SHALE TECTONICS

AAPG Memoir 93

Edited by Lesli J. Wood

ISBN13: 978-0-89181-375-0

AAPG Editor: Stephen E. Laubach
AAPG Geoscience Director: James B. Blankenship

ON THE COVER: Back cover photograph taken at the Galfa Point outcrop on Trinidad's southwest coast illustrating the proximity of the fault zone to the layer of intruded mobile mud (Henry et al., 2010, this volume); Front cover 3-D views of the basement surface and the overlying Miocene-age shales in the Alboran Sea, western Mediterranean region (Soto et al., 2010, this volume).

This publication is available from:

The AAPG Bookstore
P.O. Box 979
Tulsa, OK U.S.A. 74101-0979
Phone: 1-918-584-2555 or 1-800-364-AAPG (U.S.A. only)
E-mail: bookstore@aapg.org
www.aapg.org

Canadian Society of Petroleum Geologists
600, 640 – 8th Avenue S.W.
Calgary, Alberta T2P 1G7
Canada
Phone: 1-403-264-5610
Fax: 1-403-264-5898
E-mail: reception@cspg.org
www.cspg.org

Geological Society Publishing House
Unit 7, Brassmill Enterprise Centre
Brassmill Lane, Bath BA13JN
United Kingdom
Phone: +44-1225-445046
Fax: +44-1225-442836
E-mail: sales@geoloc.org.uk
www.geolsoc.org.uk

Affiliated East-West Press Private Ltd.
G-1/16 Ansari Road, Darya Gaaj
New Delhi 110-002
India
Phone: +91-11-23279113
Fax: +91-11-23260538
E-mail: affiliate@vsnl.com

About the Editor

Dr. Lesli J. Wood is a senior research scientist at the Bureau of Economic Geology in the University of Texas, Jackson School of Geosciences, where she directs the Quantitative Clastics Laboratory research program.

She teaches field geology, sequence stratigraphy, seismic geomorphology, and seismic interpretation courses for both industry and academia, and has published numerous peer-reviewed publications, federal research reports, and university consortium documents pertaining to both freshwater and marine clastic systems, hydrates, martian sedimentology, and shale tectonics.

Wood holds a Bachelor of Science in Geology from Arkansas Tech University, and a Master of Science in Geology from the University of Arkansas, as well as a Ph.D. in Earth Resources from Colorado State University.

Table of Contents

*NOTE TO THE READER: Chapters 1, 3, and 8 contain references to animations. These animations can be found on the CD located in the inside back cover of this book. In addition, all chapters of the book can be found on that CD, and where animations are referred to in Chapters 1, 3, and 8, there are direct links to those animations within the chapter .pdf file on the CD.

1

Wood, Lesli J., 2010, Shale tectonics: A preface, *in* L. Wood, ed., Shale tectonics: AAPG Memoir 93, p. 1–4.

Shale Tectonics: A Preface

Lesli J. Wood
Bureau of Economic Geology, Jackson School of Geosciences, University of Texas at Austin, Austin, Texas, U.S.A.

INTRODUCTION

The phenomenon of rocks moving under their own means has always fascinated both scientists and the nonscientists. Salt is known to extrude and flow as a result of differences in density of the material and surrounding sediments. However, movement of fine-grained clastics as intrusive injectites or diapirs or as extrusive eruptive sand blows or mud volcanoes has captured the public's imagination and given scientists the impetus to reconsider the physics of how sediments behave in the subsurface.

The 2006 AAPG Hedberg Conference on Mobile Shale Basins was held in response to a need to gather industry and academic communities in a common forum to address the very existence of mobile shales. Because this question involved integrated understanding of argillokinesis from the grain scale to the basin-gravity scale, the forum attracted a broad cross section of the geoscience community. These attendees presented a wide variety of topical presentations that ranged from geochemistry of modern fluid-mud extrusions to gravity studies in basins characterized by mobile shales. An ongoing topic of debate at this forum was the very term "mobile shales." The attendees decided at the meeting that this term was a misnomer and that a more appropriate term to be used to describe this phenomena was shale tectonics, thus the title of this volume. Stimulating and informative discussions at the Hedberg conference led to this special volume on shale tectonics, with contributions from researchers in industry, academia, and government covering diverse aspects of shale tectonics from grain to basin scale.

In addition, to works from authors who attended the Hedberg conference, this book also contains articles from authors who did not have an opportunity to present their work at the meeting. We hope this volume presents a representative cross section of the most current research and understanding regarding shale tectonics. The volume documents shale tectonics from a variety of basins around the world, including the southern Beaufort Sea (Elsley and Tieman, 2010), the Krishna-Godavari Basin, India (Choudhuri et al., 2010), the Niger Delta (Wiener et al., 2010), eastern offshore Trinidad (De Landro Clarke, 2010), offshore Brunei (Warren et al., 2010), and along the Spanish arm of the Mediterranean Sea (Soto et al., 2010).

Publication of this memoir coincides with a growing interest in shales as hydrocarbon reservoirs. Because of the burgeoning of shale gas and shale oil research, geoscientists are gaining a better understanding of the petrographic framework of shales, as well as their behavior under various pressure and temperature regimes and the manner in how fluids move thorough these strata (for a review, see Day-Stirrat et al., 2010). Advances in seismic imaging and processing technologies illuminate stratal geometries associated with shale tectonics that have led to a new understanding of the processes responsible for the geometries we observe in shale strata (see Day-Stirrat et al., 2010; Elsley and Tieman, 2010). In addition, advances in modeling and understanding of how both muds and shales behave after burial have led to new geodynamic models for interpreting process from response reflected in stratal packages (Albertz et al., 2010). Field geoscientists have added to our understanding of the geochemistry and physical character of extrusive mud features and their relationship to the overall basin hydrocarbon system (Battani et al., 2010;

DOI:10.1306/13231305M93730

Delisle et al., 2010; McNeil et al., 2010; Warren et al., 2010). As with mobile salt, shale-cored structures are commonly closely associated with hydrocarbons in many basins around the world. In Henry et al. (2010), information on the character of these strata can be found as well as documentation on drilling into mobile mud-cored anticlinal features (diapirs) in southern Trinidad.

TERMINOLOGY: THE QUESTION THAT WON'T GO AWAY

To develop a working classification of shale tectonics, recognition that two primary classes of features occur is critical: (1) those associated with extrusion of fluids and material that does not involve grain-to-grain contact (Battani et al., 2010; Delisle et al., 2010), and (2) those associated with larger scale deformation of apparent highly overpressured mud or shale substrates involving grain-to-grain plastic flow (Elsley and Tieman, 2010; Wiener et al., 2010). In addition, several varieties of highly fractured seep features exist (Warren et al., 2010). These features allow fluid leakage from overpressured beds, but they do not always involve plastic or fluid-mud extrusion. Shales form a variety of nonextrusive structures, many of which resemble features generated through mobile salt movement, such as domes, welds, and walls. In contrast, muds form a variety of extrusive features, such as volcanoes, ponds, and flows and can erupt explosively as seen in the Piparo mud volcano video included on the CD attached to the back cover of this book.

Several terms are used to describe the processes and features of shale mobility. These include shale tectonics, mud diapirism, shale diapirism and diapirs, mobile shales, mud volcanoes, mud diatremes, mud flows, and finally, from the Greek, argillokinesis and pellitokinesis. Although many scientists continue to argue against the existence of mobile shale, the term is well entrenched in the lexicon and unlikely to go away. A GeoRef search on the topic of mobile shale resulted in 34 instances of the combined term present in the peer-reviewed literature. Mobile shale and mud are the primary subject of a 2003 publication from the Geological Society of London, *Subsurface Sediment Mobilization* (Van Rensbergen et al., 2003). This single publication accounts for 36 additional articles on the subject, quadrupling the previous literature offering. Likewise, the 2006 AAPG Hedberg Conference on the subject of mobile shale basins resulted in 33 abstracts and extended abstracts on the subject of shale tectonics and mobile muds (Wood et al., 2006). Papers in the resulting publication of full-length manuscripts (this volume) show pervasive use of the term mobile shale(s).

Argillokinesis is a broadly applied, all-encompassing term used to describe the dynamics of uncompacted flexible clays. With broad consideration of previous literature, as well as numerous conversations with the scientific community of subsurface researchers, we propose herein to use the term shale tectonics to define the structuring within a basin associated with shale or mudstone plasticity or mobility, either as the cause of such mobility or as a result of such mobility. Mobile shales are defined as any manifestation of clay constituents (indurated or not) that show evidence of microscopic-scale fluid or plastic movement. We acknowledge that such mobility is currently poorly understood and may be some manifestation of microscopic shearing or as a reconstitution of partly indurated muds through diagenetic alteration.

HOW IS MOBILE SHALE IDENTIFIED IN THE SUBSURFACE?

Setting is a very important factor in the possible occurrence of mobile shales. Because shale mobility is caused by an imbalance of hydrostatic and lithostatic pressures, anything that varies the balance of forces within or around a shale mass, such as compressive tectonics, or rapid loading, will affect this balance and may initiate movement. Undeniably, the most influential variable in shale tectonics is fluid pressure. Two conditions commonly exist in regions of tectonically active muds or shales: (1) the presence of undercompacted mud at depth within the basin (this mud is presumed overpressured and held at depth by an overlying low permeability material) and (2) a triggering of overburden breach by some process that passes a critical threshold, causing sudden or gradual release of fluids, either with or without sediments. Triggers might be the gradual buildup of pressure beyond confining stresses of the overburden, sudden mass failure downslope of the overburden, regional compression, or seismicity.

Graue (2000) published a set of criteria that, when met, are conducive to shale tectonics and mud extrusion. These criteria are as follows.

1) Rocks that have low mechanical strength, that are commonly undercompacted, and that have a low degree of consolidation.
2) Compressive tectonic stress to provide ductile deformation and increased pore pressures.
3) Structural culminations that focus decompacted water and hydrocarbons.
4) Carrier beds for migrating fluids and gases.

In AAPG Memoir 8, *Diapirs and Diapirism*, Musgrave and Hicks (1968) provided a set of characteristics for what appear to be displaced shale masses in the Gulf of Mexico:

(1) low-velocity sound transmission, in the range of 6500–8500 ft/s (1981.2–2590.8 m/s) with very little increase in velocity with depth; (2) low density, estimated to be in the range of 2.1–2.3 g/cm^3; (3) low resistivity, approximately 0.5 ohm m; and (4) high fluid pressure, about 90% of the overburden pressure. Other authors (Henry et al., 2010) have documented sonic velocities in near surface (~0–3500 ft [~0–1070 m] below surface) mobile muds of onshore Trinidad to be lower than that of water! Authors from a range of disciplines have attributed these behaviors to high fluid pressure within the shales. At times, these densities are reduced to the point that the shales will rise as a mass in a diapiric fashion that is similar to that of salt. However, such behavior is not widely documented.

Improved seismic technologies have allowed a step change forward in the interpretation of shale tectonics in the subsurface with many previously interpreted mobile masses now much better defined. In some instances, improved imaging techniques have shown features previously interpreted as shale diapirs to be tightly folded anticlinal cores (Elsley and Tieman, 2010). However, other instances exist in which shale substrates do appear to show inflation and upward mobility (Wiener et al., 2010). Because criteria for interpreting mobile shales are not well documented in literature, many of the characteristics that were developed for interpreting mobile salts have been applied to shale basins, albeit with mixed success. In addition, in basins where both salt and shale occur, geoscientists commonly fail to differentiate mobile shales from mobile salts in seismic images. To rectify this deficiency, the following criteria for differentiating salt and shale are provided.

1) Shale welds are less prevalent in shale systems than salt welds are in salt systems.
2) Shales have a much slower velocity than salt. Salt will be underlain by seismic pull-up versus no pull-up beneath shale features.
3) Salt requires a pipe several kilometers wide to facilitate vertical migration. In contrast, fluids migrating through shales will crack and hydraulically fracture a zone through which they will migrate upward.
4) Shale may flow laterally but will develop overhangs of less than 6 km (3.7 mi). In contrast, salt overhangs may be on the order of tens of kilometers.
5) Ductile behavior of shales is unlikely above 80°C (176°F), and brittle behavior is more likely. Therefore, this temperature will provide a plasticity basement within a basin to constrain interpretation of tectonically active shales.
6) Salt volumes tend to be underestimated in interpretations of seismic data. In contrast, shale mass volumes are commonly overestimated by geophysical interpreters (Van Rensbergen and Morley, 2000) because of the manner in which geophysical data are processed. The apparent pull-up of beds around shale diapirs is caused by overmigration of the shale mass and treating its velocity as one would a salt.

FUTURE DIRECTIONS

Areas remain in which additional advances can be made in the study of shale tectonics. As exploration and development of hydrocarbons move into deeper waters along continental margins, our ability to seismically image shale versus salt must improve if we are to deduce the role that these very different materials play in deep-water fold-belt evolution. Improved understanding of how muds behave at grain-to-grain scale when buried will inform modelers and improve our understanding of diagenetic processes in shales.

At present, many intriguing questions remain regarding shale tectonics and mobile muds. How do we get fluid and plastic muds extruded at the sea floor that appear on seismic data to have originated well below the lithifaction depth of shales (see Day-Stirrat et al., 2010, for a review)? What is the physical interaction between mud-volcano pipes and underlying hydrocarbon reservoirs? Some researchers have shown that no hydraulic links between these crosscutting strata exist (DeVille et al., 2006). Do shales truly inflate as salts do above a regional stratigraphic datum? Evidence in Nigeria seems to suggest that they do (see Wiener et al., 2010). Why have we not identified more basins in which such regional inflation of shale occurs? Is there truly such a thing as a shale diapir or is this term a misdirected application of terminology? And of course, there is always the terminology itself. Although a laborious task to standardize, because this is the language by which scientific communities communicate their ideas, some attention must be paid to its clarification.

CONCLUSIONS

Hollis Hedberg was fascinated by mud volcanoes. He spent a good part of his career in eastern Venezuela and Trinidad examining and publishing on the phenomenon that was mobile shale. Today, the same phenomenon is documented in basins and settings all over our planet, as well as on other planets in our solar system. Like any geologic phenomenon, understanding will come with examination, and examination will come with a growing realization of the role that this phenomenon plays in the evolution of a planet's systems. We hope that this volume will add to the body of literature that significantly addresses both extrusive and intrusive phenomena and generates as many new questions as the answers it provides. These questions are what will drive the next generation of understanding.

ACKNOWLEDGMENTS

The author acknowledges the contributions that innumerable geoscientists interested in the study of shale tectonics have made to these discussions over the past decade. Although to list individuals would be unreasonable because of the space needs, the author is particularly indebted to the participants in the 2006 Hedberg Research Conference on Mobile Shale Basins, held in Port of Spain Trinidad, and its co-convenors, Richard Davies, Stanley Wharton, Curtis Archie, and Fazal Hosein. In addition, Andy Weinzaphel, Martin Jackson, Bruno Vendeville, and Mark Rowan are all thanked for their knowledge on salt and shale tectonics. Appreciation is extended to the AAPG for hosting several sessions in shale tectonics and mobile muds at their annual meetings, as well as support for this topic as a Hedberg Conference and the subsequent support for this publication. Angela McDonald and Ruarri Day-Stirrat are thanked for their reviews of this manuscript. Lana Dietrich is thanked for her professional editing skills.

REFERENCES CITED

Albertz, M., C. Beaumont, and S. J. Ings, 2010, Geodynamic modeling of sedimentation-induced overpressure, gravitational spreading, and deformation of passive margin mobile shale basins, *in* L. Wood, ed., Shale tectonics: AAPG Memoir 93, p. 29–62.

Battani, A., A. Prinzhofer, E. Deville, and C. J. Ballentine, 2010, Trinidad mud volcanoes: The origin of the gas, *in* L. Wood, ed., Shale tectonics: AAPG Memoir 93, p. 223–236.

Choudhuri, M., D. Guha, A. Dutta, S. Sinha, and N. Sinha, 2010, Spatiotemporal variations and kinematics of shale mobility in the Krishna-Godavari basin, India, *in* L. Wood, ed., Shale tectonics: AAPG Memoir 93, p. 91–109.

Day-Stirrat, R. J., A. McDonnell, and L. J. Wood, 2010, Diagenetic and seismic concerns associated with interpretation of deeply buried 'mobile shales', *in* L. Wood, ed., Shale tectonics: AAPG Memoir 93, p. 5–27.

De Landro Clarke, W., 2010, Geophysical aspects of shale mobilization in the deep water off the east coast of Trinidad, *in* L. Wood, ed., Shale tectonics: AAPG Memoir 93, p. 111–118.

Delisle, G., M. Teschner, E. Faber, B. Panahi, I. Guliev, and C. Aliev, 2010, First approach in quantifying fluctuating gas emissions of methane and radon from mud volcanoes in Azerbaijan, *in* L. Wood, ed., Shale tectonics: AAPG Memoir 93, p. 209–222.

Deville, E., S.-H. Guerlais, Y. Callec, R. Griboulard, P. Huyghe, S. Lallemant, A. Mascle, M. Noble, and J. Schmitz, 2006, Liquefied vs. stratified sediment mobilization processes: Insight from the south of the Barbados accretionary prism: Tectonophysics, v. 428, p. 33–47, doi:10.1016/j.tecto.2006.08.011.

Elsley, G. R., and H. Tieman, 2010, A comparison of prestack depth and prestack time imaging of the Paktoa complex, Canadian Beaufort MacKenzie Basin, *in* L. Wood, ed., Shale tectonics: AAPG Memoir 93, p. 79–90.

Graue, K., 2000, Mud volcanoes in deepwater Nigeria: Marine and Petroleum Geology, v. 17, p. 959–974.

Henry, M., M. Pentilla, and D. Hoyer, 2010, Observations from exploration drilling in an active mud volcano in the southern basin of Trinidad, West Indies, *in* L. Wood, ed., Shale tectonics: AAPG Memoir 93, p. 63–78.

McNeil, D. H., J. R. Dietrich, D. R. Issler, S. E. Grasby, J. Dixon, and L. D. Stasiuk, 2010, A new method for recognizing subsurface hydrocarbon seepage and migration using altered foraminifera from a gas chimney in the Beaufort-Mackenzie basin, *in* L. Wood, ed., Shale tectonics: AAPG Memoir 93, 195–208.

Musgrave, A. W., and W. G. Hicks, 1968, Outlining shale masses by geophysical methods, *in* E. Braunstein and G. D. O'Brien, eds., Diapirism and diapirs: AAPG Memoir 8, p. 122–136.

Soto, J. I., F. Fernández-Ibáñez, A. R. Talukder, and P. Martínez-García, 2010, Miocene shale tectonics in the northern Alboran Sea (western Mediterranean), *in* L. Wood, ed., Shale tectonics: AAPG Memoir 93, p. 119–143.

Van Rensbergen, P., and C. K. Morley, 2000, 3D seismic study of a shale expulsion syncline at the base of the Champion Delta, offshore Brunei and its implications for the early structural evolution of large delta systems: Marine and Petroleum Geology, v. 17, p. 861–872.

Van Rensbergen, O., R. R. Hillis, A. J. Maltman, and C. K. Morley, 2003, Subsurface sediment mobilization: Introduction, *in* P. Van Rensbergen, R. R. Hillis, A. J. Maltman, and C. K. Morley, eds., Subsurface sediment mobilization: Geological Society (London) Special Publication 216, p. 1–8.

Warren, J. K., A. Cheung, and I. Cartwright, 2010, Organic geochemical, isotopic, and seismic indicators of fluid flow in pressurized growth anticlines and mud volcanoes in modern deep-water slope and rise sediments of offshore Brunei Darussalam: Implications for hydrocarbon exploration in other mud- and salt-diapir provinces, *in* L. Wood, ed., Shale tectonics: AAPG Memoir 93, p. 161–194.

Wiener, R. W., M. G. Mann, M. T. Angelich, and J. B. Molyneux, 2010, Mobile shale in the Niger Delta: Characteristics, structure, and evolution, *in* L. Wood, ed., Shale tectonics: AAPG Memoir 93, p. 145–160.

Wood, L., S. Wharton, R. Davies, F. Hosein, and C. Archie, conveners, 2006, Mobile shale basins—Genesis, evolution and hydrocarbon systems: Proceedings of the AAPG/Geological Society of Trinidad and Tobago Hedberg Conference, Port-of-Spain, Trinidad, 1 p.

2

Day-Stirrat, R. J., A. McDonnell, and L. J. Wood, 2010, Diagenetic and seismic concerns associated with interpretation of deeply buried 'mobile shales', *in* L. Wood, ed., Shale tectonics: AAPG Memoir 93, p. 5–27.

Diagenetic and Seismic Concerns Associated with Interpretation of Deeply Buried 'Mobile Shales'

R. J. Day-Stirrat

Bureau of Economic Geology, John A. and Katherine G. Jackson School of Geosciences, University of Texas at Austin, Austin, Texas, U.S.A.

A. McDonnell

Bureau of Economic Geology, John A. and Katherine G. Jackson School of Geosciences, University of Texas at Austin, Austin, Texas, U.S.A.

L. J. Wood

Bureau of Economic Geology, John A. and Katherine G. Jackson School of Geosciences, University of Texas at Austin, Austin, Texas, U.S.A.

ABSTRACT

The study of mudstones is becoming increasingly significant in both conventional and unconventional hydrocarbon exploration. Starting at the microscale, we discuss the chemical and physical properties of mudstones and how they alter with burial and diagenesis. Moving to the macroscale, we address interpretation risks from seismic reflection data in 'shale' tectonic terrains (see Terminology section below). We focus on deeply buried mudstones that experienced temperatures in excess of 50 to 80°C (122 to 176°F). In these deeply buried settings, large-scale ductile deformation is commonly invoked. Above 80°C (176°F), chemical burial compaction processes reduce the ability of a mudstone to deform in a ductile manner, even under overpressure conditions. At temperatures above 80°C (176°F), we envisage 'shale tectonics' as mostly a brittle process; thus, deeply sourced 'shale diapirs' are highly unlikely, as is lateral flowage of 'shale' along deep 'shale' detachments. At the seismic scale, interpretation of 'mobilized shale' at depth is inherently controlled by the geophysical quality or limits of the seismic reflection data. One must consider the survey design and acquisition parameters, migration and velocity issues, and data vintage. Recent advances in seismic technology lead to the conclusion that the actual mobility of deeply buried mudstone features has been overinterpreted in several classic 'mobile shale' basins. Deeply buried mudstones most likely deform by a brittle process, with ductile deformation being highly improbable. When working

DOI:10.1306/13231306M93730

in a 'shale' tectonic province, workers need to consider (1) mudstone composition and its burial history; (2) chemical compaction and diagenesis; (3) depth, temperature, and timing of the invoked mobilization; and (4) geophysical risks.

INTRODUCTION

'Shale tectonics' (see Terminology section below) and its close kinship with hydrocarbon exploration are by no means a new topic. As far back as 1866, we can find an interpretation of a relationship between 'mobile shale' phenomena and hydrocarbons (Ansted, 1866). Hollis Hedberg, one of the earliest researchers on 'mobile shale' systems, held a lifelong fascination with the interrelationships of clays and 'shale,' overpressure, and hydrocarbons. He produced some of the earliest works on this topic (Hedberg, 1936) and continued to explore the topic in his later works (Hedberg, 1974). Hedberg's contemporary, Hans Kugler, discussed mud volcanism as a "saturated stratum of gas, petroleum or water, under pressure with enough energy to produce a similar geological phenomenon that causes magmatic intrusion and eruption" (Kugler, 1933, p. 743). As resource exploration intensified in the 1960s and technology allowed drilling to deeper and deeper strata, the need to understand and predict overpressured mudstones in basins around the world became more imperative. Several seminal works on 'mobile shale' masses were published in *AAPG Memoir 8, Diapirs and Diapirism* (Braustein and O'Brien, 1965). More recently, conferences in 2003 and 2006 either focused on 'mobile shales' (van Rensbergen et al., 2003; AAPG Hedberg Conference, 2006 [Wood et al., 2006]) or contained a substantial scientific component on 'mobile shale' (Bob F. Perkins Research Conference, 2004 [Post et al., 2004]). Although these activities have increased the discussion of the 'mobile-shale' phenomenon, many questions remain regarding exactly how mobile is shale (or, instead, mudstone), what causes and maintains its mobility, what are the functions of fluids and temperature in the mobility or ductility, and what are the implications of mobile mudstone for a basin's tectonic and hydrocarbon condition.

In the study of 'shale tectonics,' one needs to recognize that two primary classes of 'shale mobility' features occur. The first is associated with extrusion of fluids and material that does not involve grain-to-grain contact, and the second is associated with larger scale deformation of apparent, highly overpressured mud substrates involving grain-to-grain ductile flow. Mobile mudstone systems that are not deeply buried have been addressed (e.g., Morgan et al., 1968; Hovland et al., 1998; Haskell et al., 1999), as have 'mobile shale' belts that are deeply buried (e.g., Bruce, 1973; Bradshaw and Watkins, 1994; Cohen and McClay, 1996; Huh et al., 1996; Morley and Guerin, 1996; Brown et al., 2004). Fluid-driven processes (e.g., mud volcanoes) have also been studied extensively (e.g., Kopf, 2002; Deville et al., 2003; Davies and Stewart, 2005; Davies and Clark, 2006; Deville, 2006; Deville et al., 2006) and are frequently, although not always, extrusive features. The earliest scientific attention by academic researchers was placed on extrusive expression of mobile mud systems (mud volcanoes), with relatively minor evaluation of the character and form of nonexplosive 'mobile' mud features. The advent of improved seismic imaging and the ability to drill into more overpressured depths in later decades of the 20th century resulted in a need for increased understanding of deeply buried intrusive 'mobile' mud systems. Fluid-expulsion features, such as mud volcanoes, pipes or chimneys, and shallow ductile deformation processes are herein considered separate processes and are not discussed.

In this chapter, we discuss the probability of plastic sediment deformation at deep burial depths from a microscale (pore and grain size) to a macroscale (seismic reflection data). We first describe physical properties and heterogeneity of fine-grained sediment deposits and their evolution through burial and diagenesis. Next we discuss implications of mechanical, chemical, and geophysical properties in 'shale tectonic' basins to infer that 'mobile-shale' tectonics appears to be overinterpreted in the deep subsurface. We then discuss questions pertinent to those working in a 'shale-tectonic' province: from microscale physical and chemical properties of the buried sediment to macroscale recognition of these phenomena in seismic data.

TERMINOLOGY

No good lexicon exists for 'mobile-shale' features. Morley (2003) suggested a comprehensive classification of features associated with 'mobile' mud systems, recognizing mud diapirs, true 'shale' diapirs, fluidized mud-feeder pipes, and gas clouds and chimneys. He stressed the close association between shale state and structural regime and framework but lamented the large number of unanswered questions about these systems. Although in the shallow subsurface muds can certainly deform in a ductile or plastic manner, the degree of ductility that is retained or evolves as the mudstone is buried and enters the realm of mechanical and chemical compaction is unknown. In this chapter, as in van Rensbergen and Morley (2003), several terms (e.g., 'mobile shale' and 'shale diapir') will be used in single quotation marks because these terms are unqualified and they inherently imply plastic or ductile deformation. The first, simplistic, nomenclature for 'mobile shales' was proposed by Kugler (1933), when he defined a diapiric structure as a competent mass of mobile pelite with rock fragments; a mud extrusion as being characterized by large transported blocks of rock and mud volcanoes; and finally, a mud volcano as being an explosive feature characterized by clay, lime, sand, and small subordinate blocks. Maltman and Bolton (2003) defined a process-based structural classification of shale- and mud-associated features (see their figure 1). In their classification, structures (i.e., ball and pillows, flames, volcanoes, surface diapirs, shale diapirs, folds, etc.) were linked to the sediment state (i.e., unconfined cataclastic flow, brecciation, fluidization, liquefaction, etc.), which was inherently tied to the burial depth and degree of lithifaction. Stewart and Davies (2006) created a terminology for mud volcanoes that focused on the internal architecture of these features. This terminology included a mud source, an intrusive domain, an extrusive domain, and a roof domain, along with a terminology that reflected their seismic data imaging of the subsurface morphology. L. Cartwright (2004, personal communication) suggested a classification of faults, intrusions (sand, mud, salt, and igneous), diapirs (mud and salt), and pipes in these systems, each linked to its flux state as dynamic, episodic, or steady.

In discussing fine-grained siliciclastic deposits, the terms used to describe the size and composition from Wentworth (1922) and Folk (1974) are implicit. In strict terms, shale is a fine-grained rock that exhibits fissility (the tendency to split into thin sheets) that develops as a consequence of the alignment of particles through burial and, possibly, surface weathering. As such, the term 'mobile shale' is an unfortunate oxymoron entrenched in the literature. The work of Weaver (1989) allows for a full appreciation of the complexity of mudstone composition and mineral evolution with increasing temperature. This compositional evolution has important implications for the physical properties of a mudstone as it is buried. In this chapter, we try to avoid the term 'shale' because, strictly speaking, shale implies fissility. Instead, we employ the word 'mudstone' to describe fine-grained siliciclastic rocks, either lithified or semilithified, composed of clay or silt-size material, which may or may not have fissility and may or may not be dominated by clay minerals or quartz. In contrast, we define 'mud' as a material found in a liquid state composed of particles smaller than 0.062 mm (0.39 in.) (4 on the phi scale). The mineralogical composition of a mudstone at deposition is fundamentally different from its composition at the boundary with metamorphism (~6–7 km [~20,000–23,000 ft] of vertical burial). Mechanical and chemical alterations that occur during this burial transition from mud to mudstone to shale are complicated, with many variables, all of which have consequences as to whether deformation occurs in a brittle or ductile manner.

PHYSICAL AND CHEMICAL PROCESSES DURING MUDSTONE BURIAL

Mudstone Diagenesis

Mud is transformed to mudstone through diagenesis, which is a set of processes occurring between deposition and metamorphism encompassing both physical and mineralogical changes (e.g., Harrison and Tempel, 1993; Battaglia et al., 2004). Diagenesis, under its broadest definition, is the transformation of unconsolidated sediments into lithified rock. This lithification promotes fundamental differences in composition (mineralogical assemblages) (e.g., Savin and Epstein, 1970; Hower et al., 1976; Yeh and Savin, 1977; Land and Milliken, 1981; Awwiller, 1993; Bjorlykke and Hoeg, 1997; Land et al., 1997; Clauer et al., 1999; Land and Milliken, 2000; Merriman, 2005; Srodon et al., 2006; McCarty et al., 2008) and physical properties (e.g., Athy, 1930; Baldwin, 1971; Baldwin and Butler, 1985; Burland, 1990; Hansen, 1996; Bjorlykke and Hoeg, 1997; Dewhurst et al., 1998; Yang and Aplin, 1998; Nygard et al., 2004; Worden et al., 2005; Yang and Aplin, 2007; Day-Stirrat et al., 2008) of a mudstone at depth, in comparison with that of its depositional precursor. Although mineral diagenesis and compaction are inherently linked, this section of the chapter deals with mineralogical changes associated with diagenesis. Compaction and other physical parameters are addressed later.

Burial diagenesis is regularly split into shallow- and deep-burial settings, although the boundary between these two settings is commonly ambiguous. Shallow-burial settings have a fairly restricted time-temperature-pressure interval, a lower limit commonly being placed at the point where biologically mediated processes cease, at about 75°C (167°F) (e.g., Hesse, 1986). In such a shallow setting, the open-pore network of a noncompacted sedimentary deposit allows a ready flux of dissolved components that are out of equilibrium with the environment and highly reactive. Such early diagenetic processes can involve reactive solids, organic matter, volcanic glass, iron oxide, aluminum oxide, and amorphous silica. In early diagenesis, biological processes mediate a series of redox reactions, exploiting the ready availability of oxidants in an idealized series: oxygen-nitrate-manganese oxide, iron oxide, and sulfate carbonate (e.g., Hesse, 1986). Sulfate reduction to form pyrite is commonly characteristic of many 'black shales,' for example, the Kimmeridge Clay Formation of the North Sea and southern United Kingdom (MacQuaker and Gawthorpe, 1993; MacQuaker et al., 1997). However, early carbonate cementation has the most impact on mudstone physical properties. It can radically change the mud and mudstone response to loading and transform rock behavior from ductile to brittle (e.g., Bjorlykke and Hoeg, 1997).

Silica diagenesis at shallow depths (<2 km [<6600 ft]) is best understood in terms of the conversion of amorphous opal (opal-A) to cristobalite and tridymite (opal-CT) and ά-quartz (Bloch, 1998). Opal-A is transformed with burial to opal-CT and, finally, ά-quartz by approximately 50°C (122°F) (Knauth, 1994). The opal-A to opal-CT conversion liberates bound water that rapidly reduces porosity, thereby driving overpressure formation (Davies et al., 2006). This liberated fluid may also prime adjacent porous sandstones for injectite processes by hydraulic fracturing.

As muds are progressively buried, increasing temperature and pressure cause mineral reactions to move to silicate phases (quartz, feldspar, and clay minerals). This results in a variety of reaction mechanisms, including transformation, recrystallization, dissolution, or precipitation. These reactions cause progressive reorganization of the rock fabric (e.g., Ho et al., 1999; Day-Stirrat et al., 2008) and mineralogy (e.g., Srodon et al., 2006; van de Kamp, 2008). In most sedimentary basins, clay-mineral diagenesis is particularly significant in terms of reactive volume at any temperature and pressure condition, although mudstone suites have heterogeneous compositions. As metamorphism is approached, an idealized mineralogical assemblage comprising illite, chlorite, quartz, and feldspar is typically formed, with associated mineralogical flux, increase in crystal size, and changes in anisotropy and compaction and permeability (see later section).

Smectite-to-illite transformation (Burst, 1959; Perry and Hower, 1970; Hower et al., 1976), along with its associated increase in crystallinity, is the most ubiquitous and volumetrically significant clay-mineral reaction in the evolution and mobilization of mudstones. These processes begin to occur at about 50°C (122°F) (Guggenheim et al., 2002). With main controls being temperature and chemical variability, the reaction proceeds through to the mixed-layer-phase illite-smectite. Depending on temperature, this mixed-layer-phase illite-smectite may have either a random or ordered interstratification (e.g., Srodon, 1980). As illitization proceeds, changes occur in the stacking pattern of smectite and illite layers in the mixed-layer-phase illite-smectite. A change in ordering from random to ordered illite layers at approximately 100°C (212°F) is observed. This ordering of layers is accompanied by significant elemental composition changes within the clay structure: gain in aluminum; loss of silica, iron, and magnesium; uptake of nonexchangeable potassium at the expense of calcium in the phyllosilicate interlayer; and significant water release from the smectite interlayer as it converts to illite (e.g., Hower et al., 1976; Boles and Franks, 1979; Ahn and Peacor, 1986). Diagenetic reactions (Hower et al., 1976; Boles and Franks, 1979) release large amounts of silica into the system, with the potential for this to form quartz cement. Recent calculations (van de Kamp, 2008) point to as much as 17 to 28 wt.% SiO_2 being liberated during the smectite-to-illite transformation. Distribution and occurrence of this quartz cement are an area of active research. Petrographic evidence in mudstones of quartz cement is lacking because the resolution of cathodoluminescence is insufficient to resolve the finest fraction in a mudrock (Milliken, 1994).

This review of diagenetic changes in mud suggests that early burial of muds is dominated by an open-pore network, with mud being highly reactive or reacting with organic material, biological input, iron and aluminum oxides, and silica. However, early carbonate cementation seems to be a strong control on the transition of muds from ductile to brittle behavior. Clay-mineral reactions are volumetrically the most important during diagenesis. In particular, the release of silica can lead to significant quartz cementation, and the release of water leads to overpressure. Mineralogical diagenesis via cementation is critical as to whether a mudstone will react in a brittle or ductile manner. If cementation is retarded, under high overpressures, a rock may act in a ductile manner longer in its burial history.

Mudstone Compaction

At deposition, clays and muds typically have porosities ranging from 60 to 80% (e.g., Magara, 1968; Dzevanshir et al., 1986; Mondol et al., 2007). The term consolidation is used in soil mechanics literature to describe reduction in this depositional porosity by burial processes. In geological literature, the term compaction is used to describe both mechanical and chemical processes. Complex interplay of mechanical and chemical compaction processes causes physical changes in sediment porosity, permeability, density, velocity, and strength (Figure 1). Conceptual models adopted for compaction (Figure 2) commonly comprise an initial porosity reduction associated with mechanical processes, as well as later porosity reduction governed by chemical diagenetic reactions

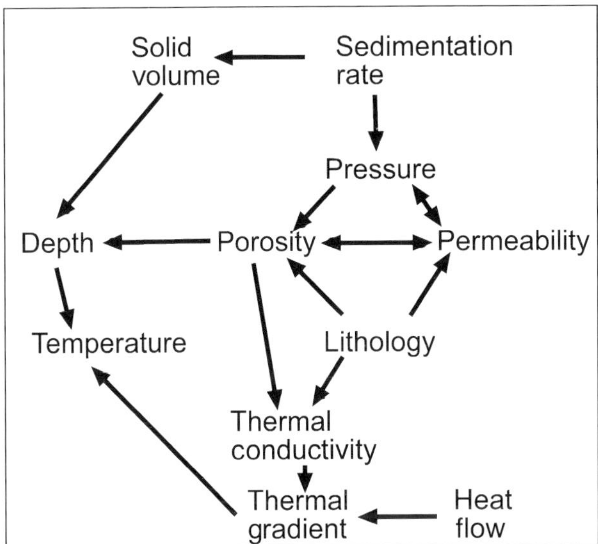

Figure 1. Interconnectedness of burial regime parameters.

causing dissolution and new crystal growth (Hedberg, 1936; Bjorlykke and Hoeg, 1997; Bjorlykke, 1999).

Initial stages of compaction are controlled by mechanical rearrangement of grains and ductile deformation. Early mechanical compaction is more pronounced in fine-grained rocks (clays and muds) because of their initially high porosity relative to sand and carbonates. Porosity-depth curves are numerous (see Figure 3 and curves presented by Mondol et al., 2007) and show a rapid initial decrease down to 2 to 2.5 km (6600 to 8200 ft) of burial (Figure 3). Deeper, the rate of porosity reduction is greatly reduced as mineral diagenesis begins to dominate. Although chemical compaction may occur in early burial (Bjorlykke and Hoeg, 1997), it generally dominates the deeper section, at depths of 2 to 5 km (~6600 to 16,500 ft), depending on the geothermal gradient (80–100°C [176–212°F]). At these depths, cementation will retard the ability of mudstones to move in a ductile manner.

Early mechanical compaction is related to effective stress (σ_e), which is the difference between overburden stress caused by loading (σ_v) and initial fluid pressure (P_p).

$$\sigma_e = \sigma_v - P_p \qquad (1)$$

Mechanical compaction cannot occur if fluid cannot escape, but chemical precipitation of minerals is still possible. A variety of empirical relationships have been proposed to describe the compaction of mudstones in response to the increase of effective stress (e.g., Giles et al., 1998; Goulty, 1998; Broichhausen et al., 2005), but unfortunately for basin and thermal modeling, no unique solution is globally applicable, and compaction is still the great unknown (Giles et al., 1998). Sedimentary composition, however, is a key factor in mudstone compressibility, with empirical equations relating clay content to void ratio (e, ratio of volume of voids to volume occupied by sediment grains) at known effective stresses (σ_e) (Aplin and Larter, 1995; Yang and Aplin, 1998). Recent work on experimental mechanical compaction (Mondol et al., 2007) points to the importance of composition and grain size to the porosity-depth relationship (Figure 4). Further factors include differences in sedimentation rate, depositional fabrics, and permeability characteristics. Additionally, some differences associated with porosity-depth curves may be related to strata uplift and associated overburden unloading on the compaction-curve profile (Figure 5).

Compaction generally reduces sedimentary deposit permeability. Katsube and Williamson (1994) suggested that mudrock permeability decreases 6 to 10 orders of magnitude during burial compaction, with modal pore-size distributions (strictly, pore-throat distributions) reducing from 200 nm at about 1 km (~3300 ft) (30% porosity) to 10 to 40 nm at 4.5 km (~14,500 ft) (>5% porosity). Permeability may vary by several orders of magnitude for any given porosity (Neuzil, 1994) and is probably a function of differences in grain size and grain-size distribution (Dewhurst et al., 1999b). Therefore, defining grain-size distributions at a given porosity may be key to defining permeability and, hence, the degree of overpressure a rock can withstand before it fractures (Moore, 2005). One must consider questions of scale of phenomena when discussing grain-size distribution, permeability variations, unit thickness, and

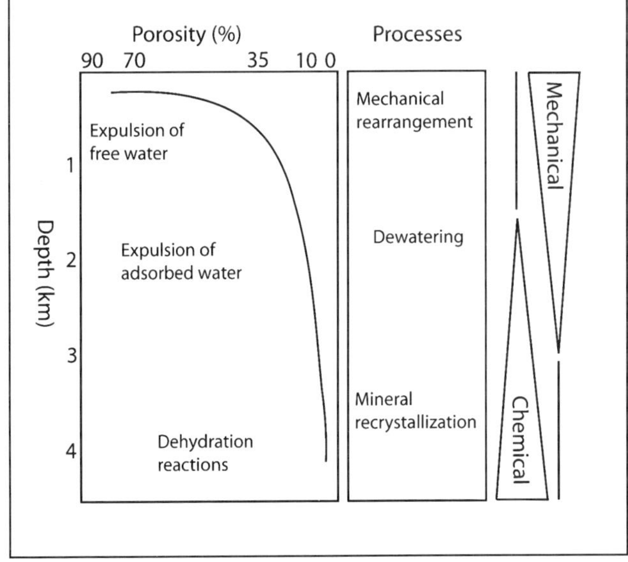

Figure 2. Model showing processes that drive porosity reduction as a function of depth. 1 km = 3281 ft.

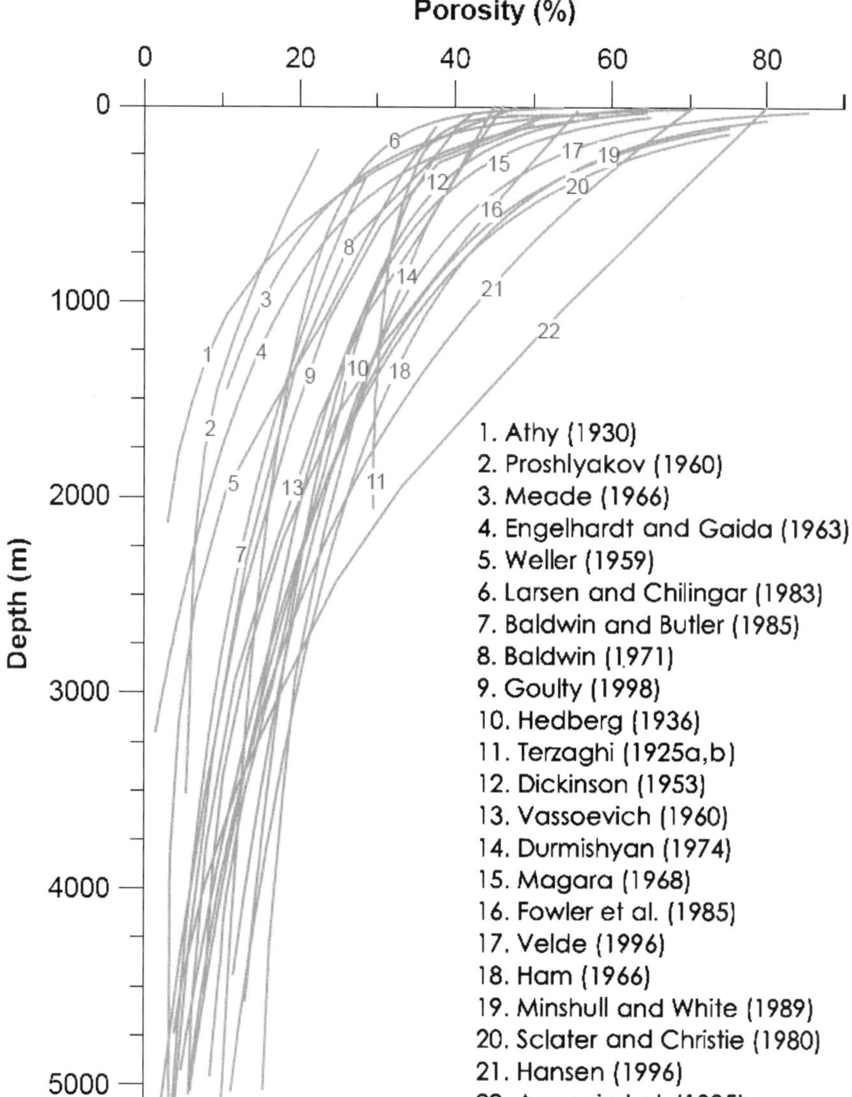

Figure 3. Selected porosity-depth curves for mudstone (from Mondol et al., 2007; reprinted with permission of Elsevier). 1000 m = 3281 ft.

any permeability pathway (such as fractures that may open in response to uplift or tectonic compression). Note also that physical properties of a recently deposited mud are quite different from those of a mudstone, which is formed as a result of porosity reduction and mineral diagenesis from 2 to 5 km (~6600 to 16,500 ft) of burial.

In summary, mudstone composition (i.e., percent clay minerals) is strongly linked to compressibility or, in other words, mudstone compaction, which significantly alters porosity and permeability with depth.

Overpressure

In basins with high sedimentation rates, long drainage pathways, and low permeabilities, excess pore pressures may become elevated because of the excessive amount of time it takes for fluids to dissipate. In addition, overpressure can result from fluid expansion caused by heating, mineral dehydration and water release, and hydrocarbon generation (Osborne and Swarbrick, 1997). Such expansion leads to transient excess pore pressures. Because mechanical compaction cannot occur if fluid cannot escape, overpressured rocks will commonly have anomalously high porosity (Dickinson, 1953; Waples and Couples, 1998; Bolas et al., 2004; Broichhausen et al., 2005; Dugan and Germaine, 2008). Therefore, porosity is a function of effective stress and the effect of permeability in reducing effective stress. Overpressured sections that retain high fluid contents for extended periods may have a higher thermal gradient than they would under normal consolidation conditions. These above-normal-trending thermal gradients can likewise influence the chemical compaction of sediments. Waples and Couples (1998) showed that a sequential series of

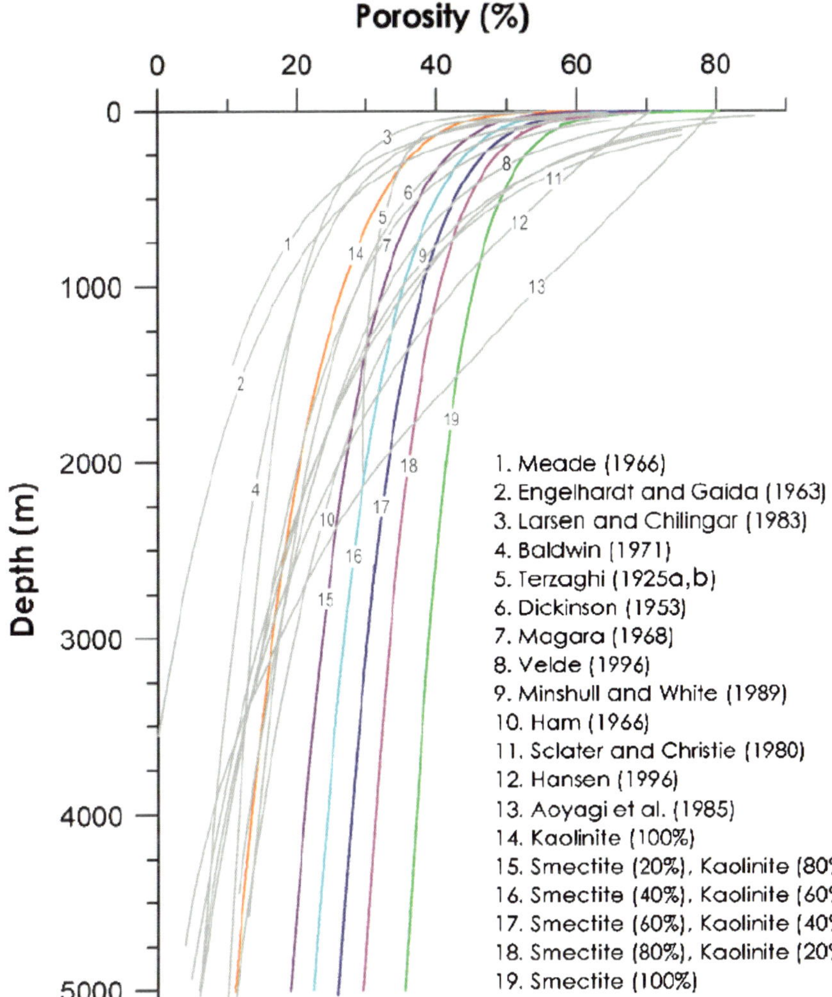

Figure 4. Selected porosity-depth curves for mudstone (from Mondol et al., 2007; reprinted with permission of Elsevier), with experimental mineral assemblages superimposed. 1000 m = 3281 ft.

steps occur in a compaction regime to heavily influence overpressure formation. These steps in order of occurrence are (1) sediment accumulation leading to the application of stress to the system, (2) response to this applied stress in the grain framework, (3) transfer of the stress in the grain framework, and (4) fluid flow out of the rock, lowering overpressure and permitting further mechanical compaction. The final step in the process of overpressure formation is the rate-limiting function. The degree of overpressure for a given lithology is probably limited by porosity and associated permeability. Thus, a silt-rich mudstone at a given porosity will have a higher permeability (Neuzil, 1994; Dewhurst et al., 1999a) than a clay-rich mudstone at the same porosity. An overburdened clay-rich mudstone will therefore preferentially develop an overpressure response. In summary, not all mudstones are created equal, with large heterogeneities being possible in a variety of physical parameters that can influence their behavior under different temperature and stress conditions.

Fabric Anisotropy

The importance of mineral diagenesis to sediment compaction cannot be overlooked (Bjorlykke and Hoeg, 1997; Nygard et al., 2004). The soil-mechanics approach to compaction used at shallow depth (e.g., Burland, 1990) is insufficient to explain compaction fully. Laboratory compaction tests between two Kimmeridge Clay samples from differing maximum burial depths, an uncemented sample (50% porosity at 500-m [1640-ft] maximum burial) and a cemented sample (22% porosity at 1.7-km [~5600 ft] maximum burial), indicate that mechanical compaction and burial depth cannot explain differences in hydromechanical behavior between the two samples (Nygard et al., 2004). Odometer and triaxial compression tests (in which a load is applied to a specimen, which is also subjected to a lateral pressure or confining pressure at right angles to the vertical load) were able to reduce porosity (and void ratio) significantly in the uncemented sample (Figures 6, 7), whereas

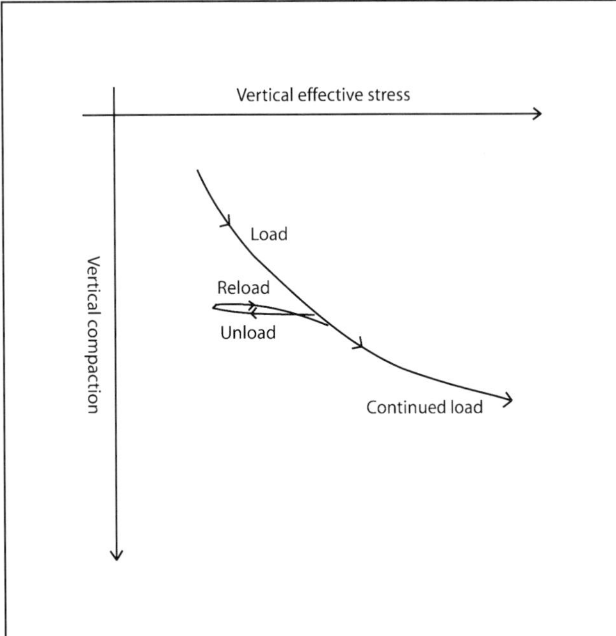

Figure 5. Conceptual model showing typical stress-strain behavior during simple loading and unloading of a compacting mudstone.

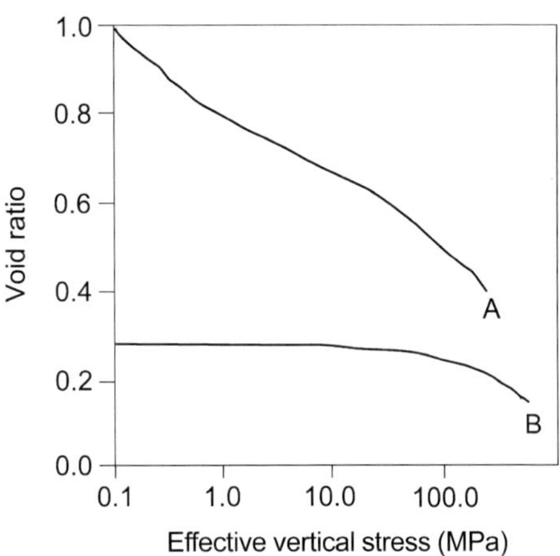

Figure 7. Compaction curves for (A) shallowly buried Kimmeridge Clay and (B) more deeply buried Kimmeridge Clay as a function of void ratio and log-effective vertical stress from odometer and triaxial tests (Nygard et al., 2004; reprinted with permission of Elsevier). 1 MPa = 145 psi.

only small amounts of porosity reduction were observed at 50 to 100 MPa (~7250 to 14,500 psi) of effective vertical stress in the deeper cemented sample because of precipitation of diagenetic phases in the sediment's pores. Cementation processes also radically reduced the compressibility index (C_c, compression index; C_c^*, intrinsic compression index). The stiffer response of the cemented sample could not be explained

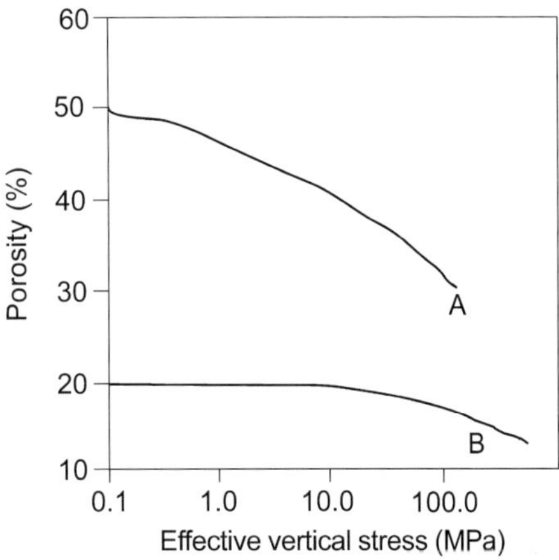

Figure 6. Compaction curves for (A) shallowly buried Kimmeridge Clay and (B) more deeply buried Kimmeridge Clay as a function of porosity (%) and log-effective vertical stress from odometer and triaxial tests (Nygard et al., 2004; reprinted with permission of Elsevier). 1 MPa = 145 psi.

by mechanical compaction alone. Permeability differed by six orders of magnitude between the samples for the same void ratio because of pore-throat reduction and increased clay-mineral alignment (anisotropy). The deeper sample had significantly less permeability than the shallow sample. These detailed observations have far-reaching consequences because they demonstrate that permeability is not a unique function of porosity. Permeability variations as a function of composition and diagenesis may set up permeability anisotropy, which may contribute to overpressure formation.

Fabric anisotropy (Figure 8) in a diagenetic context was also explored by Ho et al. (1999) and Day-Stirrat et al. (2008), who demonstrated a correlation between preferred alignment generation in illite-smectite and chlorite, as well as diagenetic alteration of smectite to illite. Illitization is ubiquitous in most basins around the world (e.g., Burst, 1959; Dunoyer de Segonzac, 1970; Perry and Hower, 1970; Hower et al., 1976; Srodon and Eberl, 1984; Srodon et al., 2006) and is commonly attributed to a mechanism of dissolution-reprecipitation (e.g., Boles and Franks, 1979; Inoue et al., 1987). Clay-mineral alignment anisotropy has not been explored fully in terms of its effects on permeability and rock-mechanic properties (see Arch and Maltman, 1990; Bolton et al., 2000; Nygard et al., 2004, for a preliminary review). However, diagenetic alignment studies prove that mechanical compaction is not the sole driver in creating mudstone anisotropy. This fact is important, given that phyllosilicates align with diagenesis (Ho et al., 1999; Day-Stirrat et al., 2008). Such alignment induces a permeability anisotropy (e.g., Bolton et al., 2000; Nygard et al., 2004); therefore, it follows that anisotropy will occur in mechanical properties.

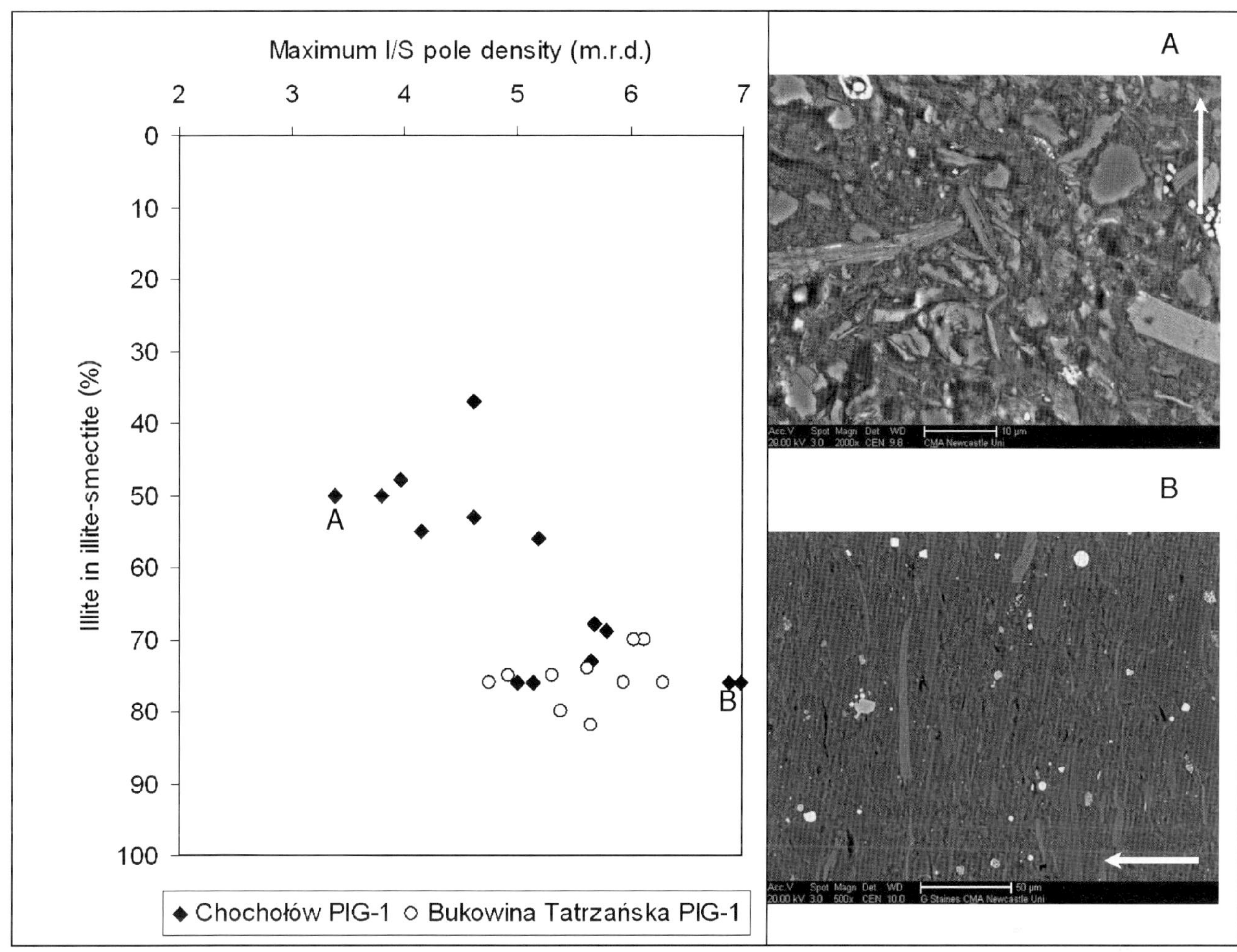

Figure 8. Quantification of illite-smectite alignment (m.r.d.) plotted as a function of the percentage of illite in the mixed-layer-phase illite-smectite (I/S) and associated backscattered scanning electron images (A, B) showing the difference in alignment between selected samples (Day-Stirrat et al., 2008; reprinted with permission of Clay Minerals Society).

LABORATORY TESTS

Ibanez and Kronenberg (1993) systematically tested the mechanical strengths of lithified samples that had been compressed parallel with perpendicular to, and at 45° to bedding (phyllosilicates were preferentially aligned parallel to bedding at a maximum burial depth of 4 km [~13,000 ft]). These workers found that lower Tertiary Wilcox Formation samples (from Louisiana, United States) were stronger when compressed perpendicular with and parallel to bedding and weaker when compressed at 45° to bedding. This observation has implications for explaining the nature of mudstone response to stress, which may be either ductile or brittle. Ductile behavior is characterized by gradual deformation until failure, whereas brittle deformation has a sudden failure at well-defined peak shear strengths. According to Ibanez and Kronenberg (1993), mudstone response to compressive stress is a brittle or semibrittle process that leads to macroscopic fracturing and semibrittle shear zones with entrained, nondeformed blocks and kink-band formation in tested materials. These results confirm those of Swan et al. !(1989), who provided backscattered scanning electron images of triaxial tests on undrained Kimmeridge Bay (Dorset, United Kingdom) mudstones at varying strain rates. The study by Swan et al. (1989) documents changes in the deformation mechanism from macroscopic fault-plane formation at low strain rates to distributed shear microstructures at high strain rates. The key conclusion is that movement is localized, both in terms of rock movement and fluid flow, at high strain rates and that nucleation and growth of a crack or fracture in one sample locale do not change the pore pressure distribution in the bulk sample. Note that experiments (Nygard et al., 2006) on undrained samples show that, under normal consolidation (compaction) conditions, materials exhibit a ductile response to increased load, but in cases where chemical diagenesis and overpressure buildup occur (>80°C [<176°F]), mudstone responds to loading by brittle deformation.

SUMMARY

Immediately after deposition, siliciclastic deposits are muds with certain mineralogical and physical characteristics, which radically alter with burial and diagenesis. A mudstone residing in a burial regime at 80°C (176°F) will behave significantly differently with applied forces than a mudstone at 140 or 200°C (284 or 392°F). Ductile deformation is unlikely above 80°C (176°F), and more brittle behavior can be expected.

SEISMIC IMAGING

The preceding review of mudstone response to burial suggests that ductile deformation is unlikely above 80°C (176°F); this temperature cannot be tied to a specific depth because it depends on several factors, including sedimentation-burial rate, geothermal gradient, and basement type, and the depth will be unique to each basin. Ductile deformation processes have been widely invoked (e.g., 'shale diapir or mobile shale') (e.g., Bradshaw and Watkins, 1994; Cohen and McClay, 1996; Huh et al., 1996; Morley and Guerin, 1996; Brown et al., 2004), yet burial temperatures are generally not referred to in studies of 'mobile-shale' belts. Because ductile deformation is highly unlikely at burial temperatures above 80°C (176°F), seismic phenomena that are commonly interpreted as 'mobile shales' in deeply buried settings need to be examined, along with a possible reinterpretation of the seismic appearance. This geophysical review will focus mostly on risks associated with the interpretation of 'mobile shale' from seismic reflection data at depths at which mechanical and chemical compaction will have altered the shallowly buried mud to a mudstone.

Seismic data, gravity modeling, multibeam, and 3.5-kHz profiles are geophysical techniques that have all been used to interpret regions of 'mobile shale,' ranging from deeply buried 'shale ridges' or 'diapirs' to shallower subsurface mud chambers to extrusive mud volcanoes. We will focus on seismic reflection data because loss of seismic reflection signal is one of the most widely used criteria for interpretation of mobile sediment masses, not only in the shallow subsurface (e.g., Hovland et al., 1998, upper 1500 m [4921 ft]), but particularly also in the deep subsurface (e.g., Cohen and McClay, 1996; Morley and Guerin, 1996). Interpretation of 'mobilized sediment' at depth is controlled inherently by geophysical quality or limits of seismic reflection data. With poor data quality, interpreters may be more likely to interpret a shale mass than with better quality, higher resolution data. Areas previously interpreted as regions of 'shale mobilization' are being reexamined because newer technology and processing provide better quality seismic data, and new conclusions differ from those of previous interpretations (e.g., van Rensbergen and Morley, 2003; Elsley and Tieman, 2010).

UNCERTAINTIES

In basins where conditions (i.e., fine-grained sediments, compressional tectonics, rapid burial, etc.) suggest the likelihood of overpressured shales, loss of seismic reflection signal is commonly assumed to result from a distorted mobilized sedimentary mass (e.g., Damuth, 1994; Cohen and McClay, 1996). However, seismic character is not unique and is certainly not direct evidence that a deeply buried mudstone is acting in a mobile manner (van Rensbergen and Morley, 2003). Seismic quality is strongly influenced by survey design and processing, as well as inherent physical properties of the geological system. Numerous geologic conditions may limit our ability to clearly resolve bedding and other geologic information using seismic methods. Issues with signal-to-noise ratio, migration, velocity gradients, and fault shadow can result in apparent zones of low image quality. In contrast, some zones of low image quality (i.e., chaotic reflectivity) are produced by homogenous lithology, mass transport, and hydrofracturing (van Rensbergen and Morley, 2003). When studying a region of highly overpressured mudstones, a worker should consider (1) the survey design (a survey should be designed with the idea of imaging the mudstone zone in mind, which is unlikely to be the case), (2) data migration strategies, (3) velocity and the occurrence of overpressure, (4) sedimentologic facies of the section, and (5) vintage and quality of data under examination. Some of these considerations warrant further elaboration.

Migration

Seismic migration compensates for distortion introduced by wave propagation and acquisition geometry. Migration was traditionally applied after data stacking; however, prestack migration is now more common and allows for better imaging of complex structural settings. Prestack depth migration (PSDM) is also increasingly popular for resolving more complex structural settings

Before migration

Two-way traveltime (s)

1
2
3
4

After migration

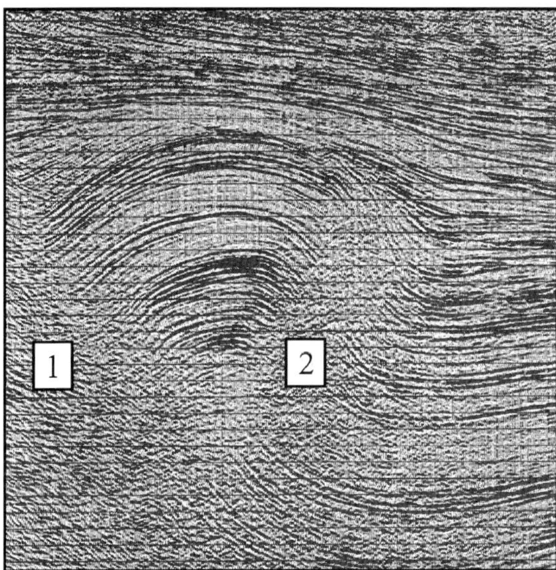

Figure 9. The common-midpoint stack across a thrust fault, before and after migration. Following migration, lack of reflectivity in area (1) is caused by short line length, whereas wipeout in area (2) occurs because the migration algorithm cannot resolve steep dips (from Yilmaz, 2001; reprinted with permission of Society of Exploration Geophysicists).

or areas where imaging problems exist. Lack of seismic reflectivity seen in some basins could be a processing artifact caused by inaccurate migration of steeply dipping reflections (Yilmaz, 2001; van Rensbergen and Morley, 2003; Lansley, 2004; Dooley et al., 2008). In compressional settings, overly steepened thrust limbs (or overturned beds) can be imaged only using appropriate migration algorithms that can handle steep dips (Yilmaz, 2001) and only if the steep events exist in the original recorded data. If normal migration algorithms are applied to seismic data acquired over steeply dipping beds, then a wipeout zone (zone of low image quality) results (Figure 9). Similarly, in extensional tectonic settings, bed angles in the hanging wall of a fault block can be oversteepened, resulting in a zone of poor seismic imaging. This reduced quality may take the form of a triangular zone in the footwall of a fault block that can be misinterpreted as a 'shale roller.' Such wipeout zones may also be a consequence of inadequate seismic aperture. Fault-shadow effects caused by lateral velocity changes are another possibility for poor imaging across faults (see later discussion).

Velocity and Overpressure

Highly overpressured intervals can cause a decrease in acoustic-impedance velocity contrast between sand and mudstone, thereby causing a dimming of reflections from lack of impedance contrast (Maucione and Surdam, 1997). Velocity decreases as fluids, instead of rock matrix, begin to support the overburden weight significantly. When the hydrostatic pressure increases to 0.9 psi/ft (~20.5 kPa/m), the differential or effective pressure is reduced significantly, thereby causing the velocity to drop to a smaller value. Wiener et al. (2006) described low-velocity sags beneath low-velocity, overpressured 'shale structures' in the Niger Delta. Failure to take such low-velocity zones into account during time-to-depth conversion can cause significant errors in seismic processing. A decrease in acoustic velocity, accompanied by seismic wipeout, may indicate an overpressured interval that is in normal stratigraphic position; it should not be assumed to represent a mobilized mass. Highly overpressured mudstone is frequently characterized on logs by low resistivity (0.5 ohm m or less), low density (2.1 to 2.3 g/cm^3), and low acoustic velocity (6500–8500 ft/s [~1980–2590 m/s]) (Musgrave and Hicks, 1968), but these parameters do not indicate whether the mudstone is part of a diapiric mass or is in normal stratigraphic position.

In the Champion delta (Brunei), van Rensbergen and Morley (2000) inferred that an overpressure hydraulic front (i.e., front of high overpressure and hydraulic fracturing) caused a decrease in reflection continuity and amplitude. The impedance contrast at the transition between this front and adjacent lithologies was caused by mudstone compaction, diagenesis, and cementation at the front. This hydraulic front could be misinterpreted as a fault because it represents a relatively sharp boundary between a zone of good reflection continuity and one of poor reflectivity. Crosscutting reflections observed on good-quality three-dimensional (3-D) seismic data acquired over the area emphasize that this boundary was not a fault.

Although overpressure is normally characterized by low sonic velocity, it can occasionally be associated with higher velocities. Offshore Brunei, overpressure in the outer-shelf of the Baram and Champion deltas can be associated with low sonic velocity and a gradually increasing effective vertical stress path (van Rensbergen et al., 2006). Overpressures are attributed to disequilibrium compaction. However, on the inner-shelf, overpressure is associated with high sonic velocities and a sharp decrease in effective vertical stress. In shelfal areas, overpressures are attributed to inflation caused by lateral or vertical fluid transfer, not in-situ disequilibrium compaction. This difference suggests that the cause of the overpressure influences the velocity response, further suggesting that the image quality in 'shale masses' may be influenced strongly by the cause of fluid overpressure instead of the degree of organization of stratigraphy.

Distortion of seismic reflections by hydrofracturing leads to seismic signal loss. Hydraulic fracturing is a frequent occurrence in overpressured basins, in which overpressure increase approaches or exceeds lithostatic pressure (Deville et al., 2003; Deville et al., 2006). Pore pressures exceed fracture strength, thereby causing natural hydraulic fracturing of the sedimentary overburden (Engelder, 1993). Hydrofractures can vertically penetrate several kilometers (Engelder, 1993; Davies, 2007) toward the surface (Deville, 2006) to form mud pipes, sills, volcanoes, etc. Where hydraulic fractures are present, poor seismic resolution results, and a danger exists that the amount of 'mobile shale' will be overestimated as a consequence (e.g., in the Baram and Champion deltas; van Rensbergen et al., 2006).

Fault Shadow

Seismic discontinuity and false structures can occur in the shadow of normal faults, created by lateral velocity contrasts or velocity inversion, across the fault. Such velocity anomalies can create false structures, with the footwall appearing to roll into the fault, although no such roll occurs in reality (Allen and Bruso, 1989). These anomalies can cause ineffective imaging in the time domain (Gochioco et al., 2002). A risk exists that, in shallow-dipping faults, such fault-shadow anomalies may be misinterpreted as 'shale ridges.' Allen and Bruso (1989) presented an example from the Oligocene Frio Formation of the Texas Gulf Coast (United States) with more than 200 ms of apparent seismic dip in the shadow of the growth fault. However, this dip appeared to be false. They also showed how minor faulting behind the major growth fault was a false effect caused by velocity contrasts across the growth fault. These velocity effects are caused by the juxtaposition of a downthrown high-velocity zone in the footwall against a lower velocity zone in the hanging wall (Figure 10).

Prestack depth migration offers the best approach to correct for the fault-shadow effect (Trinchero, 2000; Gochioco et al., 2002) because depth migration can account for strong velocity variations across fault planes. Prestack depth migration can be used along with traveltime tomography, depth focusing analysis, and migration velocity scans to further constrain the depth image, generally removing the fault-shadow effect (Gochioco et al., 2002).

Facies, Sedimentology, and Lateral Changes

A thick homogeneous lithology (e.g., thick mudstone or thick sandstone) with no internal impedance contrasts can produce a reflection-free or low-amplitude seismic facies. If this lithology transitions laterally and vertically to a more heterogeneous facies, the wipeout zone may appear as a discreet, anomalous, reflection-free zone that could be misinterpreted as a 'shale diapir' or 'shale mass,' although the lithology is in normal stratigraphic position (Davies and Stewart, 2005). 'Shale' detachments are typically of low seismic amplitude, in places imaged as diffuse zones of poor reflectivity (e.g., Morley and Guerin, 1996, their figure 3; Bilotti and Shaw, 2005, their figure 4; Corredor et al., 2005, their figure 3). Typical low-impedance contrast may arise not only from weak lithologic contrast across the detachment surface, but also as a response to intense shearing and hydraulic fracturing along the decollement. Prodelta mudstones are generally invoked as mobile substrate in many delta systems (e.g., Niger Delta, Baram-Champion deltas, and Oligocene-aged deltas of the Texas coast). Because a low-impedance contrast is most likely to be expected in prodelta mudstones, these deposits are already predisposed to being characterized by low-reflection continuity and amplitude. For instance, the Akata Formation of marine clays in the Niger Delta is mostly devoid of internal reflections (Corredor et al., 2005, their figure 5).

Similarly, postdepositional alteration (e.g., liquefaction, slumping, and mass transport) can eradicate original bedding to produce a homogeneous mass that appears seismically transparent (Posamentier, 2004; Moscardelli et al., 2006). Such homogenous masses should be distinguishable from potential 'shale mobilization' zones based on regional mapping and stratigraphic position. Because overpressure varies laterally and vertically, chaotic seismic facies associated with 'mobile

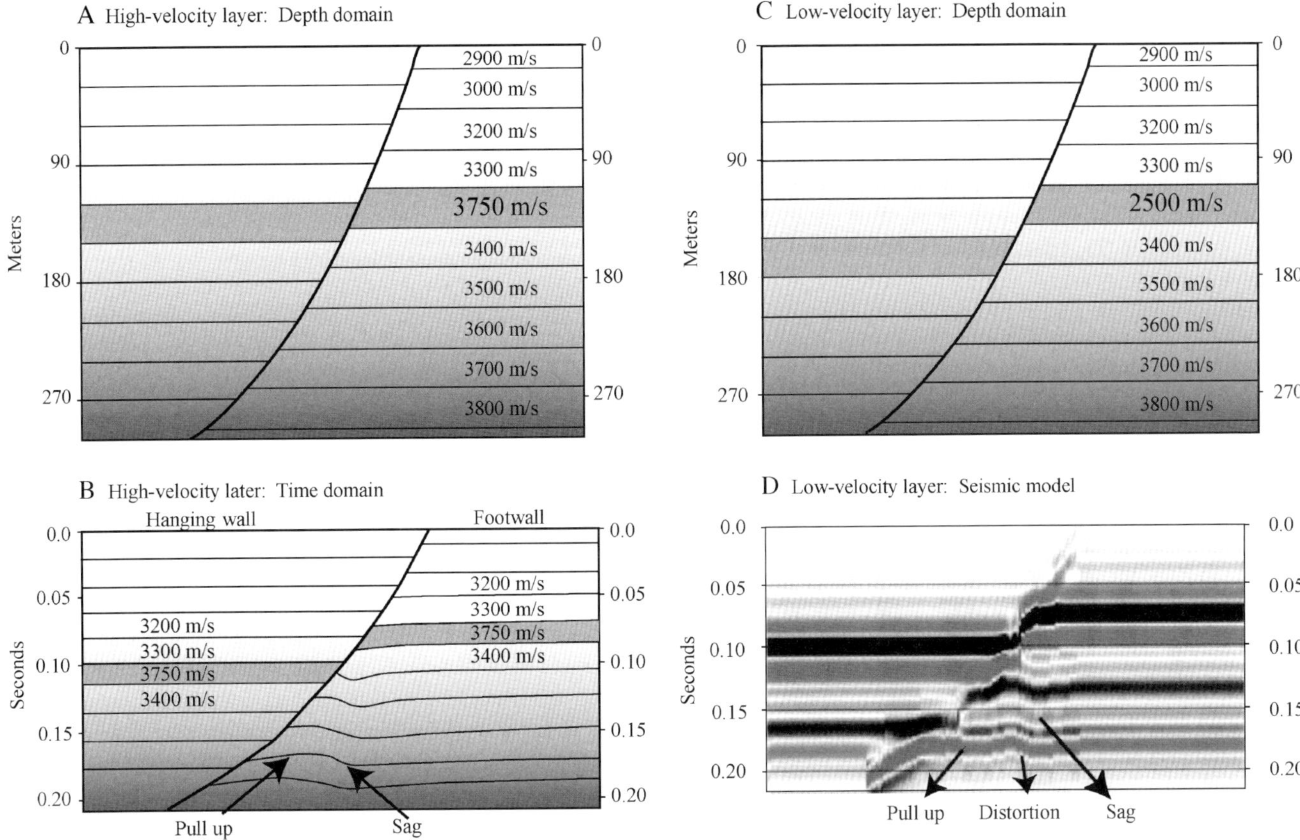

Figure 10. (A) Depth-domain geological model of a normal fault. Velocity increases with depth. A high-velocity layer (3750 m/s [12,303 ft/s]) is present partway down the fault. (B) In the time domain, velocity anomaly across the fault creates an apparent structure in the footwall caused by juxtaposition of a high-velocity zone in the footwall against a low-velocity zone in the hanging wall. (C) Depth-domain geological model of a normal fault with a low-velocity layer (2500 m/s [8202 ft/s]) present partway down the fault. (D) Variable-density display of model C. Footwall pull-up, sag, and disruption in reflection continuity also result when a low-velocity layer is modeled. Modified from Trinchero (2000), used with permission of the Society of Exploration Geophysicists. 10 m = 33 ft.

shale' should not necessarily be confined to a specific stratigraphic unit. Instead, the transition between overpressured mobilized 'shale' and overburden is expected to be diffuse and progressive (Vendeville, 2006). Consequently, the 'mobile shale' mass is typically mapped with a diffused irregular geometry.

Data Vintage and Quality

Importantly, an interpreter has to appreciate the limitations of the seismic data he or she works with. Assuming that the processed image you see is the 'real' geology is easy, but processing results are not unique. Because two-dimensional (2-D) seismic data have generally poorer resolution than 3-D seismic data, higher confidence can be placed in 3-D data, if 2-D data are ambiguous. Three-dimensional seismic data provide more complete surface and subsurface coverage and allows for better visualization. With advances in 3-D processing (e.g., PSDM), imaging has even greater clarity, and more realistic depth conversions are possible. Different survey designs and processing flows between different vintages of seismic data may account for much of the difference in observed quality. The following questions should be born in mind when interpreting in a 'shale' tectonic domain. (1) When was the seismic data acquired? (2) Was the survey design appropriate for optimal imaging of the 'shale tectonic' interval, e.g., optimal fold, aperture, etc.? (If insufficient seismic signal was acquired, no amount of reprocessing will improve the image.) (3) When were the data processed or reprocessed? (4) What algorithms were applied to the data; were the data processed to image a specific deep target or a shallow interval? Did processing account for steep dips? All of these factors allow interpreters to assess geophysical risk to the data clarity, with older, 2-D, low-fold seismic data generally having a higher risk than modern PSDM 3-D seismic data.

EXAMPLES: GEOPHYSICAL REEVALUATION

Numerous articles and studies have dealt with 'mobile shales' in stratigraphically deep basins around the world (for a review, see van Rensbergen et al., 2003; Wood et al., 2006). If, as we have indicated here, mudstones do not act in a ductile manner when buried to temperatures above 80°C (176°F), then some of these key 'mobile-shale' basins may warrant reexamination. Key basins discussed next include the Niger Delta, the Mackenzie delta, the western Gulf of Mexico, the Champion delta, the Columbus Basin of Trinidad, and the Bight Basin of Australia.

Niger Delta

The Niger Delta is a broad Tertiary delta prism up to 12 km (~39,400 ft) thick resting on overpressured marine clays of the Akata Formation (Cohen and McClay, 1996). These clays are thought to act as a mobile substrate that deforms in response to stepped progradation and loading (Doust and Omatsola, 1990) to form buried 'shale' structures on the upper slope and outer-shelf. The deep-water Niger Delta contains two fold and thrust belts that are the downdip compressional expression of gravity-driven extension occurring updip on the continental shelf (Corredor et al., 2005). Between the extensional province on the shelf and the contractional deep-water province sits a zone of interpreted 'shale ridges' and mud diapirs (Corredor et al., 2005). The Akata Formation contains multiple detachments linking contractional and extensional provinces. Zones of chaotic reflections have been interpreted as 'shale' structures, including deeply buried (>3-s two-way traveltime) triangular features, fingerlike intrusions, and broad swells (Cohen and McClay, 1996).

In the Niger Delta, 'plastic shales' are commonly invoked as a mechanism for folding, but improved seismic imaging and balanced cross sections of the deep-water fold complexes indicate that deep structures in this part of the delta are mostly duplexes of the mechanically competent, deeply buried Akata shale (Corredor et al., 2005; Gilbert, 2006). High-quality seismic data reveal primarily shear-fault-bend folds, with localized and less-common fault-bend and fault-propagation folds (Corredor et al., 2005). In support, Connors et al. (2006) used a regional 2-D seismic data set to show that because most complex contractional structures are, in essence, imbrications, they are difficult to resolve and could be misinterpreted as 'shale diapirs' in the deep fold belt. Wiener et al. (2009) used gravity and seismic velocity data to indicate that a thick zone of overpressured mudstone lies in the middle of the delta that thins basinward. This thickness change of the overpressured mudstone results in a changing structural style basinward, from high-relief detachment folds to low-relief detachment folds to shear-fault-bend fold systems. Gravity data and a seismic-based interpretation of the top 'mobile shale' are used to model the density of the mudstone. Wiener et al. (2010) used restorations to show that 'mobile-shale' withdrawal was an important sediment accommodation mechanism and that mudstone withdrawal zones were balanced downdip by inflation of thick 'mobile-shale'-cored contractional anticlines. The Niger Delta therefore remains enigmatic as to how mudstones are interpreted. This basin clearly warrants further examination regarding the origin and tectonics of these structures.

Mackenzie Delta

The Mackenzie Basin on the Beaufort Sea shelf in Arctic Canada contains more than 9 km (29,500 ft) of Tertiary and Quaternary deltaic sediments (Dixon et al., 1992; Kroeger et al., 2008). A 'mobile-shale' system is reported from the Mackenzie delta (Elsley and Graham, 2006), with 'shale diapirs,' mud volcanoes, and detachment folds all described. However, Elsley and Tieman (2010) demonstrate that structures previously interpreted as 'shale diapirs' on conventional time-domain data become resolved as tightly folded anticlines on beam PSDM data (compare Elsley and Tieman, 2010, their figure 8A, B). This new interpretation not only demonstrates the overestimation of 'mobile shale' in the region, but also has significant implications for reservoir distribution in the structure. The new interpretation features intact strata broadly distributed within a tightly folded anticline, instead of the anticline being filled with 'mobile shales.' Elsley and Tieman (2010) suggest applying such PSDM techniques to other 'shale-tectonic' basins to improve the resolution of high-dipping, folded beds that may have been interpreted previously as chaotic mobile 'shale.'

Prestack depth migration on 2008 vintage, regional 2-D data from the Mackenzie Basin shows a similar resolution of intact, highly dipping strata that had previously been interpreted as 'mobile-shale'-cored anticlines (M. Dinkelman, 2008, personal communication). These newly interpreted, faulted compressional folds form a fold train within a shallow- and deep-marine fault and thrust belt resulting from the northeastward movement of the Brooks Range (M. Dinkelman, 2008, personal communication).

Champion Delta

Offshore Brunei is composed of thick Miocene to Holocene deltaic deposits up to 10 km (6.2 mi) thick

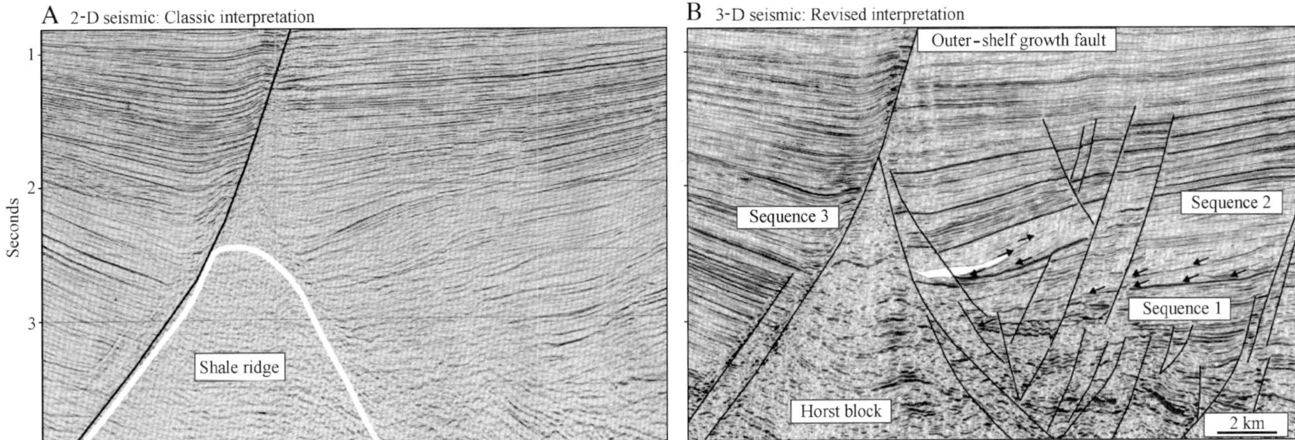

Figure 11. (A) Two-dimensional (2-D) seismic section with classic "shale ridge" interpreted in the footwall of a growth fault, from Champion delta, offshore Brunei. (B) New interpretation, based on three-dimensional (3-D) data, showing reflection stratification (horst block) where a shale ridge was previously interpreted (from van Rensbergen and Morley, 2003; reprinted by permission of the Geological Society [London] Publishing House).

(van Rensbergen and Morley, 2000) that display gravity-driven deformation belts of extensional faults and compressional folds and thrusts. The middle to upper Miocene Champion delta sits on the eastern part of the Brunei shelf, and the younger Pliocene–Pleistocene Baram delta lies on the western part of the shelf. Many of the extensional counterregional faults have chaotic seismic facies at their base (conventionally interpreted as 'mobile shale'), which was assumed to have deformed syndepositionally in a ductile manner as the sedimentary deposit 'flowed' from an area of high pressure to an area of lower pressure in response to loading. Van Rensbergen and Morley (2003) compared the conventional 'mobile shale' interpretation from 2-D data with interpretations from 3-D data and showed that the function of 'mobile shale' in the structural evolution of the Baram and Champion deltas has been significantly overestimated mostly because of poor seismic image resolution. A diapiric 'shale ridge' in the footwall was reimaged as a horst block of well-stratified prodelta clays (Figure 11). The chaotic seismic facies masked early prodelta synclines, growth faults, thrust faults, and narrow pipes of fluidized sediment (van Rensbergen and Morley, 2003). Fluid-expulsion phenomena are thought to result in large regions of compaction, creating compaction synclinal folding of the rocks (van Rensbergen and Morley, 2000).

Western Gulf of Mexico

Oligocene and lower Miocene, gravity-driven extensional fault systems of the Texas coastal and shelf zones in the westernmost Gulf of Mexico were thought to have regionally detached on a shale decollement (Bruce, 1973; Bradshaw and Watkins, 1994; Brown et al., 2004; Palmes, 2004). Triangular-shaped 'shale rollers' or 'sediment ridges' were interpreted in the growth-fault footwalls (Brown et al., 2004; Hammes et al., 2007). As seismic capabilities improve, the image resolution of these hanging-wall blocks also improves, and unpublished prestack time migration (PSTM) seismic data provide greater clarity (Seismic Exchange Inc., Texas Offshore OBS 3-D Survey; J. Sposato, 2008, personal communication). Coherent seismic reflections can be imaged in several of the footwall triangular-shaped zones. An even better resolution is provided by PSDM data (Figure 12), which show previously interpreted zones of incoherent reflectivity to actually exhibit good stratification, implying that the sediment within those zones is not significantly deformed (although Figure 12 compares PSTM and PSDM data, the PSTM data in itself was a significant improvement over the original, poststack seismic data). These reprocessed images suggest that the amount of mobile 'shale,' as traditionally interpreted, is most likely considerably overestimated in the western Gulf of Mexico. In addition, the Oligocene and lower Miocene faults that were thought to have detached on 'shale' are now mostly thought to have detached on salt (Peel et al., 1995; Radovich, 2006; McDonnell et al., 2009), which raises further uncertainty over the interpretation of large 'shale ridges' across the area.

Columbus Basin

In the Columbus Basin in offshore Trinidad, West Indies, triangular zones of relatively poor seismic imaging occur at depths of nearly 12,200 m (40,000 ft). These regions were previously interpreted as masses of 'mobile shale'

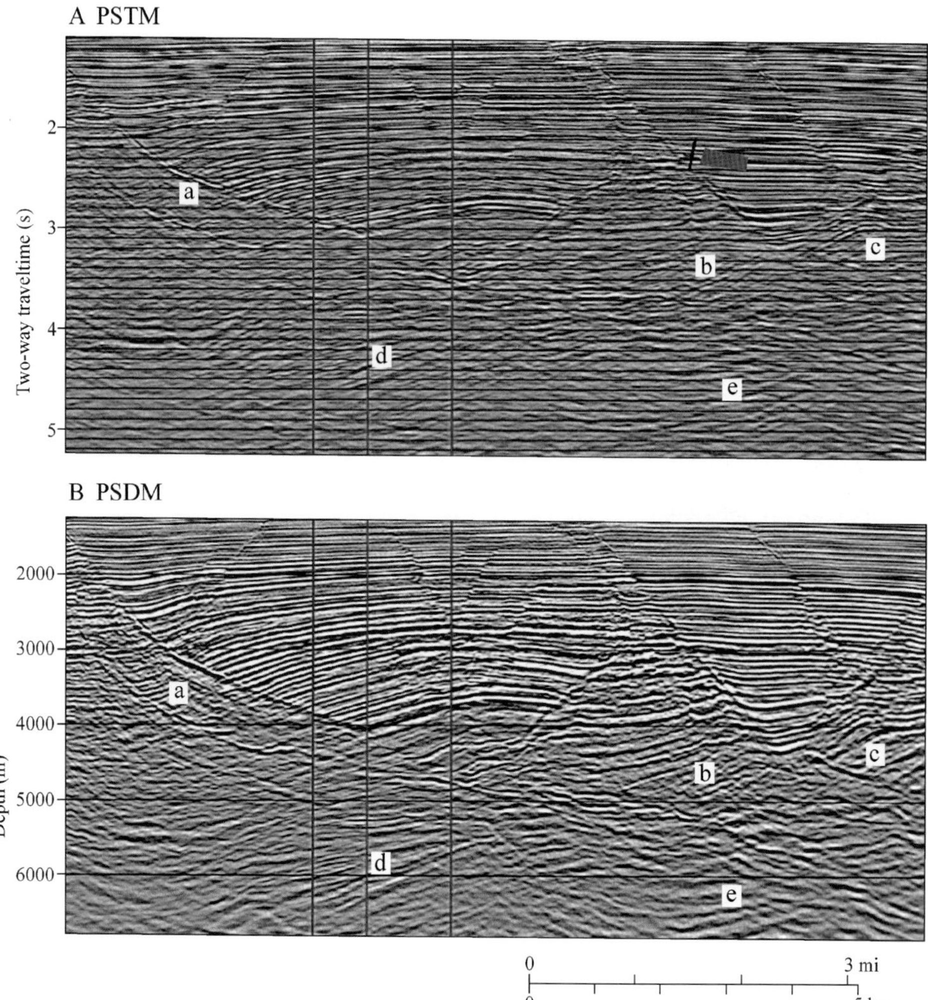

Figure 12. (A) Prestack time-migrated (PSTM) 3-D seismic data from the Mustang Island protraction area, western Gulf of Mexico. (B) Prestack depth-migrated (PSDM) version of the same data. The PSDM data have enhanced resolution and clarity. (a–e) Hanging-wall blocks with little reflectivity in the PSTM data. In contrast, these areas display significant reflectivity in the PSDM data. Seismic data are courtesy of CGGVeritas.

(Wood, 2000), but they have been reinterpreted as rafted blocks of slope mudstones bounded by (1) relatively young, basinward-dipping normal faults and (2) faulted segments of older, counterregional, normal faults (Gibson et al., 2006).

Bight Basin

Another example of 'mobile shale' occurs in the sparsely drilled Bight Basin, off southern Australia (Totterdell and Krassay, 2003). In this basin, the Cretaceous White Pointer delta shows extensive growth faults above a regionally extensive decollement. Reactive diapirs are interpreted in the footwall of each fault where a triangular zone of poor reflectivity with chaotic reflections is observed (Figure 13) (Totterdell and Krassay, 2003). Little seismic penetration occurs beneath this zone, which is characteristic of overpressured facies. Given the previously discussed examples from the Gulf of Mexico, Columbus Basin of Trinidad, and the Niger and Mackenzie deltas, we suggest that an alternative interpretation of highly folded, intact overpressured and fractured mudstone is possible. Such an interpretation could be tested using new, properly designed 3-D survey data and/or reprocessed older data.

DISCUSSION

Examples presented here from a variety of basins around the world show that improved seismic acquisition and processing have enabled new interpretations. Areas of deeply buried (>2 km [>6550 ft]) 'shale diapirism or mobile shale' are commonly revealed to be intact, steeply dipping, highly fractured, or overpressured strata. Collaboration between geophysicists and geologists will best determine if seismic incoherent zones are real or virtual. An interpreter who maps a wipeout zone as 'mobile shale' without considering other geophysical or structural reasons for the wipeout may be

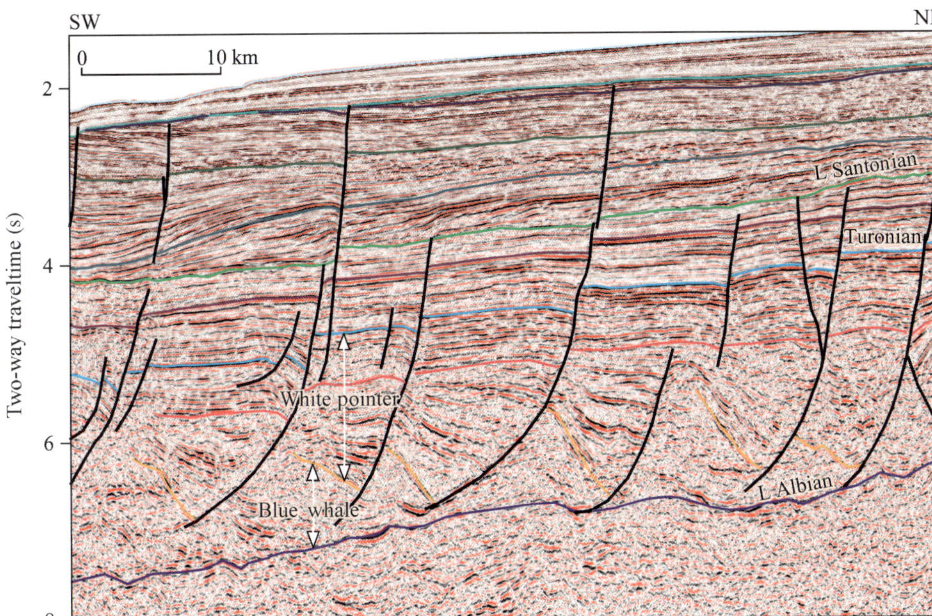

Figure 13. Two-dimensional seismic section with classic triangular wipeout zones in the footwall of growth faults, from the White Pointer delta, Bight Basin, offshore southern Australia. Wipeout zones would be classically interpreted as reactive diapirs; however, the wipeout zones may partly result from poor data quality and low resolution of the 2-D seismic data (Totterdell and Krassay, 2003; reprinted by permission of the Geological Society [London] Publishing House). 1 km = 3281 ft.

introducing an inappropriate or misleading geological model for an exploration target. The chance of success is reduced because an incorrect model impacts trap geometry, bed- and fault-seal risking, and reservoir distribution associated with structural development. Advances in seismic technology and improved understanding of mudstone behavior can combine to resolve the question of exactly how mobile the mudstone substrate is. However, several questions remain.

1) How long did overpressure persist, and could it have been transient? Overpressure needs to be sustained indefinitely for 'mobilization' to continue. As soon as fluid escapes, the system freezes, and mechanical compaction proceeds as normal. Aside from mechanical compaction, chemical compaction will continue to occur, thereby retarding the ability to deform in a plastic manner.
2) What are the proposed depths, temperatures, and timing or age of the 'shale tectonic' event? We caution against the interpretation of 'mobile shale' above approximately 2-km (6550 ft) depth because diagenesis in this shallow zone has made the rock more brittle than surface sedimentary deposits and any movement would involved brittle deformation. Because diagenetic mineral reactions are kinetic reactions that are a function of both time and temperature, 2 km (6550 ft) is a general guide. Temperature is the key differentiator, with ductile deformation unlikely above 80°C (176°F).
3) What material is present, is it capable of movement, and, if so, what type of movement? Consider the lithology of deeply buried deposits. Sediments dominated by silt-rich mudstone will exhibit different porosity and permeability than sediments dominated by clay-rich mudstone. This difference has implications for the ability of the lithology to dissipate internal pore pressure.
4) What are the geophysical uncertainties?

Whereas we cannot rule out 'shale diapirism' as a phenomenon that exists in nature (see Wiener et al., 2010), our view is that it has been overattributed and overinterpreted in the subsurface. Instances of mobile movement in shallowly buried, young, and highly overpressured sediments are observed (Hovland et al., 1998), but 'shale mobility', or rather mudstone mobility, in a deeply buried system seems to be physically impossible.

CONCLUSIONS

In conclusion, the processes of deposition, burial, compaction, diagenesis, and movement are complex interplays of many variables. Composition and mechanical behavior of fine-grained sediments at deposition are fundamentally different from the composition and mechanical behavior of the resultant rock at 80, 140, and 200°C (176, 284, or 392°F) in the deep subsurface. Consequently, responses of these rocks to processes that create a seismic image in the subsurface are dynamic and variable. Variability also exists in composition, which has profound implications for physical responses to changes in temperature and pressure (porosity, permeability, and strength) in the subsurface.

We pose several questions to consider from a seismic interpretation, as well as a chemical and physical perspective, when 'diapiric structures,' 'mud diapirs,' 'shale diapirs,' 'mobile shale,' 'shale ridges,' and 'shale tectonics' are interpreted. In general, we caution against overinterpretation of these features in the subsurface. Questions concerning the timing and nature of overpressure, temperature regime, type of tectonic movement being invoked, and geophysical risks should all be considered when working in a 'shale tectonic' setting.

ACKNOWLEDGMENTS

The authors thank Richie Miller and John Girgis for access to CGGVeritas Gulf of Mexico seismic data and Jennifer Totterdell for access to Bight Basin seismic data. Ken Thies, Paul Murray, Mark Hollanders, Hermann Lebit, and members of the Deep Shelf Gas Industry-Funded Consortium are thanked for valuable discussion and exchange of ideas. We also acknowledge Ursula Hammes and members of the State of Texas Advanced Oil and Gas Resource Recovery (STARR) research team at the Bureau of Economic Geology for discussion and valuable debate. We are grateful to Richard Davies for a thoughtful review of the manuscript. We thank Lana Dieterich for editing support and Nancy Cottington for graphic support. Publication was authorized by the director, Bureau of Economic Geology, Jackson School of Geosciences, University of Texas at Austin.

REFERENCES CITED

Ahn, J. H., and D. R. Peacor, 1986, Transmission and analytical electron microscopy of the smectite-to-illite transition: Clays and Clay Minerals, v. 34, p. 165–179, doi:10.1346/CCMN.1986.0340207.

Allen, J. L., and J. M. Bruso, 1989, A case history of velocity problems in the shadow of a large growth fault in the Frio Formation, Texas Gulf Coast: Geophysics, v. 54, p. 426–439, doi:10.1190/1.1442668.

Ansted, D. T., 1866, On the mud volcanoes of the Crimea and on the relation of these similar phenomena to deposits of petroleum: Proceeding of the Research Institute of Great Britain, v. 4, p. 628–640.

Aoyagi, K., T. Kazama, H. Sekiguchi, and G. V. Chilingarian, 1985, Experimental compaction of Na-montmorillonite clay mixed with crude oil and seawater: Water-Rock Interaction, v. 49, p. 385–392.

Aplin, A. C., and S. R. Larter, 1995, Fluid flow, pore pressure, wettability, and leakage in mudstone cap rocks, *in* P. Boult and J. Kaldi, eds., Evaluating fault and cap seals: AAPG Hedberg Series 2, p. 1–12.

Arch, J., and A. Maltman, 1990, Anisotropic permeability and tortuosity in deformed wet sediments: Journal of Geophysical Research, v. 95, p. 9035–9045, doi:10.1029/JB095iB06p09035.

Athy, A. F., 1930, Density, porosity and compaction of sedimentary rocks: AAPG Bulletin, v. 14, p. 1–24.

Awwiller, D. N., 1993, Illite smectite formation and potassium mass-transfer during burial diagenesis of mudrocks—A study from the Texas Gulf-Coast Paleocene–Eocene: Journal of Sedimentary Petrology, v. 63, p. 501–512.

Baldwin, B., 1971, Ways of deciphering compacted sediments: Journal of Sedimentary Petrology, v. 41, p. 293–301.

Baldwin, B., and C. O. Butler, 1985, Compaction curves: AAPG Bulletin, v. 69, p. 622–626.

Battaglia, S., L. Leoni, and F. Sartori, 2004, The Kubler index in late diagenetic to low-grade metamorphic pelites: A critical comparison of data from 10 A and 5 A peaks: Clays and Clay Minerals, v. 52, p. 85–105, doi:10.1346/CCMN.2004.0520109.

Bilotti, F., and J. H. Shaw, 2005, Deep-water Niger Delta fold and thrust belt modeled as a critical-taper wedge: The influence of elevated basal fluid pressure on structural styles: AAPG Bulletin, v. 89, p. 1475–1491, doi:10.1306/06130505002.

Bjorlykke, K., 1999, Principle aspects of compaction and fluid flow in mudstones, *in* A. C. Aplin, A. J. Fleet, and J. H. S. MacQuaker, eds., Muds and mudstones: Physical and fluid flow properties: Geological Society (London) Special Publication 158, p. 73–78.

Bjorlykke, K., and K. Hoeg, 1997, Effects of burial diagenesis on stresses, compaction and fluid flow in sedimentary basins: Marine and Petroleum Geology, v. 14, p. 267–276, doi:10.1016/S0264-8172(96)00051-7.

Bloch, J., 1998, Shale diagenesis: A currently muddied view, *in* J. Schieber, W. Zimmerle, and P. S. Sethi, eds., Basin studies, sedimentology, and paleontology: Stuttgart, E. Schweizerbart'sche Verlagsbuchhandlung, p. 131–146.

Bolas, H. M. N., C. Hermanrud, and G. M. G. Teige, 2004, Origin of overpressures in shales: Constraints from basin modeling: AAPG Bulletin, v. 88, p. 193–211, doi:10.1306/10060302042.

Boles, J. R., and S. G. Franks, 1979, Clay diagenesis in Wilcox sandstones of southwest Texas: Journal of Sedimentary Petrology, v. 49, p. 55–70.

Bolton, A. J., A. J. Maltman, and Q. Fisher, 2000, Anisotropic permeability and bimodal pore-size distributions of fine-grained marine sediments: Marine and Petroleum Geology, v. 17, p. 657–672, doi:10.1016/S0264-8172(00)00019-2.

Bradshaw, B. E., and J. S. Watkins, 1994, Growth-fault evolution in offshore Texas: Gulf Coast Association of Geological Societies Transactions, v. 44, p. 103–110.

Braustein, J., and G. D. O'Brien, 1965, Diapirism and diapirs: AAPG Memoir 8, 444 p.

Broichhausen, H., R. Littke, and T. Hantschel, 2005, Mudstone compaction and its influence on overpressure generation, elucidated by a 3D case study in the North Sea: International Journal of Earth Sciences, v. 94, p. 956–978, doi:10.1007/s00531-005-0014-1.

Brown, L. F., R. G. Loucks, R. H. Trevino, and U. Hammes,

2004, Understanding growth-faulted, intraslope subbasins by applying sequence-stratigraphic principles: Examples from the south Texas Oligocene Frio Formation: AAPG Bulletin, v. 88, p. 1501–1522, doi:10.1306/07010404023.

Bruce, C., 1973, Pressured shale and related sediment deformation: mechanism for development of regional contemporaneous faults: AAPG Bulletin, v. 57, p. 878–886.

Burland, J. B., 1990, 30th Rankine Lecture—On the compressibility and shear-strength of natural clays: Geotechnique, v. 40, p. 329–378.

Burst, J. F., 1959, Postdiagenetic clay mineral environmental relationships in the Gulf Coast Eocene: Clays and Clay Minerals, v. 6, p. 327–341, doi:10.1346/CCMN.1957.0060124.

Clauer, N., T. Rinckenbach, F. Weber, F. Sommer, S. Chaudhuri, and J. R. O'Neil, 1999, Diagenetic evolution of clay minerals in oil-bearing neogene sandstones and associated shales, Mahakam Delta Basin, Kalimantan, Indonesia: AAPG Bulletin, v. 83, p. 62–87.

Cohen, H. A., and K. McClay, 1996, Sedimentation and shale tectonics of the northwestern Niger Delta front: Marine and Petroleum Geology, v. 13, p. 313–328, doi:10.1016/0264-8172(95)00067-4.

Connors, C. D., B. Radovich, A. Danforth, and S. Venkatraman, 2006, Long offset, PSDM imaging advances interpretation of offshore Nigeria Akata mobile shale: Stratigraphy, contractional and diapiric structural styles, and processes, *in* L. Wood, S. Wharton, R. Davies, F. Hosein, and C. Archie, conveners, Mobile shale basins—Genesis, evolution and hydrocarbon systems: Proceedings of the AAPG/Geological Society of Trinidad and Tobago Hedberg Conference, Port-of-Spain, Trinidad, 1 p.

Corredor, F., J. H. Shaw, and F. Bilotti, 2005, Structural styles in the deep-water fold and thrust belts of the Niger Delta: AAPG Bulletin, v. 89, p. 753–780, doi:10.1306/02170504074.

Damuth, J., 1994, Neogene gravity tectonics and depositional processes on the deep Niger Delta continental margin: Marine and Petroleum Geology, v. 11, p. 320–346, doi:10.1016/0264-8172(94)90053-1.

Davies, R. J., 2007, Birth of a mud volcano: East Java: GSA Today, v. 17, p. 4–9.

Davies, R. J., and I. R. Clark, 2006, Submarine slope failure primed and triggered by silica and its diagenesis: Basin Research, v. 18, p. 339–349.

Davies, R. J., and S. A. Stewart, 2005, Emplacement of giant mud volcanoes in the South Caspian Basin: 3D seismic reflection imaging of their root zones: Journal of the Geological Society, v. 162, p. 1–4, doi:10.1144/0016-764904-082.

Davies, R. J., M. Huuse, P. Hirst, J. Cartwright, and Y. S. Yang, 2006, Giant clastic intrusions primed by silica diagenesis: Geology, v. 34, p. 917–920, doi:10.1130/G22937A.1.

Day-Stirrat, R. J., A. C. Aplin, J. Srodon, and B. A. van der Pluijm, 2008, Diagenetic reorientation of phyllosilicate minerals in Paleogene mudstones of the Podhale Basin, southern Poland: Clays and Clay Minerals, v. 56, p. 100–111, doi:10.1346/CCMN.2008.0560109.

Deville, E., 2006, Shale mobilization processes: An overview from Trinidad and the Barbados accretionary prism, *in* L. Wood, S. Wharton, R. Davies, F. Hosein, and C. Archie, conveners, Mobile shale basins—Genesis, evolution and hydrocarbon systems: Proceedings of the AAPG/Geological Society of Trinidad and Tobago Hedberg Conference, Port-of-Spain, Trinidad, 5 p.

Deville, E., A. R. Battani, S. Griboulard, J. P. Guerlais, J. P. Herbin, C. Houzay, and A. Prinzhofer, 2003, The origin and processes of mud volcanism: New insights from Trinidad, *in* P. van Rensbergen, R. R. Hillis, A. J. Maltman, and C. K. Morley, eds., Subsurface sediment mobilization: Geological Society (London) Special Publication 216, p. 475–490.

Deville, E., S.-H. Guerlais, Y. Callec, R. Griboulard, P. Huyghe, S. Lallemant, A. Mascle, M. Noble, and J. Schmitz, 2006, Liquefied vs. stratified sediment mobilization processes: Insight from the south of the Barbados accretionary prism: Tectonophysics, v. 428, p. 33–47, doi:10.1016/j.tecto.2006.08.011.

Dewhurst, D. N., A. C. Aplin, J. P. Sarda, and Y. L. Yang, 1998, Compaction-driven evolution of porosity and permeability in natural mudstones: An experimental study: Journal of Geophysical Research, v. 103, p. 651–661, doi:10.1029/97JB02540.

Dewhurst, D. N., A. C. Aplin, and J. P. Sarda, 1999a, Influence of clay fraction on pore-scale properties and hydraulic conductivity of experimentally compacted mudstones: Journal of Geophysical Research, v. 104, p. 29,261–29,274, doi:10.1029/1999JB900276.

Dewhurst, D. N., Y. Yang., and A. C. Aplin, 1999b, Permeability and fluid flow in natural mudstones, *in* A. C. Aplin, A. J. Fleet, and J. H. S. MacQuaker, eds., Muds and mudstones: Physical and fluid flow properties: Geological Society (London) Special Publication 158, p. 23–43.

Dickinson, G., 1953, Geological aspects of abnormal reservoir pressures in the Gulf Coast Louisiana: AAPG Bulletin, v. 37, p. 410–432.

Dixon, J., J. R. Dietrich, L. R. Snowdon, G. Morrell, and D. H. McNeil, 1992, Geology and petroleum potential of Upper Cretaceous and Tertiary strata, Beaufort-Mackenzie area, north-west Canada: AAPG Bulletin, v. 76, p. 927–947.

Dooley, T. P., M. P. Jackson, M. R. Hudec, and J. A. Cartwright, 2008, Modeling of strain partitioning during gravity-driven deformation of multilayered evaporites an overburden: AAPG Annual Convention, v. 17, p. 46.

Doust, H., and E. Omatsola, 1990, Niger Delta, *in* J. D. Edwards and P. A. Santogrossil, eds., Divergent passive margin basins: AAPG Memoir 48, p. 201–238.

Dugan, B., and J. T. Germaine, 2008, Near-sea floor overpressure in the deep water Mississippi Canyon, northern Gulf of Mexico: Geophysical Research Letters, v. 35, p. 1–5.

Dunoyer de Segonzac, G., 1970, The transformation of clay minerals during burial diagenesis and low-grade metamorphism: Sedimentology, v. 15, p. 281–346.

Durmishyan, A. G., 1974, Compaction of argillaceous rocks: International Geology Review, v. 16, p. 650–653.

Dzevanshir, R. D., L. A. Buryakovskiy, and G. V. Chilingarian, 1986, Simple quantitative-evaluation of porosity of argillaceous sediments at various depths of burial: Sedimentary Geology, v. 46, p. 169–175, doi:10.1016/0037-0738(86)90057-6.

Elsley, G., and P. Graham, 2006, Mobile shale in the MacKenzie

Delta—The Paktoa Shale diapir complex, *in* L. Wood, S. Wharton, R. Davies, F. Hosein, and C. Archie, conveners, Mobile shale basins—Genesis, evolution and hydrocarbon systems: Proceedings of the AAPG/Geological Society of Trinidad and Tobago Hedberg Conference, Port-of-Spain, Trinidad, 3 p.

Elsley, G. R., and H. Tieman, 2010, A comparison of prestack depth and prestack time imaging of the Paktoa Complex, Canadian Beaufort MacKenzie Basin, *in* L. Wood, ed., Mobile shale basins: AAPG Memoir 93, p. 79–90.

Engelder, T., 1993, Stress regimes in the lithosphere: Princeton, New Jersey, Princeton University Press, 457 p.

Engelhardt, W. V., and K. H. Gaida, 1963, Concentration changes of pore solutions during compaction of clay sediments: Journal of Sedimentary Petrology, v. 33, no. 4, p. 919–930.

Folk, R. L., 1974, Petrology of sedimentary rocks: Austin, Texas, Hemphill, 159 p.

Fowler, S. R., R. S. White, and K. E. Louden, 1985, Sediment dewatering in the Makran accretionary prism: Earth and Planetary Science Letters, v. 75, p. 427–438, doi:10.1016/0012-821X(85)90186-4.

Gibson, R., K. Meisling, and J. Bhajan, 2006, Tectonically-driven Plio-Pleistocene structural development of the Columbus Basin, offshore Trinidad, West Indies, *in* L. Wood, S. Wharton, R. Davies, F. Hosein, and C. Archie, conveners, Mobile shale basins—Genesis, evolution and hydrocarbon systems: Proceedings of the AAPG/Geological Society of Trinidad and Tobago Hedberg Conference, Port-of-Spain, Trinidad, 2 p.

Gilbert, E., 2006, Structural styles of shale-dominated gravity-driven thrusting, southern Atlantic margins of Africa and Brazil, *in* L. Wood, S. Wharton, R. Davies, F. Hosein, and C. Archie, conveners, Mobile shale basins—Genesis, evolution and hydrocarbon systems: Proceedings of the AAPG/Geological Society of Trinidad and Tobago Hedberg Conference, Port-of-Spain, Trinidad, 1 p.

Giles, M. R., S. L. Indrelid, and D. M. D. James, 1998, Compaction-the great unknown in basin modeling, *in* S. J. Dueppenbecker and J. E. Iliffe, eds., Basin modeling: Practice and progress: Geological Society (London) Special Publication 141, p. 15–43.

Gochioco, L. M., I. R. Novianti, and R. V. Pascual, 2002, Resolving fault shadow problems in Irian Jaya (Indonesia) using prestack depth migration: The Leading Edge, p. 911–912.

Goulty, N. R., 1998, Relationships between porosity and effective stress in shales: First Break, v. 16, p. 413–419, doi:10.1046/j.1365-2397.1998.00698.x.

Guggenheim, S., et al., 2002, Report of the Association Internationale pour l'Etude des Argiles (AIPEA) Nomenclature Committee for 2001: Order, disorder and crystallinity in phyllosilicates and the use of the "Crystallinity Index": Clay Minerals, v. 37, p. 389–393, doi:10.1180/0009855023720043.

Ham, H. H., 1966, New charts help estimate formation pressures: Oil & Gas Journal, v. 65, no. 51, p. 58–63.

Hammes, U., H. Zeng, R. Loucks, and F. Brown, 2007, All fill-no spill: Slope-fan sand bodies in growth-faulted sub-basins: Oligocene Frio Formation, South Texas Gulf Coast: Gulf Coast Association of Geological Societies Transactions, v. 91, p. 361–371.

Hansen, S., 1996, A compaction trend for Cretaceous and Tertiary shales on the Norwegian shelf based on sonic transit times: Petroleum Geoscience, v. 2, p. 159–166.

Harrison, W. J., and R. N. Tempel, 1993, Diagenetic pathways in sedimentary basins: AAPG Studies in Geology 36, p. 69–86.

Haskell, N., et al., 1999, Delineation of geologic drilling hazards using 3-D seismic attributes: The Leading Edge, v. 18, p. 373–382, doi:10.1190/1.1438301.

Hedberg, H. D., 1936, Gravitational compaction of clays and shales: American Journal of Sciences, v. 31, p. 241–287.

Hedberg, H. D., 1974, Relation of methane generation to undercompacted shales, shale diapirs, and mud volcanoes: AAPG Bulletin, v. 58, p. 661–673.

Hesse, R., 1986, Diagenesis 11. Early diagenetic pore water/sediment interactions: Modern offshore basins: Geoscience Canada, v. 13, p. 165–196.

Ho, N. C., D. R. Peacor, and B. A. van der Pluijm, 1999, Preferred orientation of phyllosilicates in Gulf Coast mudstones and relation to the smectite-illite transition: Clays and Clay Minerals, v. 47, p. 495–504, doi:10.1346/CCMN.1999.0470412.

Hovland, M., E. Nygaard, and S. Thorbjornsen, 1998, Piercement shale diapirism in the deep-water Vema Dome area, Voring Basin, offshore Norway: Marine and Petroleum Geology, v. 15, p. 191–201, doi:10.1016/S0264-8172(98)80004-4.

Hower, J., E. V. Eslinger, M. E. Hower, and E. A. Perry, 1976, Mechanism of burial metamorphism of argillaceous sediment: 1. Mineralogical and chemical evidence: Geological Society of America Bulletin, v. 87, p. 725–737, doi:10.1130/0016-7606(1976)87<725:MOBMOA>2.0.CO;2.

Huh, S., J. S. Watkins, B. E. Bradshaw, J. Xi, and R. Kasande, 1996, Regional structure and stratigraphy of the Texas shelf, Gulf of Mexico: AAPG Bulletin, v. 80, p. 1504–1535.

Ibanez, W. D., and A. K. Kronenberg, 1993, Experimental deformation of shale: Mechanical properties and microstructural indicators of mechanisms: International Journal of Rock Mechanics and Mining Sciences and Geomechanics Abstracts, v. 30, p. 723–734, doi:10.1016/0148-9062(93)90014-5.

Inoue, A., N. Kohyama, R. Kitagawa, and T. Watanabe, 1987, Chemical and morphological evidence for the conversion of smectite to illite: Clays and Clay Minerals, v. 42, p. 276–287.

Katsube, T. J., and M. A. Williamson, 1994, Effects of diagenesis on shale nano-pore structure and implications for sealing capacity: Clay Minerals, v. 29, p. 451–461, doi:10.1180/claymin.1994.029.4.05.

Knauth, P. L., 1994, Petrogenesis of chert, *in* P. J. Heaney, C. T. Prewitt, and G. V. Gibbs, eds., Silica physical behavior, geochemistry, and minerals applications: Mineralogical Society of America, Reviews in Mineralogy, v. 29, p. 233–258.

Kopf, A. J., 2002, Significance of mud volcanism: Reviews of Geophysics, v. 40, p. 1–52.

Kroeger, K. F., R. Ondrack, R. di Primio, and B. Horsfield, 2008, A three-dimensional insight into the Mackenzie Basin (Canada): Implications for the thermal history and hydrocarbon generation potential of Tertiary deltaic sequences: AAPG Bulletin, v. 92, p. 225–247, doi:10.1306/10110707027.

Kugler, H. G., 1933, Contribution to the knowledge of sedimentary volcanism in Trinidad: Journal of the Institute of Petroleum Technology, v. 19, p. 743–760.

Land, L. S., and K. L. Milliken, 1981, Feldspar diagenesis in the Frio Formation, Brazoria County, Texas Gulf Coast: Geology, v. 9, p. 314–318, doi:10.1130/0091-7613(1981)9<314:FDITFF>2.0.CO;2.

Land, L. S., and K. L. Milliken, 2000, Regional loss of SiO_2, and gain of K_2O during burial diagenesis of Gulf Coast mudrocks, U.S.A., *in* R. H. Worden and S. Morad, eds., Quartz cementation in sandstones: International Association of Sedimentologists Special Publication 34, p. 183–197.

Land, L. S., L. E. Mack, K. L. Milliken, and F. L. Lynch, 1997, Burial diagenesis of argillaceous sediment, south Texas Gulf of Mexico sedimentary basin: A reexamination: Geological Society of America Bulletin, v. 109, p. 2–15, doi:10.1130/0016-7606(1997)109<0002:BDOASS>2.3.CO;2.

Lansley, R. M., 2004, CMP fold: A meaningless number?: The Leading Edge, v. 23, p. 1038–1041, doi:10.1190/1.1813358.

Larsen, G., and G. V. Chilingar, 1983, Diagenesis in sediments and sedimentary rocks: 2. Introduction: Developments in sedimentology 25B: New York, Elsevier Scientific Publishing Company, 572 p.

MacQuaker, J. H. S., and R. L. Gawthorpe, 1993, Mudstone lithofacies in the Kimmeridge Clay Formation, Wessex Basin, southern England—Implications for the origin and controls of the distribution of mudstones: Journal of Sedimentary Petrology, v. 63, p. 1129–1143.

MacQuaker, J. H. S., C. D. Curtis, and M. L. Coleman, 1997, The role of iron in mudstone diagenesis: Comparison of Kimmeridge Clay Formation mudstones from onshore and offshore (UKCS) localities: Journal of Sedimentary Research, v. 67, p. 871–878.

Magara, K., 1968, Compaction and migration of fluids in Miocene mudstones, Nagaoka plain, Japan: AAPG Bulletin, v. 52, p. 2466–2501.

Maltman, A. J., and A. Bolton, 2003, How sediments become mobilized, *in* P. van Rensberger, R. R. Hillis, A. J. Maltman, and C. K. Morley, eds., Subsurface sediment mobilization: Geological Society (London) Special Publication 216, p. 9–20.

Maucione, D. T., and R. C. Surdam, 1997, Seismic response characteristics of a regional scale pressure compartment boundary, Alberta Basin, Canada, *in* R. C. Surdam, ed., Seals traps and the petroleum system: AAPG Memoir 67, p. 269–281.

McCarty, D. K., B. A. Sakharov, and V. A. Drits, 2008, Early clay diagenesis in Gulf Coast sediments: New insights from XRD profile modeling: Clays and Clay Minerals, v. 56, p. 359–379, doi:10.1346/CCMN.2008.0560306.

McDonnell, A., M. R. Hudec, and M. P. A. Jackson, 2009, Distinguishing salt welds from shale detachments on the inner Texas shelf, western Gulf of Mexico: Basin Research, v. 21, p. 47–59.

Meade, R. H., 1966, Factors influencing the early stages of the compaction of clays and sands-review: Journal of Sedimentary Petrology, v. 36, no. 4, p. 1085–1101.

Merriman, R. J., 2005, Clay minerals and sedimentary basin history: European Journal of Mineralogy, v. 17, p. 7–20, doi:10.1127/0935-1221/2005/0017-0007.

Milliken, K. L., 1994, Cathodoluminescent textures and the origin of quartz silt in Oligocene mudrocks, South Texas: Journal of Sedimentary Research, v. A64, p. 567–571.

Minshull, T. A., and R. White, 1989, Sediment compaction and fluid migration in the Makran accretionary prism: Journal of Geophysical Research, v. 94, p. 7387–7402, doi:10.1029/JB094iB06p07387.

Mondol, N. H., K. Bjorlykke, J. Jahren, and K. Hoeg, 2007, Experimental mechanical compaction of clay mineral aggregates—Changes in physical properties of mudstones during burial: Marine and Petroleum Geology, v. 24, p. 289–311, doi:10.1016/j.marpetgeo.2007.03.006.

Moore, J., 2005, Integration of the sedimentological and petrophysical properties of mudstone samples: Ph.D. thesis, University of Newcastle-upon-Tyne, Newcastle-upon-Tyne, UK, 300 p.

Morgan, J. P., J. M. Coleman, and S. M. Gagliano, 1968, Mudlumps: Diapiric structures in Mississippi delta sediments, *in* E. Braunstein and G. D. O'Brien, eds., Diapirism and diapirs: AAPG Memoir 8, p. 145–161.

Morley, C. K., 2003, Mobile shale related deformation in large deltas developed on passive and active margins, *in* P. van Rensberger, R. R. Hillis, A. J. Maltman, and C. K. Morley, eds., Subsurface sediment mobilization: Geological Society (London) Special Publication 216, p. 335–357.

Morley, C. K., and G. Guerin, 1996, Comparison of gravity-driven deformation styles and behavior associated with mobile shales and salt: Tectonics, v. 15, p. 1154–1170, doi:10.1029/96TC01416.

Moscardelli, L., L. Wood, and P. Mann, 2006, Mass-transport complexes and associated processes in the offshore area of Trinidad and Venezuela: AAPG Bulletin, v. 90, p. 1059–1088, doi:10.1306/02210605052.

Musgrave, A. W., and W. G. Hicks, 1968, Outlining shale masses by geophysical methods, *in* E. Braunstein and G. D. O'Brien, eds., Diapirism and diapirs: AAPG Memoir 8, p. 122–136.

Neuzil, C. E., 1994, How permeable are clays and shales: Water Resources Research, v. 30, p. 145–150, doi:10.1029/93WR02930.

Nygard, R., M. Gutierrez, R. Gautam, and K. Hoeg, 2004, Compaction behavior of argillaceous sediments as function of diagenesis: Marine and Petroleum Geology, v. 21, p. 349–362, doi:10.1016/j.marpetgeo.2004.01.002.

Nygard, R., M. Gutierrez, R. K. Bratli, and K. Hoeg, 2006, Brittle-ductile transition, shear failure and leakage in shales and mudrocks: Marine and Petroleum Geology, v. 23, p. 201–212, doi:10.1016/j.marpetgeo.2005.10.001.

Osborne, M. J., and R. E. Swarbrick, 1997, Mechanisms for generating overpressure in sedimentary basins: A reevaluation: AAPG Bulletin, v. 81, p. 1023–1041.

Palmes, S. L., 2004, Shale-detached deformation along the Texas shelf: 3D analyses, recognition criteria, and development of duplex structures in extensional settings, *in* P. J. Post, D. L. Olson, K. T. Lyons, S. L. Palmes, P. F. Harrison, and N. C. Rosen, eds., Salt-sediment interaction and hydrocarbon prospectivity: Concepts, applications, and case studies for the 21st century: Gulf Coast Section SEPM 24th Annual B. F. Perkins Research Conference (CD-ROM), p. 464–482.

Peel, F. J., C. J. Travis, and J. R. Hossack, 1995, Genetic structural provinces and salt tectonics of the Cenozoic offshore United States Gulf of Mexico: A preliminary analysis, *in* M. P. A. Jackson, D. G. Roberts, and S. Snelson, eds., Salt tectonics: A global perspective: AAPG Memoir 65, p. 153–175.

Perry, E., and J. Hower, 1970, Burial diagenesis in Gulf Coast pelitic sediments: Clays and Clay Minerals, v. 18, p. 165–177, doi:10.1346/CCMN.1970.0180306.

Posamentier, H., 2004, Stratigraphy and geomorphology of deep-water mass transport complexes based on 3D seismic data: Proceedings of the 2004 Offshore Technology Conference, Houston Texas, Paper OTC 16740, 7 p.

Post, P. J., D. L. Olson, K. T. Lyons, S. L. Palmes, P. F. Harrison, and N. C. Rosen, eds., 2004, Salt-sediment interactions and hydrocarbon prospectivity: Concepts, applications, and case studies for the 21st century: 24th Annual Gulf Coast Section SEPM: Gulf Coast Section B. F. Perkins Research Conference, CD-ROM.

Proshlyakov, B. K., 1960, Reservoir properties of rocks as a function of their depth and lithology: Geologia Nefti i Gaza, v. 12, p. 24–29.

Radovich, B. J., 2006, West Cameron 76 field and deep lower Tertiary framework revealed using sequence stratigraphy integrated with "SPICE" (abs.): Houston Geological Society Luncheon Talk: http://www.hgs.org/en/cev/500 (accessed March 25, 2010).

Savin, S. M., and S. Epstein, 1970, The oxygen and hydrogen isotope geochemistry of clay minerals: Geochimica et Cosmochimica Acta, v. 34, p. 25–42, doi:10.1016/0016-7037(70)90149-3.

Sclater, J. G., and P. A. F. Christie, 1980, Continental stretching—An explanation of the post-mid-Cretaceous subsidence of the central North-Sea Basin: Journal of Geophysical Research, v. 85, no. NB7, p. 3711-3739.

Srodon, J., 1980, Precise identification of illite/smectite interstratifications by x-ray powder diffraction: Clays and Clay Minerals, v. 28, p. 401–411, doi:10.1346/CCMN.1980.0280601.

Srodon, J., and D. D. Eberl, 1984, Illite, *in* S. W. Bailey, ed., Micas: Reviews in Mineralogy, v. 13, p. 495–544.

Srodon, J., M. Kotarba, A. Biron, P. Such, N. Clauer, and A. Wojtowicz, 2006, Diagenetic history of the Podhale-Orava Basin and the underlying Tatra sedimentary structural units (western Carpathians): Evidence from XRD and K-Ar of illite-smectite: Clay Minerals, v. 41, p. 751–774, doi:10.1180/0009855064130217.

Stewart, S. A., and R. J. Davies, 2006, Structure and emplacement of mud volcano systems in the South Caspian Basin: AAPG Bulletin, v. 90, p. 753–770, doi:10.1306/11210505090.

Swan, G., J. Cook, S. Bruce, and R. Meehan, 1989, Strain rate effects in Kimmeridge Bay Shale: International Journal of Rock Mechanics and Mining Sciences and Geomechanics Abstracts, v. 26, p. 135–149, doi:10.1016/0148-9062(89)90002-8.

Terzaghi, C., 1925a, Principles of soil mechanics: I. Phenomena of cohesion of clays: Engineering News-Record, v. 95, no. 19, p. 742–746.

Terzaghi, C., 1925b, Principles of soil mechanics: IV. Settlement and consolidation of clay: Engineering News-Record, v. 95, no. 19, p. 874–878.

Totterdell, J. M., and A. A. Krassay, 2003, The role of shale deformation and growth faulting in the late Cretaceous evolution of the Bight Basin, offshore southern Australia, *in* P. van Rensberger, R. R. Hillis, A. J. Maltman, and C. K. Morley, eds., Subsurface sediment mobilization: Geological Society (London) Special Publication 216, p. 429–442.

Trinchero, E., 2000, The fault shadow problem as an interpretation pitfall: The Leading Edge, v. 19, p. 132–135, doi:10.1190/1.1438549.

van de Kamp, P. C., 2008, Smectite-illite-muscovite transformation, quartz dissolution, and silica release in shales: Clays and Clay Minerals, v. 56, p. 66–81, doi:10.1346/CCMN.2008.0560106.

van Rensbergen, P., and C. K. Morley, 2000, 3D seismic study of a shale expulsion syncline at the base of the Champion Delta, offshore Brunei and its implications for the early structural evolution of large delta systems: Marine and Petroleum Geology, v. 17, p. 861–872, doi:10.1016/S0264-8172(00)00026-X.

van Rensbergen, P., and C. K. Morley, 2003, Re-evaluation of mobile shale occurrences on seismic sections of the Champion and Baram deltas, offshore Brunei, *in* P. van Rensberger, R. R. Hillis, A. J. Maltman, and C. K. Morley, eds., Subsurface sediment mobilization: Geological Society (London) Special Publication 216, p. 395–409.

van Rensbergen, O., R. R. Hillis, A. J. Maltman, and C. K. Morley, 2003, Subsurface sediment mobilization: Introduction, *in* P. van Rensbergen, R. R. Hillis, A. J. Maltman, and C. K. Morley, eds., Subsurface sediment mobilization: Geological Society (London) Special Publication 216, p. 1–8.

van Rensbergen, P., M. Tingay, C. Morley, J. Warren, L. Quoc Lap, I. Cartwright, and H. Darman, 2006, Overpressure and fluid migration in the evolution of the Baram and Champion deltas, Brunei, *in* L. Wood, S. Wharton, R. Davies, F. Hosein, and C. Archie, conveners, Mobile shale basins—Genesis, evolution and hydrocarbon systems: Proceedings of the AAPG/Geological Society of Trinidad and Tobago Hedberg Conference, Port-of-Spain, Trinidad, 3 p.

Vassoevich, N. B., 1960, Experiment in constructing typical gravitational compaction curve of clayey sediments: Novosti Neftianoi Tekhniki, Geologiia Seriia, v. 4, p. 11–15.

Velde, B., 1996, Compaction trends of clay-rich deep sea

sediments: Marine Geology, v. 133, p. 193–201, doi:10.1016/0025-3227(96)00020-5.

Vendeville, B. C., 2006, Similarities and differences between salt and shale tectonics, *in* L. Wood, W. Wharton, R. Davies, F. Hosein, and C. Archie, conveners, Mobile shale basins—Genesis, evolution and hydrocarbon systems: Proceedings of the AAPG/Geological Society of Trinidad and Tobago Hedberg Conference, Port-of-Spain, Trinidad, 1 p.

Waples, D. W., and G. D. Couples, 1998, Some thoughts on porosity reduction-rock mechanics, overpressure and fluid flow, *in* S. J. Düppenbecker and J. E. Iliffe, eds., Basin modeling: Practices and progress: Geologic Society (London) Special Publication 141, p. 73–81.

Weaver, C. E., 1989, Clays, muds, and shales: Developments in sedimentology 44: Amsterdam, Elsevier, 623 p.

Weller, J. M., 1959, Compaction of sediments: AAPG Bulletin, v. 43, no. 2, p. 273–310.

Wentworth, C. K., 1922, A scale of grade and class terms for clastic sediments: Journal of Geology, v. 30, p. 377–392, doi:10.1086/622910.

Wiener, R. W., M. G. Mann, D. M. Advocate, M. T. Angelich, and S. A. Barboza, 2006, Mobile shale characteristics and impact on structural and stratigraphic evolution of the Niger Delta, *in* L. Wood, S. Wharton, R. Davies, F. Hosein, and C. Archie, conveners, Mobile shale basins—Genesis, evolution and hydrocarbon systems: Proceedings of the AAPG/Geological Society of Trinidad and Tobago Hedberg Conference, Port-of-Spain, Trinidad, 2 p.

Wiener, R. W., M. G. Mann, M. T. Angelich, and J. B. Molyneux, 2010, Mobile shale in the Niger Delta: Characteristics, structure, and evolution, *in* L. Wood, ed., Shale tectonics: AAPG Memoir 93, p. 145–160.

Wood, L. J., 2000, Chronostratigraphy and tectonostratigraphy of the Columbus Basin, eastern offshore Trinidad: AAPG Bulletin, v. 84, p. 1905–1928.

Wood, L., S. Wharton, R. Davies, F. Hosein, and C. Archie, conveners, 2006, Mobile shale basins—Genesis, evolution and hydrocarbon systems: Proceedings of the AAPG/Geological Society of Trinidad and Tobago Hedberg Conference, Port-of-Spain, Trinidad, http://www.searchanddiscovery.com/abstracts/html/2006/hedberg_intl/index.htm (accessed October 5, 2010).

Worden, R. H., D. Charpentier, Q. J. Fisher, and A. C. Aplin, 2005, Fabric development and the smectite to illite transition in Upper Cretaceous mudstones from the North Sea: An image analysis approach, *in* R. P. Shaw, ed., Understanding the micro to macro behavior of rock-fluid systems: Geological Society (London) Special Publication 249, p. 103–114.

Yang, Y. L., and A. C. Aplin, 1998, Influence of lithology and compaction on the pore size distribution and modeled permeability of some mudstones from the Norwegian margin: Marine and Petroleum Geology, v. 15, p. 163–175, doi:10.1016/S0264-8172(98)00008-7.

Yang, Y. L., and A. C. Aplin, 2007, Permeability and petrophysical properties of 30 natural mudstones: Journal of Geophysical Research, v. 112, doi:10.1029/2005JB004243

Yeh, H. W., and S. M. Savin, 1977, Mechanism of burial metamorphism of argillaceous sediments: 3. O-isotope evidence: Geological Society of America Bulletin, v. 88, p. 1321–1330, doi:10.1130/0016-7606(1977)88<1321:MOBMOA>2.0.CO;2.

Yilmaz, Ö., 2001, Seismic data analysis: Processing, inversion and interpretation of seismic data: Investigations in geophysics: Tulsa, Society of Exploration Geophysicists Investigations in Geophysics 10, 2027 p.

3

Albertz, Markus, Christopher Beaumont, and Steven J. Ings, 2010, Geodynamic modeling of sedimentation-induced overpressure, gravitational spreading, and deformation of passive margin mobile shale basins, *in* L. Wood, ed., Shale tectonics: AAPG Memoir 93, p. 29–62.

Geodynamic Modeling of Sedimentation-induced Overpressure, Gravitational Spreading, and Deformation of Passive Margin Mobile Shale Basins

Markus Albertz[1]

Department of Oceanography, Dalhousie University, Halifax, Nova Scotia, Canada

Christopher Beaumont

Department of Oceanography, Dalhousie University, Halifax, Nova Scotia, Canada

Steven J. Ings

Department of Earth Sciences, Dalhousie University, Halifax, Nova Scotia, Canada

ABSTRACT

We investigate the differential loading and pore-fluid pressure required for failure and subsequent prolonged gravitational spreading of passive margin shale basins using two-dimensional analytical limit analysis and plane-strain finite element modeling. The limit analysis, supported by the models, indicates that narrow margins (slope regions that are approximately 50 to 100 km [31 to 62 mi] wide) require pore-fluid pressures that are 80–94% of the overburden weight for failure to occur, whereas for wider margins (~400 km [~250 mi] wide) like the Niger Delta, the corresponding values are 95–99%; these ranges depend on the intrinsic strength of sediments.

In the large deformation models, gravitational spreading in response to sedimentation-induced overpressure caused by delta progradation is investigated. Shale is modeled as a visco-plastic Bingham fluid that is frictional-plastic below yield and has a yield criterion that depends on the effective pressure (mean stress minus pore-fluid pressure). The velocity of the postyield flow of the shale is limited by the viscosity of the Bingham fluid, chosen for this study to be 10^{18} Pa·s. Pore-fluid pressure is predicted parametrically to be proportional to the local sedimentation rate during progradation, where the proportionality constant, k_c, depends inversely on the hydraulic conductivity. Varying the sediment progradation rate, the depth of

[1]*Present address*: Exxon Mobil Upstream Research Company, Houston, Texas, U.S.A.

DOI:10.1306/13231307M933417

onset of excess pore pressure, and k_c produces model deformation patterns consistent with seaward-directed squeeze-type flow of overpressured shale (Poiseuille flow) or wholesale seaward motion of the shale and overburden (Couette flow), depending on the overall mobility of the model.

INTRODUCTION

Weak substrates, such as salt and overpressured shale, facilitate gravity-induced deformation in sedimentary basins worldwide (e.g., Mohriak et al., 1995; Morley and Guerin, 1996; Van Rensbergen and Morley, 2000; Rowan et al., 2001; Tari et al., 2002; Morley, 2003; Modica and Brush, 2004). Particularly in passive margin settings, the overall structural style observed in response to gravitational spreading in salt and shale basins is commonly remarkably similar, including near-shelf extensional provinces, deep-water contraction, and translation of the intervening overburden (e.g., Cohen and McClay, 1996; Tari et al., 2003; Corredor et al., 2005; Camerlo and Benson, 2006). Despite such resemblance at the large scale, individual structures typically exhibit considerable differences (Figure 1). For example, in salt basins, the primary salt commonly becomes reorganized into multiple allochthonous levels and forms complex primary and secondary structures (e.g., tongues, diapirs, glaciers, canopies, and roho systems, including minibasins; e.g., Diegel et al., 1995; Jackson, 1995; Rowan, 1995; Hamiter et al., 1997; Rowan et al., 1999). Furthermore, in many salt basins, the transport of primary salt beyond the depositional limit can exceed several tens of kilometers (e.g., Fletcher et al., 1995; Wu and Bally, 2000; Ings and Shimeld, 2006). In contrast, seismic data acquired in mobile shale basins commonly display only primary structures, such as seaward-dipping and -younging normal faults, stratigraphically distinct depocenters, and deep-water fold and thrust belts (Knox and Omatsola, 1989; Haack et al., 2000; Bilotti and Shaw, 2005). Although many examples of fluid escape structures at the sea floor and the Earth's surface, such as pock marks and mud volcanoes, respectively, show that shale can locally be highly mobile (e.g., Graue, 2000;

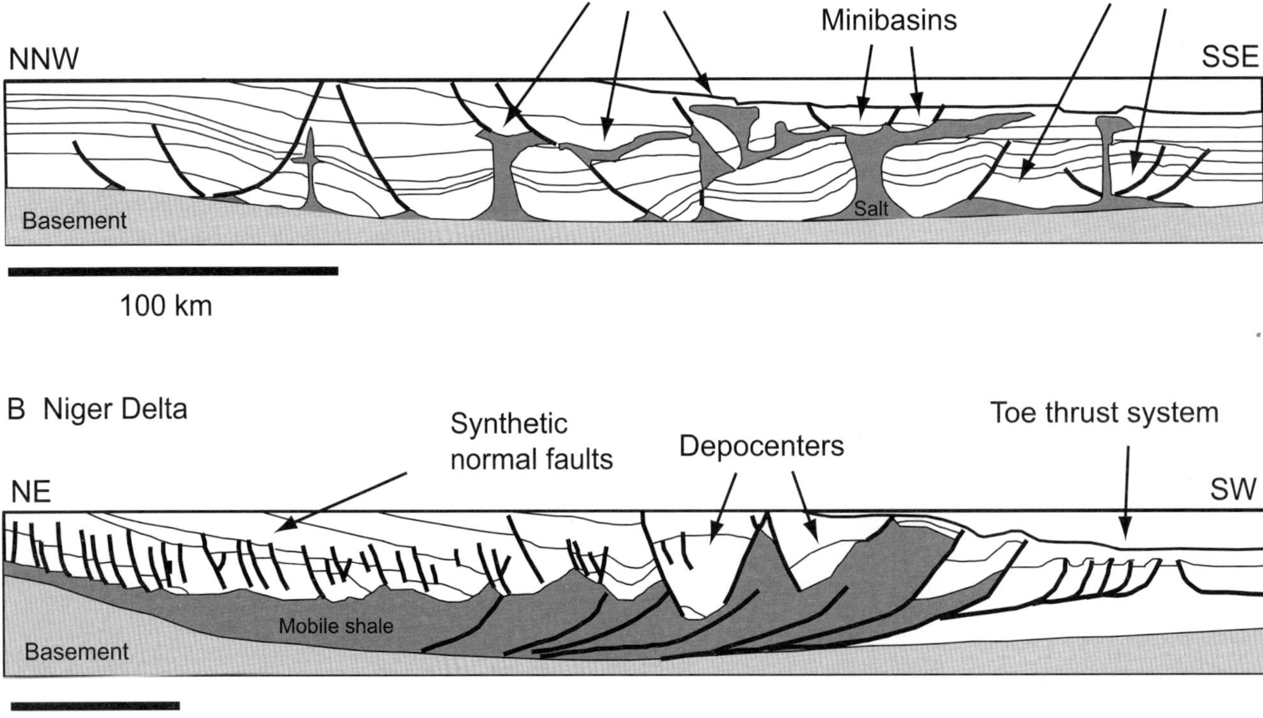

Figure 1. Structural styles typically observed in salt- and shale-bearing passive margin sedimentary basins. Both modified from Wu and Bally (2000). (A) Salt-related structures in the northeastern Gulf of Mexico. (B) Shale-related structures in the Niger Delta.

Sullivan et al., 2004), the bulk of the sediments in mobile shale basins typically remains autochthonous or parautochthonous (e.g., Wu and Bally, 2000; Van Rensbergen and Morley, 2003), implying that, at larger scales, shale seems to be much less mobile compared to salt.

Whereas advances in seismic imaging (e.g., Ratcliff, 1993), laboratory experiments (e.g., Spiers et al., 1990; Ter Heege et al., 2005a, b), and analog modeling (e.g., Ge et al., 1997; Ge and Jackson, 1998; Fort et al., 2004; Dooley et al., 2005; Gaullier and Vendeville, 2005; Vendeville, 2005) as well as numerical modeling (e.g., Tuncay and Ortoleva, 2001; Gemmer et al., 2004, 2005; Ings et al., 2004) have markedly improved our understanding of the mechanics and dynamics of salt tectonic systems, much remains to be learned concerning shale tectonic systems. For example, even the most basic concerns, such as the approximate rheological behavior of mobile shale and the processes leading to wholesale mobilization of shale and thus deformation of the overburden sediments, are rather poorly constrained. This seems surprising, given that major productive shale tectonic provinces exist around the world (e.g., southern Gulf of Mexico, Caspian Sea, northeastern Venezuela, Beaufort-Mackenzie delta, and offshore Nigeria) and that, in recent years, hydrocarbon exploration has been increasingly focused on the technically and exploratively challenging deep-water contractional provinces. Hence, both fundamental and economic benefits are to be gained from a more comprehensive understanding of shale tectonic systems.

Several studies indicate that shale mobilization is commonly associated with pore-fluid pressures that exceed hydrostatic levels (e.g., Wu and Bally, 2000; Morley, 2003; Rowan et al., 2004), suggesting that sufficiently high fluid pressures are required before shale will yield, mobilize, and thus deform overburden material. The purpose of this article is to investigate what levels of differential loading and pore-fluid pressures are required to cause failure and subsequent prolonged, large-scale gravitational spreading of passive margin shale basins. We contrast this deep-seated failure and deformation with the shallow shelf-edge and slope failure (slumping) and surficial fluid-escape phenomenon (e.g., pock marks and mud volcanoes). Furthermore, we examine how variation in sediment progradation rate, depth of onset of excess fluid pressure, and sensitivity of excess pore-fluid pressure to sedimentation rate may affect the structural evolution in mobile shale basins. We address these problems with a twofold approach: (1) use of limit analysis using analytical theory and verification by numerical experiments, and (2) use of two-dimensional, large deformation, numerical models in which pore-fluid pressures are parametrically coupled to sedimentation rates. Here, our goal is not to simulate natural systems, because this would require accurate observational constraints on various parameters, such as the physical dimensions, structure, stratigraphy, and sediment loading, as well as a thorough understanding of the processes at work in a given system. Instead, we use numerical models based on simplifying assumptions to investigate the nature of the fundamental processes, their interactions, and the structural styles that result.

AN APPROXIMATE RHEOLOGICAL MODEL OF SHALE

The analytical and numerical modeling of shale tectonics requires that we adopt a model for the rheology of shale. Sediment from continental slopes and mud volcanoes are typically composed of a large variety of clasts and rock fragments in a clay-mineral-rich matrix; the entire spectrum from clast-supported to clast-free homogeneous mud systems is covered (e.g., Lance et al., 1998; Dugan et al., 2003). The rheology of mud and shale is principally controlled by the fluid content as particle interaction becomes less important when fluid contents increase (e.g., Kopf, 2002). Although beyond the scope of our study, the friction between grains can be entirely lost when the local fluid content is extremely high, such that the shear strength and viscosity of mud are lowered greatly, causing small-scale fluidization and venting via diatremes (Dimitrov, 2002).

At the margin scale, the mechanical complexities noted above can be approximated, to first order, by a simple rheological model for shale that parametrically couples strength to pore-fluid pressure. Fluid pressure controls the transition from the undeformed frictional-plastic state to viscoplastic flow on yielding. Conceptually, this rheological model can be visualized in one dimension as a frictional block and slider in parallel with a linear dashpot (Figure 2), and is termed a nonelastic Bingham viscoplastic fluid (commonly termed a Bingham fluid; Bingham, 1922). Prior to yielding, the material does not deform because of the finite frictional-plastic strength of the block and slider. Upon yielding, the material flows viscously (controlled by the dashpot) and the bulk rheology is controlled by both the frictional block and slider and the viscous dashpot. Given the same force F and the same magnitude of viscosity η_B, the viscoplastic flow of a Bingham fluid will be slower than a comparable purely viscous flow because the total force $F = F_P + F_V$ acting on the system is shared between the plastic and viscous components F_P and F_V, respectively, whereas the total force for purely viscous flow is simply $F = F_P$. Therefore, during Bingham flow, the force F_V acting on the dashpot

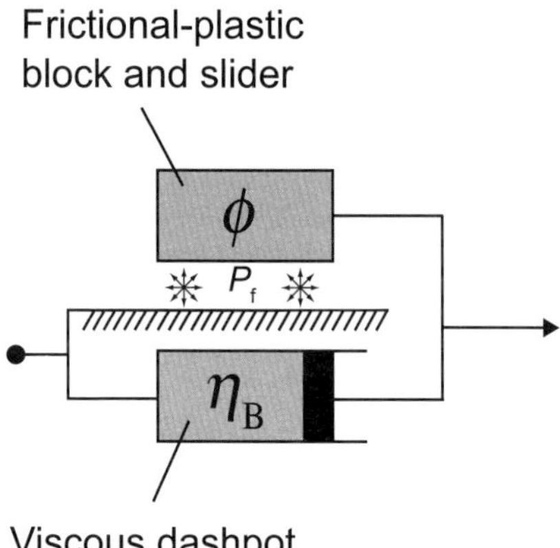

Figure 2. One-dimensional conceptual model of a Bingham viscoplastic fluid. Prior to yielding, the material does not deform because of the finite frictional-plastic strength of the block and slider. On yielding, the bulk flow of the material is determined by the plastic and viscous components where the linear viscous dashpot limits the flow velocity. ϕ = internal angle of friction; η_B = viscosity of Bingham fluid; P_f = pore-fluid pressure.

is the amount in excess of the force needed to cause yielding or $F_V = F - F_P$.

We use Terzaghi's (1923, 1943) effective stress principle to calculate the effect of pore-fluid pressure, P_f, on the Drucker-Prager yield criterion (Drucker and Prager, 1952) for the incompressible frictional-plastic shale matrix. In terms of the scalar invariants, the yield stress, σ_y, is given by

$$\begin{aligned}\sigma_y = J_{2D}^{1/2} &= (P - P_f)\sin\phi + C\cos\phi \\ &= (P - \lambda_{HR}P_o)\sin\phi + X\cos\phi \qquad (1)\end{aligned}$$

where $J_{2D}^{1/2}$ is the second invariant of the deviatoric stress, P is the dynamic pressure (mean stress), P_o is the weight of the overburden, ϕ is the internal angle of friction of the matrix, C is the cohesion, and $\lambda_{HR} = P_f/P_o$ is the Hubbert-Rubey pore-fluid pressure ratio (Hubbert and Rubey, 1959). Equation 1 states that the yield stress depends on the effective pressure $P_e = P - P_f$ such that when $P_f = 0$, the yield strength will be a maximum and as λ_{HR} approaches 1 and P approximates P_o (i.e., the fluid pressure is approximately the weight of the overburden), the yield strength tends to zero. As shown in Appendix 1, this equation applies even in the case where tectonic stresses exist, such that the mean stress is not equal to the weight of the overburden.

This simple approach allows a numerical representation of the transition from a shale system at rest with sufficiently strong grain-to-grain friction (below-yield state) to a viscoplastically deforming mobile shale system where intergrain friction is reduced by increased pore-fluid pressures (postyield state). A more complex rheological model of a Bingham fluid with a history-dependent yield strength has been successfully applied to simulate submarine slides (Gauer et al., 2005). In the remainder of this article, we refer to the Bingham fluid as Bingham shale and assume $C \sim 0$, which is reasonable for deforming continental margin sediments.

LIMIT ANALYSIS: CONDITIONS FOR YIELDING

The structural zoning typically observed in shale tectonic provinces (i.e., landward extension, seaward contraction, and intermediary translation) requires failure of the overburden. In this section, we determine the minimum amounts of differential loading (differential pressure) and pore-fluid pressure required for a shale system to yield and become unstable. We also analyze the effects of varying the depth d of the onset of excess pore-fluid pressure, the dry internal angle of friction ϕ_o in the overpressured region, and the width W across which the differential pressure is applied.

Approximate Analytical Theory

Lehner (2000) and Gemmer et al. (2004, 2005) investigated the circumstances required for yielding in salt systems using a horizontal force balance based on lubrication theory (e.g., Lobkovsky and Kerchman, 1991). We adopt this approach here and modify it for the effects of pore-fluid pressures in shale and local isostatic adjustment. We consider a simple passive margin shale system comprising a thick continental shelf, a gradually seaward-thinning section beneath the slope, and a deep-water basin with smaller sediment thickness (Figure 3A). The horizontal force balance is given by equation 2, and when the system is on the verge of failure such that the region between x_1 and x_2 is about to translate seaward, these forces assume their limiting values

$$F_t = F_c + F_w + F_b + F_i \qquad (2)$$

where F_t and F_c result from the landward and seaward horizontal stresses, respectively; F_w is the horizontal component of the water loading pressure (i.e., a buttress force); F_b is the shear traction at the base of the system caused by the weight of the overlying material; and F_i is the landward-directed horizontal component of the downdip component of the weight of the rock that arises because the base is tilted as a result of isostatic adjustment. The force F_t acts to destabilize the system whereas F_c, F_w, F_b, and F_i oppose F_t and hence tend to maintain

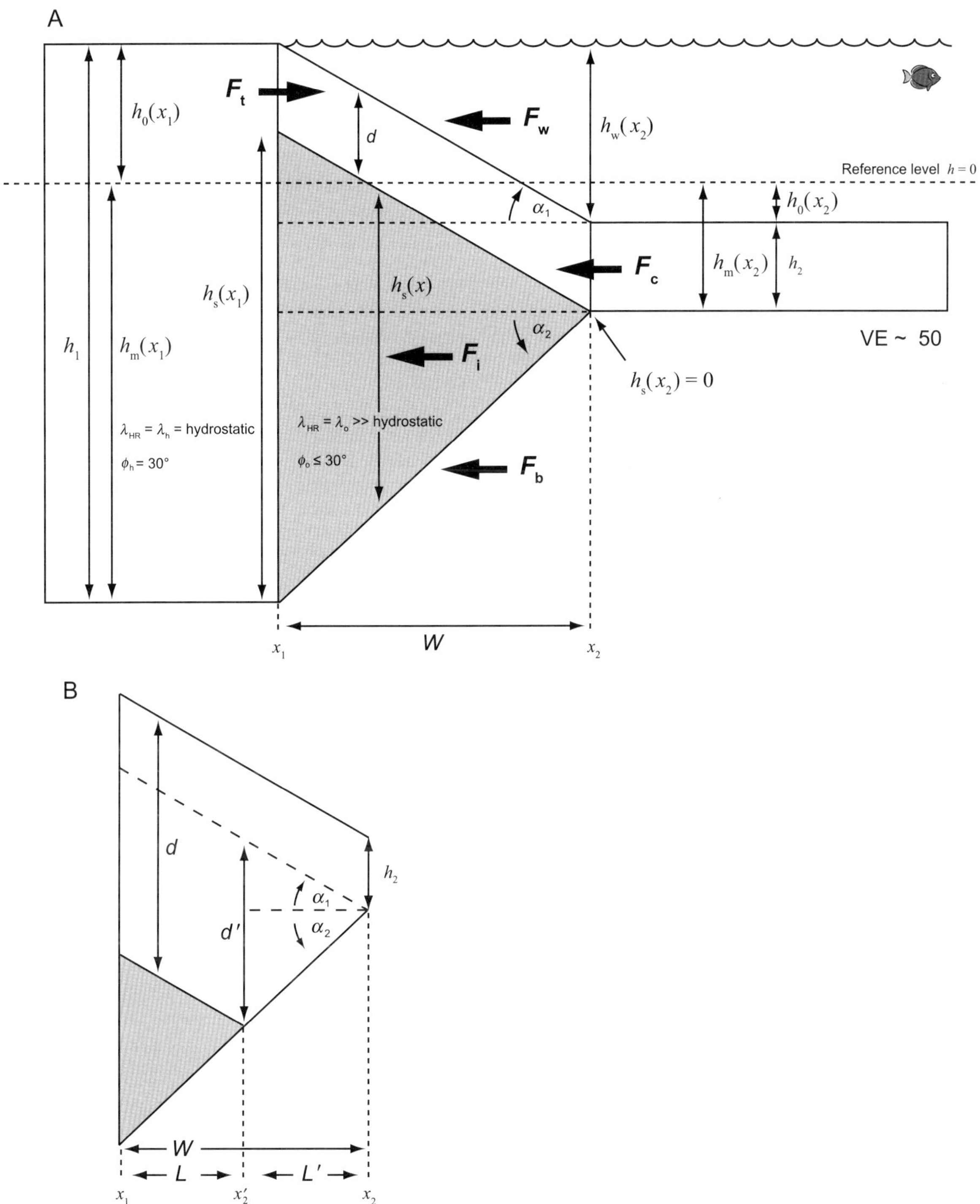

Figure 3. (A) Schematic cross section through the sedimentary section of a passive margin showing horizontal forces acting on a shale system. Gray shading indicates region of excess pore-fluid pressures. h_1 = landward total sediment thickness; h_2 = seaward total sediment thickness; h_m (x_1) and h_m (x_2) = heights of displaced mantle at locations x_1 and x_2, respectively; $h_0(x_1)$ = height of landward surface measured from reference level $h = 0$; $h_0(x_2)$ = height of seaward surface measured from reference level; $h_s(x)$ = thickness of overpressured shale in the x direction; $h_s(x_1)$ and $h_s(x_2)$ = thickness of overpressured shale at locations x_1 and x_2, respectively; d = depth of onset of excess pore-fluid pressure; $h_w(x_2)$ = maximum water depth; α_1 and α_2 = angles at surface and base, respectively, when in isostatic equilibrium. λ_h and λ_o are the Hubbert-Rubey pore-fluid pressure ratios (λ_{HR}) above and below depth d, respectively. ϕ_h and ϕ_o are the internal angles of friction in the hydrostatic and overpressured sediments, respectively. Horizontal forces: F_t = force in extensional domain; F_c = force in contractional domain; F_w = force caused by the weight of the water column; F_b = basal traction force; F_i = force arising from isostatically induced tilt of the base. (B) Illustration of adjustment of width (w) of overpressured region L when $d > h_2$. L' = amount by which W is decreased. $d' = d - h_2$. VE = vertical exaggeration.

stability. The limiting values of the individual terms are given approximately by the following equations (see Appendix 1 for the derivation):

$$F_t = \frac{1}{2}\rho g\left[k_w d^2 + k_{ws}\left(h_1^2 - d^2\right)\right] \tag{3}$$

$$F_c = -\frac{1}{2}g\left[\rho\left(k'_w d^2 + k'_{ws}\left(h_2^2 - d^2\right)\right) + \rho_w\left(h_w(x_2)\left(2d + 2h_s(x_2)\right)\right)\right] \tag{4}$$

$$F_w = -\frac{1}{2}\rho_w g\left(h_w(x_2)\right)^2 \tag{5}$$

$$F_b = -\frac{1}{2}(1 - \lambda_o)\sin\phi_o g L\left[\rho(h_1 + h_2) + \rho_w h_w(x_2)\right] \tag{6}$$

$$F_i = -\frac{1}{2}gW\ \tan\alpha_2\left[\rho(h_1 + h_2) + \rho_w h_w(x_2)\right] \tag{7}$$

where

$$k_w = \frac{1 - (1 - 2\lambda_h)\sin\phi_h}{1 + \sin\phi_h} \tag{8}$$

$$k'_w = \frac{1 + (1 - 2\lambda_h)\sin\phi_h}{1 - \sin\phi_h} \tag{9}$$

$$k_{ws} = \frac{1 - (1 - 2\lambda_o)\sin\phi_o}{1 + \sin\phi_o} \tag{10}$$

$$k'_{ws} = \frac{1 + (1 - 2\lambda_o)\sin\phi_o}{1 - \sin\phi_o} \tag{11}$$

Equations 8 to 11 are the limit state coefficients of the Rankine passive and active earth pressure modified to include the effect of fluid pressure (λ). In these equations ρ denotes the density of the sediments and ρ_w is the water density. Above and below depth d, the Hubbert-Rubey pore-fluid-pressure ratio represents what are to be assumed to be hydrostatic ($\lambda_{HR} = \lambda_h$) and overpressured conditions ($\lambda_{HR} = \lambda_o$), respectively. Correspondingly, ϕ_h and ϕ_o are the dry internal angles of friction in the hydrostatic and overpressured sediments, respectively. Other parameters and variables are defined in Figure 3A and Table 1. The maximum water depth is given by

$$h_w(x_2) = h_O(x_1) \quad h_O(x_2) \tag{12}$$

and the depths to the surfaces $h_O(x_1)$ and $h_O(x_2)$ at positions x_1 and x_2 measured from the reference level $h = 0$ (this level marks the base of models without isostastic adjustment) are derived using Archimedes' principle for local isostatic adjustment, $h(x)\rho + h_w(x)\rho_w = h_m(x)\rho_m$

$$h_O(x_1) = h_1\left(1 - \frac{\rho}{\rho_m}\right) \tag{13}$$

and

$$h_O(x_2) = h_2\left(\frac{\rho_m - \rho}{\rho_m - \rho_w}\right) - h_O(x_1)\left(\frac{\rho_w}{\rho_m - \rho_w}\right) \tag{14}$$

where ρ_m is the density of the mantle and L is the width of the overpressured region and is less than the width of the transition zone W when $d > h_2$ (Figure 3B). Width L can be calculated according to

$$L = W - \left[\frac{(d - h_2)}{(\tan\alpha_1 + \tan\alpha_2)}\right] \tag{15}$$

The case where overpressured conditions occur at shallow depths below the sediment surface ($d < h_2$) is not considered in this study because overpressures in natural deltas are generally not encountered at burial depths

Table 1. Symbols and units.

Variable	Symbol	Unit	Value
Hydraulic conductivity	K	m/s	
Permeability	κ	m^2	
Normal stress	σ	Pa	
Shear stress	τ	Pa	
Dynamic pressure (mean stress)	P	Pa	
Effective mean pressure	P_e	Pa	
Pore-fluid pressure	P_f	Pa	
Hubbert-Rubey pore-fluid pressure ratio*	λ_{HR}	–	
Sedimentation parameter	λ'	–	
Coupling coefficient	k_c	s/m	
Bingham viscosity	η_B	Pa·s	10^{18}
Cohesion	C	Pa	
Internal angle of friction*	ϕ	Degrees	
Width of transition zone	W	m	
Width of overpressured zone	L	m	
Progradation rate	V_P	cm/yr	
Sedimentation rate	V_S	cm/yr	
Density of solid	ρ	kg/m^3	2300
Density of water	ρ_w	kg/m^3	1000
Density of fluid	ρ_{fl}	kg/m^3	
Density of mantle	ρ_m	kg/m^3	3300
Fluid viscosity	η_{fl}	Pa·s	
Flexural rigidity	D	Nm	10^{23}

*Subscript h denotes hydrostatic sediments, o denotes overpressured.
1 Pa = 145 × 10^{-6} psi; 1 cm = 0.4 in.; 1 m = 3.3 ft; 1 kg = 2.2 lb; 1 Nm = 0.74 ft-lb.

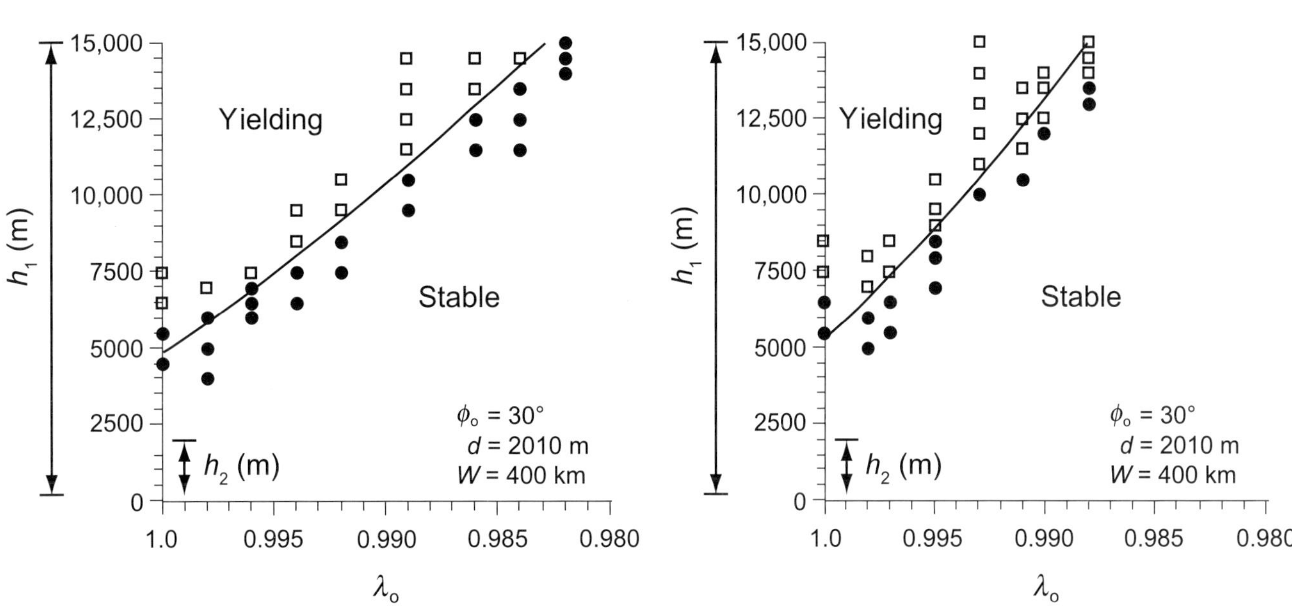

Figure 4. Results of limit analysis showing analytically determined boundary (solid line) between stable and unstable (yielding) shale systems for variation in h_1 (landward sediment thickness) and λ_o (Hubbert-Rubey pore-fluid pressure ratio in overpressured sediments). Solid circles and open squares are stable and unstable numerical models, respectively. d = depth of onset of excess pore-fluid pressure; ϕ_o and ϕ_h = internal angle of friction in overpressured and hydrostatic sediments, respectively; W = width of transition zone; h_2 = seaward sediment thickness; 1 m = 3.3 ft; 1 km = 0.6 mi. (A) Shale system without effects of water loading. (B) Same system with effects of water loading included.

shallower than ca. 2000 m (6562 ft) (e.g., Mann and Mackenzie, 1990; Yassir and Bell, 1994; Kooi, 1997).

The total landward sediment thickness h_1 decreases linearly across the width of the transition zone W to the seaward thickness h_2 (Figure 3A). Although bathymetric profiles in nature are geometrically more complex, we choose this linear geometry to keep the equations as simple as possible yet explain the basic mechanics. The equivalent equations can be derived for a more naturally shaped margin bathymetry, but they tend to obscure the basic relationships because of their complexity. Equations 3 to 15 were combined using Maple© software (version 10) and solved for the limiting value of h_1, which places the system on yield.

Results of the Analytical Theory

Figure 4 (solid line) shows the on-yield variation of h_1 for the case where the seaward sediment thickness h_2 is held constant at 2010 m (6594 ft) and the Hubbert-Rubey pore-fluid pressure ratio λ_{HR}, which equals λ_o, is varied systematically below depth d. Figure 4A (solid line) illustrates the equivalent result when the effects of water loading are removed by setting $\rho_w = 0$ (equations 4–7, 14). When water loading is included (Figure 4B), the curve rotates slightly counterclockwise because the weight of the water stabilizes the system. Both plots indicate that, for the extreme case where the yield stress is zero (i.e., $\lambda_o = 1$), a landward sediment thickness of approximately 5000 m (~16,500 ft) or a differential height of approximately 3000 m (~9900 ft) is required to cause yielding (Figure 4). As h_1 increases, progressively lower values of λ_o are necessary for yielding. For example, yielding with the thickest landward sediment thickness investigated (h_1 = 15,000 m [49,213 ft]) corresponds to λ_o values of approximately 0.983–0.988. The results presented below include the effects of water loading.

Increasing the depth of the onset of excess pore-fluid pressure d will stabilize the system because less of the sediment is overpressured and the net strength therefore increases. This effect is illustrated in Figure 5A. The curve representing d = 2010 m (6594 ft) is the same as the one in Figure 4B. With increasing d, the curves showing the on-yield condition shift to greater h_1 and λ_o values. When d = 4000 m (13,123 ft) (i.e., about twice the value of h_2) and $\lambda_o = 1$, a landward sediment thickness of 10,000 m (32,808 ft) is required for yielding. However, if h_1 = 15,000 m (49,213 ft), yielding for the same value of d requires that $\lambda_o \geq 0.997$. For simplicity, in the following section, we set $d = h_1$ = 2010 m (6594 ft).

The previous results assume a dry internal angle of friction ϕ of 30°. However, marine deposits appear to

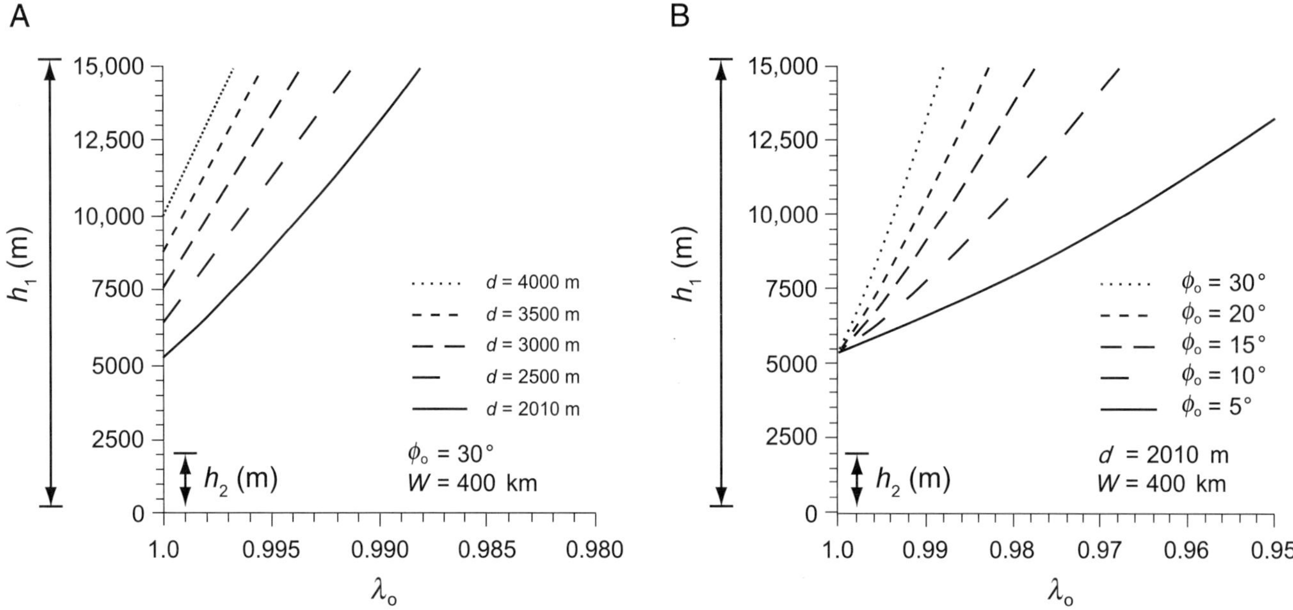

Figure 5. Results of limit analysis. (A) Analytically determined boundaries (lines) between stable and unstable (yielding) shale systems for variation in h_1 (landward sediment thickness), λ_o (Hubbert-Rubey pore-fluid pressure ratio in overpressured rocks) and d (depth of onset of excess pore-fluid pressure). (B) Analytically determined boundaries between stable and unstable (yielding) shale systems (lines) for variation in h_1, λ_o, and ϕ_o (internal angle of friction in overpressured sediment). h_2 = seaward sediment thickness; 1 m = 3.3 ft; 1 km = 0.6 mi.

be intrinsically weaker than continentally derived sediments. For example, laboratory deformation experiments indicate that ϕ values as low as approximately 4–16° may be characteristic for some marine shales, particularly when smectite dominates the composition (Saffer and Marone, 2003). Such low ϕ values have been attributed to alignment of clay minerals with platy morphology (Saffer et al., 2001). To bracket a realistic range of the internal angle of friction in nature, we vary ϕ_o in the overpressured region and plot the on-yield condition as contour lines for various ϕ_o values (Figure 5B). With decreasing ϕ_o, the stability curve rotates in a clockwise manner, indicating that, for a given landward sediment thickness, progressively lower λ_o values are required to achieve yielding. For example, a system with h_1 = 12,000 m (39,370 ft) and ϕ_o = 30° will yield when $\lambda_o \sim$ 0.992. When ϕ_o = 5°, however, yielding of a system with the same geometry requires that λ_o = 0.952 only.

Lastly, we investigate the effect of the model geometry on yielding. Figure 6A and B show the results for ϕ_o end members of 30 and 5°, respectively, where the width of the transition zone W is varied between 50 and 400 km (31 and 249 mi). To demonstrate the effect of W, consider a landward thickness of 12,000 m (39,370 ft) and an internal angle of friction of the overpressured region of 30°. Reducing W weakens the system substantially. When W is reduced from 400 to 50 km (249 to 31 mi), the corresponding λ_o required for yielding decreases from approximately 0.992 to 0.942 (Figure 6A). By comparison, replacing ϕ_o in the overpressured region by 5° and decreasing W renders the system much weaker. Notice that the critical λ_o values for transition zone widths of 400 and 50 km (249 and 31 mi) are 0.952 and 0.825, respectively. The strong weakening effect of reducing W occurs because the basal traction force F_b (equation 6), which must be exceeded to achieve yielding, decreases correspondingly. Despite the theoretical possibility of $\lambda_o < 0.9$, yielding in natural systems may still require levels of pore-fluid pressures that approach the weight of the overburden P_o. Using h_1 and W values estimated from published bathymetric profiles of the Niger Delta (Bilotti and Shaw, 2005), we can plot the geometry of the Niger Delta and ascertain that even when ϕ_o is only 5°, yielding will not occur unless $\lambda_o \sim 0.952$ (Figure 6B, bold dot).

Are such high fluid pressures, $\lambda_{HR} \geq 0.95$, attainable when hydrofracturing occurs when fluid pressure reaches the minimum principal stress, which can be less than the overburden weight in regions of extension in passive margin settings? In Appendix 2, we show that the effect of increasing pore-fluid pressure is to reduce the frictional-plastic yield strength such that as λ_{HR} approaches 1, the maximum and minimum principal stresses converge toward the weight of the overburden. Under these circumstances, the pore-fluid pressure must equal the weight of the overburden for

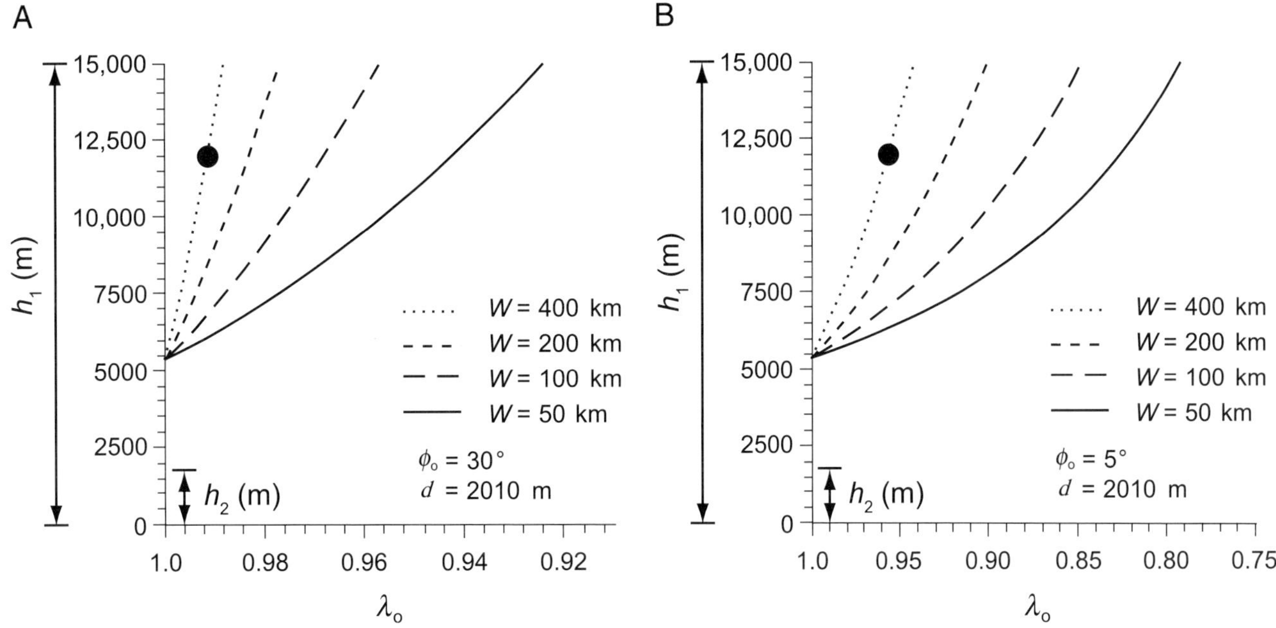

Figure 6. Results of limit analysis showing analytically determined boundaries (lines) between stable and unstable (yielding) shale systems for variation in h_1 (landward sediment thickness), λ_o (Hubbert-Rubey pore-fluid pressure ratio in overpressured sediments) and W (width of transition zone). d = depth of onset of excess pore-fluid pressure; h_2 = seaward sediment thickness; 1 m = 3.3 ft; 1 km = 0.6 mi. (A) Case where $\phi_o = 30°$. (B) Case where $\phi_o = 5°$. Bold dots indicate the approximate geometry of the Niger Delta and inferred λ_o for ϕ_o = 30 and 5°.

hydrofracturing to occur and, therefore, no basic theoretical concern exists that pore-fluid pressures cannot approach $\lambda_{HR} \sim 1$ in regions of extension in passive margin settings.

Verification of Approximate Theory by Numerical Experiments

We employ a two-dimensional plane-strain finite element model to test the accuracy of the simplified analytical limit analysis. The model domain is 1000 km (621 mi) wide and consists entirely of Bingham shale (Figure 7). The density and the internal angle of friction in the model domain are uniform ($\rho = 2300$ kg/m^3, $\phi = \phi_o = 30°$) and the initial pore-fluid pressures are hydrostatic ($\lambda_{HR} = (\rho_w g d)/(\rho g d) = 0.43$). The numerical model uses local (Airy) isostatic compensation. The total weight, including the weights of the Bingham shale and the overlying water column, rests on a fluid mantle substratum with density ρ_m = 3300 kg/m^3. In keeping with the simple

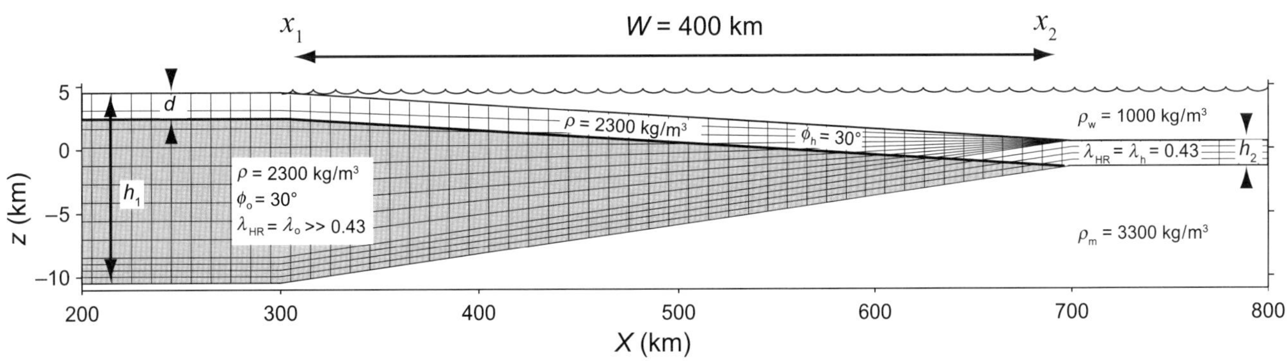

Figure 7. Configuration of numerical model for limit analysis. h_1 = landward total sediment thickness; h_2 = seaward total sediment thickness; λ_h and λ_o = Hubbert-Rubey pore-fluid pressure ratios λ_{HR} in hydrostatic and overpressured regions, respectively; ρ = density of sediments; ρ_w = density of water; ρ_m = density of mantle; W = width of transition zone between locations x_1 and x_2; d = depth of onset of excess pore-fluid pressure; 1 km = 0.6 mi; 1 m = 3.3 ft; 1 kg = 2.2 lb. The Eulerian grid is shown at a reduced resolution.

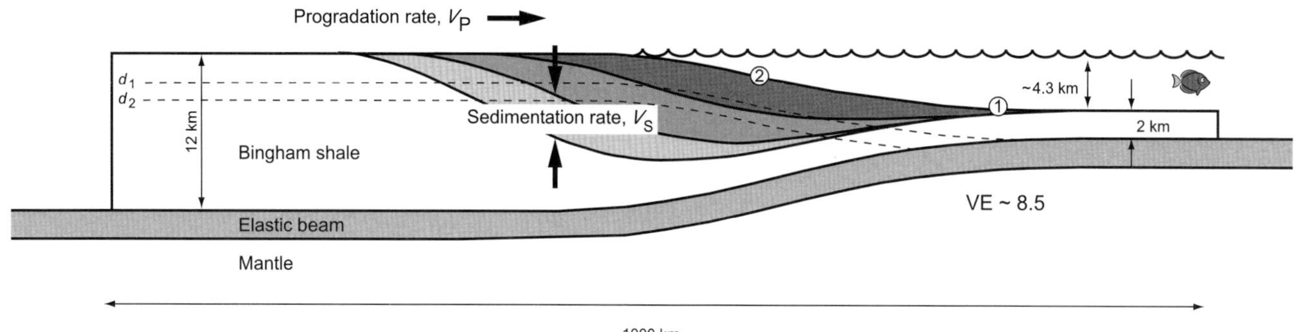

Figure 8. Configuration of a large deformation finite element model. The Hubbert-Rubey pore-fluid pressure ratio, λ_{HR}, is coupled to the vertical sedimentation rate V_S. V_P = horizontal progradation rate; d_1 = depth of onset of excess pore-fluid pressures; d_2 = depth of onset of maximum excess pore-fluid pressure. 1 and 2 indicate regions of low and high sedimentation rates, respectively. VE = vertical exaggeration; 1 km = 0.6 mi.

model geometry, we assume that the water level is the same as the surface level h_1 of the landward overburden when water loading is used. Therefore, models with larger h_1 values are marked by greater water depths. Between the model surface and depth d and $\lambda_{HR} = \lambda_h$ remains hydrostatic and $\lambda_{HR} = \lambda_o$ is varied only below d. Systematic variation of h_1 and λ_o separates stable from unstable model configurations. For a given value of h_1, we seek a value of λ_o that gives a completely stable system. We then increase λ_o until failure occurs and continue the procedure with the next h_1 value. The finite element method used is briefly described in Appendix 3.

Comparison of the results of the approximate analytical limit analysis with the complete finite element results shows a reasonably good agreement. The data match the prediction quite well for the case without water loading, although some discrepancy at extremely high λ_o values is observed when the yield stress tends to zero (Figure 4A). A similar trend is observed when water loading is included, but an additional small mismatch is observed when h_1 is larger than approximately 10,000 m (~33,000 ft) (Figure 4B). Given that the theory is an approximate solution, and that the limit analysis examines extremely high pore-fluid pressures ($\lambda_o \sim 0.980$ to 1), we believe that the limit analysis results presented above are sufficiently accurate to demonstrate the main consequences of varying the selected parameters.

LARGE DEFORMATION MODELS WITH COUPLING OF SEDIMENTATION RATE AND PORE-FLUID PRESSURE

The limit analysis provides estimates of the necessary levels of differential loading between the landward and seaward regions of a passive continental margin and pore-fluid pressures for yielding to occur and unstable flow to ensue. In addition, we examine some of the major controls on the large-deformation structural style. The objective of this section is to investigate the interaction among sediment progradation, the effect of the sedimentation rate on the generation of excess pore pressure, and the finite structures that develop when the system becomes unstable and the shale and overburden are mobilized.

We use the terms stable and unstable to refer to the state of failure of a shale system. When the rocks in a shale system remain below their yield stress, the system is stable. In contrast, when a shale system is at yield, the system fails and is referred to as unstable. We use the terms mobile and immobile to characterize the flow velocities. Hence, a mobile shale system is one that is on yield and that deforms with relatively large velocities resulting in distinctive finite structures. Conversely, an immobile shale system can either be below yield (stable) or at yield yet deforming with only very small velocities, so that only limited finite deformation occurs on geological time scales.

Configuration of Numerical Model

We use a similar two-dimensional plane-strain finite element model to that of the limit analysis, with five major modifications: (1) a more realistic bathymetric profile, (2) flexural isostasy, (3) a sediment progradation model, (4) a transition zone where λ_{HR} is increased linearly from the hydrostatic to a maximum prescribed value, and (5) a simple model that links sedimentation rate to the generation of excess pore-fluid pressure. As before, h_1 and h_2 are the total landward and seaward thicknesses of the model sediments. The model domain is 1000 km (621 mi) wide, h_1 is 12,000 m (39,370 ft), and h_2 is 2010 m (6594 ft) (Figure 8). To achieve a more realistic,

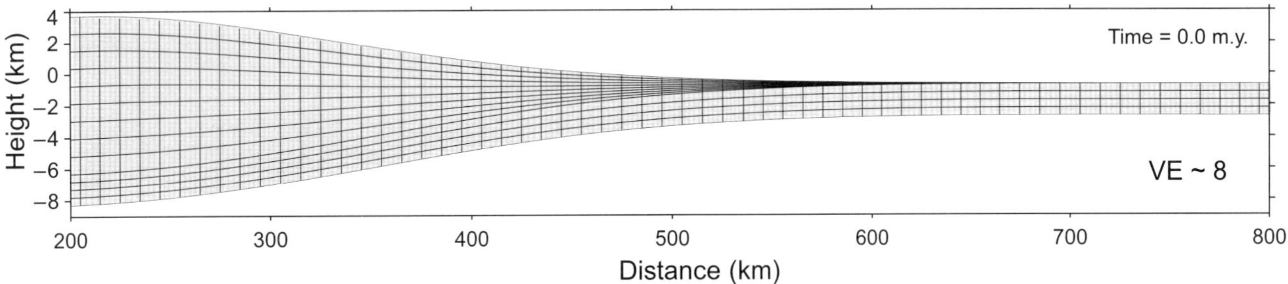

B Configuration after sediment progradation and isostatic adjustment

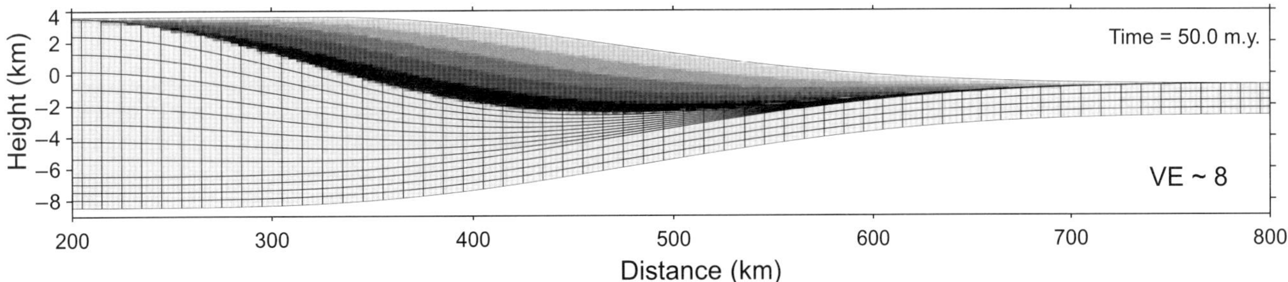

Figure 9. Stable reference Model R. (A) Initial configuration. (B) Configuration after 50 m.y. sediment progradation and isostatic adjustment. VE = vertical exaggeration; 1 km = 0.6 mi.

smooth transition from the shelf to the deep-water basin, a half-Gaussian function is used, such that

$$h(x) = \begin{cases} h_1 & \text{if } x < x_1 \\ h_2 + (h_1 - h_2)\exp\left(-\frac{4[x - x_1]^2}{W^2}\right) & \text{if } x \geq x_1 \end{cases} \quad (16)$$

and $W = 400$ km (249 mi) was chosen to approximate the current bathymetric profile of the Niger Delta (Bilotti and Shaw, 2005). When flexurally isostatically adjusted, the maximum water depth is approximately 4.3 km (~14,100 ft). All model sediments are Bingham shale. Sediment progradation is simulated by seaward translation of the half-Gaussian function across the model domain at a prescribed progradation rate V_P and by filling the area between the current model surface and the half-Gaussian curve with additional Bingham shale. Material is neither added nor removed where the current model surface is above the progradation profile. The gray-shaded bands depicted in Figure 8 represent chronostratigraphic layering, but the material properties (except the pore-fluid pressure) are uniform.

The numerical model includes flexural isostatic compensation. The model is underlain by an elastic beam (Figure 8), flexural rigidity $D = 10^{23}$ Nm, resting on a fluid mantle substratum, and density $\rho_m = 3300$ kg/m^3. Changes in the isostatic deflection at the base depend on the difference in the distribution of material (weight) between two time steps. The total weight includes the pre-existing material, material added or removed during this time step by motion (deformation), new sediments added by progradation during this time step, and the weight of the overlying water column.

Figure 9 illustrates an example of a reference model result for a shale basin where pore-fluid pressures always remain hydrostatic. The model remains stable, and flexural isostasy is the sole response to loading by the prograding sediments (Figure 9 represents Animation R at 50 m.y. model time). In the models presented below, coupling between pore-fluid pressures and sedimentation rate is modeled parametrically to generate excess pore-fluid pressures and thus cause shale mobilization and deformation.

Parametric Model for the Coupling Between Sedimentation Rate and Excess Pore-Fluid Pressure

Borehole measurements in natural deltas commonly reveal a systematic increase of pore-fluid pressures with depth (e.g., Mann and Mackenzie, 1990; Yassir and Bell, 1994): at depth d_1, pore-fluid pressures start to deviate from hydrostatic and gradually approach the weight of the overburden at depth d_2 and continue at similar levels at greater depths (Figure 10A). We represent this type of pore-fluid pressure profile parametrically. At depths less than d_1, pore-fluid pressures are hydrostatic. At depths greater than d_2, the pore-pressure ratio is

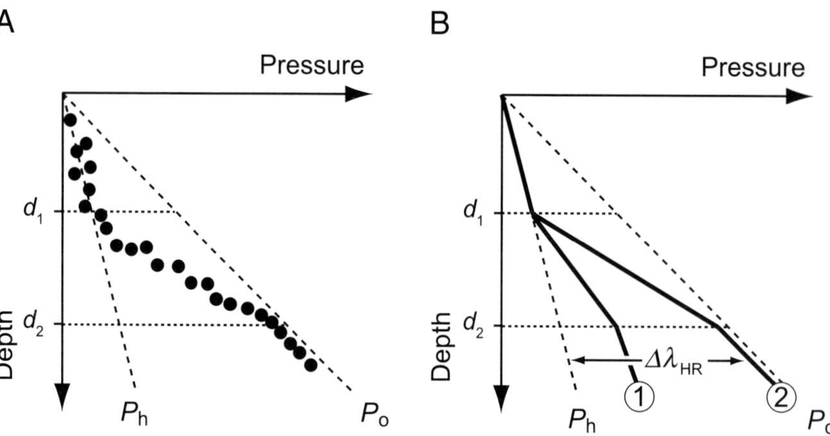

Figure 10. Diagrams showing pore-fluid pressure as a function of depth. P_h = hydrostatic pore-fluid pressure. P_o = overburden weight. (A) Typical distribution of pore-fluid pressure in basins (e.g., Mann and Mackenzie, 1990). (B) Schematic illustration of increasing pore-fluid pressure by the Hubbert-Rubey pore-fluid pressure ratio $\Delta\lambda_{HR}$ in regions of low (1) and high (2) vertical sedimentation rates. d_1 = depth at which pore-fluid pressures start to deviate from hydrostatic pressure; d_2 = depth at which they equal the weight of the overburden.

augmented with respect to the hydrostatic level by an amount

$$\Delta\lambda_{HR} = \kappa_c{}'S \tag{17}$$

where V_S is the local vertical sedimentation rate (derived from the isostatically adjusted horizontal delta progradation rate V_P) and k_c is a coupling coefficient explained later. Where V_S is moderate, the pore-fluid pressure is increased to a value (point 1 in Figures 8, 10B) and in cases of high V_S to a correspondingly higher value (point 2 in Figures 8, 10B). The maximum value λ_{HR}^{MAX} is limited to one that is close to the weight of the overburden. In the depth range between d_1 and d_2, the pore pressure increases linearly from the hydrostatic value at d_1 to the fully augmented value at d_2. This formulation allows newly added sediments to become overpressured when buried sufficiently deeply. Hence, the top of the overpressured shale is not bound to a particular stratigraphic horizon but evolves diachronously.

The coupling coefficient k_c can be shown to be inversely proportional to the hydraulic conductivity $K = \kappa\rho_{fl}g/\eta_{fl}$, where κ is the permeability, g is the acceleration due to gravity, ρ_{fl} is the fluid density, and η_{fl} is the fluid viscosity, in the following way. Audet and McConnell (1992) and Fowler and Yang (1999), among others, defined the sedimentation parameter $\lambda' = K/V_S$ where V_S is the vertical sedimentation rate. Consider a parameter $p = 1/\lambda'$, such that $p = V_S/K$ measures the ratio of the sedimentation rate to the hydraulic conductivity (which controls the rate of dissipation of excess pore-fluid pressure; Kooi, 1997). When $p >> 1$ the sediment loading is too fast for normal fluid expulsion and compaction, and the converse. The change in λ_{HR} (equation 17) is also a measure of the same tendency to generate excess pore-fluid pressure during sedimentation.

It follows that if $\Delta\lambda_{HR} \propto p$, then $k_cV_S \propto V_S/K$ and $k_c \propto 1/K$, that is, the coupling coefficient is inversely proportional to the hydraulic conductivity. A high value of k_c implies a low hydraulic conductivity and the converse. This implies that, in the parametric pore-fluid pressure model, the choice of the value of k_c corresponds to choosing a value of the hydraulic conductivity. If $k_cV_S \propto V_S/K >> 1$, the parametric model will become strongly overpressured. Correspondingly, $\lambda' = K/V_S << 1$ will yield the same result. In summary, the sedimentation rate modulates the pore-fluid pressure through the parameter k_c; the pore-fluid pressure regulates the effective strength of the model sediment column through equation 1.

Overview of the Numerical Models

Below we describe the results of six numerical models organized in two sets of three. In the first set, the coupling between sedimentation rate and pore-fluid pressure is instantaneous. That is, the excess pore-fluid pressure is always proportional to the current rate of sedimentation at a given location. When sedimentation ceases, the excess pressure immediately decays. This formulation represents an end member mechanical behavior where it is assumed that sedimentation can maintain excess pore-fluid pressures consistent with the choice of k_c, but that fluid pressures adjust rapidly to changes in the sedimentation rate. Model 1 illustrates the effects of relatively high hydraulic conductivity (small coupling coefficient k_c) and a sediment progradation rate V_P of 0.25 cm/yr (0.1 in./yr). Model 2 is identical with model 1 except a higher value of k_c is used, which implies lower hydraulic conductivities. Model 3 is also based on model 1 but demonstrates how doubling V_P to 0.5 cm/yr (0.2 in./yr) and choosing an intermediate value for k_c affect the results.

In the second set of models, excess pore-fluid pressures are maintained. That is, excess pressures remain at their maximum values. This behavior represents an end member system in which compaction reduces the

Table 2. Parameters used in large deformation models.

Model	V_P (cm/yr)	k_c (s/m)	ϕ (degrees)	λ_{HR}^{MAX}	d_1 (m)	d_2 (m)	η_B (Pa·s)	W (km)
R*	0.25	0	30	0.998	2500	4000	10^{18}	400
1	0.25	2.52×10^{11}	30	0.998	2500	4000	10^{18}	400
2	0.25	5.05×10^{11}	30	0.998	2500	4000	10^{18}	400
3	0.50	1.58×10^{11}	30	0.998	2500	4000	10^{18}	400
4	0.25	2.52×10^{11}	30	0.998	2500	4000	10^{18}	400
5	0.25	2.52×10^{11}	30	0.998	2000	4000	10^{18}	400
6	0.25	2.52×10^{11}	$30 \rightarrow 19.47$**	0.998	2500	4000	10^{18}	400

*R denotes stable reference model; 1 cm = 0.4 in.; 1 m = 3.3 ft; 1 Pa = 145×10^{-6} psi; 1 km = 0.6 mi.
**Value for ϕ representing strain softening.

hydraulic conductivity such that once generated, pore pressures are not dissipated on the time scale of the model run. All models in this set have the same V_P and k_c as model 1. Model 4 is equivalent to model 1 apart from maintaining excess pore-fluid pressures. Model 5 shows the consequences of reducing the depth d_1 of onset of excess pore-fluid pressures from 2500 to 2000 m (8202 to 6562 ft). Finally, in model 6, we explore how strain softening affects finite deformation. The model parameters are summarized in Table 2. The value of k_c is constant during each model run. Each of these models should be compared to the stable reference model (Figure 9), model R. The Lagrangian tracking grid in Figures 11–16 is shown at reduced resolution for better visualization of the deformation.

Model Results

Model Set 1: Excess Pore-Fluid Pressures Proportional to the Current Sedimentation Rate

Model 1: $V_p = 0.25$ cm/yr, $k_c = 2.52 \times 10^{11}$ s/m ($V_p = 0.09$ in./yr, $k_c = 7.68 \times 10^{10}$ s/ft). With time, new material progrades from landward to seaward across the model domain. We illustrate sediment progradation (Figure 11A–C, Animation 1), the spatial distribution of pore-fluid pressures (Figure 11D), and the accumulated finite strain after 50 m.y. (Figure 11E). At any given time, the areas receiving the highest amounts of sediment loading at the steepest section of the progradation profile are subject to generation of maximum excess pore-fluid pressures as indicated by λ_{HR}, the Hubbert-Rubey pore-fluid pressure ratio (Figure 11D). Recall that as λ_{HR} tends to 1, pore-fluid pressures approach the weight of the overburden. As the sediments prograde across the model from landward to seaward, the overpressured domain translates with the progradation profile and is augmented by rapid sedimentation in the depocenters that develop (e.g., near 400 km [249 mi], Figure 11C). Notice that the width of the actual zone attaining λ_{HR} values sufficiently large to induce failure (i.e., $\lambda_{HR} \sim 0.998$) is only ca. 15% of the width of the progradation profile. As a result of attaining maximum λ_{HR} values, two shear zones form; one lies at the base of the model, the other approximately coincides with the top of the overpressured region (Figure 11E). The material sandwiched between the two shear zones is squeezed and transferred seaward. This squeeze flow leads to a minor shale withdrawal beneath the upper slope and correspondingly inflation beneath the middle of the slope. The finite strain plots indicate that most of the finite deformation is accommodated by the two shear zones and the remaining material remains relatively undeformed. This viscoplastic plug-type flow can also be seen from the deformed Lagrangian mesh (Figure 11B, C). As the model continues to evolve, the inactive sections of the upper shear zone are deformed by subsequent extension of the overlying sediments (Figure 11E). Extension and contraction generally occur in a distributed fashion, but sporadically, the deformation also extends from the upper shear zone to the model surface via localized high strain zones, such as normal faults beneath the shelf and folds and thrusts near the toe of the slope (Figure 11E).

During some time steps, the horizontal shale transfer is large enough to raise the surface of the overburden above the progradation profile near the toe of the slope where shale is inflated vertically. Because of the progradation model and the instantaneous coupling between λ_{HR} and loading, these areas cease to receive sediments and the excess fluid pressures dissipate very rapidly and return to the hydrostatic state. As a consequence, hydrostatically pressured sediment columns divide and laterally bound two or more distinct overpressure compartments (Figure 11D).

Overall, model 1 is characterized by a restricted region of very high pore-fluid pressure (e.g., Figure 11D) and therefore a relatively low level of finite deformation. The most distinguishing structure in this model is an approximately 2-km (1.2-mi)-deep and approximately 40-km

Model 1

A Tracking grid
Time = 7.0 m.y.
VE ~ 8
Height (km)
Distance (km)

B Tracking grid
Time = 41.0 m.y.
VE ~ 8
Height (km)
Distance (km)

C Tracking grid
Time = 50.0 m.y.
VE ~ 8
Height (km)
Distance (km)

D Pore-fluid pressure
Time = 41.0 m.y.
VE ~ 8
Height (km)
Distance (km)
Velocity
0.1 cm/yr
λHR
0.40 0.50 0.60 0.70 0.80 0.90 0.99

E Finite strain
Time = 50.0 m.y.
VE ~ 8
Height (km)
Distance (km)
2nd invariant of finite strain
5x10−3 10−2 5x10−2 10−1 5x10−1 1 5 10¹ 5x10¹

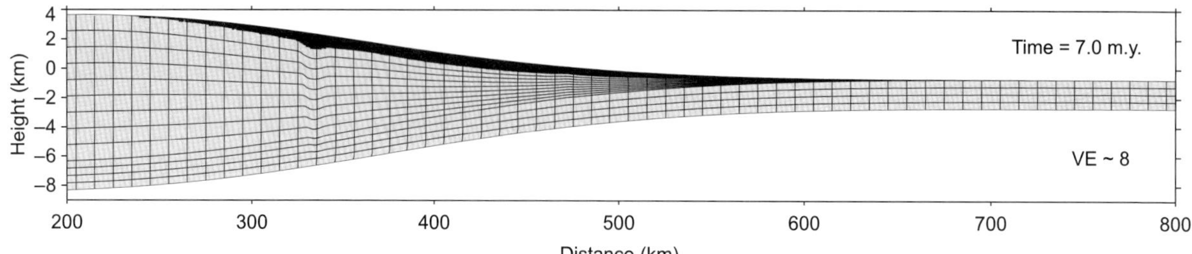

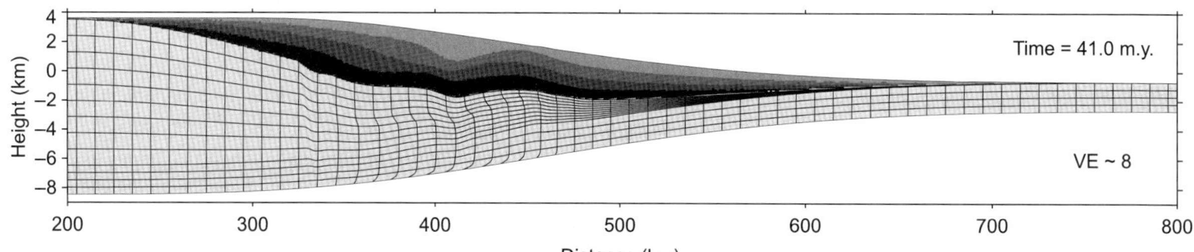

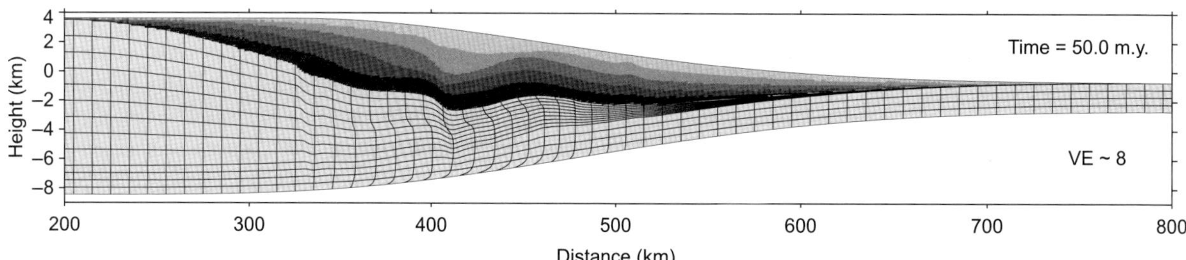

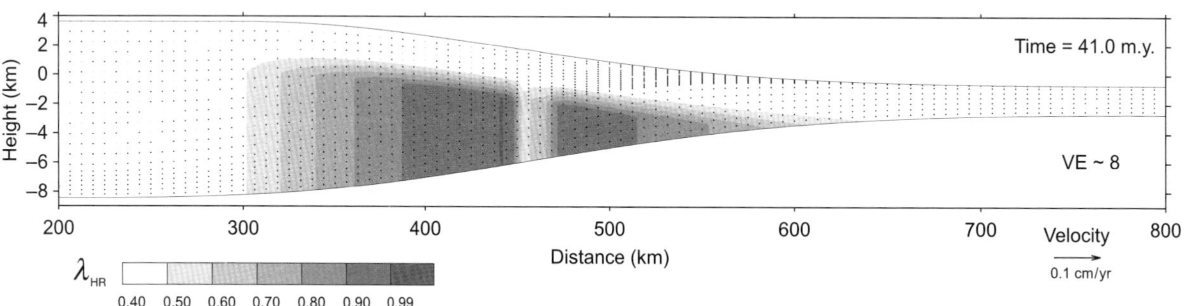

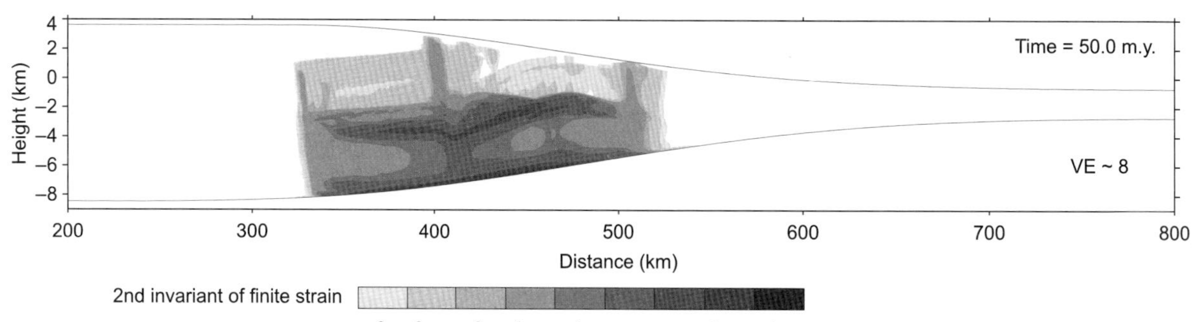

(25-mi)-wide basin, which develops within ca. 10 m.y. (Figure 11C). Regions of enhanced sedimentation also develop landward and seaward of the main basin (Figure 11B, C). These differences with respect to model R (Figure 9B) can be accounted for by the squeeze flow of the overpressured shale (Figure 11C, E). In other respects, model 1 develops mostly intact sedimentary wedges closely resembling those of model R (compare Figures 9B, 11C).

Model 2: $V_p = 0.25$ cm/yr, $k_c = 5.05 \times 10^{11}$ s/m ($V_p = 0.09$ in./yr, $k_c = 1.54 \times 10^{11}$ s/ft). Model 2 (Figure 12A–C, Animation 2) is the same as model 1 except that the coupling coefficient is doubled, which has the effect of doubling $\Delta\lambda_{HR}$ for the same sedimentation rate. As a result, the zone of maximum excess pore-fluid pressures is much wider than in model 1 (Figure 12D), leading to a behavior that differs sharply from the one observed in model 1.

Similar to model 1, two subhorizontal shear zones develop in model 2, and shale is mobilized in seaward-directed squeeze flows (Figure 12B, C, E). Throughout the model run, strongly localized high-strain zones breach the model surface, resulting in structures that can be interpreted as normal faults and folds and/or thrust faults beneath the shelf and the lower slope, respectively. The strain plots in Animation 2 indicate that these zones of extension and contraction are generally linked by the basal shear zone. Displacement on normal faults in model 2 is considerably larger than in model 1 and individual normal faults sole into the basal shear zone (Figure 12E). Consequently, strata deposited on the progressively downward displaced hanging wall thicken toward the normal faults and thus display growth character. As the sediment progradation migrates seaward, the currently active normal fault ceases to accommodate displacement and is replaced by a younger normal fault farther seaward, resulting in a progressive seaward downstepping of the upper shear zone (Figure 12E). Hence, stratigraphically distinct depocenters form, which are bounded on either side by the diachronous seaward-dipping normal faults (Figure 12C, E). Each currently active normal fault marks the seaward termination of the current depocenter and simultaneously bounds the next younger depocenter that forms seaward. As a result, model 2 is marked by the episodic development of stratigraphically distinct, migrating-and younging-seaward depocenters akin to a stepwise outbuilding of a delta. The depocenters are approximately 25–40 km (~16–25 mi) wide and approximately 2–3 km (~6600–9800 ft) deep (Figure 12C).

The formation of depocenters and corresponding fold belts can be subdivided into four individual deformation periods. Although depocenter A develops during the first deformation period, a corresponding fold belt develops, in which individual folds become younger landward (Figure 12A). As the next seaward depocenter develops during deformation period B, the corresponding contraction shifts seaward to initiate fold belt B where individual folds again are younger landward, and the cycle begins anew (Figure 12B, C). In a similar way to Model 1, shale squeeze flow, inflation, and folding cause local uplift of the model surface above the progradation profile. Until these regions are covered again by prograding material, sedimentation ceases ($V_S = 0$) and the pore-fluid pressure beneath these areas is reset to hydrostatic pressure. For this reason, model 2 is subject to extensive compartmentalization of excess pore-fluid pressure (λ_{HR} plots in Animation 2), and the hydrostatic zones limit the lengths of the active associated subhorizontal shear zones that bound the squeeze flow.

Model 3: $V_p = 0.5$ cm/yr, $k_c = 1.58 \times 10^{11}$ s/m ($V_p = 0.09$ in./yr, $k_c = 4.82 \times 10^{10}$ s/ft). Model 3 was designed to examine how larger amounts of more rapidly deposited sediments may affect the structural evolution of a mobile shale system (Figure 13A–C, Animation 3). Keeping all other parameters the same as in model 1, the sediment progradation rate V_P was increased by a factor of 2, and the value of the coupling coefficient k_c was chosen so that the width of the zone of maximum excess pore-fluid pressures approximates that in model 2 during most time steps.

In a similar manner to model 2, model 3 develops two shear zones between which overpressured Bingham shale is squeezed seaward (Figure 13B–E). Likewise, the basal shear zone accommodates significantly more finite strain, and the upper shear zone is reworked during progressive deformation by shear zones that link the surface and the basal shear zone (Figure 13E). Model 3 is characterized by more efficient shale withdrawal than model 2. For example, model 2 develops a shale bulge beneath the central slope (Figure 12C),

Figure 11. Model 1: $V_P = 0.25$ cm/yr (0.09 in./yr), $k_c = 2.52 \times 10^{11}$ s/m (7.68×10^{10} s/ft). Instantaneous coupling between V_S and $\Delta\lambda_{HR}$. (A–C) The pale gray area with a Lagrangian tracking grid denotes preexisting Bingham shale. Gray-shaded bands are prograded Bingham shale. (D) Plot of the Hubbert-Rubey pore-fluid pressure (λ_{HR}) ratio and instantaneous flow velocities at 41 m.y. model time. (E) Plot of the second invariant of the finite strain. VE = vertical exaggeration; 1 km = 0.6 mi or 3281 ft; 0.1 cm = 0.04 in. See Table 1 for explanations of symbols.

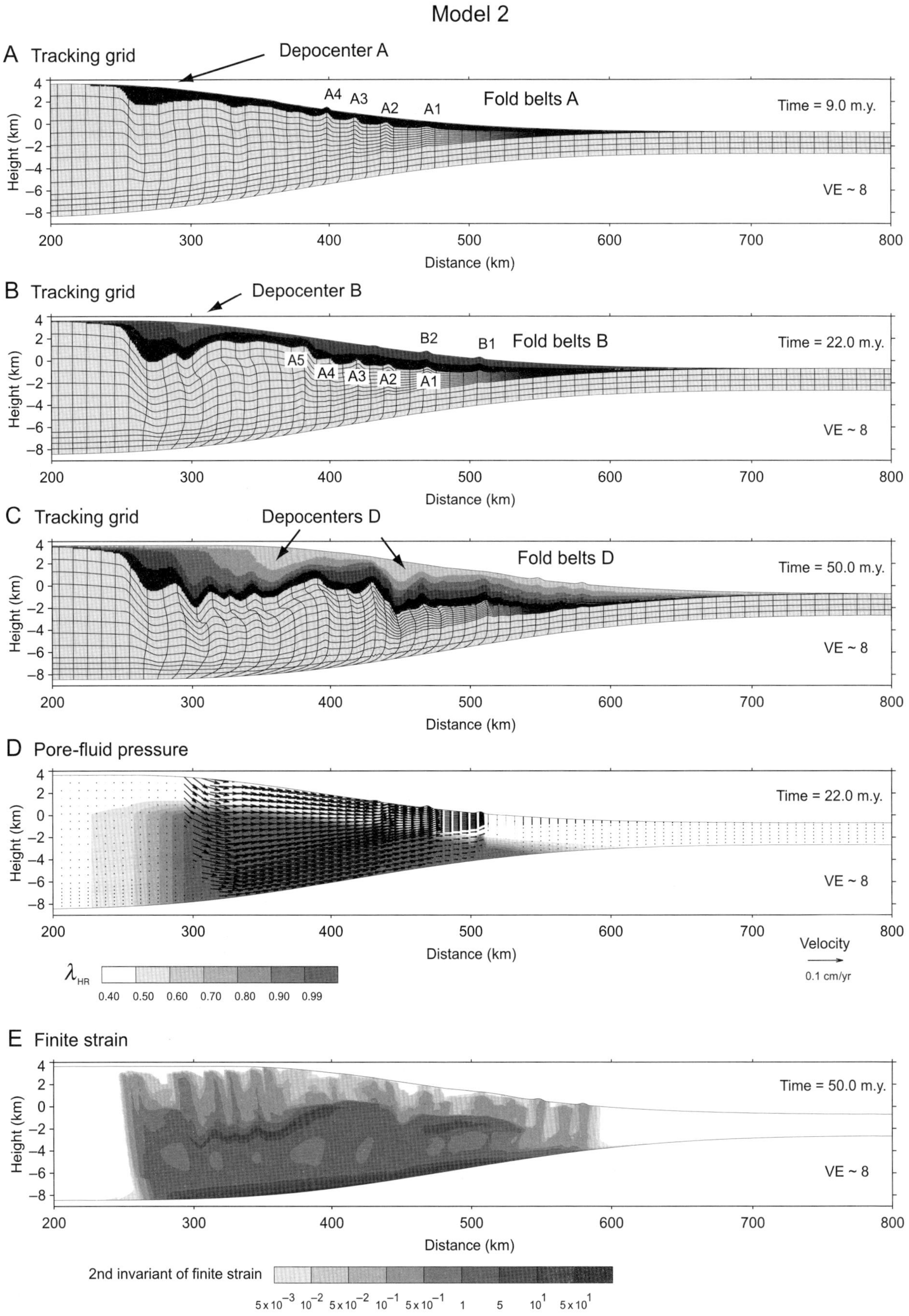

Model 2
A Tracking grid
Depocenter A
A4 A3 A2 A1
Fold belts A
Time = 9.0 m.y.
VE ~ 8
Height (km)
Distance (km)
B Tracking grid
Depocenter B
B2 B1 Fold belts B
A5 A4 A3 A2 A1
Time = 22.0 m.y.
VE ~ 8
Height (km)
Distance (km)
C Tracking grid
Depocenters D
Fold belts D
Time = 50.0 m.y.
VE ~ 8
Height (km)
Distance (km)
D Pore-fluid pressure
Time = 22.0 m.y.
VE ~ 8
Height (km)
Distance (km)
Velocity
0.1 cm/yr
λ_{HR}
0.40 0.50 0.60 0.70 0.80 0.90 0.99
E Finite strain
Time = 50.0 m.y.
VE ~ 8
Height (km)
Distance (km)
2nd invariant of finite strain
5×10^{-3} 10^{-2} 5×10^{-2} 10^{-1} 5×10^{-1} 1 5 10^{1} 5×10^{1}

whereas in model 3, the subslope region is significantly more depleted of shale at the same time (Figure 13C). This difference mostly reflects the faster progradation in model 3. Both extensional and contractional provinces develop and share many characteristics with those observed in model 2. For example, the locations of normal-fault-bounded depocenters migrate seaward diachronously, similar to the stepwise outbuilding of the delta in model 2. Furthermore, during the life span of a given depocenter, the corresponding fold belts initiate beneath the lower slope and are marked by landward-younging individual folds. Doubling the progradation and local vertical sedimentation rates augments the structural development. Compared to model 2, the extensional basins in model 3 are wider (~40 to 60 km [~25 to 37 mi]) and deeper (~3 to 5 km [~1.9 to 3.1 mi]), and the fold amplitudes are generally larger (Figure 13C). Furthermore, the number of individual structures in model 3 exceeds that in model 2 (e.g., 12 high strain extensional shear zones in model 3 [Figure 13E] exist versus 6 in model 2 [Figure 12E]).

Model Set 2: Excess Pore-Fluid Pressures Proportional to the Maximum Sedimentation Rate

Model 4: $V_p = 0.25$ cm/yr, $k_c = 2.52 \times 10^{11}$ s/m ($V_p = 0.09$ in./yr, $k_c = 7.68 \times 10^{10}$ s/ft). Model 4 is equivalent to model 1 except that the maximum excess pore-fluid pressure is maintained even when local sedimentation rates decline. This change produces two main effects. The large-scale overpressured region no longer migrates with the prograding zone of the highest sedimentation rate. Instead, the overpressured region expands seaward during progradation, leaving an overpressured region beneath the shelf (compare Figures 11D, 14D, Animation 4). Second, local uplift of the shale above the progradation profile does not diminish the associated pore-fluid pressure. The pore-fluid pressure is maintained despite the decrease or cessation of sedimentation. Both of these effects contribute to a more efficient squeeze flow of the shale. The first effect creates a potentially wider zone in which the squeeze flow can operate, and the second effect removes the compartmentalization of the excess pore-fluid pressure, thereby maintaining the longer length scale of the squeeze flow.

Extension is immediately localized in the form of a depocenter that remains active throughout the model run (Figure 14A–C). Massive shale withdrawal below the depocenter results in a thick shale bulge that rolls over landward beneath the slope, resulting in a landward overhang of pre-existing shale (Figure 14C). Because of the inflation of the slope region with shale, the top of the pre-existing shale remains very close to the progradation profile during the first ca. 30 m.y. model time (Figure 14B). Initially, the corresponding contraction is spread across an approximately 30-km (~19-mi) area below the lower slope (Animation 4). Later this deformation becomes localized in the form of a fold that continues to grow until ca. 27 m.y. model time (Figure 14B). When covered with sediments, the active fold locks and is replaced by a new one farther down the slope (Figure 14C). Overall, the deformation in model 4 is much more continuous compared to the previous models, and the formation of characteristic episodic depocenters is completely lacking.

Model 5: $V_p = 0.25$ cm/yr, $k_c = 2.52 \times 10^{11}$ s/m, Reduced Depth d_1 of Onset of Excess Pore-Fluid Pressures ($V_p = 0.09$ in./yr, $k_c = 7.68 \times 10^{10}$ s/ft). Model 5 is the same as model 4 except that the effect of a reduction of d_1 from 2500 to 2000 m (8202 to 6562 ft) was investigated. This reduction results in pore-fluid pressures that exceed hydrostatic values at shallower levels than in the previous models and as a result, the model will be more unstable because the limit values of the F_t and F_c forces (Figure 3) are reduced (see section titled Limit Analysis: Conditions for Yielding).

Up to ca. 10 m.y. model time, model 5 behaves similarly to model 4. A landward depocenter is associated with highly localized deformation and is balanced by more distributed contractional deformation downslope (Figure 15A). However, whereas model 4 is dominated by a squeezing channel type of flow of the overpressured shale (Figure 14D), in model 5, the entire overburden participates in seaward translation (Figure 15A–D). Hence, the entire sedimentary package in the central translational zone moves on the basal shear zone in a seaward direction at an approximately uniform velocity with little internal deformation (Figure 15A, B, D, Animation 5). Updip extension and downdip contraction continue as the model evolves. Contrary to model 4, contraction is accommodated by the formation

Figure 12. Model 2: $V_P = 0.25$ cm/yr (0.09 in./yr), $k_c = 5.05 \times 10^{11}$ s/m (1.54×10^{11} s/ft). Instantaneous coupling between V_S and $\Delta\lambda_{HR}$. (A–C) The pale gray area with a Lagrangian tracking grid denotes preexisting Bingham shale. Gray-shaded bands are prograded Bingham shale. (D) Plot of the Hubbert-Rubey pore-fluid pressure ratio (λ_{HR}) and instantaneous flow velocities at 22 m.y. model time. (E) Plot of the second invariant of the finite strain. VE = vertical exaggeration; 1 km = 0.6 mi or 3281 ft; 0.1 cm = 0.04 in. See Table 1 for explanation of symbols.

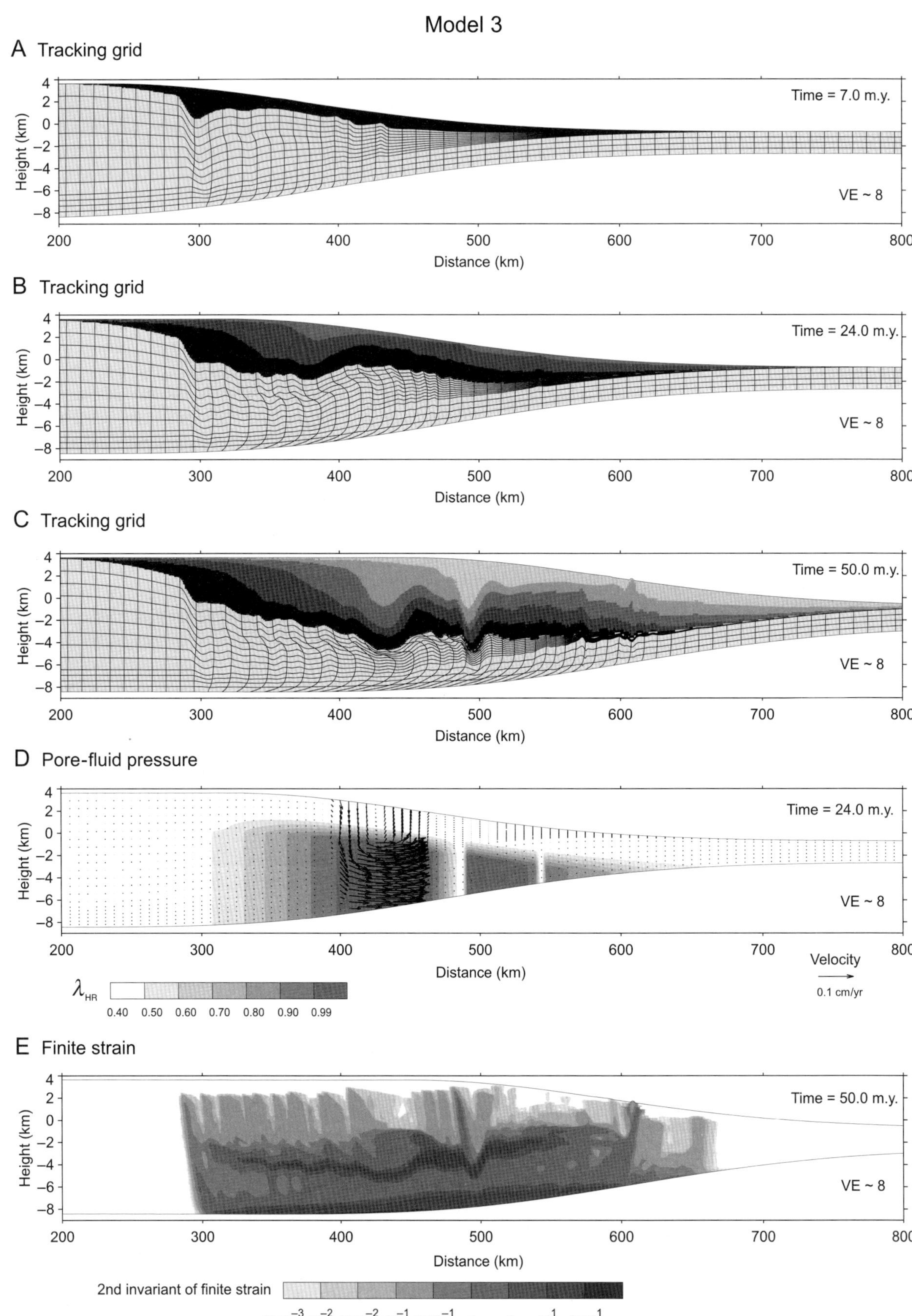

Model 3
A Tracking grid
Time = 7.0 m.y.
VE ~ 8
Height (km)
Distance (km)
B Tracking grid
Time = 24.0 m.y.
VE ~ 8
C Tracking grid
Time = 50.0 m.y.
VE ~ 8
D Pore-fluid pressure
Time = 24.0 m.y.
VE ~ 8
Velocity
0.1 cm/yr
λ_{HR}
0.40 0.50 0.60 0.70 0.80 0.90 0.99
E Finite strain
Time = 50.0 m.y.
VE ~ 8
2nd invariant of finite strain
5×10^{-3} 10^{-2} 5×10^{-2} 10^{-1} 5×10^{-1} 1 5 10^{1} 5×10^{1}

of a synchronous fold belt instead of a series of individual, seaward-propagating folds. Sustained extension, translation, and sedimentation widen the overpressured region, ultimately causing wholesale rapid seaward translation above the basal shear zone (Figure 15C). During this phase, the translation velocity reaches approximately 1 cm/yr (0.4 in./yr) and is only limited by the 10^{18} Pa·s (145×10^{12} psi·s) Bingham viscosity. Eventually, the deformation and large-scale material transfer reduce the slope, thus stabilizing the model, and sediment progradation proceeds seaward (Figure 15C). Comparison of Figures 14E and 15E illustrates that the area affected by high levels of finite strains in model 5 is approximately twice as large as that in model 4, and that most of the deformation in model 5 is focused toward the basal shear zone.

The final sedimentation pattern of model 5 contrasts strongly with that of the more stable delta models, model R and model 1, in that most of the sediment is deposited in the extensional basins beneath the shelf and upper slope. Correspondingly, less sediment is deposited on the lower slope because contraction maintains the top of the sediment near or above the progradation profile. The combined effect of prolonged normal faulting in the same vicinity (~280 to 300 km [~174 to 186 mi], Figure 15A–C) and the seaward translation of the basins produced by this faulting (~130 km [~81 mi] migration of depocenter A, Figure 15A–C) results in sedimentation that becomes younger in the landward direction (Figure 15C). In the absence of an understanding that this is a consequence of the seaward translation of the depocenters, the pattern could erroneously be interpreted to be a consequence of sediment progradation in a landward direction from a seaward source.

Model 6: V_p = 0.25 cm/yr, $k_c = 2.52 \times 10^{11}$ s/m, Strain Softening in Solid Matrix by a Factor of 1.5 (V_p = 0.09 in./yr, $k_c = 7.68 \times 10^{10}$ s/ft). In nature, it is likely that progressive deformation leads to weakening of frictional-plastic sediments by faulting, fracturing, and/or grain size reduction (e.g., Kirby, 1985), processes that produce weak fault gouge, for example. The purpose of model 6 is to examine the effects of this type of strain softening of the solid matrix. Strain softening is simulated by linearly reducing the internal angle of friction ϕ (equation 1) from $\phi_I = 30°$ to $\phi_E = 19.47°$ as the local strain (second invariant of strain) increases from 0.5 to 1. The imposed decrease in ϕ corresponds to a strength reduction ($\sin \phi_I / \sin \phi_E$) by a factor of 1.5. As can be seen from equation 1, this strain softening compounds the effect of high pore-fluid pressure. The latter reduces the effective pressure ($P_e = P - P_f$), and the former further reduces the yield strength by $\sin \phi_I / \sin \phi_E$ for fully strain-softened shale. In this regard, model 6 is similar to model 5 in that it explores the effect of small, but significant, additional weakening of the shale with respect to the other models. The relative effect of strain softening is greatest for those regions that remain close to the hydrostatic state (equation 1). By contrast, highly overpressured regions are already so weak that strain softening has a minor effect. Because strain softening is progressive, its initial effects are small, but they become larger as the model evolves.

Model 6 is initially the same as model 4; sedimentary loading generates excess pore-fluid pressures beneath the slope, thus mobilizing shale and squeezing it seaward between two shear zones (Figures 14, 16). Up to ca. 21 m.y. model time, the bulk evolution of model 6 is quite similar to the structural development of model 4. For example, at any given time, extension and contraction are linked via the basal shear zone, and the corresponding structures comprise landward normal faults and lower slope folds (Figure 16, Animation 6). Likewise, normal faulting localizes almost immediately, whereas focused contractional strain does not occur before ca. 10 m.y. model time (Figure 16A).

In terms of the structural geometry, however, model 6 differs significantly from model 4. Whereas both the shale withdrawal basin and the deep-water folds in model 4 are rather symmetric (Figure 14B), the equivalent structures in model 6 are distinctively asymmetric (Figure 16B). For instance, shale withdrawal beneath the shelf is predominantly accommodated by a well-developed listric normal fault that soles into the basal shear zone and allows the characteristic deposition of landward-thickening growth strata (Figure 16B). By the same token, the deep-water fold belt in Model 6 assumes a prominent seaward vergence (Figure 16B).

As model 6 continues to evolve between 10 and 21 m.y., the local strain increases from 0.5 to 1 or more in increasingly larger areas, and these areas correspondingly become progressively weaker. At ca. 21 m.y., the model behavior switches abruptly from relatively

Figure 13. Model 3: V_P = 0.5 cm/yr (0.2 in./yr), $k_c = 1.58 \times 10^{11}$ s/m (4.82×10^{10} s/ft). Instantaneous coupling between V_S and $\Delta\lambda_{HR}$. (A–C) The pale gray area with a Lagrangian tracking grid denotes preexisting Bingham shale. Gray-shaded bands are prograded Bingham shale. (D) Plot of the Hubbert-Rubey pore-fluid pressure ratio (λ_{HR}) and instantaneous flow velocities at 24 m.y. model time. (E) Plot of the second invariant of the finite strain. VE = vertical exaggeration; 1 km = 0.6 mi or 3281 ft; 0.1 cm = 0.04 in. See Table 1 for explanation of symbols.

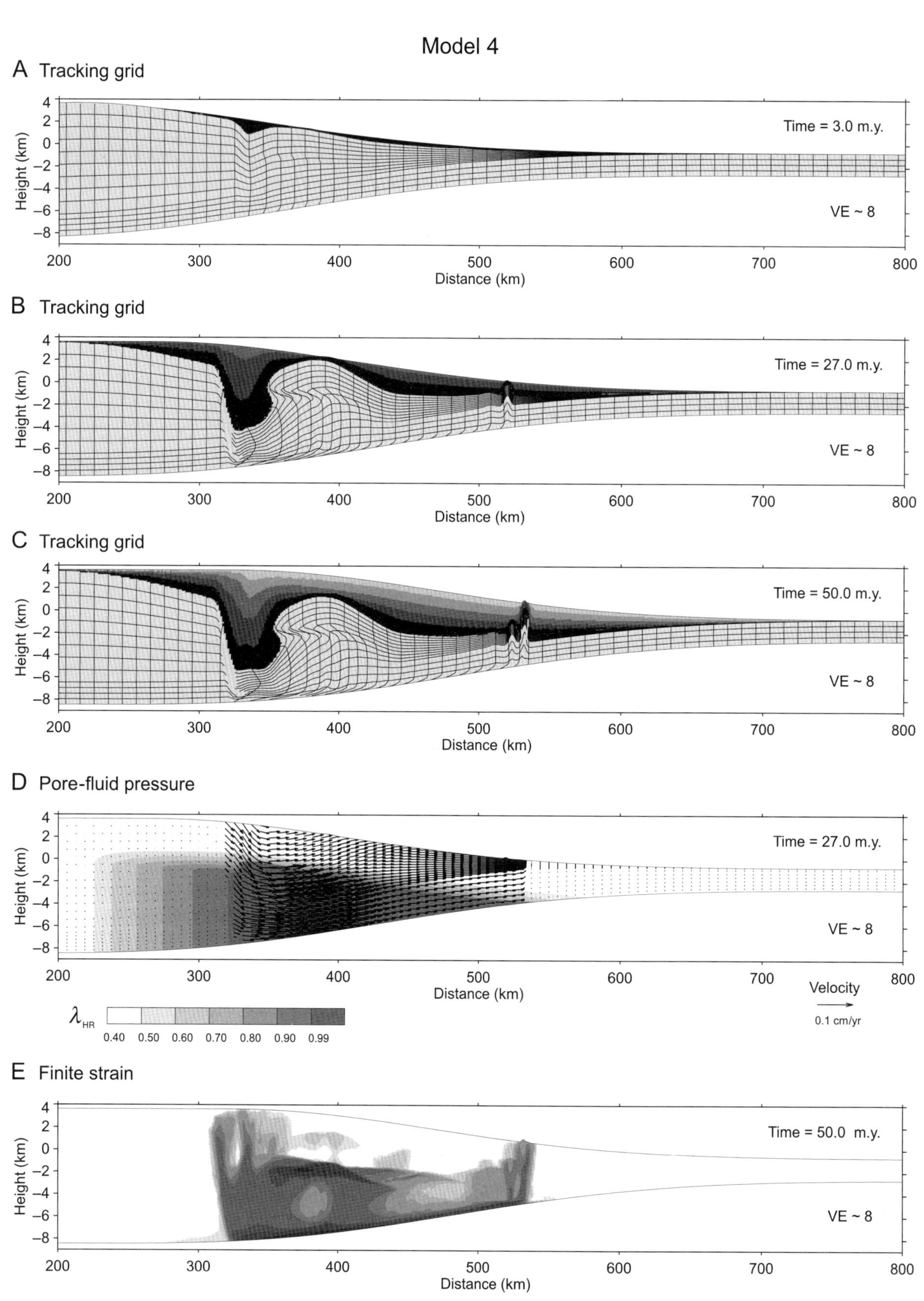

Model 4
A Tracking grid
Time = 3.0 m.y.
VE ~ 8
Height (km)
Distance (km)
B Tracking grid
Time = 27.0 m.y.
VE ~ 8
C Tracking grid
Time = 50.0 m.y.
VE ~ 8
D Pore-fluid pressure
Time = 27.0 m.y.
VE ~ 8
Velocity
0.1 cm/yr
λ_{HR}
0.40 0.50 0.60 0.70 0.80 0.90 0.99
E Finite strain
Time = 50.0 m.y.
VE ~ 8
2nd invariant of finite strain
5x10^-3 10^-2 5x10^-2 10^-1 5x10^-1 1 5 10^1 5x10^1

slow and steady shale withdrawal, inflation, and associated squeeze flow deformation to much more rapid, very large lateral and vertical displacements, and complex deformation dominated by the translation above the basal shear zone. The change in behavior is associated with major differences in structural style. The previously asymmetric shale withdrawal basin marked by listric normal faulting is overprinted by more symmetric extension. Finite strain patterns indicate that the initial seaward-dipping listric normal fault is replaced by a set of conjugate shear zones with somewhat more shear on the landward-dipping set (Figure 16E). These shear zones bound relatively symmetric grabens. This wide, rapid extension is balanced by thrust faulting in the deep-water domain with as much as 3 cm/yr (1.2 in./yr) velocities, which creates substantial bathymetric relief (Figure 16C).

DISCUSSION

Limitations and Applicability of the Numerical Models

The model results presented here will be useful if the approximations are consistent with the level of the investigation. Little is to be gained by developing a complex numerical model with multiple interactions, couplings and feedbacks when there are few constraints on the levels of excess pore fluid pressure that develop deep in these systems at the spatial scale of a margin and over the timescale of its evolution. In this study, we are concerned with the most basic stability and deformation problem, and several simplifications are made to make the problem tractable. Some of these simplifications are limitations of the numerical model, but others are a consequence of a lack of information concerning the properties of these systems at the large scale.

With this need for a balanced approach in mind, we have confined the modeling in the first part of the article to the limit analysis of overpressured frictional-plastic shale with a Mohr-Coulomb yield criterion based on the effective stress. The results of the analytical and numerical limit analysis determine the levels of the pore-fluid pressure that are required to reach the deep-seated gravitational stability limit in the absence of other processes that contribute to weakening of the sediments, such as seepage forces caused by fluid flowing through pore space that impart frictional forces to the solid framework (e.g., Mourgues and Cobbold, 2003). For these assumptions, the results should be robust, and they indicate that Hubbert-Rubey pore-fluid pressure ratios in the range of 0.8–0.992 are required for failure to occur at the scale of a shale basin.

In the second part of the article, the numerical models are expanded to consider the finite deformation that follows failure of a shale basin during sediment progradation. These models include the following simplifications.

1) The rheology is modeled as a Bingham fluid in which plastic yielding has the same Mohr-Coulomb criterion based on the effective pressure (equation 1) as in the limit analysis. In the absence of information on the rheology of mobile overpressured shale at a large scale, the postyield Bingham viscosity was set at 10^{18} Pa·s (145×10^{12} psi·s), a value that gives reasonable flow velocities when the system becomes unstable.
2) The development of excess pore-fluid pressures was not calculated by mechanical or viscous compaction of the sediments, as for example in Morency et al. (2007). Instead, the calculation was reduced to a parameterized problem in which excess pore-pressure development at depth is proportional to the local sedimentation rate with a proportionality constant (coupling coefficient, k_c) that is inversely proportional to a nominal hydraulic conductivity for the system. This parametric model will give approximate predictions where overpressure will develop in natural systems in response to sedimentary loading. The feedback effect of the pore-fluid pressure on the mechanics and deformation of the shale basin was reduced to a sensitivity analysis in which the system was made progressively weaker in a parametric manner by increasing k_c, increasing the sediment progradation rate, modifying the fluid pressure regime so that excess pore-fluid pressures develop at a shallower depth, and including the effect of strain softening of the shale matrix during deformation. The effect of persistence versus dissipation of fluid overpressure was addressed by considering two end member conditions: (1) excess pore pressure proportional to the current local rate of sedimentation (corresponding to rapid generation and dissipation), and (2) local excess pore pressure

Figure 14. Model 4: $V_P = 0.25$ cm/yr (0.09 in./yr), $k_c = 2.52 \times 10^{11}$ s/m (7.68×10^{10} s/ft). (A–C) Maximum values of $\Delta\lambda_{HR}$ are maintained. The pale gray area with a Lagrangian tracking grid denotes preexisting Bingham shale. Gray-shaded bands are prograded Bingham shale. (D) Plot of the Hubbert-Rubey pore-fluid pressure ratio (λ_{HR}) and instantaneous flow velocities at 27 m.y. model time. (E) Plot of the second invariant of the finite strain. VE = vertical exaggeration; 1 km = 0.6 mi or 3281 ft; 0.1 cm = 0.04 in. See Table 1 for explanation of symbols.

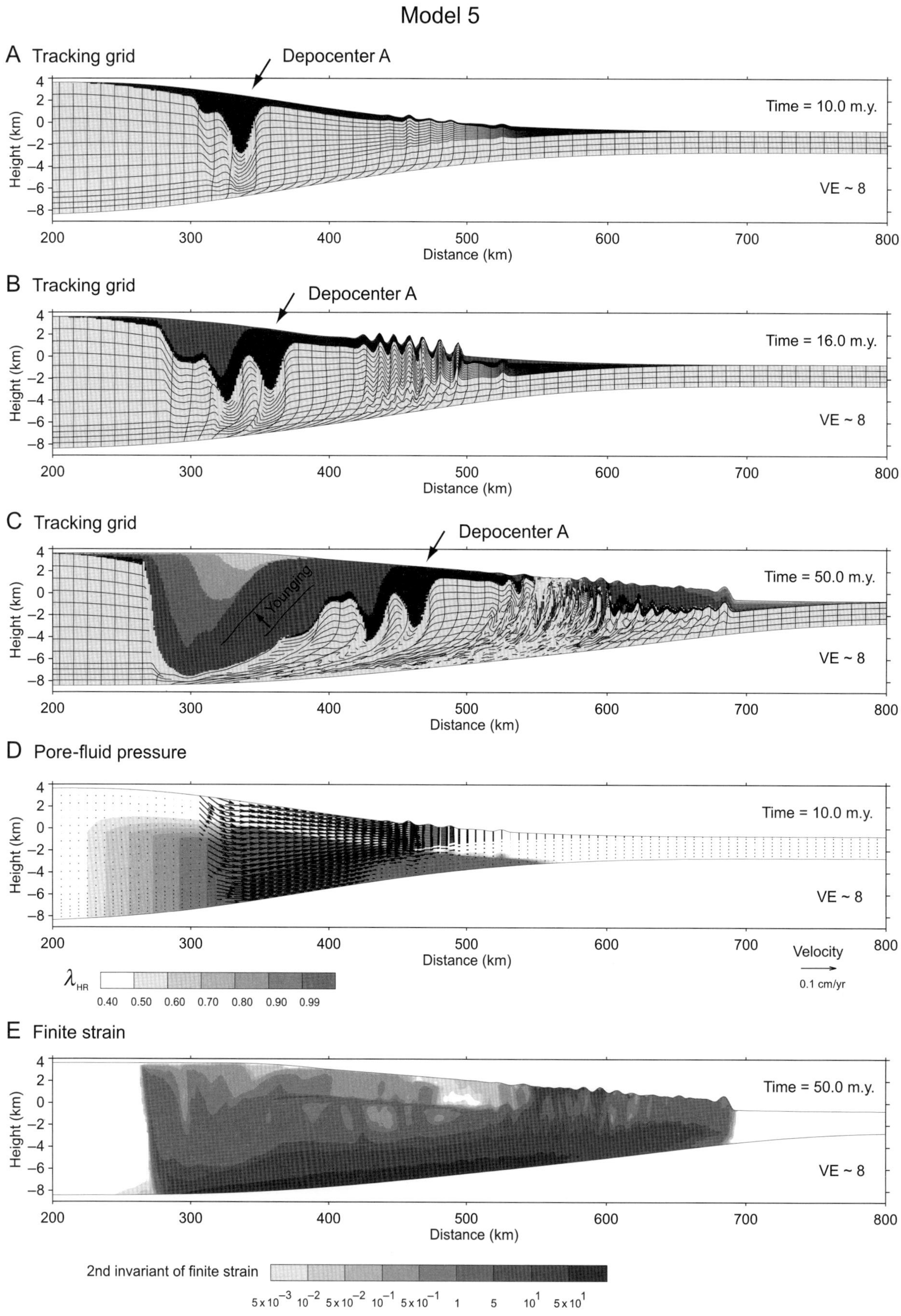

Model 5
A Tracking grid
Depocenter A
Time = 10.0 m.y.
VE ~ 8
Height (km)
4 2 0 −2 −4 −6 −8
200 300 400 500 600 700 800
Distance (km)
B Tracking grid
Depocenter A
Time = 16.0 m.y.
VE ~ 8
Height (km)
4 2 0 −2 −4 −6 −8
200 300 400 500 600 700 800
Distance (km)
C Tracking grid
Depocenter A
Younging
Time = 50.0 m.y.
VE ~ 8
Height (km)
4 2 0 −2 −4 −6 −8
200 300 400 500 600 700 800
Distance (km)
D Pore-fluid pressure
Time = 10.0 m.y.
VE ~ 8
Height (km)
4 2 0 −2 −4 −6 −8
200 300 400 500 600 700 800
Distance (km)
Velocity
0.1 cm/yr
λ_{HR}
0.40 0.50 0.60 0.70 0.80 0.90 0.99
E Finite strain
Time = 50.0 m.y.
VE ~ 8
Height (km)
4 2 0 −2 −4 −6 −8
200 300 400 500 600 700 800
Distance (km)
2nd invariant of finite strain
5×10^{-3} 10^{-2} 5×10^{-2} 10^{-1} 5×10^{-1} 1 5 10^{1} 5×10^{1}

proportional to the local maximum sedimentation rate (corresponding to persistence of excess pore pressure at the model time scale, as may be produced by viscous compaction).

These restrictions will limit the applicability of the models to shale basins where the fluid pressure regime is similar to the parametric model prediction. Similarly, the sediment rheology must be like that of a uniform Bingham fluid, where all preexisting and prograding sediments are treated as having the same yield criterion and postyield linear viscosity. In reality, the Bingham viscosity may depend on many possible variables, including the fluid content of the shale, the excess fluid pressure, and the amount of deformation, with the limiting condition that the shale may become fluidized. Developing a model with these properties would be relatively easy, but without appropriate data, calibrating for these conditions would be impossible.

3) The models are geometrically simple. They include sediment progradation according to a Gaussian sediment profile, which is a reasonable approximation for the main geometrical control, the large-scale shape. The sediment properties and the progradation velocity are uniform, and the coupling coefficient, k_c, is constant in each model experiment. No sediment compaction exists, and therefore sediment density does not change. The models are two-dimensional plane-strain vertical cross sections. Lastly, the models do not consider thermal effects. The models do, however, include the effects of flexural isostatic compensation and water loading, and therefore, the overall horizontal force balance should be close to the equivalent natural system when seepage forces are negligible.

This set of simplifications will limit the applicability of the results to uniformly prograding, approximately two-dimensional deltas, where excess pore-fluid pressure development is widely distributed and not restricted to particular formations with certain lithologies (e.g., sand versus shale). Although these requirements are quite restrictive, the models do address the first-order yield and deformation characteristics that can be anticipated, and the associated structural styles. More complexities, including initial configuration, time and space variations in sedimentation rate and type, and lithologically dependent coupling coefficients, can easily be added once the first-order behavior is understood. These additions can be justified for case-specific studies where reliable observational constraints are available. For the purposes of this study, we regard all of these additional effects to be potentially important but likely to be of second order.

Structural Styles Produced by the Numerical Models

In view of the limitations noted above, only the primary features of the model results warrant assessment and comparison to natural equivalents. We combined the key characteristics produced by the numerical models (stratigraphic tops, pore-fluid pressure (λ_{HR}), and fault patterns interpreted from finite-strain and strain-rate plots) in Figure 17.

Model 1 (Figure 17A) is consistent with conditions of high hydraulic conductivity and hence rapid dissipation of excess pore pressures. Because only the areas beneath the steepest section of the progradation profile (highest V_S) attain the required levels of λ_{HR} for yielding, the zone of squeeze flow of overpressured shale is much narrower than the progradation profile. Consequently, shale withdrawal is limited, and despite minor faulting, the strata are mostly intact and strongly resemble those of stable model R (compare Figures 9B, 17A). Model 2 (Figure 17B) is consistent with low regional hydraulic conductivity (large k_c) and local rapid dissipation of excess pore pressures in areas where sedimentation ceases. As a result, a series of diachronous, seaward-younging and normal-fault-bounded depocenters and corresponding deep-water fold and thrust belts form (Figure 17B). Significant squeeze flow of overpressured shale is indicated by the shale bulge beneath the slope (Figure 17B). Similar to model 2, model 3 produces seaward-younging depocenters and corresponding fold and thrust belts, except that doubling the sediment progradation rate increases the amplitude and number of these structures. Seaward-diachronous depocenters and shale bulges akin to the ones produced by models 2 and 3 have been observed in the Niger Delta (Knox and Omatsola, 1989; Cohen and McClay, 1996; Wu and Bally, 2000).

In model 4 (Figure 17D), excess pore pressures are maintained. Continuous shale withdrawal beneath a quasi-stationary depocenter and inflation of the adjacent slope region with shale generates an enormous ponded

Figure 15. Model 5: $V_P = 0.25$ cm/yr (0.09 in./yr), $k_c = 2.52 \times 10^{11}$ s/m (7.68×10^{10} s/ft). (A–C) Maximum values of $\Delta\lambda_{HR}$ are maintained; reduced d_2. The pale gray area with a Lagrangian tracking grid denotes preexisting Bingham shale. Gray-shaded bands are prograded Bingham shale. (D) Plot of the Hubbert-Rubey pore-fluid pressure ratio (λ_{HR}) and instantaneous flow velocities at 10 m.y. model time. (E) Plot of the second invariant of the finite strain. VE = vertical exaggeration; 1 km = 0.6 mi or 3281 ft; 0.1 cm = 0.04 in. See Table 1 for explanation of symbols.

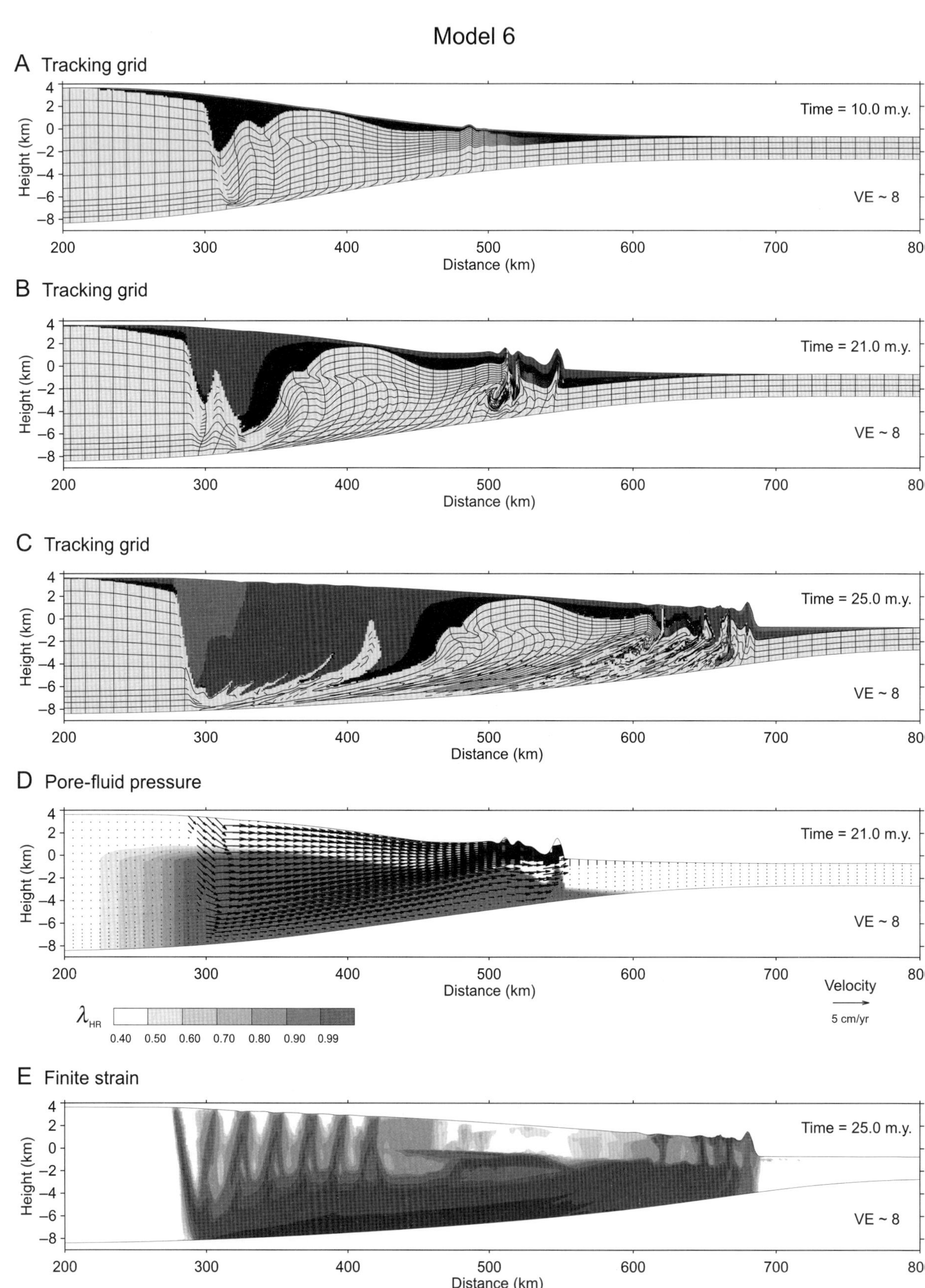

Model 6
A Tracking grid
Time = 10.0 m.y.
VE ~ 8
Height (km)
Distance (km)
B Tracking grid
Time = 21.0 m.y.
C Tracking grid
Time = 25.0 m.y.
D Pore-fluid pressure
Time = 21.0 m.y.
Velocity
5 cm/yr
λ_{HR}
0.40 0.50 0.60 0.70 0.80 0.90 0.99
E Finite strain
Time = 25.0 m.y.
2nd invariant of finite strain
5×10^{-3} 10^{-2} 5×10^{-2} 10^{-1} 5×10^{-1} 1 5 10^{1} 5×10^{1}

slope basin (~100 km [~62 mi] wide, Figure 17D). Diachroneity is almost completely suppressed, because once initiated, the extensional basin and the lower slope folds continue to be used. Consequently, the same strata that are arranged laterally in the extensional provinces of models 2 and 3 (Figure 17B, C) are stacked vertically in model 4 (Figure 17D). Because of a shallower onset of excess pore pressure, model 5 (Figure 17E) is more unstable and first marked by updip extension and simultaneous folding in the lower slope region and later followed by a wholesale seaward motion of the overburden whereby the folds are used by thrust faults. The strata assume sigmoidal, landward-dipping geometries (Figure 17E). Normal faults generally dip seaward and thrust faults landward. Likewise, model 6 is characterized by wholesale motion of the overburden. Because of continuous activity of the seaward-dipping synthetic master normal fault, symmetric grabens form in the extensional province (Figure 17F). Landward-dipping normal faults first initiate as antithetic faults to the master normal fault and then translate seaward with the moving overburden.

Models 5 and 6 are marked by nearly complete withdrawal of preexisting shale, and contrary to the previous models, most of the finite strain is focused in the basal shear zone. Finally, all models indicate that the normal and thrust faults are linked to the basal detachment when they accommodate shear (Figure 17A–F, Animations 1 to 6).

Analysis of Model Characteristics in Terms of Poiseuille and Couette Flow Modes

The overall behavior of the models and the relationships among them can be understood by reference to the equivalent salt-tectonic problem and by using the limit analysis to determine the spatial extent of the unstable region, the level of instability (i.e., mobility), and the way these quantities contribute to the finite deformation. All of the models deform in one or both of two modes: gravitationally driven Poiseuille and Couette channel flow modes (P and C modes) that characterize the expulsion of rifted margin salt under differential loading of prograding sediments (Vendeville and Jackson, 1992; Lehner, 2000; Gemmer et al., 2004, 2005). From a modeling perspective, the setting for the current shale mobilization problem (Figure 8) is exactly the same as that for salt except that mobile Bingham shale replaces the salt (Figure 18). In salt tectonics, the salt, the channel material, and overburden have intrinsically different properties. Salt is viscous and will flow under differential pressure either as a parabolic Poiseuille flow between immobile overlying or underlying strata, or as a Couette flow when the brittle (frictional-plastic) overburden also fails and moves horizontally, or as a flow that combines both modes. The shale problem is different in that the channel and overburden are not defined by two different materials. Instead, the channel corresponds to an initially unknown part of the system that self-selects, fails, and participates in the Poiseuille flow (Figure 18A). This channel flow may also have a superimposed Couette component if the overburden also fails and flows seaward (Figure 18B, C). Because shale is modeled as a visco-plastic Bingham material with a finite frictional yield strength that must be exceeded before flow can ensue, the characteristic parabolic P-mode horizontal velocity profile for linearly viscous salt is replaced by a velocity profile corresponding to uniform translation of a visco-plastic slug bounded by upper and lower shear zones.

In salt tectonics, the whole of the slope region of the margin is commonly unstable and underlain by an active channel. By contrast, in the equivalent shale problem, only regions with sufficiently high pore-fluid pressure will yield under differential loading and participate in channel flow. In the numerical models, these mobile regions will be those with high surface sedimentation rates. The shale models therefore have a range of unstable regions, and the extent of these yielding regions increases with the coupling coefficient k_c, the progradation rate V_P, a decrease in the depth of onset of excess pore-fluid pressure d_1, and strain softening.

On the basis of these concepts, we can explain the primary characteristics of the numerical models. Model R never achieves sufficient excess pore-fluid pressures to cause yielding. It remains stable during sediment progradation, and neither P nor C modes develop. This end member model represents a stable shale basin. Model 1 exhibits yielding of an approximately 100-km (~60-mi)-wide region beneath the midslope, the region with the highest sedimentation rates. The mode is primarily P, which accounts for the horizontal sluglike squeeze flow in the self-determined channel

Figure 16. Model 6: $V_P = 0.25$ cm/yr (0.9 in./yr), $k_c = 2.52 \times 10^{11}$ s/m (7.68×10^{10} s/ft). (A–C) Maximum values of $\Delta\lambda_{HR}$ are maintained; strain softening by factor 1.5. The pale gray area with a Lagrangian tracking grid denotes preexisting Bingham shale. Gray-shaded bands are prograded Bingham shale. (D) Plot of the Hubbert-Rubey pore-fluid pressure ratio (λ_{HP}) and instantaneous flow velocities at 21 m.y. model time. (E) Plot of the second invariant of the finite strain. VE = vertical exaggeration; 1 km = 0.6 mi or 3281 ft; 0.1 cm = 0.04 in. See Table 1 for explanation of symbols.

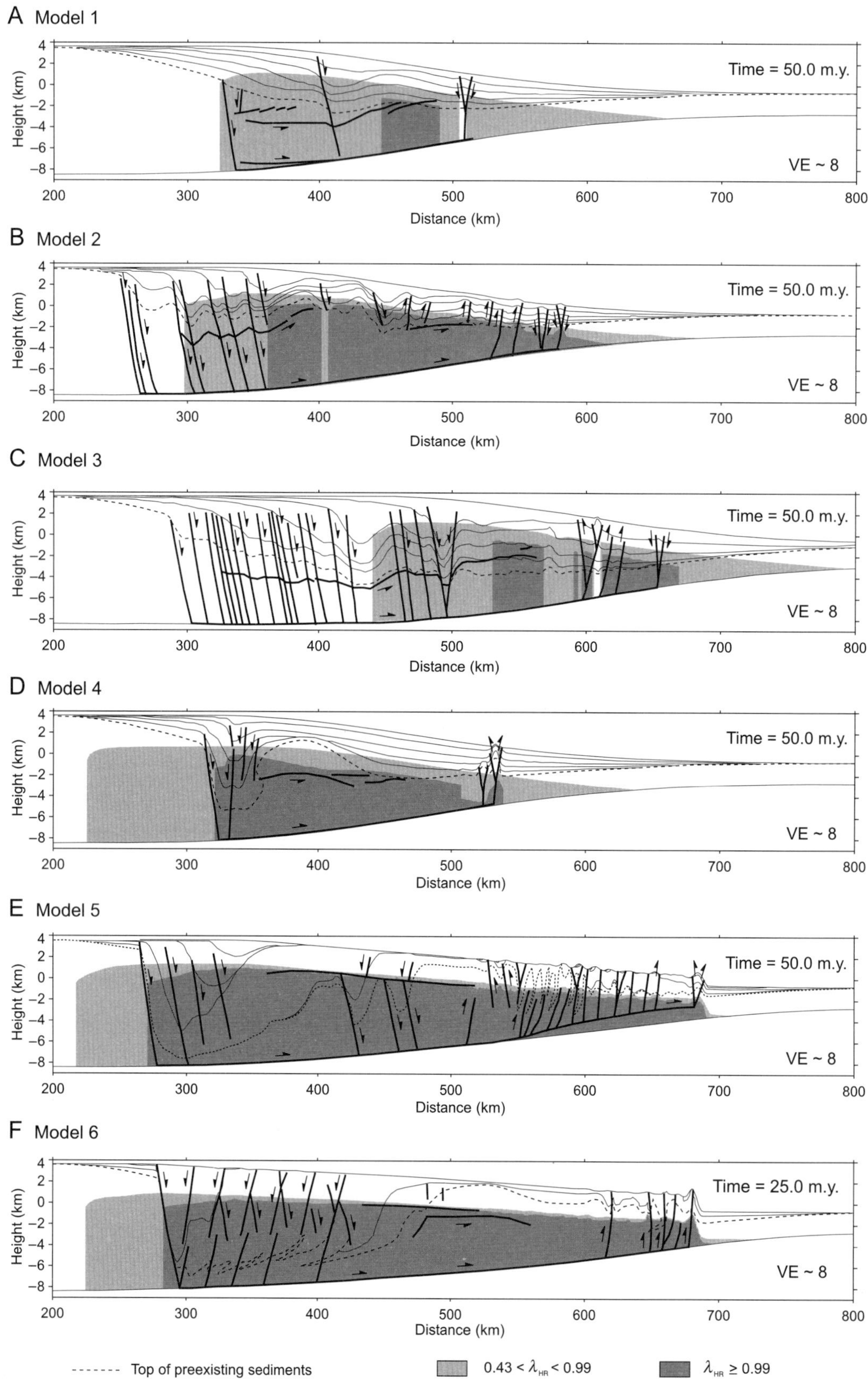

Figure 17. Summary of structural styles produced by the finite element models. The synthetic profiles combine stratigraphy, λ_{HR}, and fault patterns interpreted from strain and strain rate plots. VE = vertical exaggeration; 1 km = 0.6 mi or 3281 ft; λ_{HR} = Hubbert-Rubey pore-fluid pressure ratio.

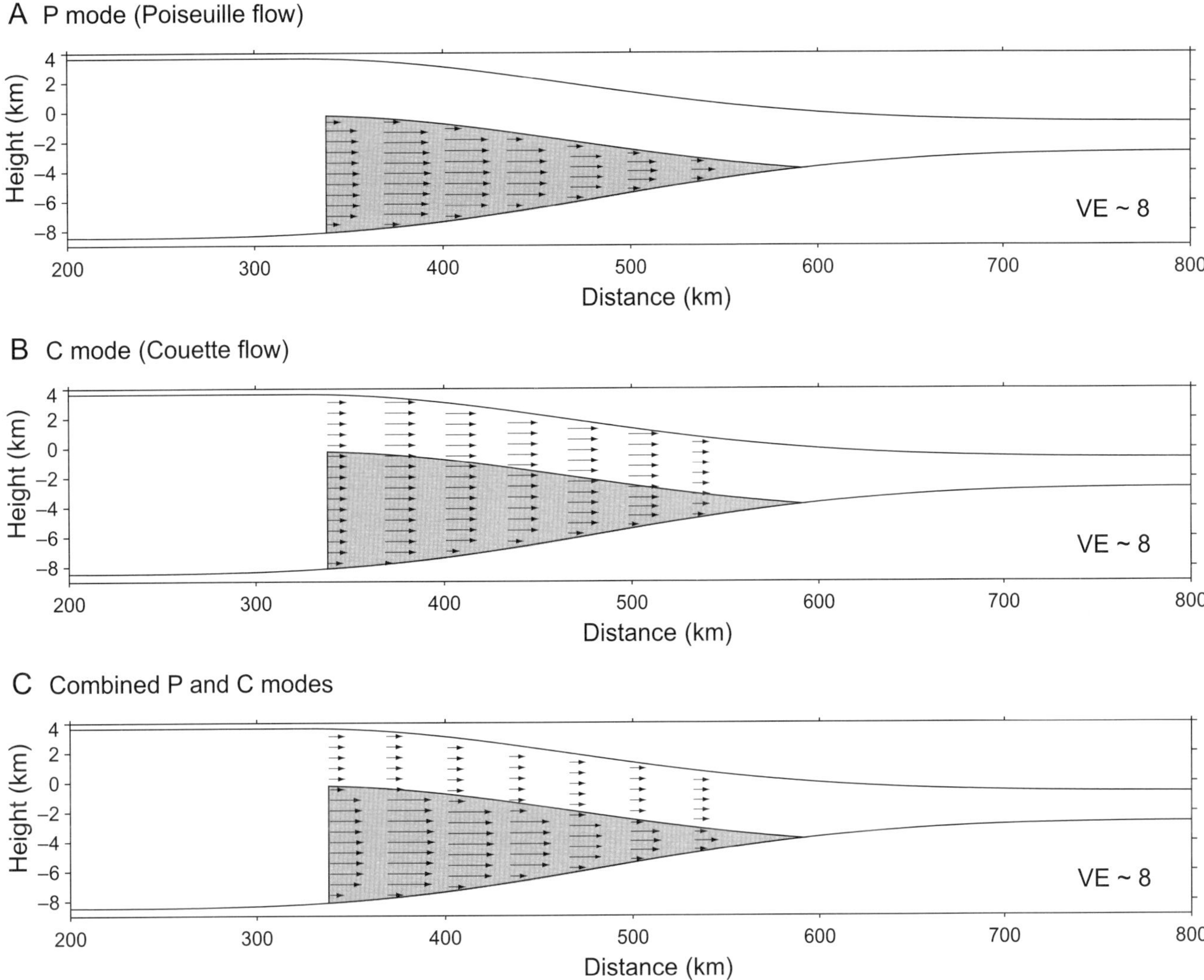

Figure 18. Schematic illustration of shale flow modes. Arrows indicate velocity vectors associated with flow of overpressured sediments (pale gray). (A) P mode (Poiseuille flow). (B) C mode (Couette flow). (C) Combined P and C modes. VE = vertical exaggeration; 1 km = 0.6 mi or 3281 ft.

and explains the development of the upper and lower shear zones that bound the channel. Only very limited Couette flow exists and therefore little horizontal transport of the overburden. Excess accommodation for sediments (i.e., excess with respect to model R) is a consequence of the forced withdrawal of the mobilized shale in the basal channel. A deficit in accommodation is caused by uplift of the overburden where the Poiseuille flow ceases laterally. Model 1 therefore corresponds to a system with limited P-mode flow. In model 2, the higher coupling coefficient leads to a wider region of failure (approximately 200 km [~125 mi] wide). The system is more unstable and the overburden as well as the channel is mobile. Overall, the P mode is stronger, which accounts for the persistence of the upper shear zone and the pluglike flow at depth. Excess accommodation is created both by vertical (channel withdrawal) and horizontal (Couette) flow. The fold belts occur where the Couette flow velocity declines. The system evolves diachronously with the migrating region of high fluid pressures. Model 3 is similar to model 2 and demonstrates that it is the product $k_c V_S$ that determines the width of the highly overpressured and yielding region for a specific time. Model 4 is a more unstable version of model 1 because of the persistence of overpressures even as sedimentation rates decline. The yielding region is wider. Deformation is characterized by landward P and seaward C modes, accounting for the seaward termination of the upper shear zone. The folds occur at the termination of the C mode flow. Model 5 is fundamentally more unstable than the earlier models. All flow appears to be C mode, although the strain plot (Figure 15E) indicates there may be a weak upper level shear zone suggestive of a P-mode component (in the earlier model evolution, Animation 5). The excess accommodation and the fold belts are produced by the spatially limited C mode. The yielding region is nearly 400 km (~250 mi) wide, much wider than in the earlier models. Model 6 is similar to model 5 and equally unstable. More evidence exists of an initial landward P mode, which developed as in model 1 before strain softening weakened the model and allowed the C mode to develop (Animation 6).

Suggestions for Comparison to Natural Examples

Despite the simplified nature of the models, their largest scale features raise some first-order questions concerning natural deltas. We suggest the following points be considered as a basis on which to develop interpretation guidelines:

1) What can be inferred about the transport of the material? Is excess accommodation produced by expulsion of the underlying material (P mode) or by lateral flow of the overburden (C mode)?
2) If the P mode is observed, at what depth does the upper shear zone occur? Does the depth coincide with the onset of overpressures? Was the seaward transport of the substrate greater than that of the overburden? This would correspond to the observation of an underlying layer of mobile shale that has been subjected to a squeeze flow and expelled. This part of the system would reveal fluidlike flow patterns, whereas the deformation of the overlying region would be more characterized by faulting.
3) If the C mode is observed, what are the horizontal length scales of the Couette flows (i.e., the distances between coeval zones of extension and contraction)? What are the time scales of the Couette flows? Is there evidence for wholesale and rapid motion of the overburden?
4) If an upper shear zone is observed, is it smooth, indicating continuous shear and shale squeeze flow of the substrate, or irregular, such as when diachronous normal faulting produces a seaward-stepping geometry?
5) Is the deformation diachronous with migrating pairs of extensional/contractional provinces or is the deformation in fixed locations at the time scale of the evolution of the system?
6) What is the level of the basal shear zone? Does it coincide with the base of a particular stratigraphic unit?
7) How unstable was the system? Compare the excess and deficit amounts of accommodation and divide by a characteristic time scale.
8) If evidence for excess pore-fluid pressure compartmentalization exists, can it be related to local bathymetric relief caused by contraction or should it be attributed to contrasting lithologies?

CONCLUSIONS

We have used analytical limit analysis, supported by numerical experiments to investigate the levels of differential loading and pore-fluid pressures required to initiate large-scale, gravitationally driven failure of passive margin shale basins. We have also used two-dimensional, large deformation, numerical models in which excess pore-fluid pressures are parametrically coupled to sedimentation rates (coupling coefficient k_c inversely proportional to the hydraulic conductivity) to study prolonged gravitational spreading in these basins.

The limit analysis demonstrates that, for the range of parameters investigated (landward sediment thickness $h_1 = 0$ to 15,000 m [49,213 ft], depth of onset of overpressures $d = 2010$ to 4000 m [6594 to 13,123 ft], internal angle of friction $\phi = 5$ to 30°, and width of transition zone $W =$ 50 to 400 km [31 to 249 mi]), the levels of pore-fluid pressures must be 80–99% of the weight of the overburden ($\lambda_{HR} \sim 0.8$–0.99) for yielding to occur. Lower levels of λ_{HR} correspond to narrow passive margins and extremely low dry internal angles of friction ($W \sim 50$ km [~30 mi], $\phi = 5°$). Highest levels ($\lambda_{HR} \sim 0.952$–0.992) correspond to geometries like the Niger Delta ($W \sim 400$ km [~250 mi]), and the range reflects the possible variation of the dry internal angle of friction. The results also show that increasing d, ϕ, and W stabilizes the system, whereas increasing h_1 tends to trigger failure. Correspondingly, decreasing these variables has the opposite effects.

In the large deformation models, we simplified the problem by treating marine shale as a viscoplastic Bingham fluid (a frictional-plastic material that flows viscously on yielding) in which the yield strength is proportional to the effective pressure (mean stress minus pore-fluid pressure). In this Bingham shale, the flow velocity is limited by assuming a postyield linear viscosity of 10^{18} Pa·s (145×10^{12} psi·s).

The large deformation models examined how the structural style varies with the value of the coupling coefficient, k_c, sediment progradation rate, V_P, depth of onset of excess fluid pressure, d_1, and strain softening. Low values of k_c are consistent with high hydraulic conductivities and rapid dissipation of excess pore-fluid pressures and the reverse. Compared to a stable model, where all prograding sediments are accommodated by isostatic adjustment and no flow of sediment exists, we can distinguish mildly unstable, mobile, and highly mobile models, depending on k_c, V_P, and additional factors that cause weakening of the system. Mildly unstable models create only insignificant excess accommodation space (where the excess is defined by comparison to the equivalent stable model), and the resulting deformation is minor. In mobile models, overpressured shale is transferred seaward by a Poiseuille-type squeeze flow between two subhorizontal shear zones. When excess pore pressures dissipate rapidly, archetypical seaward diachronous extension and contraction develop. When excess pore pressures are maintained throughout the model time, diachroneity is suppressed. Decreasing the depth of the onset of excess pore-fluid pressures or strain softening produces highly mobile models where the sediments above the overpressured zone move seaward faster than the squeeze flow beneath and the resulting deformation pattern is that of a Couette flow associated with major accommodation space near the model shelf edge. Lastly, we have proposed a list of questions based on the behavior of the models that may assist in the interpretation of data from equivalent natural systems.

ACKNOWLEDGMENTS

This research was funded by an Atlantic Canada Opportunities Agency Atlantic Innovation Fund contract (PanAtlantic Petroleum Systems Consortium), by IBM, and by Shell International Exploration and Production Inc. M. Albertz acknowledges additional support from Natural Resources Canada and the German Research Foundation. C. Beaumont was supported by the Canada Research Chair in Geodynamics. S. J. Ings was supported by a Natural Sciences and Engineering Research Council of Canada, Canada Graduate Scholarship. The numerical finite element code was developed by Philippe Fullsack, and Bonny Lee assisted with the implementation of the parametric pore-fluid pressure model. We thank Philippe Fullsack, Sofie Gradmann, Glen Stockmal, Richard Wiener, and Chris Connors for helpful comments and Dan Schultz-Ela for a technical review.

APPENDIX 1: DERIVATION OF APPROXIMATE FORCE BALANCE EQUATIONS

We assume that the principal stresses in the shale system are horizontal and vertical and apply the Coulomb criterion using the maximum shear stress $\tau_m = (\sigma_{zz} - \sigma_{xx})/2$ and mean stress $\sigma_m = (\sigma_{xx} + \sigma_{zz})/2$ (Jaeger and Cook, 1969, p. 98) to represent failure of the noncohesive frictional plastic material

$$\left(\frac{\sigma_{zz} - \sigma_{xx}}{2}\right) = \left(\frac{\sigma_{xx} + \sigma_{zz}}{2}\right) \sin \phi \tag{18}$$

where σ is stress and ϕ is the internal angle of friction. $\sigma_{zz} = \rho g d$ is the weight of the overburden where ρ is the density, g is the acceleration of gravity, and $d = h(x) - z$ is the depth below the surface, where z is measured from the bottom. At yield, σ_{xx} can be calculated using the active and passive Rankine states (e.g., Mandl, 1988, p. 301–304) for extension and contraction, respectively, for the dry and wet cases when the pore-fluid pressure $P_f = \lambda_{HR}\sigma_{zz}$ is subtracted from the principal stresses

$$\sigma_{xx,\text{extension}} = \rho g d \frac{(1 - \sin \phi)}{(1 + \sin \phi)}, \text{dry} \tag{19}$$

$$\sigma_{xx,\text{extension}} = \rho g d \frac{(1 - (1 - 2\lambda_{HR}) \sin \phi)}{(1 + \sin \phi)}, \text{wet} \tag{20}$$

and

$$\sigma_{xx,\text{contraction}} = \rho g d \frac{(1 + \sin \phi)}{(1 - \sin \phi)}, \text{dry} \tag{21}$$

$$\sigma_{xx,\text{contraction}} = \rho g d \frac{(1 + (1 - 2\lambda_{HR}) \sin \phi)}{(1 - \sin \phi)}, \text{wet} \tag{22}$$

To calculate the resulting horizontal force F_t in the yielding shale, we integrate σ_{xx} over z for the overpressured and normally pressured (i.e., hydrostatic pressure) vertical columns at position x_1 (Figure 3A)

$$F_t = \int_0^{h_1-d} \sigma_{xx,\text{extension}} dz + \int_{h_1-d}^{h_1} \sigma_{xx,\text{extension}} dz \tag{23}$$

Replacing σ_{xx} yields

$$F_t = \rho g k_{ws} \int_0^{h_1-d} (h(x) - z) dz + \rho g k_w \int_{h_1-d}^{h_1} (h(x) - z) dz \tag{24}$$

where k_{ws} and k_w account for excess pore-fluid pressures ($\lambda_{HR} = \lambda_o$) in the overpressured region and hydrostatic pressures in the overburden sediments ($\lambda_{HR} = \lambda_h$), respectively, according to

$$k_{ws} = \frac{(1 - (1 - 2\lambda_o)\sin\phi)}{(1 + \sin\phi)} \tag{25}$$

and

$$k_w = \frac{(1 - (1 - 2\lambda_h)\sin\phi)}{(1 + \sin\phi)} \tag{26}$$

After substituting h_1 for $h(x)$ and integrating, we obtain

$$F_t = \rho g k_{ws}\left[h_1 z - \frac{1}{2}z^2\right]_0^{h_1-d} + \rho g k_w\left[h_1 z - \frac{1}{2}z^2\right]_{h_1-d}^{h_1} \tag{27}$$

which, after rearranging, simplifies to

$$F_t = \frac{1}{2}\rho g\left[k_w d^2 + k_{ws}(h_1^2 - d^2)\right] \tag{28}$$

At position x_2, the horizontal force F_c can be calculated in a similar manner

$$F_c = \int_0^{h_2-d} \sigma_{xx,\text{contraction}}dz + \int_{h_2-d}^{h_2} \sigma_{xx,\text{contraction}}dz \tag{29}$$

Because the sediments at position x_2 are submerged beneath water, the weight of the water column increases the pressure and thus the strength of the seaward overpressured and overburden sediments with thicknesses $h_s(x_2)$ and d, respectively, by

$$\int_{h_2(x_2)} \rho_w g h_w(x_2)dz = \rho_w g h_w(x_2) h_s(x_2) \tag{30}$$

and

$$\int_d \rho_w g h_w(x_2)dz = \rho_w g h_w(x_2) d \tag{31}$$

respectively, where $h_w(x_2)$ is the maximum water depth and $h_s(x_2) = h_2 - d$ is the thickness of the overpressured sediments at position x_2 (Figure 3A). Rearranging and simplifying yields:

$$F_c = -\frac{1}{2}g\left[\rho(k'_w d^2 + k'_{ws}(h_2^2 - d^2)) + \rho_w(h_w(x_2)(2d + 2h_s(x_2)))\right] \tag{32}$$

where

$$k'_{ws} = \frac{(1 + (1 - 2\lambda_o)\sin\phi)}{(1 - \sin\phi)} \tag{33}$$

and

$$k'_w = \frac{(1 + (1 - 2\lambda_h)\sin\phi)}{(1 - \sin\phi)} \tag{34}$$

Furthermore, the weight of the water exerts a horizontal buttress force on the sloping overburden

$$F_w = -\int_{h_w(x_2)^2} \rho_w g z dz = -\frac{1}{2}\rho_w g(h_w(x_2)^2) \tag{35}$$

The basal traction force F_b acting at the base of the model is approximated as

$$F_b = (1 - \lambda_o)\sin\phi_o F_{GL} \tag{36}$$

where F_{GL} is the weight of the sediments and the water integrated across the width L of the overpressured zone (Figure 3B)

$$\begin{aligned} F_{GL} &= \int_{x_1}^{x'_2} (\rho g h(x) + \rho_w g h_w(x))dx \\ &= \rho g\left(\frac{1}{2}h_w(x_2)L + h_2 L + \frac{1}{2}(h_1 - h_2 - h_w(x_2))L\right) \\ &+ \frac{1}{2}\rho_w g h_w(x_2)L \end{aligned} \tag{37}$$

Substituting equation 37 in equation 36 and simplifying yields

$$F_b = -\frac{1}{2}(1 - \lambda_o)\sin\phi_o g L[\rho(h_1 + h_2) + \rho_w h_w(x_2)] \tag{38}$$

The horizontal force caused by the landward tilt of the base when the model is isostatically adjusted is estimated as

$$F_i = -F_{GW} \tan\alpha_2 \tag{39}$$

where F_{GW} is the weight of the sediments and the water integrated across the width W of the transition zone and α_2 is the angle between the landward-sloping basement and the horizontal (Figure 3A)

$$\begin{aligned} F_{GW} &= \int_{x_1}^{x_2} (\rho g h(x) + \rho_w g h_w(x))dx \\ &= \rho g\left(\frac{1}{2}h_w(x_2)W + h_2 W + \frac{1}{2}(h_1 - h_2 - h_w(x_2))W\right) \\ &+ \frac{1}{2}\rho_w g h_w(x_2)W \end{aligned} \tag{40}$$

Substituting equation 40 in equation 39 and simplifying yields

$$F_i = -\frac{1}{2}gW\tan\alpha_2[\rho(h_1 + h_2) + \rho_w h_w(x_2)] \tag{41}$$

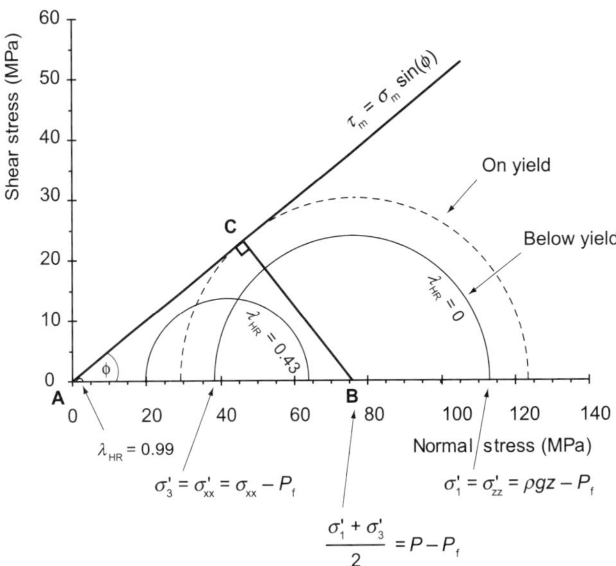

Figure 19. Mohr diagram illustrating the effect of increasing pore-fluid pressure on principal stresses. 20 MPa = 2901 psi. See text and Table 1 for explanation of symbols.

where tan α_2 can be estimated as

$$\tan\alpha_2 = \frac{h_1 - h_2 - h_w(x_2)}{W} \tag{42}$$

APPENDIX 2: BEHAVIOR OF THE PRINCIPAL STRESSES AS FLUID PRESSURES APPROACH THE WEIGHT OF THE OVERBURDEN

Our limit analysis indicates that pore-fluid pressures must be on the order of 80–99% of the weight of the overburden to induce deep-seated yielding within the sediments at the margin scale. However, the maximum level of the pore-fluid pressures in basins is also limited by the magnitude of the minimum principal stress, which is typically less than the weight of the overburden in, for example, extensional settings. Here pore pressure that is higher than the approximately horizontal minimum principal stress will likely cause hydrofracturing and thus dissipate fluid pressure in excess of the minimum principal stress, hence preventing the fluid pressure from approaching the weight of the overburden. Although these two conditions appear incompatible, they are in fact reconcilable because the effect of high pore-fluid pressures is to cause the values of the maximum and minimum principal stresses to converge, such that they are equal on yield conditions when the pore pressure reaches the minimum principal stress. Under these circumstances, the maximum and minimum principal stresses are equal and also equal to the weight of the overburden so that hydrofracturing only occurs when the pore-fluid pressure equals the weight of the overburden.

This effect of fluid pressure on the principal stresses is illustrated by the Mohr circles (Figure 19) and can also be seen by writing the two-dimensional Coulomb failure criterion in terms of the maximum shear stress, τ_m, the maximum and minimum principal stresses, σ_1 and σ_3, respectively, their mean, σ_m, and ϕ, the internal angle of friction, ($\tau_m = \sigma_m \sin\phi$; Jaeger and Cook, 1969, p. 98) modified for the pore-fluid pressure, P_f, and assuming zero cohesion

$$\begin{aligned}\tau_m &= \frac{1}{2}[(\sigma_1 - P_f) - (\sigma_3 - P_f)] \\ &= \frac{1}{2}[(\sigma_1 - P_f) + (\sigma_3 - P_f)]\sin\phi\end{aligned} \tag{43}$$

or

$$\tau_m = \frac{1}{2}(\sigma_1' - \sigma_3') = \frac{1}{2}(\sigma_1' + \sigma_3')\sin\phi, \tag{44}$$

where the primes denote effective stresses. Note that these equations are also the equations for the triangle ABC (Figure 19) (CB = ABsinϕ).

Assume that the principal directions are approximately vertical and horizontal and that $\sigma_1 = \sigma_{zz} = \rho g d$, the weight of the overburden. Under these circumstances, for given values of σ_1 and σ_3, the effect of increasing P_f is to move the center of the Mohr circle (the pressure or mean effective stress, $[\sigma_1' + \sigma_3']/2$) closer to the origin. In addition, increases in P_f are accompanied by corresponding decreases in $(\sigma_1' + \sigma_3')/2$ (the radius of the Mohr circle and side BC of the triangle ABC). This effect is seen in the way the left-hand side of equation 43 decreases as P_f on the right-hand side is increased.

The explicit effect of increasing P_f on the horizontal stress, $\sigma_3 = \sigma_{xx}$, is obtained by rearranging equation 43 to give

$$\sigma_3(1 + \sin\phi) - 2P_f\sin\phi = \sigma_1(1 - \sin\phi) \tag{45}$$

As P_f increases toward σ_3, σ_3 approaches σ_1. As a result, the diameter of the Mohr circle is reduced, and in the extreme case when $P_f \sim \sigma_{zz}$, the Mohr circle approaches a point and touches the failure envelope near the origin (Figure 19, Mohr circle for $\lambda_{HR} = 0.99$). This effect is equally true for below-yield stress conditions; in such cases, the effective stresses and mean effective stress are also smaller than the principal stresses and the Mohr circles do not touch the failure envelope (Figure 19).

The same argument can be used to show that, for compressive failure, where the horizontal stress is the maximum principal stress, the effect of the high pore-fluid pressure is exactly equivalent. As the pore-fluid pressure increases, the difference between the maximum principal stresses decreases, and they become equal when the pore-fluid pressure reaches the minimum principal stress. The overall result is that, for the extensional case at yield, the minimum principal stress converges on the maximum principal stress as the pore pressure increases (and the converse holds for compressive failure), which means the pore pressure can reach the weight of the overburden before hydrofracturing occurs.

Coupled changes of pore-fluid pressures and horizontal stress of this kind have been addressed in previous studies (e.g., Hillis, 2003) and are associated with differential tensile stress fields in nature. For example, Yassir and Bell (1994) documented increased horizontal stresses just above the onset of excess pore-fluid pressures beneath the Scotian Shelf, offshore eastern Canada.

APPENDIX 3: DESCRIPTION OF NUMERICAL METHOD

We employ the SOPALE (simplified optimized arbitrary Lagrangian Eulerian) model, a velocity-based two-dimensional plane-strain finite element model designed for fluid Stokes flow (Fullsack, 1995; Dalhousie Geodynamics Group, 2009). The model uses an arbitrary Lagrangian-Eulerian method where computations are performed on an Eulerian grid that adapts vertically to the evolving model domain, and Lagrangian particles track and update the material properties. In this approach, very large strains can be achieved without the numerically problematic distortion of the discretizing mesh, and sediments can be added to the current surface. For the models described here, the Eulerian and Lagrangian grids were chosen to comprise 104 vertical by 800 horizontal elements. The element thicknesses are adjusted to match the half-Gaussian geometry (equation 16). Hence, the vertical resolution increases seaward.

Water loading is represented by applying a boundary load normal to the upper surface seaward of position x_1 (Figure 7), which increases the solid and fluid pressures in proportion to the overlying water column. No horizontal velocities are permitted at the vertical model boundaries or the base of the model except for isostatic adjustments. Deformation is therefore driven solely by the gravitational force caused by the differential height and by the prograding sediments.

REFERENCES CITED

Audet, D. M., and J. D. C. McConnell, 1992, Forward modeling of porosity and pore pressure evolution in sedimentary basins: Basin Research, v. 4, p. 147–162, doi:10.1111/j.1365-2117.1992.tb00137.x.

Bilotti, F., and J. H. Shaw, 2005, Deep-water Niger Delta fold and thrust belt modeled as a critical-taper wedge: The influence of elevated basal fluid pressure on structural styles: AAPG Bulletin, v. 89, no. 11, p. 1475–1491, doi:10.1306/06130505002.

Bingham, E. C., 1922, Fluidity and plasticity: New York, McGraw-Hill, p. 215–218.

Camerlo, R. H., and E. F. Benson, 2006, Geometric and seismic interpretation of the Perdido fold belt; northwestern deep-water Gulf of Mexico: AAPG Bulletin, v. 90, no. 3, p. 363–386, doi:10.1306/10120505003.

Cohen, H. A., and K. McClay, 1996, Sedimentation and shale tectonics of the northwestern Niger Delta front: Marine and Petroleum Geology, v. 13, no. 3, p. 313–328, doi:10.1016/0264-8172(95)00067-4.

Corredor, F., J. H. Shaw, and F. Bilotti, 2005, Structural styles in the deep-water fold and thrust belts of the Niger Delta: AAPG Bulletin, v. 89, no. 6, p. 753–780, doi:10.1306/02170504074.

Dalhousie Geodynamics Group, 2009, http://geodynamics.oceanography.dal.ca/sopaledoc.html (accessed March 4, 2010).

Diegel, F. A., J. F. Karlo, D. C. Schuster, R. C. Shoup, and P. R. Tauvers, 1995, Cenozoic structural evolution and tectono-stratigraphic framework of the northern Gulf Coast continental margin, *in* M. P. A. Jackson, D. G. Roberts, and S. Snelson, eds., Salt tectonics: A global perspective: AAPG Memoir 65, p. 109–151.

Dimitrov, L. I., 2002, Mud volcanoes—The most important pathway for degassing deeply buried sediments: Earth-Science Reviews, v. 59, p. 49–76.

Dooley, T., K. R. McClay, M. Hempton, and D. Smit, 2005, Salt tectonics above complex basement extensional fault systems: Results from analog modeling, *in* A. G. Doré and B. A. Vinning, eds., Petroleum geology: North-west Europe and global perspective: Proceedings of the 6th Petroleum Geology Conference: Geological Society (London), p. 1631–1648.

Drucker, D. C., and W. Prager, 1952, Soil mechanics and plastic analysis of limit design: Quarterly of Applied Mathematics, v. 10, no. 2, p. 157–165.

Dugan, B., P. B. Flemings, D. L. Olgaard, and M. J. Gooch, 2003, Consolidation, effective stress, and fluid pressure of sediments from ODP site 1073, U.S. mid-Atlantic continental slope: Earth and Planetary Science Letters, v. 215, p. 13–26.

Fletcher, R. C., M. R. Hudec, and I. A. Watson, 1995, Salt glacier and composite sediment-salt glacier models for the emplacement and early burial of allochthonous salt sheets, *in* M. P. A. Jackson, D. G. Roberts, and S. Snelson, eds., Salt tectonics: A global perspective: AAPG Memoir 65, p. 77–108.

Fort, X., J. P. Brun, and F. Chauvel, 2004, Salt tectonics on the Angolan margin, synsedimentary deformation processes: AAPG Bulletin, v. 88, no. 11, p. 1523–1544, doi:10.1306/06010403012.

Fowler, A. C., and X. Yang, 1999, Pressure solution and viscous compaction in sedimentary basins: Journal of Geophysical Research, v. 104, p. 12,989–12,997.

Fullsack, P., 1995, An arbitrary Lagrangian-Eulerian formulation for creeping flows and its application in tectonic models: Geophysical Journal International, v. 120, p. 1–23, doi:10.1111/j.1365-246X.1995.tb05908.x.

Gauer, P., T. J. Kvalstad, C. F. Forsberg, P. Bryn, and K. Berg, 2005, The last phase of the Storegga Slide: Simulation of retrogressive slide dynamics and comparison with slide-scar morphology: Marine and Petroleum Geology, v. 22, p. 171–178, doi:10.1016/j.marpetgeo.2004.10.004.

Gaullier, V., and B. C. Vendeville, 2005, Salt tectonics driven by sediment progradation: Part II. Radial spreading of sedimentary lobes prograding above salt: AAPG Bulletin, v. 89, p. 1081–1089.

Ge, H., and M. P. A. Jackson, 1998, Physical modeling of structures formed by salt withdrawal: Implications for deformation caused by salt dissolution: AAPG Bulletin, v. 82, no. 2, p. 228–250.

Ge, H., M. P. A. Jackson, and B. C. Vendeville, 1997, Kinematics and dynamics of salt tectonics driven by progradation: AAPG Bulletin, v. 81, no. 3, p. 398–423.

Gemmer, L., S. J. Ings, S. Medvedev, and C. Beaumont, 2004, Salt tectonics driven by differential sediment loading: stability analysis and finite-element experiments: Basin Research, v. 16, p. 199–218, doi:10.1111/j.1365-2117.2004.00229.x.

Gemmer, L., C. Beaumont, and S. J. Ings, 2005, Dynamic modeling of passive margin salt tectonics: Effects of water loading, sediment properties and sedimentation patterns: Basin Research, v. 17, p. 383–402, doi:10.1111/j.1365-2117.2005.00274.x.

Graue, K., 2000, Mud volcanoes in deep water Nigeria: Marine and Petroleum Geology, v. 17, p. 959–974, doi:10.1016/S0264-8172(00)00016-7.

Haack, R. C., P. Sundararaman, J. O. Diedjomahor, H. Xiao, N. J. Gant, E. D. May, and K. Kelsch, 2000, Niger delta petroleum systems, Nigeria, *in* M. R. Mello and B. J. Katz, eds., Petroleum systems of South Atlantic margins: AAPG Memoir 73, p. 213–231.

Hamiter, R., A. Lowrie, M. MacKenzie, and E. Guderian, 1997, Origin of salt-withdrawal mini-basins, proven producers along present and paleo-Louisiana Slope: Transactions - Gulf Coast Association of Geological Societies, v. 47, p. 185–191.

Hillis, R. R., 2003, Pore pressure/stress coupling and its implications for rock failure, *in* P. Van Rensbergen, R. R. Hillis, A. J. Maltmann, and C. K. Morley, eds., Subsurface sediment mobilization: Geological Society (London) Special Publication 216, p. 359–368.

Hubbert, M. K., and W. W. Rubey, 1959, Role of fluid pressure in mechanics of overthrust faulting: 1. Mechanics of fluid-filled porous solids and its application to overthrust faulting: Geological Society of America Bulletin, v. 70, p. 115–166, doi:10.1130/0016-7606(1959)70[115:ROFPIM]2.0.CO;2.

Ings, S. J., and J. W. Shimeld, 2006, A new conceptual model for the structural evolution of a regional salt detachment on the northeast Scotian margin, offshore eastern Canada: AAPG Bulletin, v. 90, no. 9, p. 1407–1423, doi:10.1306/04050605159.

Ings, S. J., C. Beaumont, and L. Gemmer, 2004, Numerical modeling of salt tectonics on passive margin continental margins: Preliminary assessment of the effects of sediment loading, buoyancy, margin tilt, and isostasy: Concepts, applications and case studies for the 21st century, *in* P. J. Post, D. L. Olson, K. T. Lyons, S. L. Palmes, P. F. Harrison, and N. C. Rosen, eds., Gulf Coast Section SEPM Foundation 24th Annual Bob F. Perkins Research Conference Proceedings, p. 2–40.

Jackson, M. P. A., 1995, Retrospective salt tectonics, *in* M. P. A. Jackson, D. G. Roberts, and S. Snelson, eds., Salt tectonics: A global perspective: AAPG Memoir 65, p. 1–28.

Jaeger, J. C., and N. G. W. Cook, 1969, Fundamentals of rock mechanics: London, Chapman and Hall, 515 p.

Kirby, S. H., 1985, Rock mechanics observations pertinent to the rheology of the continental lithosphere and the localization of strain along shear zones: Tectonophysics, v. 119, p. 1–27, doi:10.1016/0040-1951(85)90030-7.

Knox, G. J., and E. M. Omatsola, 1989, Development of the Cenozoic Niger Delta in terms of the 'Escalator Regression' model and impact on hydrocarbon distribution: Proceedings of the Royal Geological and Mining Society of the Netherlands Symposium 'Coastal Lowlands, Geology and Geotechnology', 1987, p. 181–202.

Kooi, H., 1997, Insufficiency of compaction disequilibrium as the sole cause of high pore fluid pressures in pre-Cenozoic sediments: Basin Research, v. 9, p. 227–241.

Kopf, A. J., 2002, Significance of mud volcanism: Reviews of Geophysics, v. 40, p. 1–52.

Lance, S., P. Henry, X. Le Pichon, S. Lallemant, H. Chamley, F. Rostek, J. C. Faugères, E. Gonthier, and K. Olu, 1998, Submersible study of mud volcanoes seaward of the Barbados accretionary wedge: Sedimentology, structure and rheology: Marine Geology, v. 145, p. 255–292, doi:10.1016/S0025-3227(97)00117-5.

Lehner, F. K., 2000, Approximate theory of substratum creep and associated overburden deformation in salt basins and deltas, *in* F. K. Lehner and J. L. Urai, eds., Aspects of tectonic faulting: Berlin, Springer-Verlag, p. 21–47.

Lobkovsky, L. I., and V. I. Kerchman, 1991, A two-level concept of plate tectonics: Application to geodynamics: Tectonophysics, v. 199, p. 343–374, doi:10.1016/0040-1951(91)90178-U.

Mandl, G., 1988, Mechanics of tectonic faulting: Models and basic concepts: Amsterdam, Elsevier, 407 p.

Mann, D. M., and A. S. Mackenzie, 1990, Prediction of pore fluid pressures in sedimentary basins: Marine and Petroleum Geology, v. 7, p. 55–65, doi:10.1016/0264-8172(90)90056-M.

Modica, C. J., and E. R. Brush, 2004, Postrift sequence stratigraphy, paleogeography, and fill history of the deep-water Santos Basin, offshore Brazil: AAPG Bulletin, v. 88, no. 7, p. 923–945, doi:10.1306/01220403043.

Mohriak, W. U., et al., 1995, Salt tectonics and structural styles in the deep-water province of the Cabo Frio region, Rio de Janeiro, Brazil, *in* M. P. A. Jackson, D. G. Roberts, and S. Snelson, eds., Salt tectonics: A global perspective: AAPG Memoir 65, p. 273–304.

Morency, C., R. S. Huismans, C. Beaumont, and P. Fullsack, 2007, A numerical model for coupled fluid flow and matrix deformation with applications to disequilibrium compaction and delta stability: Journal of Geophysical Research, v. 112, B10407, doi:10.1029/2006JB004701.

Morley, C. K., 2003, Mobile shale related deformation in large deltas developed on passive and active margins, *in* P. Van Rensbergen, R. R. Hillis, A. J. Maltmann, and C. K. Morley, eds., Subsurface sediment mobilization: Geological Society (London) Special Publication 216, p. 335–357.

Morley, C. K., and G. Guerin, 1996, Comparison of gravity-driven deformation styles and behavior associated with mobile shales and salt: Tectonics, v. 15, no. 6, p. 1154–1170, doi:10.1029/96TC01416.

Mourgues, R., and P. R. Cobbold, 2003, Some tectonic consequences of fluid overpressures and seepage forces as demonstrated by sandbox modeling: Tectonophysics, v. 376, p. 75–97, doi:10.1016/S0040-1951(03)00348-2.

Ratcliff, D. W., 1993, New technologies improve seismic images of salt bodies: Oil & Gas Journal, v. 91, September 27, p. 41–48.

Rowan, M. G., 1995, Structural styles and evolution of allochthonous salt, central Louisiana outer shelf and upper slope, *in* M. P. A. Jackson, D. G. Roberts, and S. Snelson, eds., Salt tectonics: A global perspective: AAPG Memoir 65, p. 199–228.

Rowan, M. G., M. P. A. Jackson, and B. D. Trudgill, 1999, Salt-related fault families and fault welds in the northern Gulf of Mexico: AAPG Bulletin, v. 83, no. 9, p. 1454–1484.

Rowan, M. G., R. A. Ratliff, B. D. Trudgill, and J. Barceló Duarte, 2001, Emplacement and evolution of the Mahogany salt body, central Louisiana outer shelf, northern Gulf of Mexico: AAPG Bulletin, v. 85, no. 6, p. 947–969.

Rowan, M. G., F. J. Peel, and B. C. Vendeville, 2004, Gravity-driven fold belts on passive margins, *in* K. R. McClay, ed., Thrust tectonics and hydrocarbon systems: AAPG Memoir 82, p. 157–182.

Saffer, D. M., and C. Marone, 2003, Comparison of smectite- and illite-rich gouge frictional properties: Application to the updip limit of the seismogenic zone along subduction megathrusts: Earth and Planetary Science Letters, v. 215, p. 219–235.

Saffer, D. M., K. M. Frye, C. Marone, and K. Mair, 2001, Laboratory results indicating complex and potentially unstable frictional behavior of smectite clay: Geophysical Research Letters, v. 28, no. 12, p. 2297–2300.

Spiers, C. J., P. M. T. M. Schutjens, R. H. Brzesowsky, C. J. Peach, J. L. Liezenberg, and H. J. Zwart, 1990, Experimental determination of constitutive parameters governing creep of rocksalt by pressure solution, *in* R. J. Knipe, and E. H. Rutter, eds., Deformation mechanisms, rheology and tectonics: Geological Society (London) Special Publication 54, p. 215–227.

Sullivan, S., L. J. Wood, and M. Paul, 2004, Distribution, nature and origin of mobile mud features offshore Trinidad, *in* P. J. Post, D. L. Olson, K. T. Lyons, S. L. Palmes, P. F. Harrison, and N. C. Rosen, eds., Salt-sediment interactions and hydrocarbon prospectivity: Concepts, applications and case studies for the 21st century: Gulf Coast Section SEPM Foundation 24th Annual Bob F. Perkins Research Conference Proceedings, p. 498–513.

Tari, G., J. Molnar, and P. Ashton, 2003, Examples of salt tectonics from west Africa: A comparative approach, *in* T. T. Arthur, D. S. MacGregor, and N. R. Cameron, eds., Petroleum geology of Africa: New themes and developing technologies: Geological Society (London) Special Publication 207 p. 85–104.

Tari, G. C., P. R. Ashton, K. L. Coterill, J. S. Molnar, M. C. Sorgenfrei, W. A. Philip Thompson, D. W. Valasek, and J. F. Fox, 2002, Are west Africa deep water salt tectonics analogous to the Gulf of Mexico?: Oil & Gas Journal, v. 100, March 4, p. 73–82.

Ter Heege, J. J., J. H. P. De Bresser, and C. J. Spiers, 2005a, Dynamic recrystallization of wet synthetic polycrystalline halite: Dependence of grain size distribution on flow stress, temperature and strain: Tectonophysics, v. 396, p. 35–57, doi:10.1016/j.tecto.2004.10.002.

Ter Heege, J. J., J. H. P. De Bresser, and C. J. Spiers, 2005b, Rheological behavior of synthetic rocksalt: the interplay between water, dynamic recrystallization and deformation mechanisms: Journal of Structural Geology, v. 27, p. 948–963, doi:10.1016/j.jsg.2005.04.008.

Terzaghi, K., 1923, Die Berechnung der Durchlässigkeitsziffer des Tones aus dem Verlauf der hydrodynamischen Spannungserscheinungen: Sitzungsberichte der Akademie der Wissenschaften in Wien: Mathematisch-Naturwissenschaftliche Klasse, Abteilung IIa, v. 132, p. 125–138.

Terzaghi, K., 1943, Theoretical soil mechanics: New York, John Wiley and Sons, p. 12.

Tuncay, K., and P. Ortoleva, 2001, Salt tectonics as a self-organizing process: A reaction, transport, and mechanics model: Journal of Geophysical Research, v. 106, no. B1, p. 803–817, doi:10.1029/2000JB900107.

Van Rensbergen, P., and C. K. Morley, 2000, 3D seismic study of a shale expulsion syncline at the base of the Champion Delta, offshore Brunei and its implications for the early structural evolution of large delta systems: Marine and Petroleum Geology, v. 17, p. 861–872, doi:10.1016/S0264-8172(00)00026-X.

Van Rensbergen, P., and C. K. Morley, 2003, Re-evaluation of mobile shale occurrences on seismic sections of the Champion and Baram deltas, offshore Brunei, *in* P. Van Rensbergen, R. R. Hillis, A. J. Maltmann, and C. K. Morley, eds., Subsurface sediment mobilization: Geological Society (London) Special Publication 216, p. 395–409.

Vendeville, B. C., 2005, Salt tectonics driven by sediment progradation: Part I. Mechanics and kinematics: AAPG Bulletin, v. 89, no. 8, p. 1071–1079, doi:10.1306/03310503063.

Vendeville, B. C., and M. P. A. Jackson, 1992, The rise of diapirs during thin-skinned extension: Marine and Petroleum Geology, v. 9, p. 331–353, doi:10.1016/0264-8172(92)90047-I.

Wu, S., and A. W. Bally, 2000, Slope tectonics—Comparisons and contrasts of structural styles of salt and shale tectonics of the Northern Gulf of Mexico with shale tectonics of offshore Nigeria in Gulf of Guinea, *in* W. Mohriak and M. Talwani, eds., Atlantic rifts and continental margins: AGU Geophysical Monograph 115, p. 151–172.

Yassir, N. A., and J. S. Bell, 1994, Relationships between pore pressure, stresses, and present-day geodynamics in the Scotian Shelf offshore Eastern Canada: AAPG Bulletin, v. 78, no. 12, p. 1863–1880.

4

Henry, Maria, Michael Pentilla, and Darrell Hoyer, 2010, Observations from exploration drilling in an active mud volcano in the southern basin of Trinidad, West Indies, *in* L. Wood, ed., Shale tectonics: AAPG Memoir 93, p. 63–78.

Observations from Exploration Drilling in an Active Mud Volcano in the Southern Basin of Trinidad, West Indies

Maria Henry
Henry GeoConsulting Services, Inc., Littleton, Colorado, U.S.A
Michael Pentilla
Target Surveys, Westminster, Colorado, U.S.A
Darrell Hoyer
Hoyer Petrophysics, Inc., Fort Collins, Colorado, U.S.A

ABSTRACT

The Trinidad Exploration and Development Company drilled the Habanero 1 well within Trinidad's southern basin in an area of surface mud volcano flows and vents. The well location is along a trend of mud volcanoes that extends across northern Venezuela and southern Trinidad. The upper 3200 ft (975 m) of the well drilled through interbedded mud volcano layers as confirmed by palynology, paleontology, lithology, well log, and seismic information. Below 3200 ft (975 m), the well drilled primarily country rock deposits of the upper Miocene through Pliocene Cruse Formation. Paleontology and palynology data give an age range of Eocene through Miocene for the mud volcano material; the highest recovery of foraminifera and dinocysts were from the Oligocene–Miocene Cipero and Lengua formations.

Several drilling problems were encountered, especially in the shallow, mudflow-rich part of the hole. Drilling issues included lost circulation intervals and the necessity for high mud weights (up to 17.5 ppg), to control high pressures. The well encountered numerous oil and gas shows, but no commercial hydrocarbons were tested.

Seismic data in the area illustrate a downward-tapering cone of disruption around mud volcano vents, with numerous faults providing conduits for the flow of mud. Drilling samples confirmed this interpretation, because the zones described as faulted commonly coincided with an influx of exotic mudflow material.

Subsurface logs measured decreased resistivity, lower density, and higher interval transit time (slower velocity) in mud volcano layers relative to intervals of country rock formations. These intervals are anomalous even when compared to log data for overpressured shale zones

DOI:10.1306/13231308M933418

in neighboring wells. The Habanero 1 checkshot survey documents shallow mud volcano layers with velocities slower than the velocity of water, perhaps caused by the presence of entrained gases within the matrix of the mud. Refraction static velocities are also anomalously slow around Habanero 1 and along the trend of mud volcanoes.

INTRODUCTION

The southwest peninsula of the Caribbean island of Trinidad is located along a regional trend of mud volcanoes (Figure 1) that extends from Venezuela (Duerto and McClay, 2002) across southern Trinidad and into eastern Trinidad's deep-marine offshore areas (Sullivan, 2005). Eastern Venezuela and Trinidad are part of the Eastern Venezuela basin, a large Tertiary-aged basin located along a transpressional plate boundary zone; the Caribbean plate moving east-southeast relative to the South American plate (Figure 2). This transpression contributes to the high density of mud volcanoes in the region. Duerto et al. (2006) proposed that the diachroneity of the oblique collision between the Caribbean and South American plates caused shale deformation to become progressively younger from west to east. The Orinoco delta, the principal clastic sediment source for the region, is located to the southeast and has deposited large amounts of sediment in the area since the middle Miocene.

Triggers that initiate mud volcanism are poorly understood. In this region of the world, rapid sedimentation and tectonic loading caused by southeast-directed transpression likely caused overpressuring in deeply buried basinal shales and resultant shale mobilization. (Duerto et al., 2006). Deville et al. (2003) cited regional hydrocarbon generation as an additional mechanism for overpressure generation and reduction of rock density. He also discusses the function of hydrofracturing and liquefaction by seismic agitation. A few of Trinidad's mud volcanoes alternate between phases of slow but permanent expulsion and intermittent catastrophic events. Overpressured shales also contribute to drilling problems in the subsurface.

The Trinidad Exploration and Development Company drilled the Habanero 1 well in 2005 in southwestern Trinidad. The well drilled in an area with mapped surface mud volcano flows in close proximity to mud volcano vents (Figure 3). Data acquired while drilling show that the well drilled through approximately 3200 ft (975 m) of layered muds associated with a mud volcano before drilling out into country rock deposited by the proto-Orinoco delta. This article discusses paleontology, palynology, lithology, seismic, and well data from the Habanero 1 well and compares the character of intervals interpreted as mud volcano flows with those of the adjacent country rock. A brief description of mud volcano outcrops 1.5 mi (2.5 km) to the east of the Habanero 1 location and a discussion of the seismic character of the sections near the outcrop are also included.

REGIONAL GEOLOGY

Trinidad's southwest peninsula is bounded by the Southern Range anticline to the south, the Los Bajos fault to the north and east, and the Erin syncline to the north (Figure 3). Miocene–Pliocene sedimentary rocks dip northeastward into the Erin syncline. The faulted Southern Range anticline trends east to west along the southern coast of the southwest peninsula, coming onshore near Galfa Point. The major oil fields of Palo Seco, Southwest Soldado, and Pedernales are located nearby (Figure 1). Mud volcanoes (Figure 3) are mapped along the extent of the Southern Range anticline. Their occurrence is confirmed by seismic data (Figure 4A, B) that show these features associated with the core of the anticline. Previous authors (Duerto and McClay, 2002) documented the association of similar deep diapiric features with the frontal edge of a buried fold and thrust belt interpreted at depth. This transpressional fold and thrust belt is extrapolated to trend southwest to northeast, from Venezuela through Trinidad, parallel with the mud volcano trend shown in Figure 1.

Evidence presented by Deville et al. (2003) documents that mud material from Trinidad mud volcanoes originates from several stratigraphic levels, ranging in age from Cretaceous to Pliocene. A few mud volcanoes along this trend extrude mud on a slow, continuous basis, punctuated by episodic eruptive events. Ejected clasts include megacrystals of calcite and pyrite, and breccias from the tectonic wedge. These exotic clasts are Paleocene to Miocene in age. Many of the clasts are angular with carbonate-cemented fractures and include angular breccias of joined elements within calcite crystals. Deville et al. (2003) believed that these clasts formed because of natural hydraulic fracturing processes. A stratigraphic column (Figure 5) displays the age range for extruded mud material superimposed on the stratigraphic nomenclature for Trinidad's southern basin.

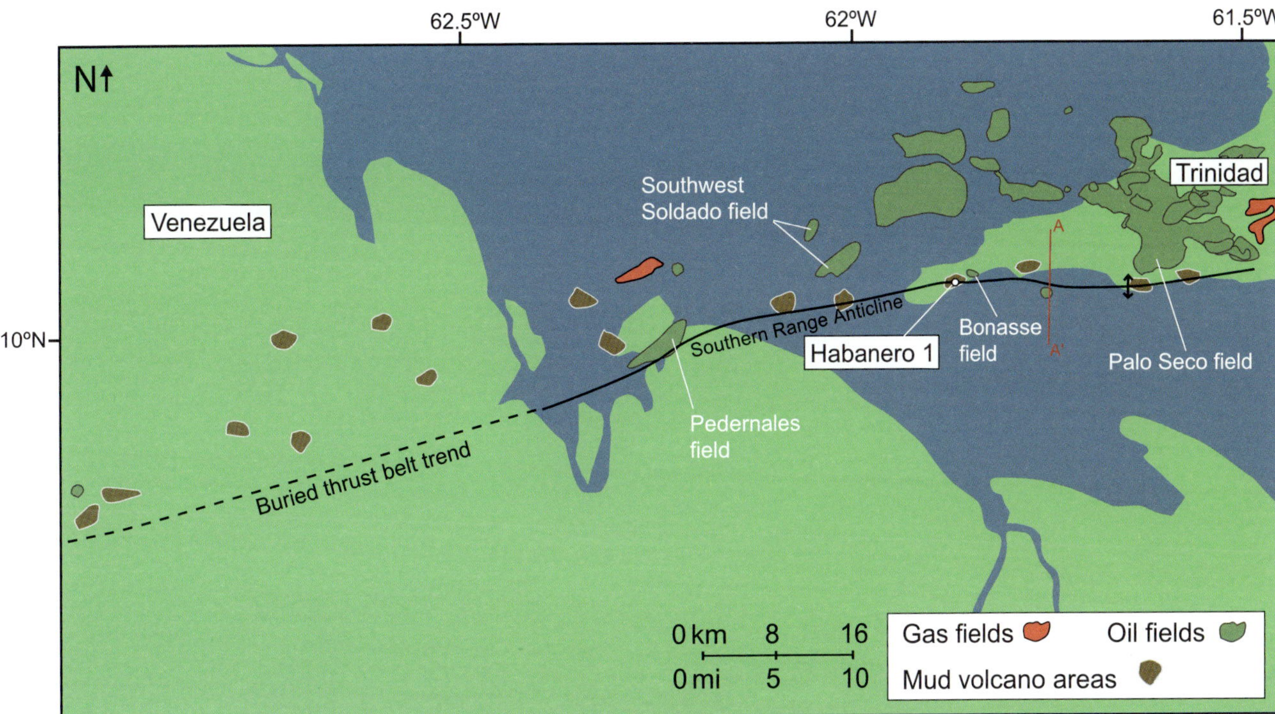

Figure 1. Map of eastern Venezuela and western Trinidad illustrating the location of Habanero 1, mud volcanoes, and oil and gas fields. The location of seismic line AA′ shown in Figure 4A is noted.

SURFACE OUTCROP OF THE SOUTHERN RANGE ANTICLINE

Strata folded into the Southern Range anticline, with associated faulting and interbedded mud volcano layers, are exposed in outcrop along Trinidad's southern coast at Galfa Point.

Figure 3 illustrates the relative locations of the Habanero 1 well, Galfa Point, and extrusive mud volcano flows and vents. Galfa Point is located on the crest of the Southern Range anticline.

Beach outcrops show a fault zone in the exposed Cruse Formation, with a layer of interbedded mudflow material adjacent to the fault (Figure 6A). Farther west along the beach, numerous clasts eroded from mud volcano layers are winnowed by waves adjacent to the outcropping mud (Figure 6B). Limestones, sandstones, and oil-stained conglomerates comprise the blocks littering the beach adjacent to the mudflows. Fractured clasts, tabular calcite, sandstones with slickensides, and crystalline pyrite are also common in the area of exposed mud and clasts at Galfa Point. Morphology of these clasts is consistent with Deville et al's (2003) description for expelled clasts from Trinidad mud volcanoes.

A high-resolution two-dimensional seismic line runs 2700 ft (822 m) east of Galfa Point (Figure 7A) and images the south flank of the Southern Range anticline. An interpreted line (Figure 7B) highlights the mud volcano vent with faulting and associated mudflows. A north-dipping main fault continues to depth, as well as small-scale, north-dipping faults. Faults and fractures provide zones of weakness for stratigraphically deep mud to flow to the surface.

HABANERO 1 DRILLING RESULTS

The Habanero 1 well was drilled in close proximity to the Southern Range anticline, approximately 1.5 mi (2.5 km) west of the Galfa Point outcrop (Figure 3). The Habanero 1 well surface location is near the crest of the Southern Range anticline in an area covered by surface mud volcano flows, just south of a mud volcano surface vent. The Bonasse field to the east produces oil from the Forest and upper Cruse formations at depths of less than 2000 ft (610 m). Upper Miocene through Pliocene Forest and Cruse formation sandstones were drilling targets for the Habanero 1 well (Figure 5).

The section from surface to 3140 ft (957 m) was dominated by Oligocene–Miocene foraminiferal microfaunas. The faunas are interpreted as reworked within mudflows composed of material derived from the Cipero and Lengua formations. A thin zone of Eocene-age

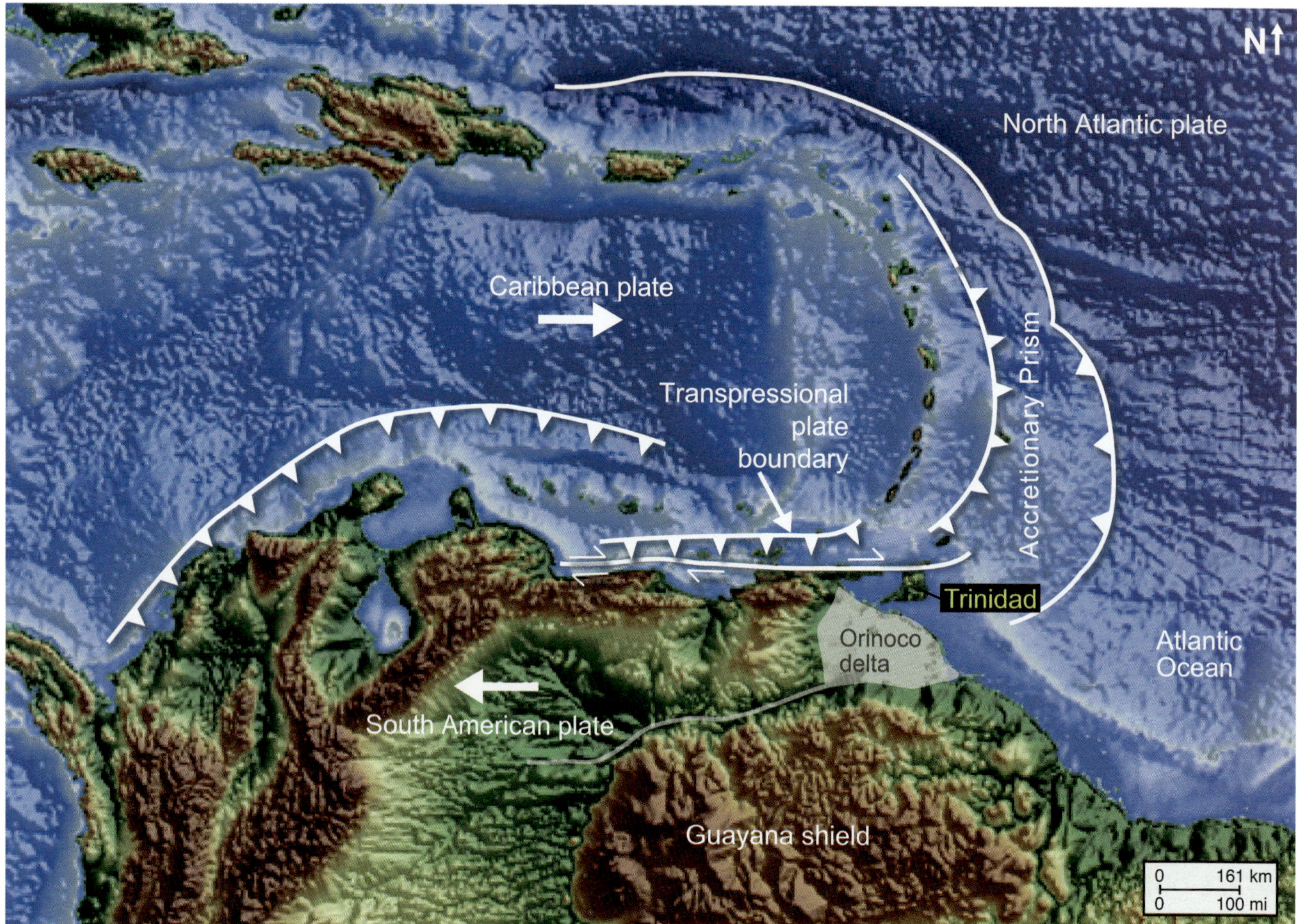

Figure 2. Geologic setting of Trinidad showing locations of Orinoco delta and key tectonic elements of the area.

Navet limestone, composed of planktonic foraminiferal oozes, was also recovered within this interval. Zones of probable in-situ microfaunas of Forest and Cruse formations were identified between 1310 and 2140 ft (399 and 652 m). Below 3300 ft (1005 m), the foraminiferal stratigraphy indicates the well drilled through primarily shelf and upper slope deposits of the middle Cruse Formation, although minor amounts of the microfaunas interpreted as reworked within exotic mudflows were found in deeper sections (Figure 8).

The upper 3200 ft (975 m) of the Habanero 1 well included a rich Oligocene through lower Miocene Cipero Formation palynological assemblage interpreted as reworked within mudflow deposits. Below 3200 ft (975 m), the palynological assemblages were typical for upper Miocene to Pliocene middle Cruse Formation deltaic deposits, with some minor dinocyst recoveries down to the total well depth of 8000 ft (2438 m). This background signal of Oligocene–Miocene dinocyst assemblages could be caused by drilling mud contamination, reworking, or an influx of mudflow material, possibly while drilling through faulted or fractured zones. Palynological and paleontological data confirmed that the Habanero 1 well did not encounter a normal stratigraphic sequence. The upper 3200 ft (975 m) of the well included flora and fauna normally associated with deeper Oligocene through lower Miocene stratigraphic units, whereas the deeper part of the well below 3200 ft (975 m) comprised a more typical sequence of upper Miocene to Pliocene deposits. Out-of-place microfaunas and floras are interpreted as reworked within exotic mudflows.

Zones of possible faults and fractures were identified by wellsite geologists through sample cutting data and are identified in Figure 8. Evidence for occurrence of faults and fractures included the presence of calcite (commonly tabular crystals), possible slickensides, crystalline pyrite, and other exotic minerals. Some zones were also associated with an influx of gas into the borehole and with lost circulation while drilling. These possible fault and fracture intervals are below the scale of seismic resolution. The influx of exotic mud material, as identified by mixed faunal and floral assemblages, commonly coincided with interpreted faulted intervals based

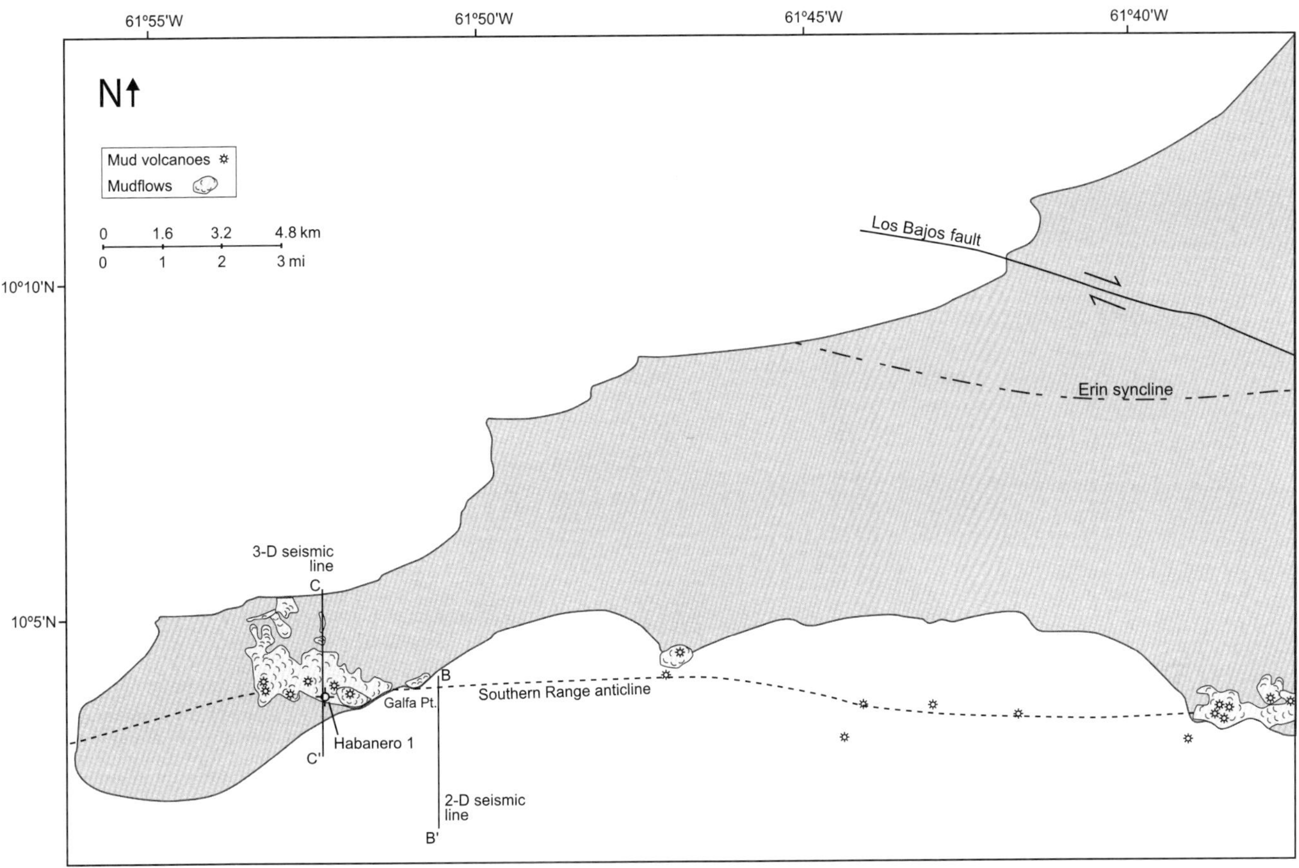

Figure 3. Map of the southwest peninsula of Trinidad showing key structural elements and mud volcano locations. The Habanero 1 well location and Galfa Point are located in mud volcano areas. The locations of seismic lines BB′ and CC′ (Figures 7, 9, respectively) are noted.

on lithologic descriptions. However, with only sample cutting data available, differentiating zones of actual faults and fractures from mudflow zones containing fractured clasts is difficult. As discussed in the previous section, fractured clasts, abundant tabular calcite, sandstones with slickensides, and crystalline pyrite are also common at the nearby Galfa Point mud volcano outcrop.

Claystones in the mudflow-rich part of the well above 3200 ft (975 m) included abundant fossils and calcite. Sandstones had variable lithologies, including gray, fine- to coarse-grained, poorly sorted to well-sorted units with black lithics and abundant quartz grains. Likely, exotic clasts from the mud volcano layers were mixed with samples from the in-situ country rock. Below the mud volcano contact at 3200 ft (975 m), lithologies were more consistent with the typical stratigraphic section in this part of the basin. Claystones were gray with few fossils and calcite laminae, whereas sandstones were gray, very fine- to fine-grained, and moderately sorted.

Palynology, paleontology, and lithology data confirm that the Habanero 1 well drilled through an interval of volcanic mudflows from the surface to 1350 ft (411 m), followed by an interval of in-situ formation (1350 to 2370 ft [411 to 722 m]). The well drilled another zone of mudflows from 2370 ft (722 m), finally emerging from the extrusive and intrusive mud body at around 3200 ft (975 m). Several technical problems were encountered while drilling the mudflow-rich part of the hole, including lost circulation intervals and requirements of high mud weights (up to 16.5 ppg) to stabilize the borehole. The well had both oil and gas shows, including a full range of heavy and light hydrocarbon indicators. Below 3200 ft (975 m), the well continued to drill through the Cruse Formation, requiring high mud weights of up to 17.5 ppg. Gas shows continued to a total depth of 8000 ft (2438 m). Three production tests recovered water with one minor show of oil.

SEISMIC DATA

A north-to-south-oriented seismic line through the Habanero 1 location illustrates the poor seismic data quality in the area where the Southern Range anticline

A

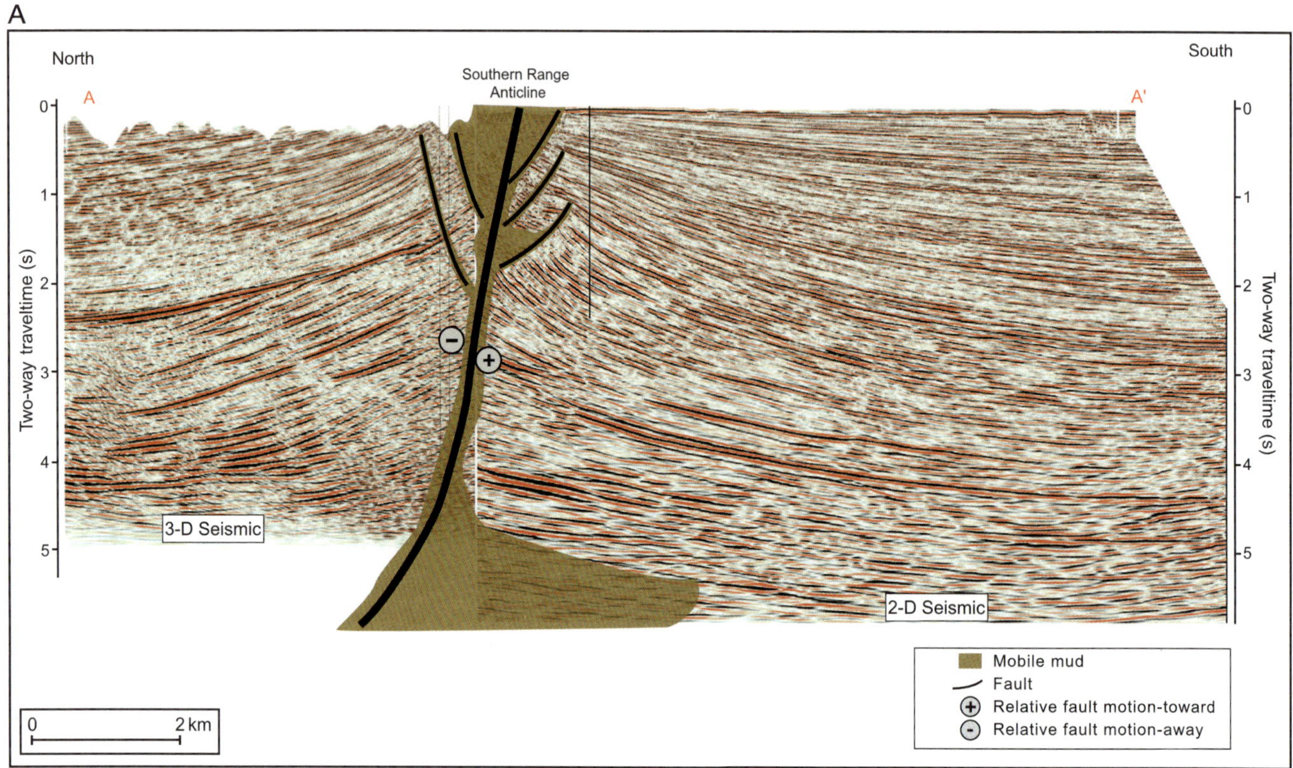

B

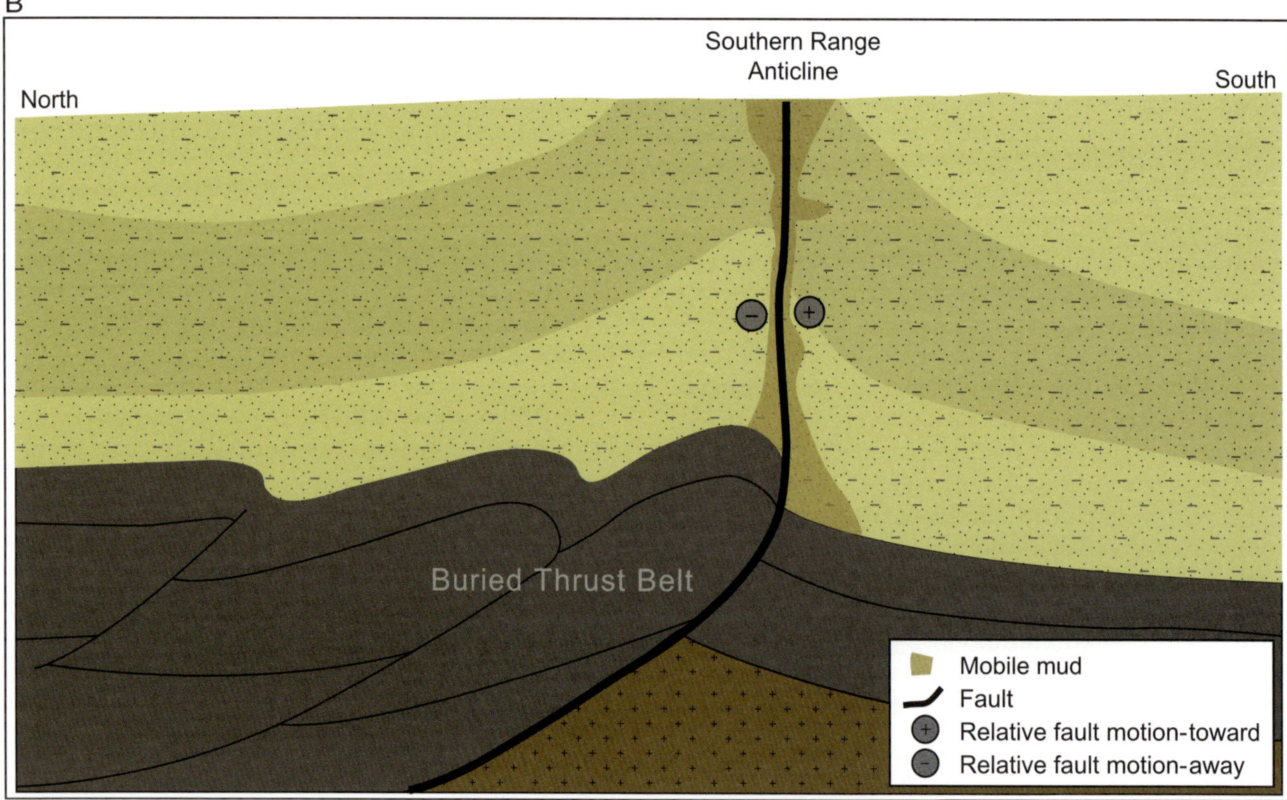

Figure 4. (A) North to south seismic line, including spliced data from three- and two-dimensional data sets. Faults associated with the Southern Range anticline provide conduits for mudflow from depth. The line location is shown on Figure 1 (seismic line courtesy of Petrotrin). 1 km = 0.6 mi. (B) North to south schematic diagram across the Southern Range anticline illustrating the proximity of mud volcanoes to the frontal edge of a buried thrust belt trend.

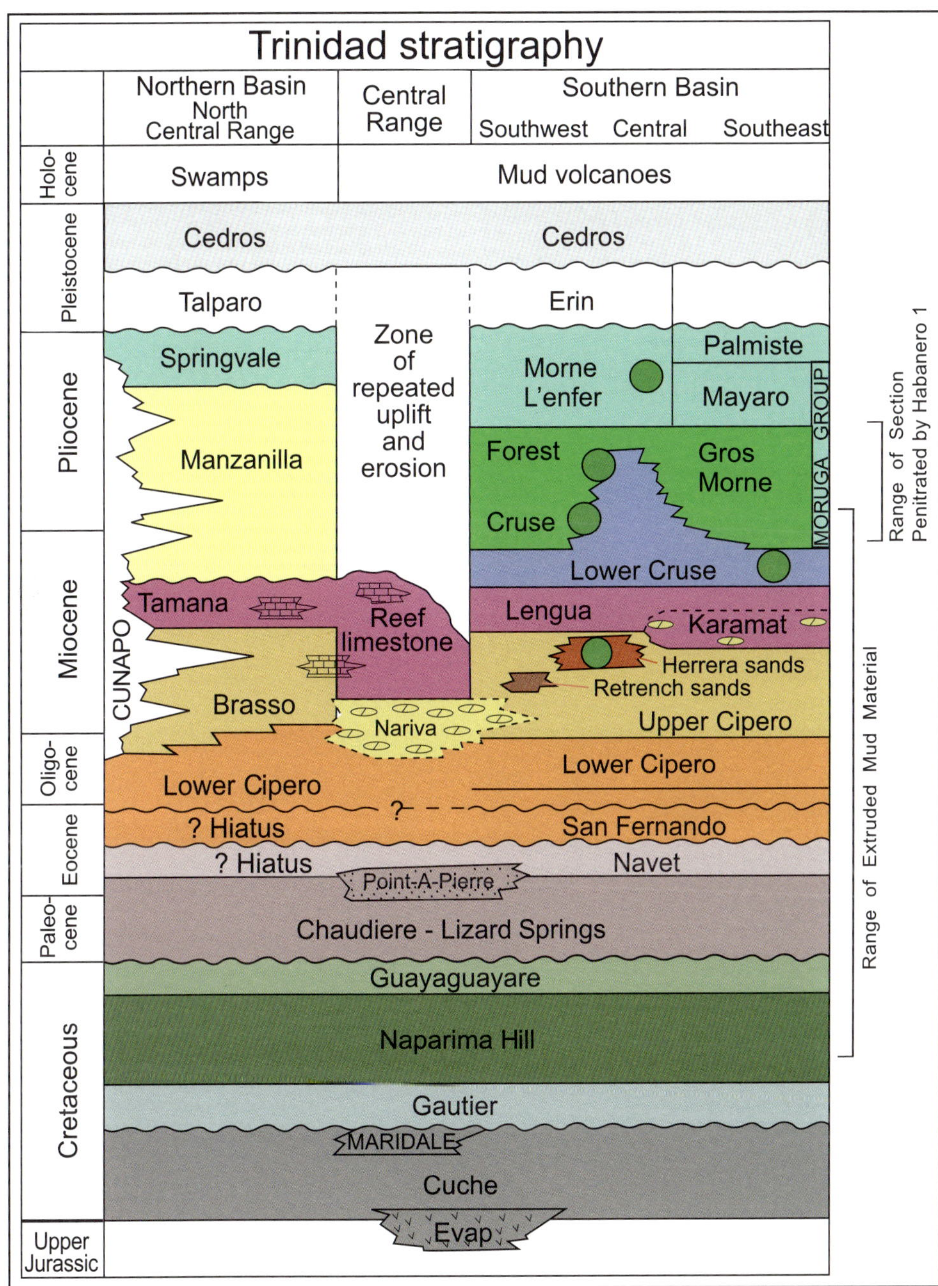

Figure 5. Trinidad stratigraphic column. Intervals of extruded mud material and penetrated section of the Habanero 1 well are noted to the right of the column.

and associated mud volcano flows are exposed along Trinidad's southwest peninsula (Figure 9). Habanero 1 was spudded in mudflow material and was drilled through several faults and mudflows down to about 1.2 s. Below 1.2 s, the well drilled through primarily bedded Cruse rocks to a total depth of 2.3 s.

ROCK PROPERTIES

Drilling difficulties necessitated casing the well bore below 3000 ft (914 m) prior to open-hole logging. Therefore, open-hole logs for Habanero 1 were run down to 3000 ft (914 m), and below 3000 ft (914 m), cased-hole logs were run. Figure 10 displays a panel with log curves to 3000 ft (914 m), including volume of sand (Vsand), formation resistivity, sonic, density, and neutron curves. The sections with increased mudflow material (0 to 1350 ft [411 m] and 2370 to 3000 ft [722 to 914 m]) show decreased resistivity and increased porosity measurements. In general, higher shale porosities are related to fluid-supported overburden and abnormally high pressures, whereas lower porosity shales or siltstones are related to normal compaction and pressure trends.

A

B

Figure 6. (A) Photograph taken at the Galfa Point outcrop on Trinidad's southwest coast illustrating the proximity of the fault zone to the layer of intruded mobile mud. (B) Photograph of the Galfa Point outcrop showing numerous clasts eroded from the mud volcano layers and winnowed by waves adjacent to the outcropping mud.

Resistivity decreases in high-porosity shale are caused by an increase in water volume. Figure 11 displays a crossplot of depth versus density for shales in wells in Trinidad's southwest peninsula area. These data show the incredible low density exhibited by mudflow shales. Figure 12 illustrates a similar trend with sonic data from shales from wells drilled in the southwest peninsula. This graph illustrates that the Habanero 1 well-log data exhibit two zones of anomalously high interval transit time (slow velocity). Both are mudflow zones.

An average checkshot velocity trend is extrapolated based on data from three wells in the southwest peninsula area (Figure 13). Points from the three wells follow similar velocity-versus-depth trends, whereas Habanero 1 data follow a similar slope but are shifted to slower average velocity values with depth. A comparison of checkshot interval velocity to average velocity from

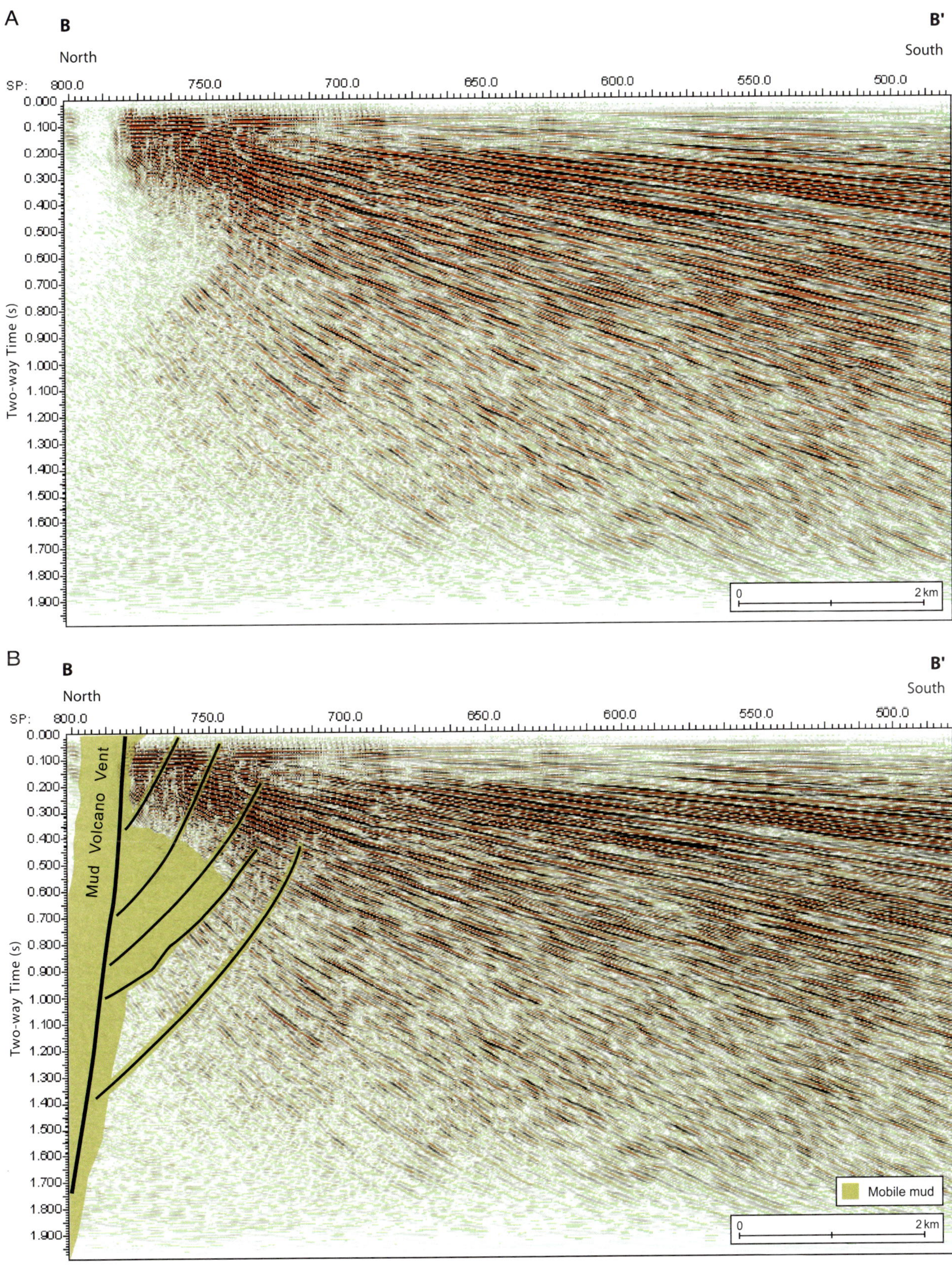

Figure 7. (A) Uninterpreted north-to-south-oriented high-resolution seismic line east of the Galfa Point outcrop. The location of the line is shown in Figure 3 (seismic line courtesy of Petrotrin). (B) Interpreted line, showing faulting patterns and associated mud volcano features. A north-dipping primary fault and numerous smaller faults likely provide conduits for mud to flow from stratigraphically deeper units. 1 km = 0.6 mi.

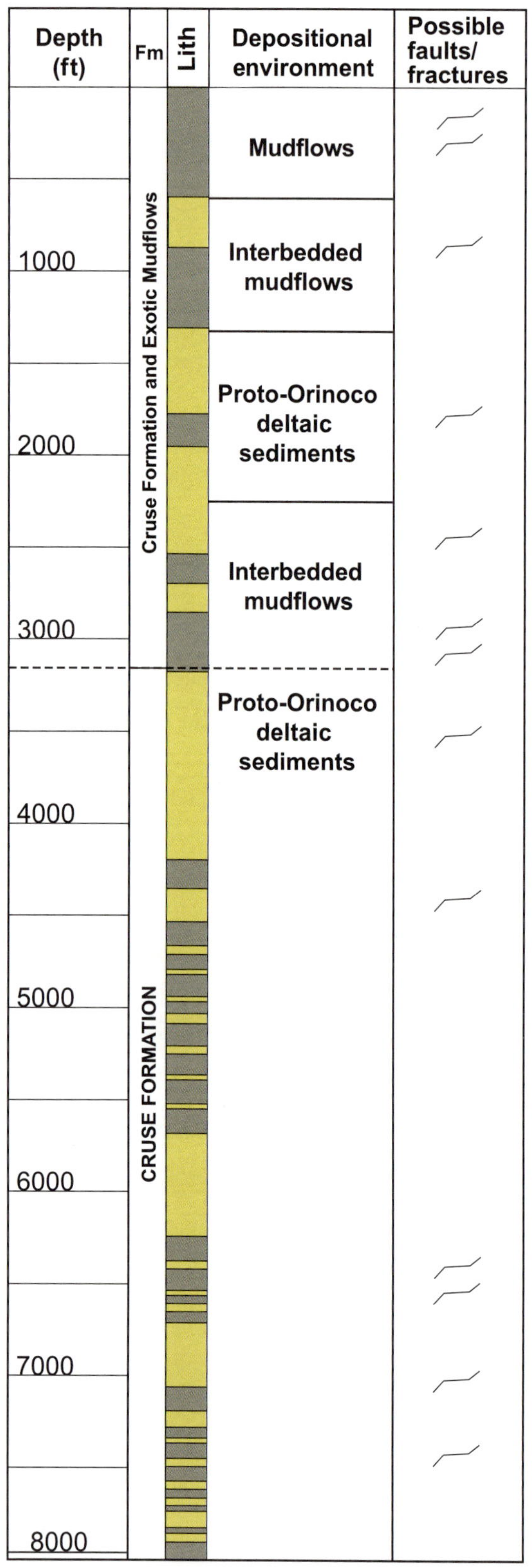

Figure 8. Column summarizing the lithology of rocks drilled by the Habanero 1 well as described from sample cuttings. The right column displays zones of possible faults and fractures, as identified by wellsite geologists. 1000 ft = 305 m.

Habanero 1 (Figure 14) illustrates that much of the velocity difference is caused by a very slow, shallow layer from the surface to around 1100 ft (335 m), which causes average velocity values for Habanero 1 to decrease. All points to the left of the blue line on Figure 14 are slower than the velocity of water (5300 ft/s [1615 m/s]). The anomalously slow velocities of the mudflow zones may be caused by entrained gases within the matrix of the mud.

A map of refraction static velocities (Figure 15) shows anomalously slow velocities in areas with surface mud volcano flows and vents. The Habanero 1 well is located on the north side of a slow velocity trend extending west to east along the Southern Range anticline.

SUMMARY

The Trinidad Exploration and Development Company's Habanero 1 well drilled within Trinidad's southern basin along a regional trend of mud volcanoes extending from southwest to northeast across northern Venezuela and the southwest peninsula of Trinidad. The well location is near the crest of the Southern Range anticline in an area of mapped mud volcano flows and surface vents.

Paleontology, palynology, lithology, well-log, and seismic data indicate that the well drilled from surface to 3200 ft (975 m) through interbedded mud volcano flows. Paleontology and palynology data give an age range of Eocene through Miocene for the exotic mudflow material, with an abundance of Oligocene–Miocene foraminera and dinocysts from the Cipero and Lengua formations. From 3200 ft (975 m) to a total depth of 8000 ft (2438 m), the well drilled primarily in-situ upper Miocene through Pliocene Cruse Formation rocks.

Drilling problems plagued the mudflow-dominated upper part of the well, with lost circulation intervals and the need for high mud weights (up to 16.5 ppg). This interval had numerous oil and gas shows, including a full range of heavy and light hydrocarbon signatures. The Cruse Formation section below 3200 ft (975 m) also encountered significant gas shows, but no commercial hydrocarbons were tested. Production tests produced water with one minor show of oil. High pressures during drilling necessitated using up to 17.5 ppg of mud to control the well in this interval.

Although seismic data through the well are poor, high-resolution seismic lines to the east of the well illustrate a downward-tapering cone of disruption around mud volcano vents near the crest of the anticline, with numerous faults providing conduits for the flow of mud from depth. Drilling observations confirm this interpretation because zones described as faulted while drilling

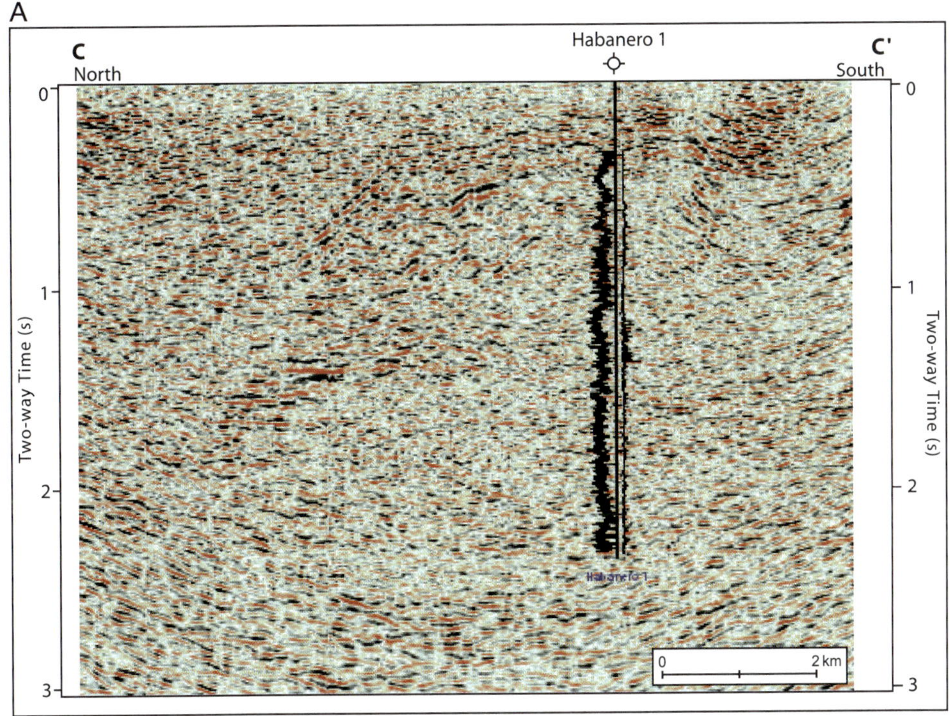

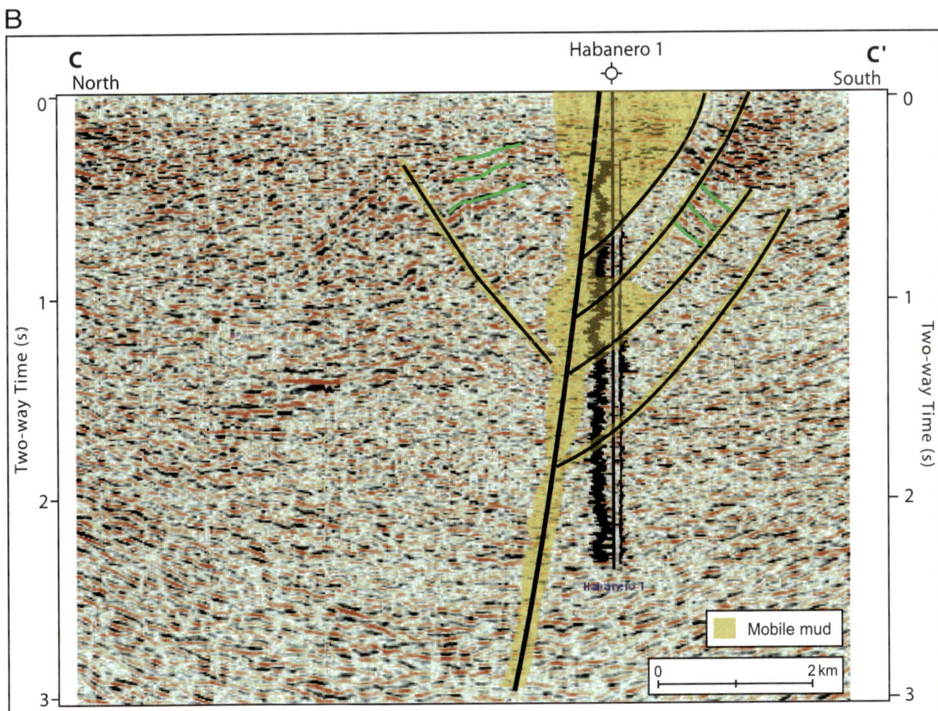

Figure 9. (A) Uninterpreted north-to-south-oriented seismic line through Habanero 1 extracted from a three-dimensional seismic volume illustrating poor seismic data quality in the area of surface mud volcano flows. The location of the line is shown in Figure 3. (B) Interpreted line illustrating north-dipping primary fault and numerous smaller faults with associated mudflows. 1 km = 0.6 mi.

commonly coincided with an influx of exotic mudflow material.

Subsurface log and seismic data from the Habanero 1 well reflect distinctive rock properties for mud volcano layers compared to shales of the country rock. Volcanic mudflow intervals in the well measured decreased resistivity, lower density, and higher interval transit time (slower velocity) relative to intervals of the in-situ formation. Sonic and density values for the mudflow zones are anomalous even when compared to shale trends for overpressured zones in neighboring wells. Density values are lower (higher porosity) and interval transit

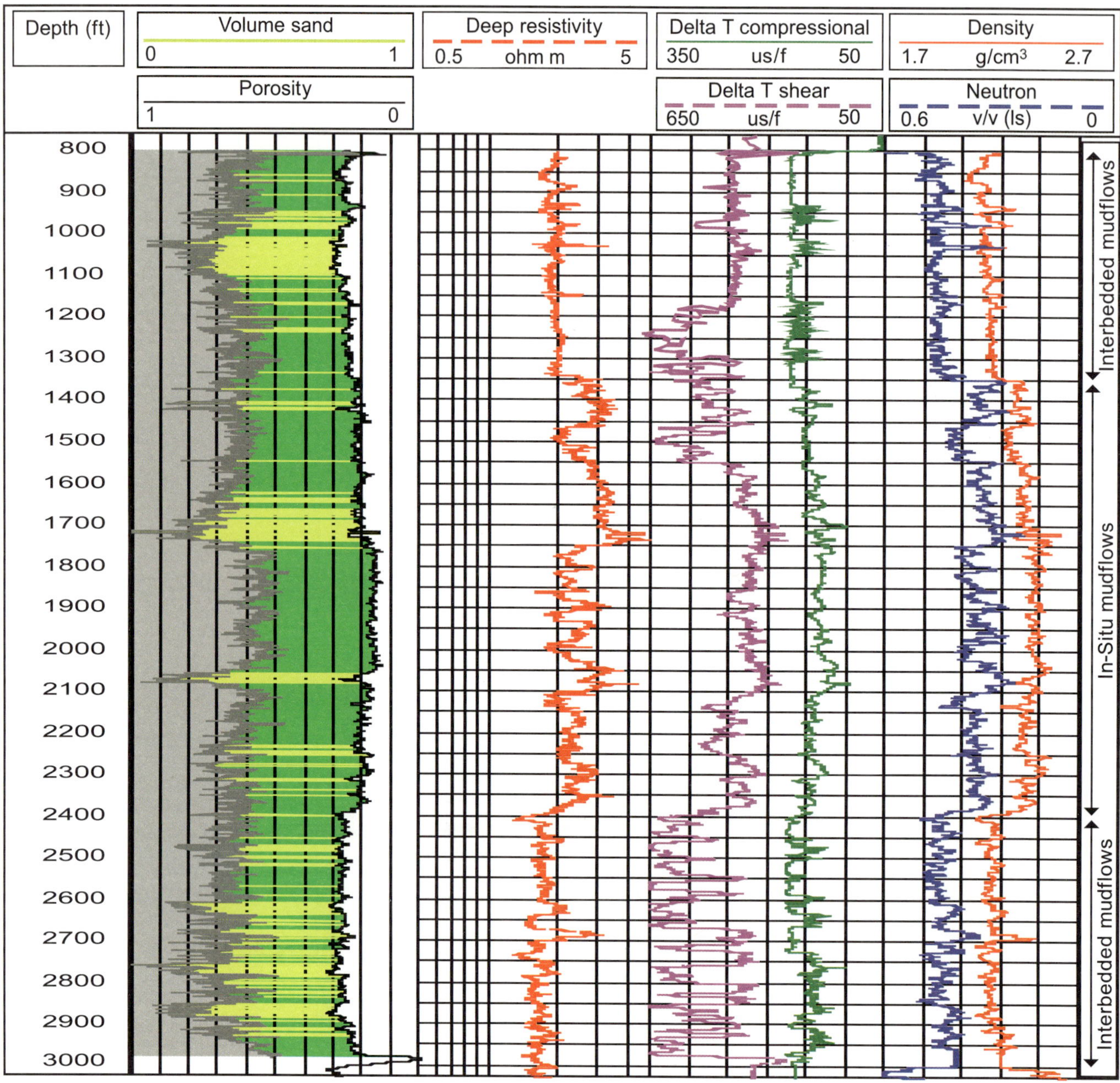

Figure 10. Panel with log curves to 3000 ft (914 m), including volume of sand (Vsand), formation resistivity, sonic (delta T), density, and neutron curves. The sections with increased mudflow material (0 to 1350 ft and 2370 to 3000 ft) show decreased resistivity and increased porosity measurements. 1000 ft = 305 m.

times are larger (slow velocity). Resistivity decreases in the high-porosity zones are caused by an increase in water volume. Higher mudflow porosities may be related to fluid-supported overburden and abnormally high pressures in those zones, whereas lower porosity shales are associated with normal compaction and pressure. Higher pressures within the mud volcano flows may be related to an active connection with the mud source at depth. Adjacent faults are zones of weakness that provide conduits into deeper overpressured units and facilitate mudflow from depth.

Checkshot velocities in Habanero 1 within shallow mud volcano layers above 1100 ft (355 m) are slower than the velocity of water (less than 5300 ft/s [1615 m/s]). This may be caused by the presence of entrained gases within the matrix of the mud. Refraction static velocities are also less than the velocity of water around the Habanero 1 well location. This slow velocity coincides

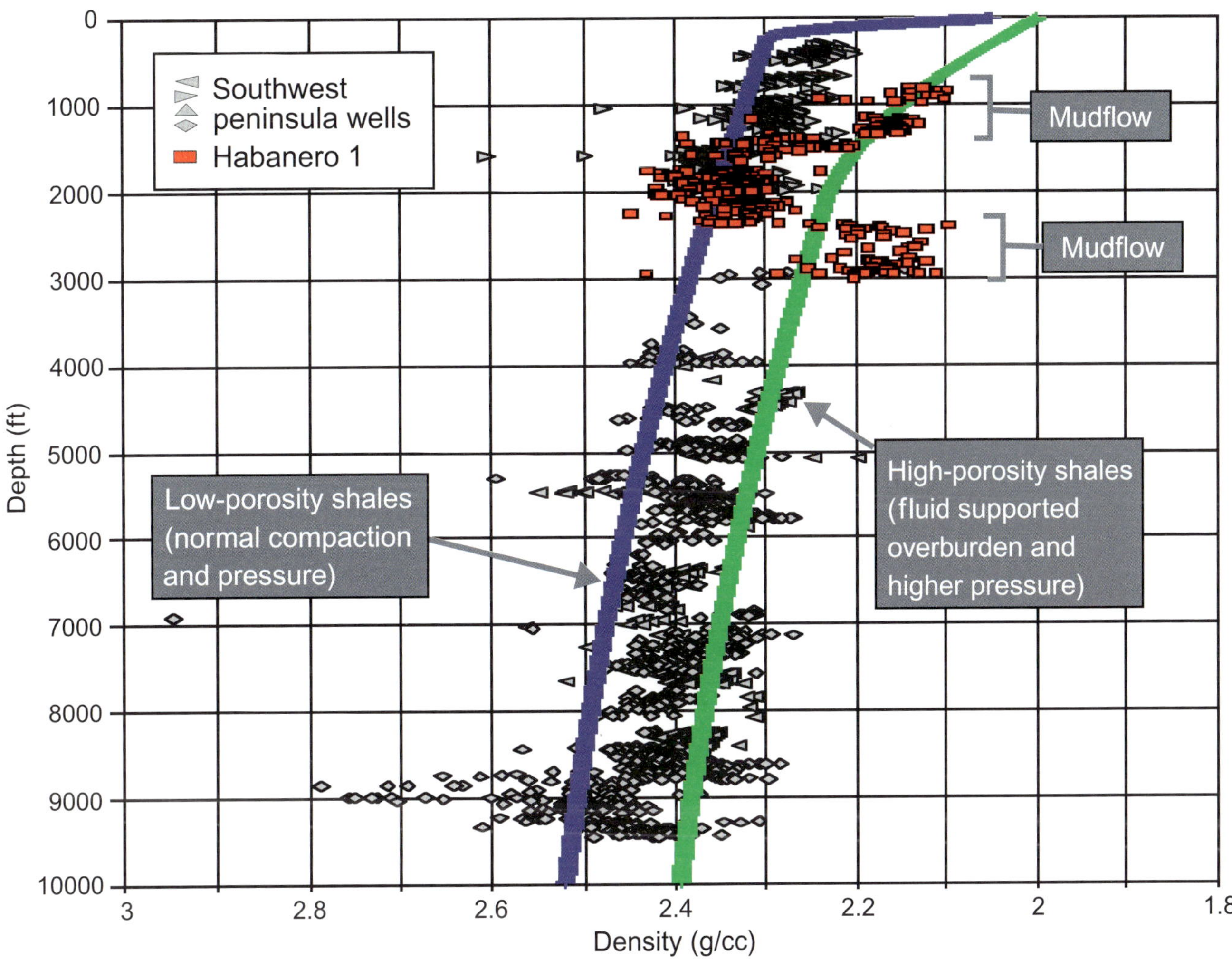

Figure 11. Crossplot of depth versus density for shales in Trinidad's southwest peninsula area wells. Data points for Habanero 1 are posted as red squares, whereas points for all other southwest peninsula wells are in gray. The blue curve defines a high-density trend through the points, and the green curve defines a low-density trend through the points. These points establish compaction trends and help identify intervals with abnormally high pore pressures. Note that the two zones with mudflow material diverge farther to the right (lower density) than even the high-porosity shale line. TED = Trinidad Exploration and Development Company. 1000 ft = 305 m.

with the trend of mud volcanoes following the west to east strike of the Southern Range anticline.

IMPLICATIONS

Empirically, the close association of petroleum accumulations and mud volcanoes within eastern Venezuela and Trinidad suggests a linkage between mud volcanoes and the local petroleum system. Faults within Trinidad mud volcanoes provide conduits for bringing up deep-seated fluids from depth, including water, gas, and liquid hydrocarbons associated with the mud (Dia et al., 1999). The faults are conduits into deeper, overpressured units in an area of active transpression.

High pressures maintained along these faults influence the petroleum system in several ways. They act as migration pathways for hydrocarbons traveling from higher pressured source rocks at depth into lower pressure reservoirs adjacent to the mud volcano. The relatively higher pressures of the mud volcano systems along these faults also act as high-pressure seals for fluids, with their sealing capacity enhanced by physical properties of the mud. Habanero 1 may be located within a mud-rich, high-pressure seal for lower-pressured downdip oil fields. Not all Trinidad mud volcanoes are highly pressured (Archie and Wach, 2007). Away from areas

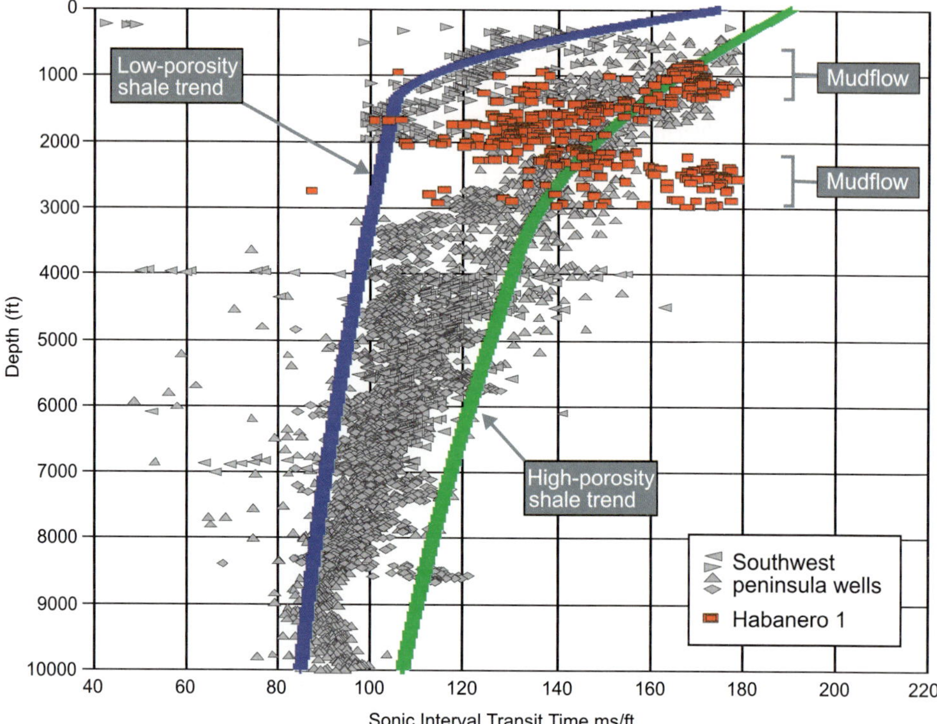

Figure 12. Crossplot of depth versus sonic log interval transit time data in shales from wells on the southwest peninsula. This graph illustrates that the Habanero 1 well-log data exhibit two zones of anomalously high interval transit time (slow velocity). Both are mudflow zones. 1000 ft = 305 m.

of active transpression, mud volcanoes may become inactive fossil mud volcanoes as pressures dissipate.

Although the inherent instability of mud volcanoes suggests the possibility for breached seals in nearby rocks, the large number of associated oil fields implies adequate seal capacity. Several fields are located near mud volcanoes trending along Trinidad's Southern Range anticline, including Pedernales, Bonasse, Palo

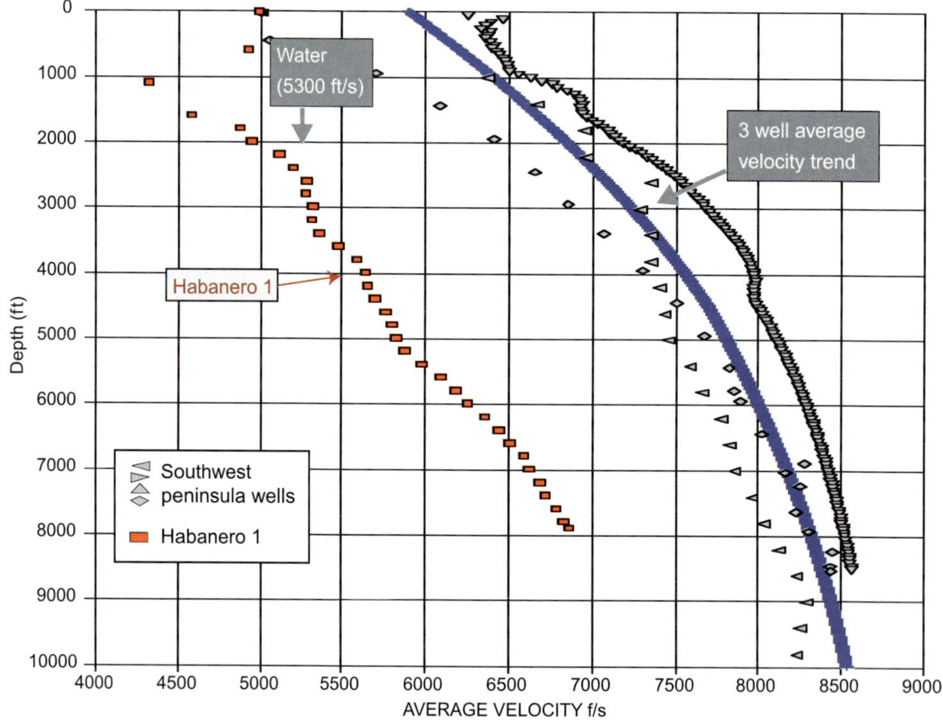

Figure 13. Crossplot of depth versus average velocity in feet per second. An average checkshot velocity trend is extrapolated (blue curve) based on data from three wells in the southwest peninsula area. Points from the three wells follow similar velocity versus depth trends, whereas Habanero 1 well data follow a similar slope but are shifted to slower average velocity values with depth. 1000 ft = 305 m.

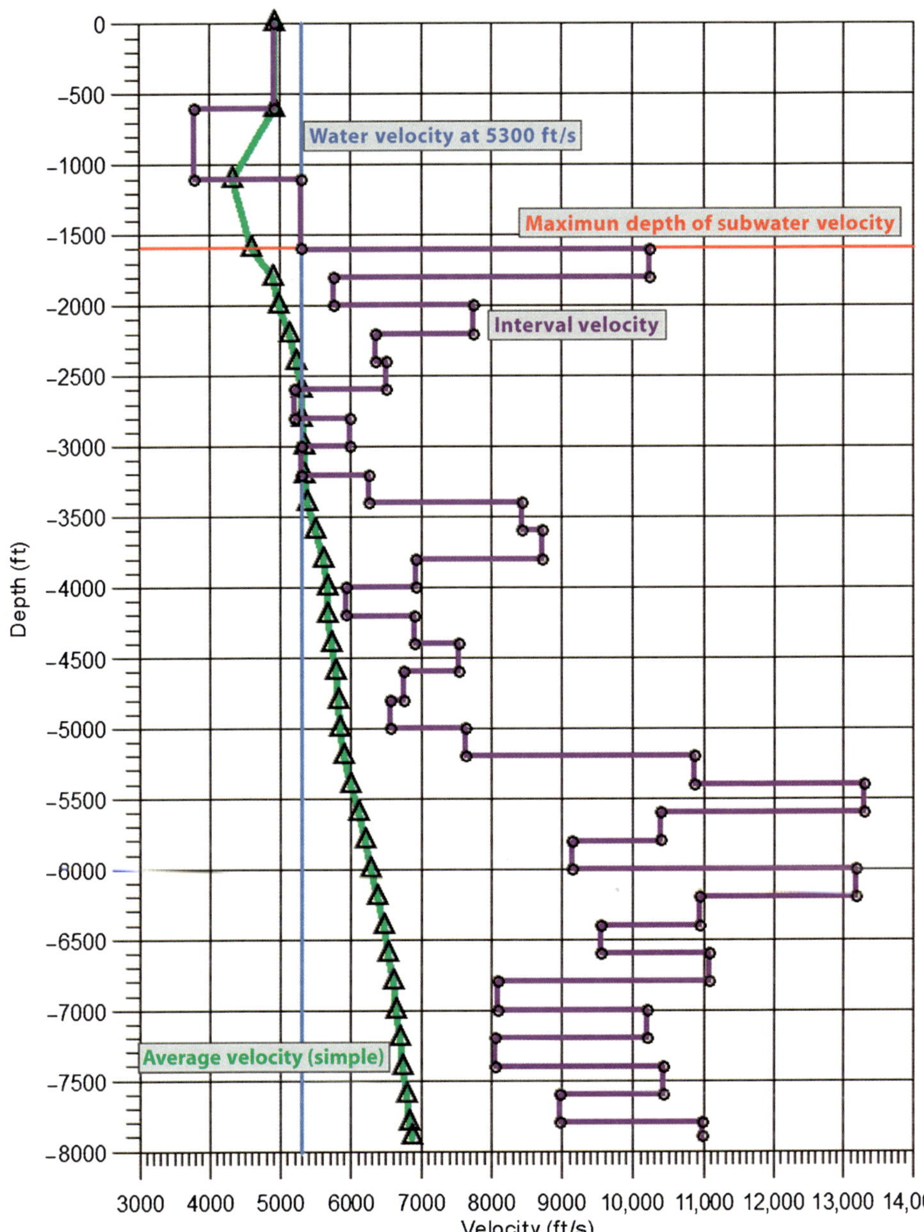

Figure 14. Crossplot of depth versus velocity data for Habanero 1. A comparison of checkshot interval velocity to average sonic velocity from the well illustrates that much of the velocity difference is caused by a very slow, shallow layer from the surface to around 1100 ft (335 m) that causes average velocity values for Habanero 1 to decrease. All points to the left of the blue line are slower than the velocity of water (5300 ft/s [1615 m/s]). 1000 ft = 305 m.

Seco, and Southwest Soldado (Figure 1). Mud volcano systems likely enhance the updip trap closure for these fields through high pressures and mud-rich fault seals.

More highly productive wells are documented away from the core of the mud volcano system in some Trinidad fields (Archie and Wach, 2007). Structural complexities in close proximity to the mud volcanoes may degrade productivity in these areas.

Poor seismic data imaging is common around mud volcano systems, and the data surrounding Habanero 1 are no exception. Developing techniques to enhance seismic data quality in these areas is important for assessing potential hydrocarbon prospectivity. With better seismic data, mapped areas of mud may shrink, and additional areas of bedded rock become prospective in the former shadow of the mud volcano.

Habanero 1 provides a better understanding of rock properties associated with mud volcano systems. This information coupled with a more accurate model of a mud volcano's physical geometry can help unravel geophysical problems, and can help to (1) refine ray trace modeling for seismic data acquisition parameters and (2) provide data for modeling velocities during prestack depth migration to better deal with the abrupt lateral and vertical changes in rock properties in the vicinity of mud volcano systems.

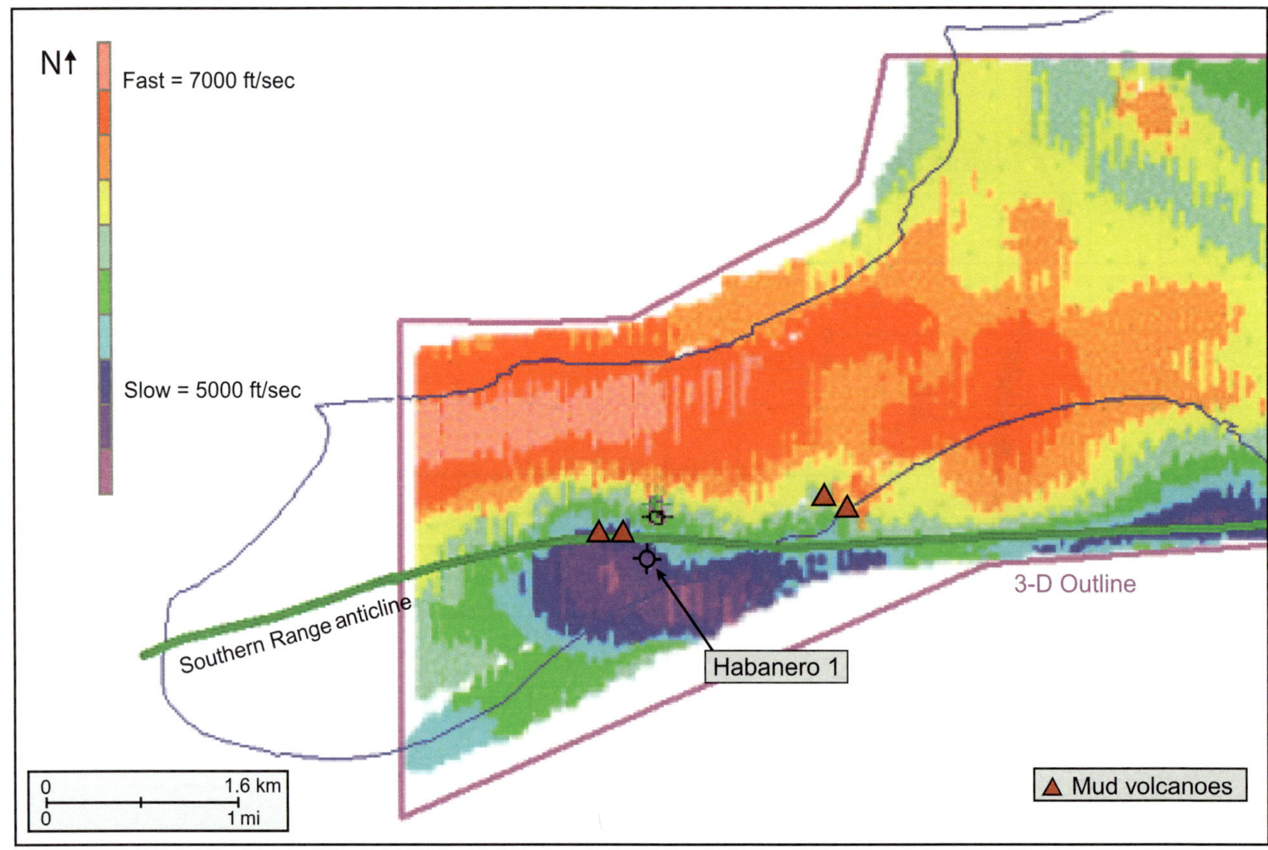

Figure 15. Map of refraction static velocities showing anomalously slow velocities in areas with surface mud volcano flows and vents. The Habanero 1 well is located on the north side of a slow velocity trend extending east to west along the Southern Range anticline. Slow velocities are shown in cool colors (around 5300 ft/s [1615 m/s]), whereas faster velocities (around 7000 ft/s [2134 m/s]) are represented with warm colors. Velocities displayed as dark blue, purple, and pink are slower than the velocity of water (5300 ft/s [1615 m/s]). The mapped data are from seismic receiver information, but seismic source information follows the same trends. 1000 ft = 305 m.

ACKNOWLEDGMENTS

The authors appreciate the support of Trinidad Exploration and Development Company Unlimited for the permission to publish this article. In addition, contributions from the following companies and individuals are acknowledged: Petrotrin, Richard Sams, Latinum/Ichron, BioStrat, Clyde Ramkhalawan, Krishna Persad, International Logging SA, Bernd Gehre, and GX Technology and Generation Services. This article benefited immensely from comments by Lesli Wood, Keith Jagiello, and Bruce Eggertson.

REFERENCES CITED

Archie, C., and G. Wach, 2007, Reservoir distribution and production in mobile shale basins: Examples from the Cruse Formation of southern Trinidad: 4th Geological Conference of the Geological Society of Trinidad and Tobago (CD-ROM).

Deville, E., A. Battani, R. Griboulard, S. Guerlais, J. P. Herbin, J. P. Houzay, C. Muller, and A. Prinzhofer, 2003, The origin and processes of mud volcanism: New insights from Trinidad, *in* P. Van Rensbergen, R. Hillis, A. Maltman, and C. Morley, eds., Subsurface sediment mobilization: Geological Society (London) Special Publication 216, p. 475–490.

Dia, A. N., M. Castrec-Rouelle, J. Boulegue, and P. Comeau, 1999, Trinidad mud volcanoes: Where do the expelled fluids come from?: Geochemica et Cosmochimica Acta, v. 63, no. 7–8, p. 1023–1038.

Duerto, L., and K. McClay, 2002, 3D geometry and evolution of shale diapirs in the Eastern Venezuelan basin (abs.): AAPG Annual Meeting Program, v. 11, p. A46.

Duerto, L., A. Escalona, M. Nunez, A. Perez, and A. Benguigui, 2006, Onshore Venezuelan shale mobilization associated to the south Caribbean plate boundary (abs.): *in* L. Wood, S. Wharton, R. Davies, F. Hossein, and C. Archie, conreners, Mobile shale basins—Genesis, evolution and hydrocarbon systems: AAPG Hedberg Conference, AAPG Search and Discovery Article 90057, 4 p.

Sullivan, S., 2005, Geochemistry, sedimentology, and morphology of mud volcanoes, eastern offshore Trinidad: M.Sc. thesis, University of Texas at Austin, Austin, Texas, U.S.A., 93 p.

5

Elsley, G. R., and H. Tieman, 2010, A comparison of prestack depth and prestack time imaging of the Paktoa complex, Canadian Beaufort MacKenzie Basin, *in* L. Wood, ed., Shale tectonics: AAPG Memoir 93, p. 79–90.

A Comparison of Prestack Depth and Prestack Time Imaging of the Paktoa Complex, Canadian Beaufort MacKenzie Basin

G. R. Elsley
Devon Canada Corporation, Calgary, Alberta, Canada

H. Tieman
Applied Geophysical Services, Calgary, Alberta, Canada

ABSTRACT

The Canadian Beaufort MacKenzie Basin (BMB) contains many examples on two-dimensional seismic sections of what appear to be shale diapirs or shale-cored anticlines. Devon Canada acquired three-dimensional (3-D) seismic data across one such feature called Paktoa and, in the winter of 2006, drilled an exploration well flanking the possible shale diapir. Although the well successfully tested oil from the Eocene Taglu unit, no clear evidence existed that the structure was a shale diapir instead of a shale-cored or fractured anticline.

Tests of a Kirchoff prestack depth migration (PSDM) algorithm prior to drilling had resulted in no significant improvement. Following the drilling of the Paktoa C-60 well, Applied Geophysical Services used their proprietary beam algorithm to image very steeply dipping reflectors in the core of the Paktoa structure. Interpretation of this new seismic volume shows that Paktoa is an inversion anticline, likely formed by the reactivation of an earlier extensional fault system, and not a shale diapir.

The beam PSDM took just three weeks to apply to 400 km^2 (154 mi^2) of marine 3-D seismic data, and the vastly improved imaging has not only changed the interpretation of the Paktoa structure but also implies that many similar features in the Canadian BMB may also be steeply dipping anticlines instead of shale diapirs or shale-cored anticlines.

INTRODUCTION

The Beaufort MacKenzie Basin (BMB), in northernmost Canada, has long been recognized to contain many of the features that are characteristic of mobile shale basins. These include mud volcanoes, pingo-like features, detachment folds, and possible shale diapirs or shale-cored anticlines interpreted on two-dimensional (2-D) seismic lines. With the advent of oil and gas exploration in the 1960s, many of the conditions thought to be

DOI:10.1306/13231309M933419

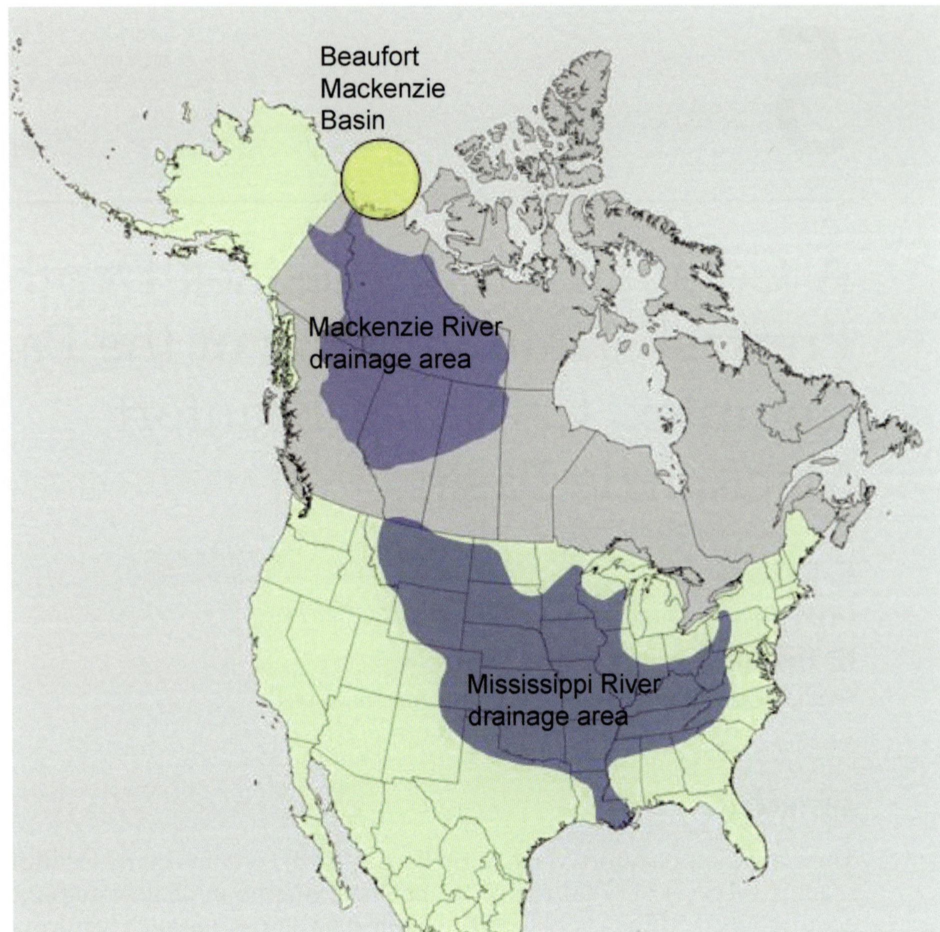

Figure 1. Map of North America showing the location of the Beaufort Mackenzie Basin study area and the location and size of the Mackenzie River drainage area relative to that of the Mississippi River drainage area to the south.

necessary for mud volcanism and shale diapirism have been noted to exist in this basin, including conditions of rapid sediment deposition, hydrocarbon generation, active tectonics, and very high subsurface geopressure.

In the winter of 2006, Devon Canada drilled, completed, and flow tested oil from the Paktoa C-60 well located in 13 m (43 ft) water depth in the Canadian Beaufort Sea part of the BMB. The predrill three-dimensional (3-D) seismic interpretation on which the Paktoa prospect was mapped was that of a three-way-dip-closed structure flanking a shale diapir. This interpretation was influenced by the fact that mobile shale is believed to have an important function in the evolution of this basin. Despite having 3-D seismic data, which covered the entire Paktoa complex, some doubt still existed as to the geometry and extent of the shale diapir, and if it was indeed a diapir instead of a shale-cored or heavily fractured anticline. This article will examine both the predrill and postdrill interpretation of the Paktoa complex and the seismic technological approaches to resolving questions regarding the nature and origin of the Paktoa feature.

Previous attempts in the application of prestack depth migration (PSDM) seismic processing of the Paktoa data had been limited to using a Kirchoff algorithm on several test lines. Because this approach had done little to change the image quality of these test lines, no attempt was made to depth migrate the entire volume before the C-60 well was drilled. Following the drilling of C-60, a beam algorithm (Sherwood et al., 2008) was applied in hopes that it could yield better results than previous PSDM methods in a shorter time frame of weeks instead of months. Previous application of this algorithm in areas of the U.S. Gulf of Mexico had been very successful at imaging steeply dipping structures similar to the Paktoa structure. This article will show the results of this technological approach to PSDM and the resulting new interpretation of the Paktoa structure.

REGIONAL OVERVIEW

The BMB is located beneath the waters of the Arctic Ocean in northwestern Canada (Figure 1). Sediments to

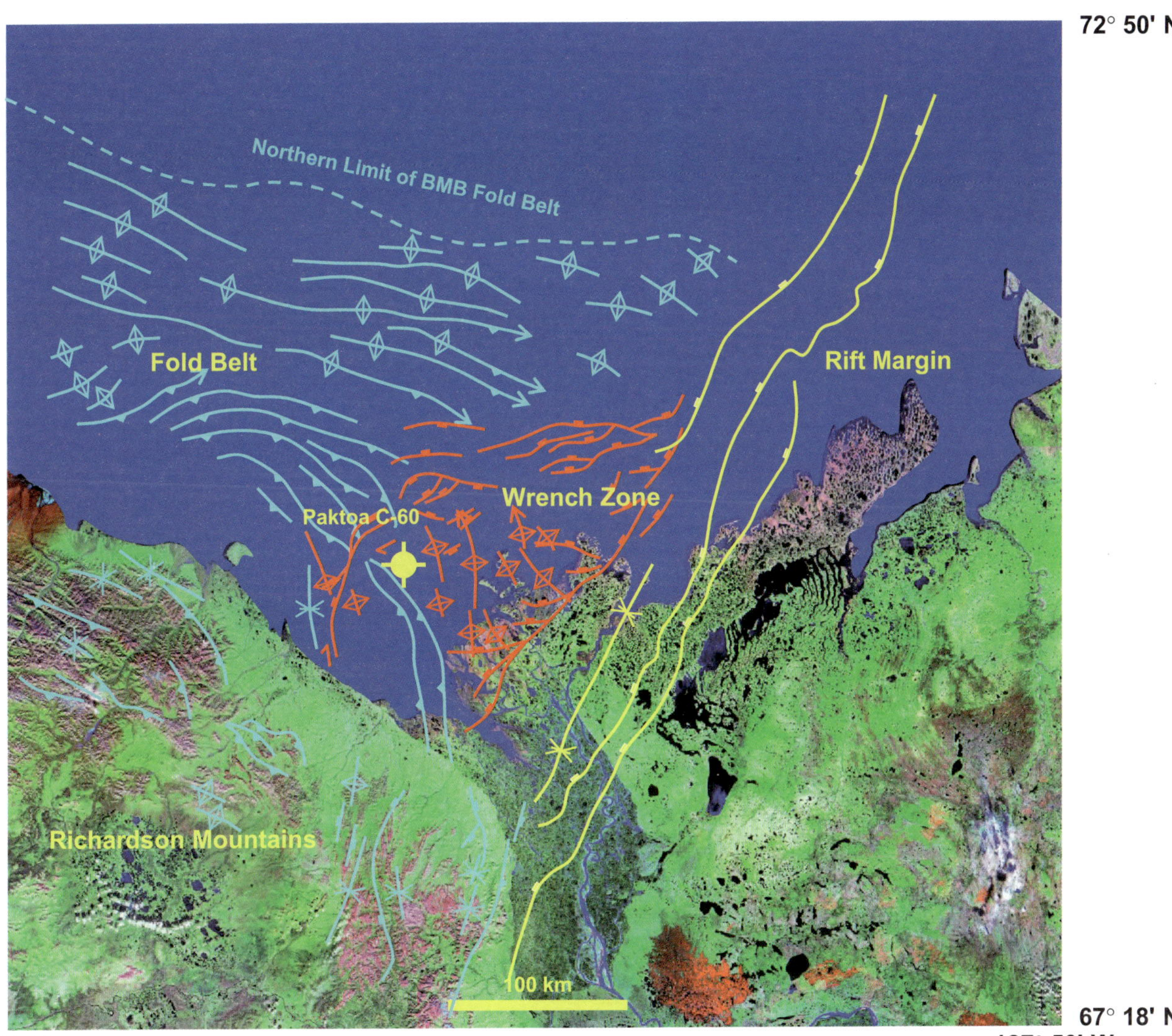

Figure 2. Structural domains of the Beaufort MacKenzie Basin (BMB) (modified from Lane and Dietrich, 1995). Blue denotes the offshore region of the Arctic Ocean. The location of the Paktoa C-60 well is shown. 100 km = 62 mi. Green = vegetation; pink, red, brown = outcrop.

the basin are supplied by the MacKenzie River draining the second largest drainage basin in North America. The MacKenzie River drains the western Canadian Cordillera, the Interior plains, the Precambrian shield, and the Arctic coastal plain.

The basin is bounded to the southeast by a rift margin that was active during the Mesozoic and to the southwest by the present-day Richardson Mountains, which are a northern equivalent to the Front Ranges of the Rocky Mountains and MacKenzie Mountains (Figure 2). The fold belt associated with the compression of the Richardson Mountains extends far into the offshore Arctic Ocean. A structurally complex area in the center of the basin is dominated by north–south-trending anticlines and east–west-trending extensional faults. These structural trends have formed as a result of right-lateral wrench movement as the fold belt interacted obliquely with the more eastward ancient passive margin. The Paktoa complex is located on the western edge of this complex wrench zone (Figure 2).

The stratigraphic fill of the basin spans a time frame from the Paleozoic to the present day, reflecting the major structural phases of basin evolution (Figure 3). The eastern rift margin is underlain by prerift Proterozoic to late Paleozoic-aged carbonates and early Mesozoic-aged clastics. Overlying this succession is a synrift sequence, which consists mainly of deep-marine organic-rich shales of the Upper Cretaceous. The drift domain is the third main succession and consists of six major progradational sequences that were deposited as the

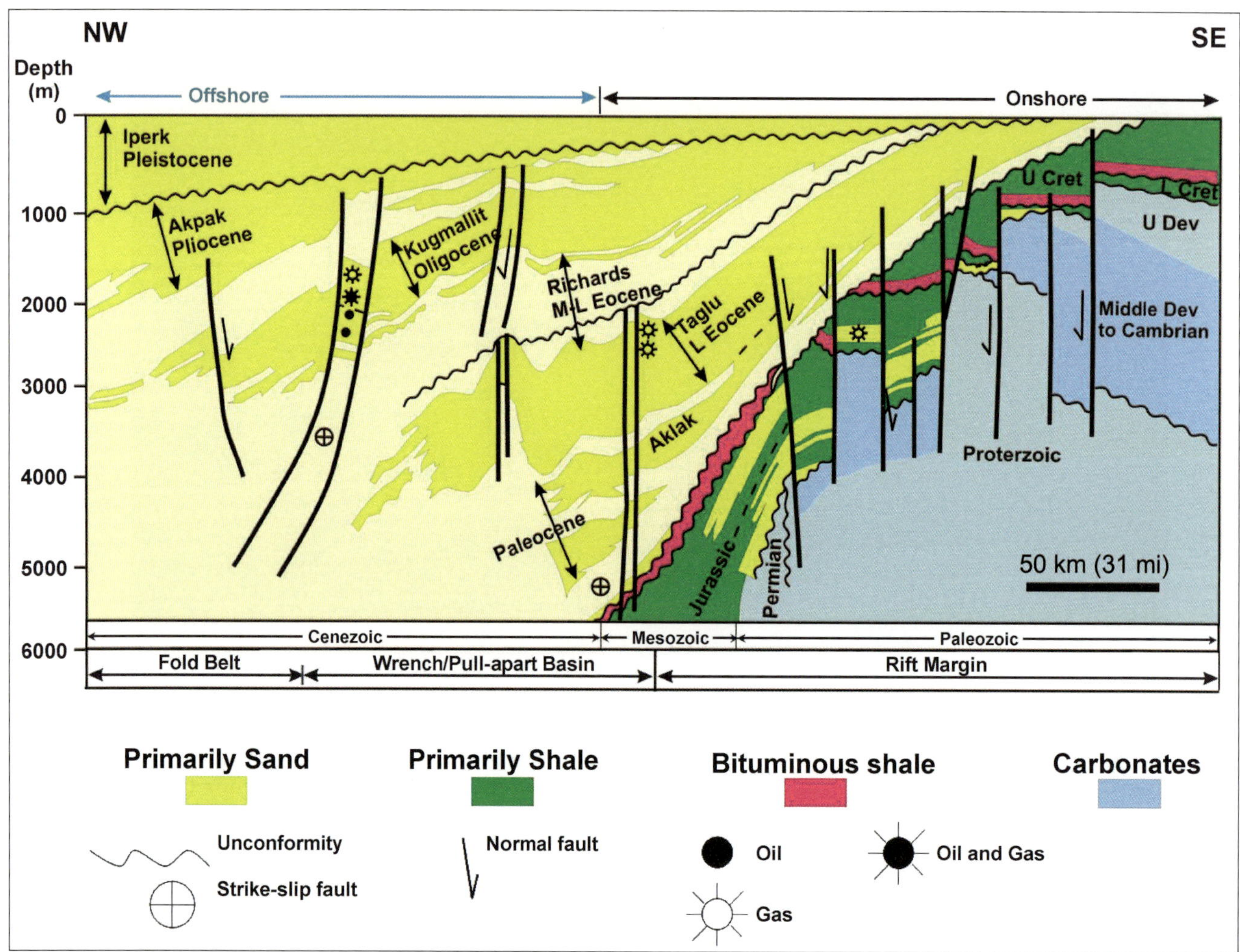

Figure 3. Schematic cross section showing the stratigraphy and structural elements of the Beaufort Mackenzie Basin. 1000 m = 3281 ft.

Laramide deformation advanced northward. These progradational sequences resulted in the deposition of a thick clastic wedge ranging in age from the Late Cretaceous to the present day. The clastic wedge is more than 10 km (6.2 mi) thick and consists of the Aklak, Taglu, Kugmallit Akpak, and Iperk units. These Tertiary sequences contain most of the discovered petroleum resources in the basin (see Bergquist et al., 2003).

The lower Eocene Taglu unit was the primary target of the Paktoa well, and the overlying Richards Formation shale formed an important seal for the play.

INTERPRETATION OF PRESTACK TIME-MIGRATED SEISMIC VOLUME

In the summers of 2001 and 2002, Devon Canada acquired approximately 1800 km^2 (695 mi^2) of 3-D marine seismic data (Figure 4) to evaluate several large exploration blocks. The Paktoa complex was identified as a primary target, whose boundaries fell within the western 3-D survey and within the 400-km^2 (154-mi^2) area shown in Figure 4 (highlighted in purple). This area was selected for seismic reprocessing in hopes of improving the imaging of this structure.

Figure 5 shows a southwest to northeast inline or structural-dip line in the northern area of the 3-D seismic volume, approximately 4 km (2.4 mi) north of the Paktoa well location. Seismic data quality is generally good; this area is not affected by permafrost because of the close proximity to the main outflow of the MacKenzie River. The top Taglu unconformity (shown on Figure 5 in yellow) is an important surface, which separates the shale-dominated Richards and uppermost Taglu units from the interbedded shales and reservoir sands of the Taglu unit. This unconformity is easily mappable, and the reflectors above the unconformity

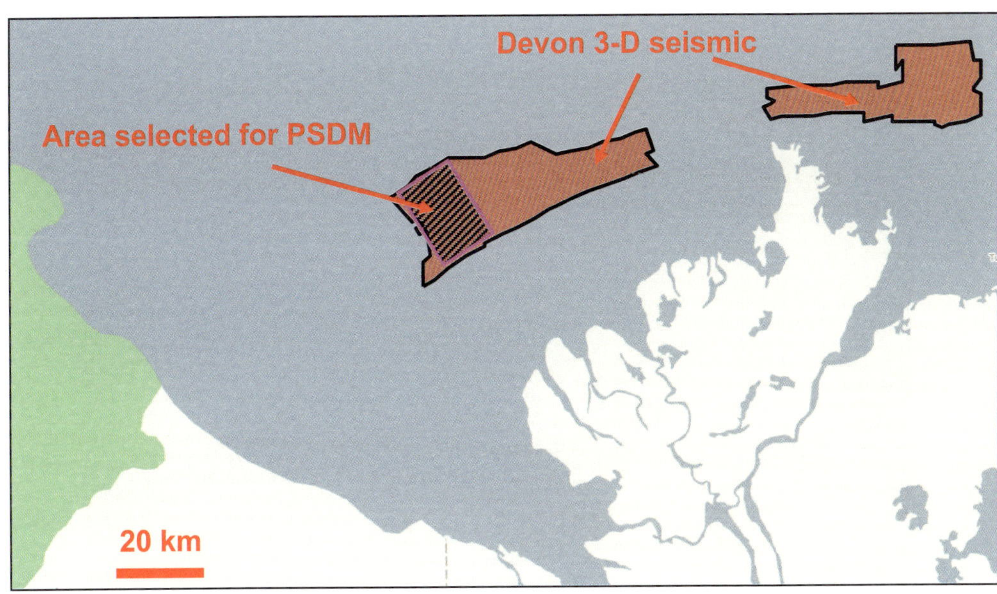

Figure 4. Map of the three-dimensional (3-D) seismic surveys used in this study. Blue denotes the offshore region of the Arctic Ocean. Onshore Canada is shown in white. Onshore Alaska is shown in green. PSDM = prestack depth migration. 20 km = 12.4 mi.

are relatively gently dipping (less than 20°) and also easily mappable. Below the unconformity, the bedding dips increase to greater than 30° at the C-60 well location, and much more structural complexity is observed. The difference in structural attitude across the unconformity implies that this surface is associated with a period of significant tectonic activity. The core of the Paktoa complex is dominated by a region with no coherent seismic reflectivity that appears to truncate the Taglu unit sands and shales.

The original interpretation of the Paktoa complex was that it was a shale diapir, consisting of Cretaceous-aged material that had become active during the middle Eocene. Although the existence and nature of shale diapirs are widely debated (see Van Rensbergen et al., 2003; Wood et al., 2006), the Paktoa feature did exhibit several characteristics to suggest it was a reactive shale diapir. These included (1) a seismically noncoherent core, (2) an apparent inflation above regional dip, and (3) an apparent deformation of the surrounding beds. An alternative interpretation was that Paktoa may be a tight, fractured fold or a shale-cored anticline, as these had been interpreted in the fold belts to the west. The core was proposed as simply an area of poor data quality, and that new techniques in seismic processing might shed some light on this area.

The Paktoa C-60 well was drilled in the winter of 2006 on the east flank of a large, four-way-dipping north–northwest-plunging anticline that is cut by east–west-trending extensional faults. The location was selected using a prestack time-migrated volume of the western 3-D. The anticline shows a four-way dip closure at the middle Eocene unconformity (Figure 6). The concept for the hydrocarbon trap configuration was that the upper Taglu sands formed a three-way-dip-closed trap with the shale diapir providing a lateral and updip seal. Although the well was a success, many questions remained about the dip of bedding in the Paktoa structure, expected pressures, volumetrics, etc. These questions necessitated further analysis of the seismic volume to better understand the nature and origin of the Paktoa structure.

PRESTACK DEPTH MIGRATION AND THE BEAM ALGORITHM

Conventional time migration of seismic data is based on the assumptions that seismic velocity only changes in the vertical direction and that changes in arrival times from increasing offsets result in hyperbolic moveout. If this is the case, then velocity analysis will flatten these hyperbolae, and the resulting stacked section will show reflectors in their true subsurface position. In structurally complex areas, this is not commonly the case. The PSDM processing differs from time-migration processing, mainly in its ability to consider lateral changes in velocity. This ability allows processors to ray trace using an earth velocity model and thus correctly position each reflection event.

Two main traditional approaches to PSDM exist. The first is a Kirchoff method, which consists of modeling a single raypath, and the second is a wave equation method, which models an entire wave front. A third method has recently been developed and applied, the

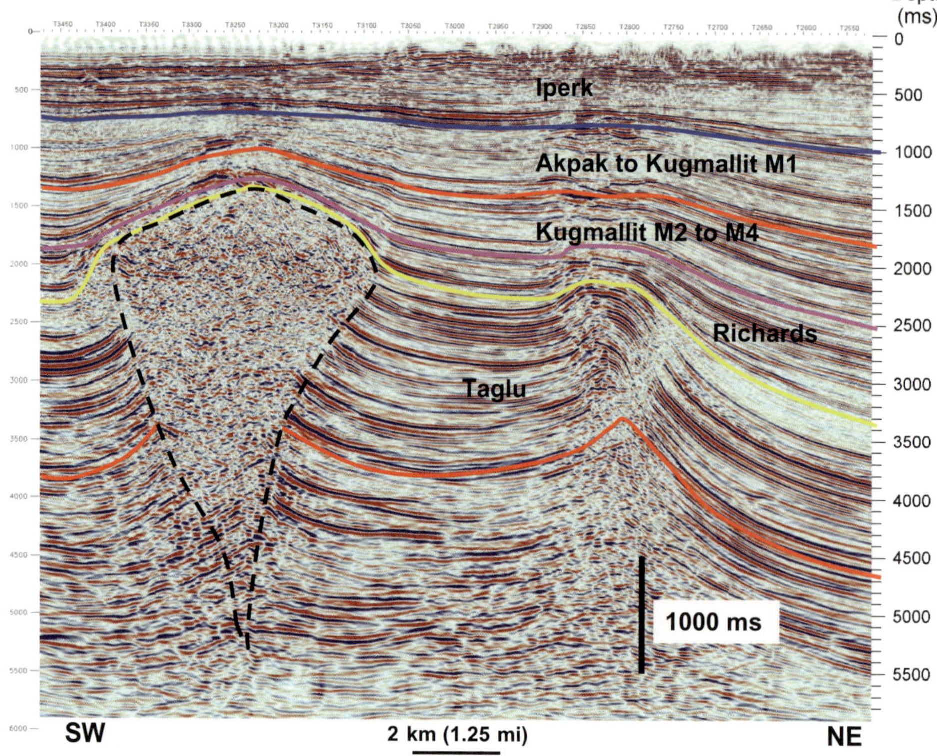

Figure 5. Northeast–southwest seismic cross section showing the mapped surfaces that define the stratigraphy of the basin, as well as the original diapir body mapped in three-dimensional seismic data. Note the apparent truncation of horizons into the diapir.

beam algorithm method, which is quite different to either the Kirchoff or the wave equation methods.

The beam PSDM algorithm analyzes a small volume of seismic data for coherent events. This data volume is centered upon a common depth point inline and crossline coordinate (x and y), a source-receiver offset, and a two-way traveltime. The data volume analyzed typically covered a time interval of about 100 to 200 ms and has spatial widths of 200 to 800 m (656 to 2625 ft), depending on the depth of interest. The time dip of a coherent event in x, y, and offset is measured, and the corresponding time wavelet is extracted. Sufficient wavelets are extracted so that when summed back together they approximate the original prestack time data minus noise and other events not of exploration interest.

A typical workflow consists of taking a suitably preprocessed time seismic volume and decomposing it into numerous wavelets. Ray tracing is then performed for each wavelet through an initial simple interval velocity model, which for the Paktoa data set was determined from a smoothed version of the stacking velocities used for the prestack time migration. This approach allows the determination of the most likely subsurface-migrated reflector location for each wavelet. The next main stage is data reconstruction and stacking to provide a depth-migrated volume. Prior to this, excluding any wavelet from the output based on a variety of criteria, such as quality of focus, is possible. This allowance provides flexibility for the reduction of coherent noise, such as multiple reflections, in the depth-migrated volume. After this initial workflow, the velocity model can be improved, and this updated version can be used for a second iteration. It is typical to repeat this process several times.

One of the advantages of beam PSDM is that there is no need to limit the aperture of the input data, so turning waves can be dealt with and very steep dips can be imaged.

RESULTS OF SEISMIC REPROCESSING

Figures 7–9 show comparisons between prestack time- and prestack depth-migrated structural dip lines. All depth lines shown in this article have been converted back to time to allow for direct comparison with the time-migrated volume. The top Taglu unconformity surface is shown in yellow in Figures 7–9.

The time-migrated version of the most northerly located of the dip lines (Figure 7A) shows a good example of the area of poor reflectivity that was interpreted as a shale diapir beneath the top Taglu unconformity. After beam PSDM (Figure 7B), very strong reflectivity is present in the core of Paktoa beneath the unconformity.

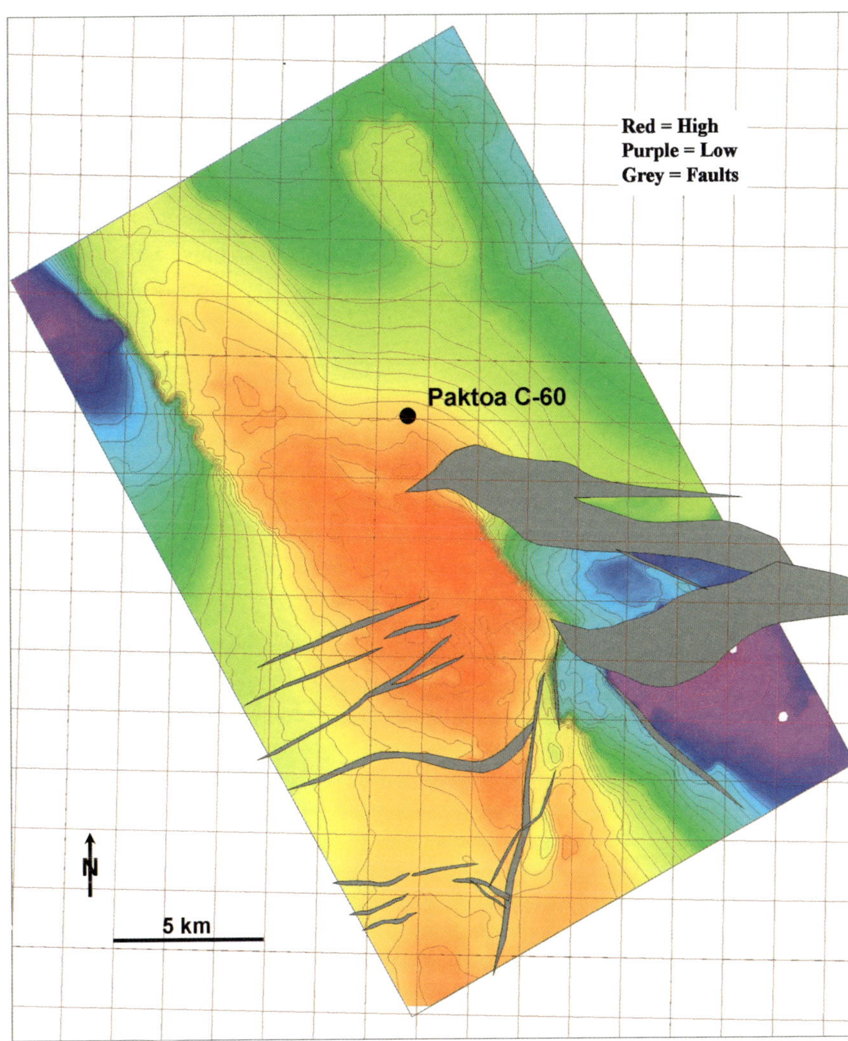

Figure 6. Top Taglu structure map. Reds and yellows denote structural highs. Blues and purples denote structurally low areas. The location of the Paktoa C-60 well is shown. 5 km = 3.1 mi.

This is very strong evidence that no discoherent core exists, but instead a coherent bedding in a tight fold with large amounts of erosion at the fold crest. The truncation of beds against the top Taglu unconformity is very clear and is consistent with the tectonic history of this part of the basin. The image quality is much better on the eastern flank of the structure than on the western flank. This degraded quality on the western flank is caused by the proximity of the western flank to the edge of the seismic data volume, so very poor seismic coverage is seen and no automatic gain control has been applied to the depth volume.

A slightly more southerly located time-migrated structural dip line (Figure 8A) shows more reflectivity on the time volume than is present on the more northerly line. However, very little indication of the steepness of the bedding dips and the complexity of the folding can be seen on the beam-processed depth version (Figure 8B). One feature that is clear on both versions is an obvious thickening of the stratigraphic section in the upper Taglu unit on the east limb of the fold, as indicated by a divergence of reflectors. This thickening is clearly a growth sequence related to the creation of accommodation space during the early Eocene. Such early thickening and later inversion of strata suggest that the inversion of earlier extensional half grabens is probably a factor in the formation of this structure. The depth-migrated line in Figure 8B shows that an inversion anticline is the most likely interpretation for the nature of this structure.

Figure 9A shows a dip line seismic cross section that was believed in earlier interpretations to lie south of the shale diapir. However, the time-migrated line (Figure 9A) indicates the presence of a large anticline, which is poorly imaged beneath the crest and on the steeply dipping eastern flank. The depth-migrated version (Figure 9B) has been much more successful at imaging the steep dips to the extent that overturned beds

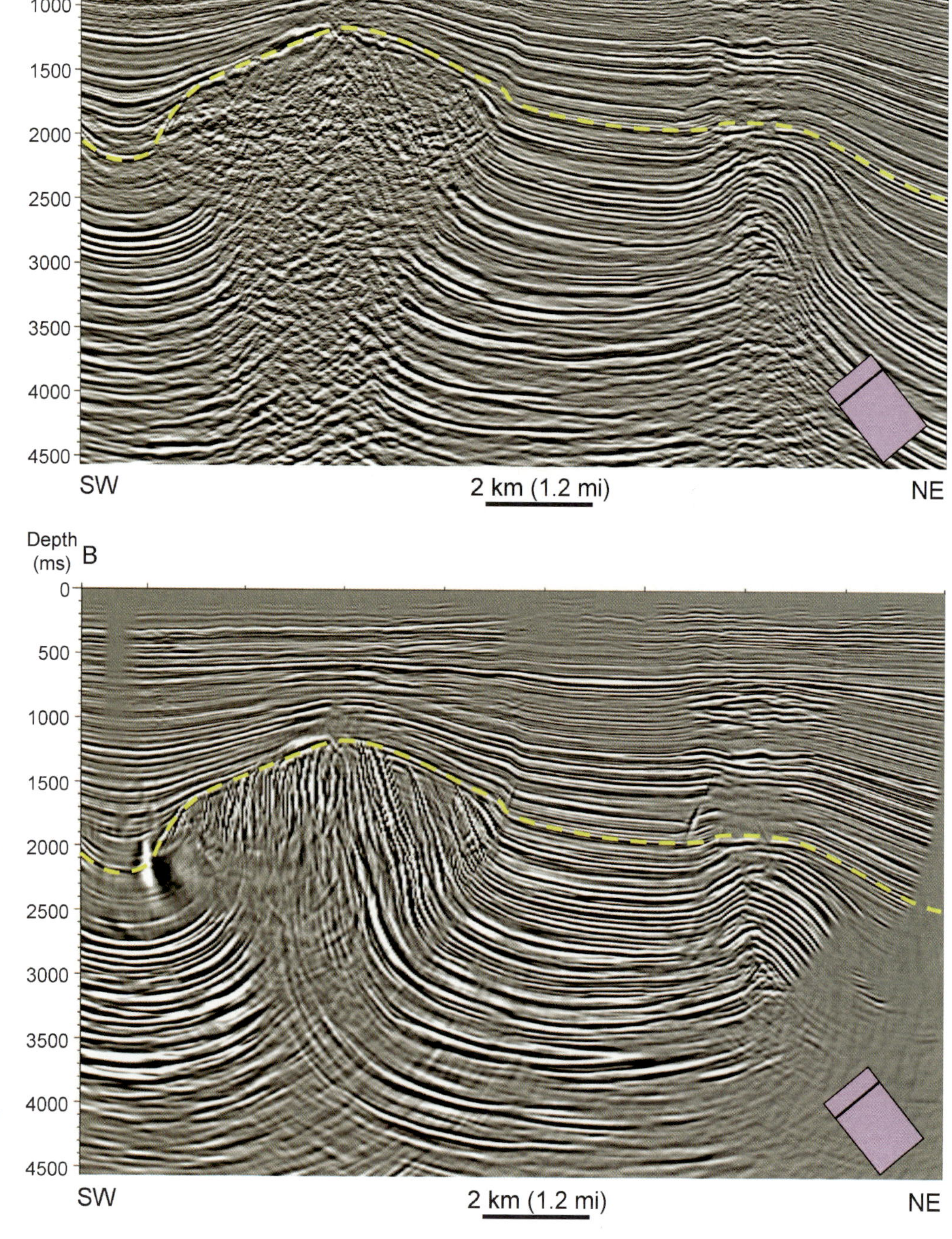

Figure 7. (A) Prestack time-migrated structural-dip-oriented seismic image across the Paktoa structure showing an apparent discoherent core material. The vertical exaggeration on this image is approximately 2:1. The yellow horizon is the top Taglu unconformity. (B) Prestack, beam-processed, depth-migrated structural-dip-oriented seismic image showing apparent steeply dipping, vertical beds forming the core of the Paktoa complex, apparently truncated beneath the unconformity.

are observed. The image becomes quite complex in the syncline along the eastern margins of the study area where the Paktoa structure has been cut by many extensional faults whose orientation parallels this line.

The three sample lines shown from the depth-migrated volume (Figures 7B, 8B, 9B) show that the Paktoa structure is a steeply dipping anticline that varies significantly along strike. Figure 7B shows a fairly symmetrical

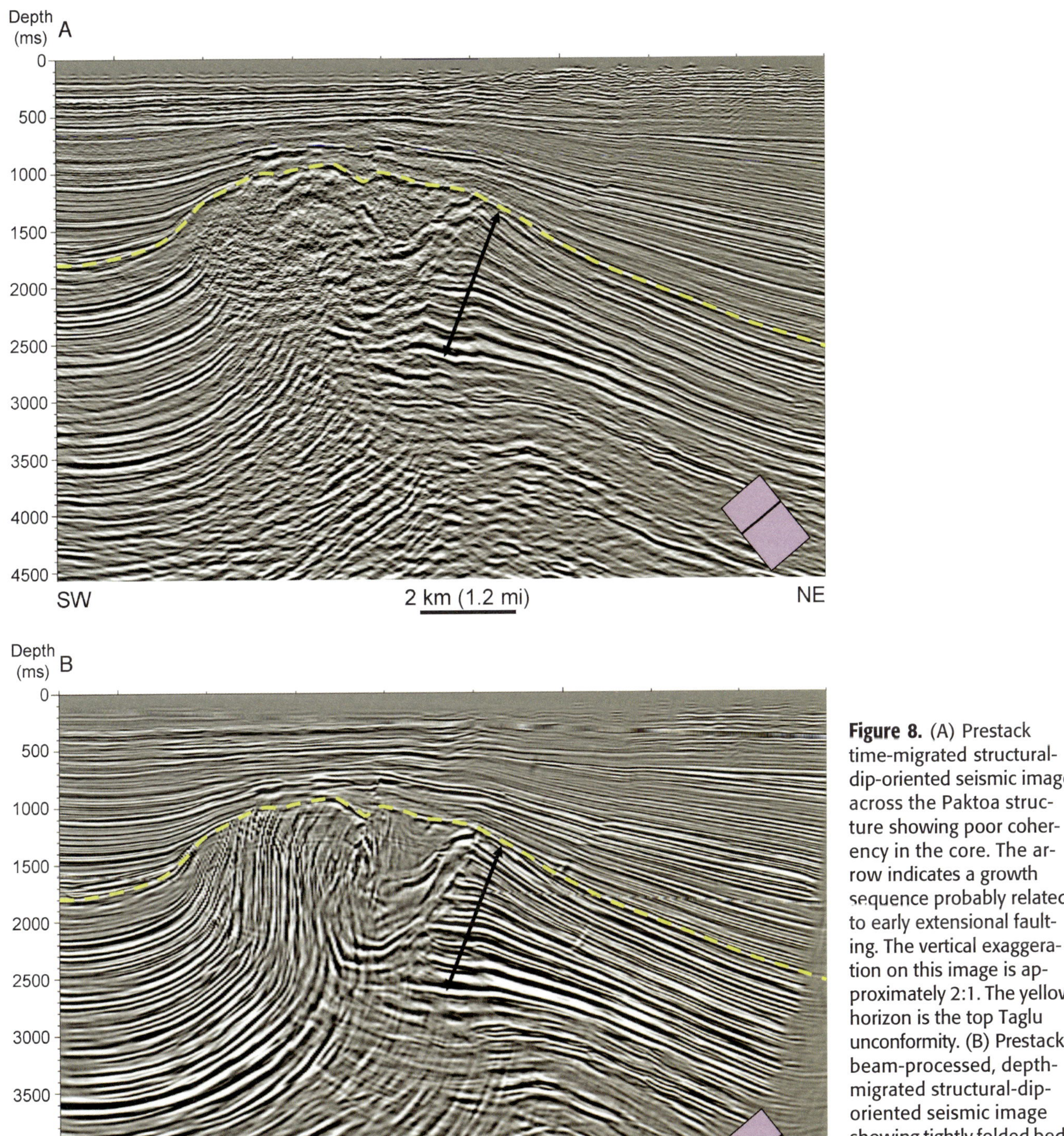

Figure 8. (A) Prestack time-migrated structural-dip-oriented seismic image across the Paktoa structure showing poor coherency in the core. The arrow indicates a growth sequence probably related to early extensional faulting. The vertical exaggeration on this image is approximately 2:1. The yellow horizon is the top Taglu unconformity. (B) Prestack, beam-processed, depth-migrated structural-dip-oriented seismic image showing tightly folded beds forming the core of the Paktoa complex, apparently truncated beneath the unconformity.

structure beneath the top Taglu unconformity, whereas Figure 8B has a much steeper west flank and Figure 9B has a steeper east flank. This apparent change in vergence along strike may be caused by the orientation of preexisting extensional faults that were later reactivated and inverted by contractional deformation.

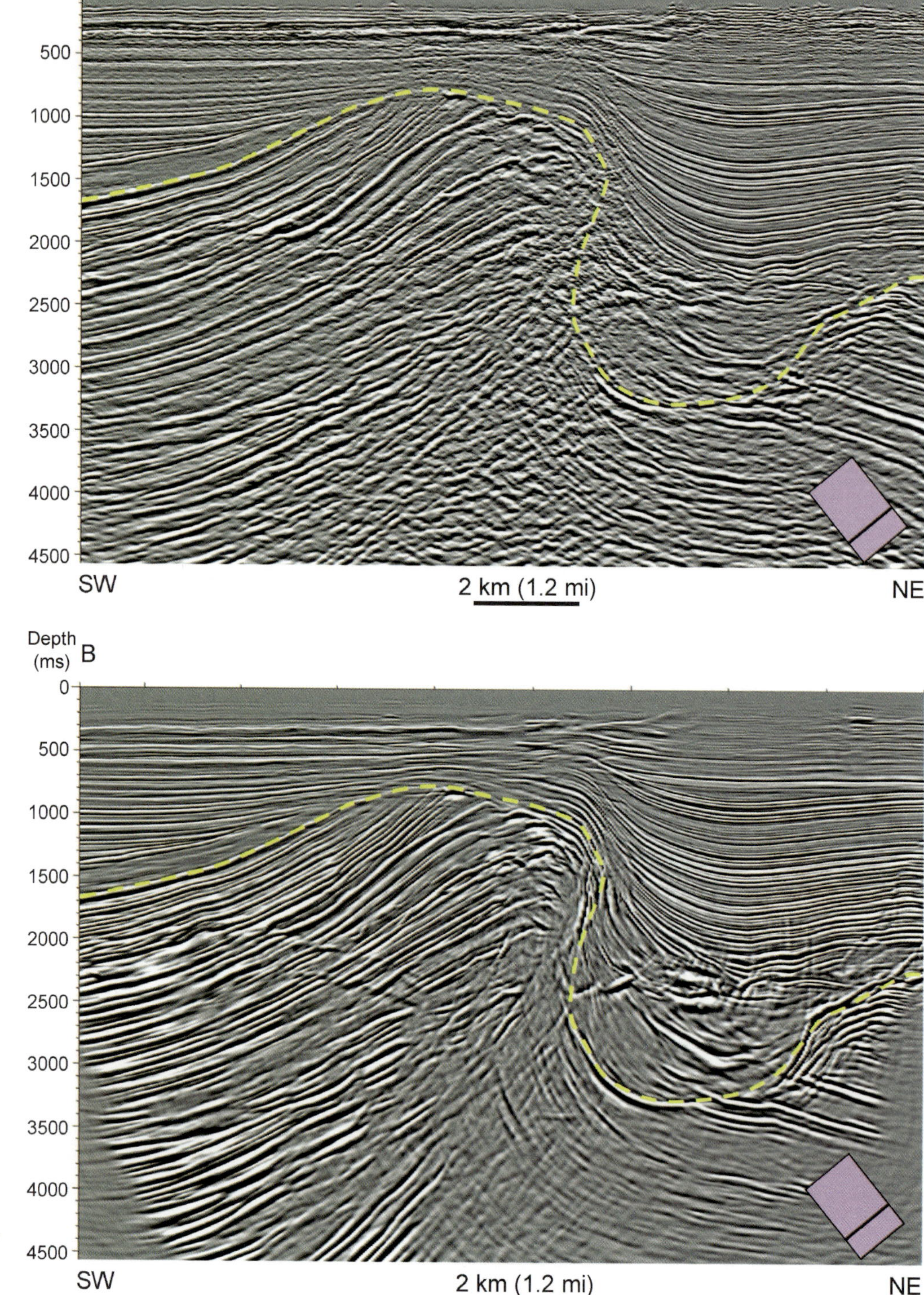

Figure 9. (A) Prestack time-migrated structural-dip-oriented seismic image across the Paktoa structure showing poor coherency on the northeast limb of the anticline. The vertical exaggeration on this image is approximately 2:1. The yellow horizon is the top Taglu unconformity. (B) Prestack, beam-processed, depth-migrated structural-dip-oriented seismic image showing overturned beds on the northeast limb of the anticline.

TECTONIC EVOLUTION OF THE PAKTOA STRUCTURE

A historical reconstruction of the evolution of the Paktoa structure was created through the interpretation of flattened seismic lines and interval isopach mapping (Figure 10). Throughout the early Tertiary (Figure 10, early Eocene), deposition of the Taglu unit was influenced by extensional faulting, probably gravity-driven

A) Early Eocene

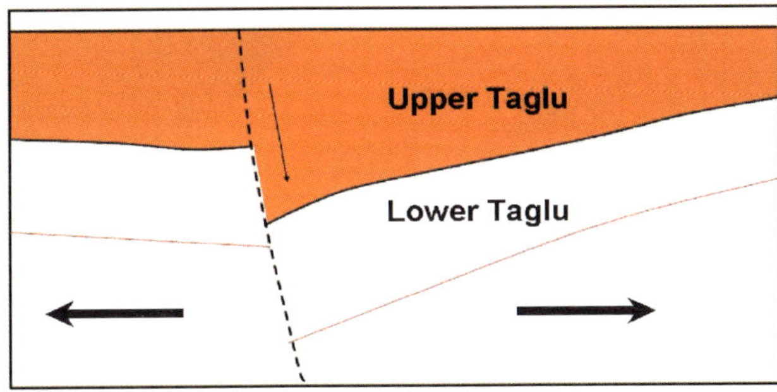

C) Late Eocene

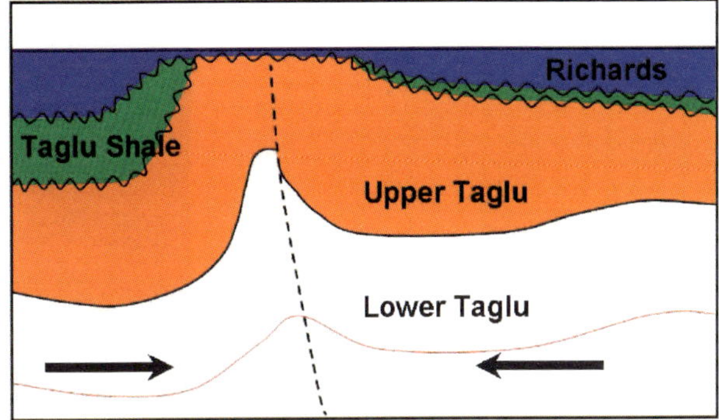

B) Middle Eocene

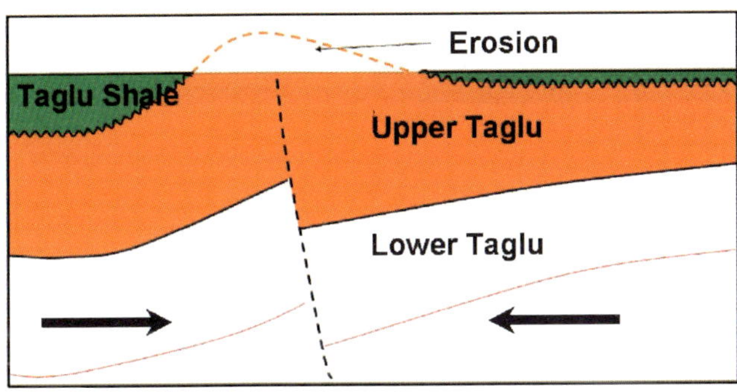

D) Present Day

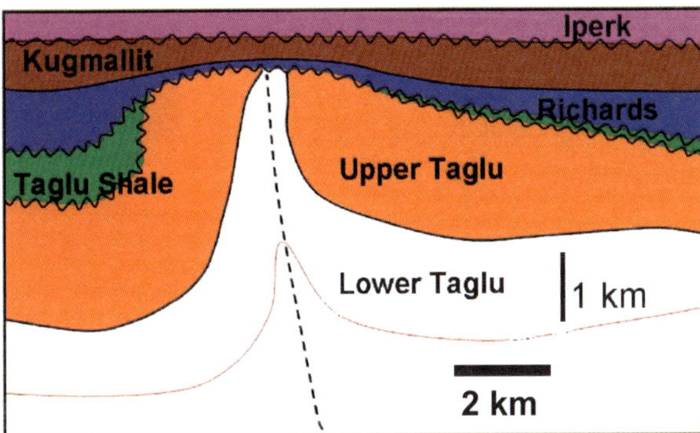

Figure 10. A series of schematic sections generated based on flattened seismic lines and isopach mapping showing a likely sequence of events from early Eocene (A), to middle Eocene (B), to late Eocene (C), and to present day (D) that have generated the present-day Paktoa structure.

listric faults that are typical of other large delta systems. By the middle Eocene, the influence of the Laramide orogeny, which had already started to form the fold belt to the west, caused compression and inversion of many of these earlier extensional faults. The Taglu beds became emergent at the crest of Paktoa, and significant erosion was observed. Upper Taglu shales were deposited during this time and onlapped the flanks of the structure.

Continued uplift through the late middle Eocene caused more erosion and the creation of several unconformity surfaces. By the latest Eocene, the rates of uplift had decreased, and the upper Richards Formation shales are preserved over the crest of the anticline. Gentle folding continued during the Oligocene and perhaps until the Pliocene, although the basinwide Pliocene–Pleistocene unconformity has removed some of the evidence for the latest folding. The result of this evolution is a tightly folded inversion anticline with relatively poorly consolidated Eocene, and likely Paleocene sedimentary rocks preserved in the core.

The nature of the final structure is well imaged in 3-D seismic data. Figure 11 shows the depth structure on the top Taglu unconformity and the clearly seismically resolvable nature of beds in the upper Taglu unit. A four-way closure at both Paktoa and the previously drilled Minuk I-53 gas discovery is observed. This is a very high-relief structure with a very thick, tightly folded, steeply dipping reservoir section.

CONCLUSIONS

The beam PSDM method was very successful at imaging steep subsurface stratigraphic dips where prestack time migration seismic processing had previously failed. These new images have resulted in a revised interpretation of the Paktoa structure as that of an inversion anticline instead of a true shale diapir. The new interpretation of the presence of reservoir rock throughout the core of the anticline has an enormous impact on the

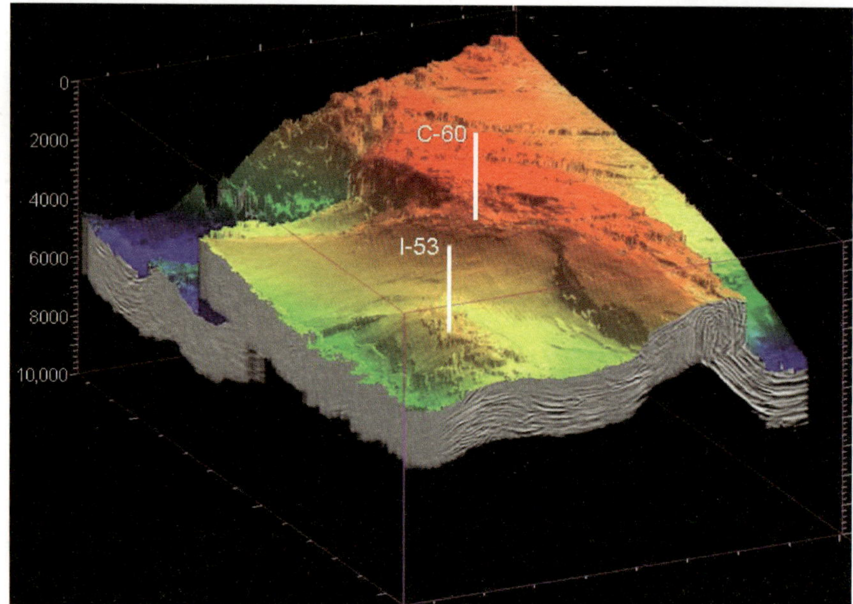

Figure 11. Depth structure on the top Taglu unconformity. The vertical exaggeration on this display is zero and the depth scale is zero to 10,000 m (32,808 ft), illustrating that this is a very high-relief structure with a very thick, tightly folded, steeply dipping reservoir section. The view is looking south. Well locations for the Paktoa C-60 and the I-53 wells are shown. Red is structurally high; blue is structurally low.

distribution and volume of hydrocarbons within the structure.

This PSDM method can be applied very rapidly to large volumes of data. The Paktoa volume is 400 km^2 (154 mi^2) in area and 10,000 m (32,808 ft) in depth. It contains almost 200,000 traces and was processed in just three weeks. This method could be very useful in the western BMB, which is dominated by fold belts; these have been interpreted on 2-D seismic lines as being shale-cored anticlines. Given the commonly poor data quality and possible steep dips in many mobile shale basins, this is a technique that should be applied in areas where there is some doubt as to the nature of the presence of mobile shale features.

ACKNOWLEDGMENTS

We would like to thank Peter Graham of Niko Resources (formerly of Devon Canada) for his geological contributions to this paper, and we would also like to thank Paul Feldman of Applied Geophysical Services for his work on the depth imaging of the seismic volume.

REFERENCES CITED

Bergquist, C. L., P. P. Graham, D. H. Johnston, and K. R. Rawliinson, 2003, Canada's MacKenzie Delta: A fresh look at an emerging basin: Oil & Gas Journal, v. 101 (November 3, 2003), p. 42–46.

Lane, L. S., and J. R. Dietrich, 1995, Tertiary structural evolution of the Beaufort Sea-MacKenzie delta region, Arctic Canada: Bulletin of Canadian Petroleum Geology, v. 43, no. 3, p. 293–314.

Sherwood, J. W. C., K. Sherwood, H. Tieman, and K. Schleicher, 2008, 3D beam prestack depth migration with examples from around the world: European Association of Geoscientists and Engineers Proceedings, Rome, Italy, unpaginated.

Van Rensbergen, P., R. R. Hillis, A. J. Maltman, and C. K. Morley, 2003, Subsurface sediment mobilization: Geological Society (London) Special Publication 216, p. 1–8.

Wood, L., S. Wharton, R. Davies, F. Husein, and C. Archie, convenors, 2006, Mobile shale basins— Genesis evolution and hydrocarbon systems: Proceedings of the AAPG/Geological Society of Trinidad and Tobago Hedberg Conference, Port of Spain, Trinidad, http://www.searchanddiscovery.com/abstracts/html/2006/hedberg_intl/index.htm (accessed October 5, 2010).

6

Choudhuri, Mainak, Debajyoti Guha, Arindam Dutta, Sudipta Sinha, and Neeraj Sinha, 2010, Spatiotemporal variations and kinematics of shale mobility in the Krishna-Godavari basin, India, *in* L. Wood, ed., Shale tectonics: AAPG Memoir 93, p. 91–109.

Spatiotemporal Variations and Kinematics of Shale Mobility in the Krishna-Godavari Basin, India

Mainak Choudhuri
Reliance Industries Limited, Petroleum Business (E&P), Navi Mumbai, Maharashtra, India

Debajyoti Guha
Shell Technology India Private Limited, Bangalore, Karnataka, India

Arindam Dutta
Reliance Industries Limited, Petroleum Business (E&P), Navi Mumbai, Maharashtra, India

Sudipta Sinha
Reliance Industries Limited, Petroleum Business (E&P), Navi Mumbai, Maharashtra, India

Neeraj Sinha
Reliance Industries Limited, Petroleum Business (E&P), Navi Mumbai, Maharashtra, India

ABSTRACT

Tertiary sediment loading on the Late Jurassic–Early Cretaceous passive-margin rift fills in the offshore Krishna-Godavari basin generated different episodes and patterns of mobile, shale-cored structures from the Paleocene to Pliocene. The Paleocene–Eocene shales in the deeper shelf and upper slope areas moved as a series of thrust slices, whereas the younger Miocene–Pliocene shales moved as individual bulges (diapirs). Lithological variability and shifting depocenters, prevailing tectonic conditions, and available space in the system for translation in space and time all influenced the spatiotemporal distribution of the shale-cored structures. A relation is also observed between the location of the toe thrusts and shale diapirs, and the basement highs and escarpments. Two-dimensional palinspastic restorations, incorporating all the above variables, confirm the linkages between the sediment depocenters, growth faults, and mobile shales in the Krishna-Godavari basin.

DOI:10.1306/13231310M933420

INTRODUCTION

Shale-cored thrusts and detachment folds are found worldwide in different deep-water settings where sediment loading and basin-margin uplifts are key factors in the localization of these features. The extent, locale, and intensity of these shale-cored structures are strongly controlled by the basin geohistory, especially the sediment loading pattern, the shifting of depocenters, the shelf width, and the slope geometry.

Our goal is to describe and address some of these factors and their function in the development of shale bulges or toe thrusts through geologic time in a major offshore basin of India, the Krishna-Godavari basin. Basin-scale isochore maps have been used to understand the sediment dispersal pattern through geologic time, and episodes of shale movement and localization in the Krishna-Godavari basin. Structural restoration studies along key profiles have helped in detailing the thin-skinned tectonic episodes and will aid in discussing the spatial and temporal variation of these episodes.

Study Area

The study area is located in the offshore Krishna-Godavari basin, which is a Jurassic–Early Cretaceous-aged passive-margin rift basin along the east coast of India. It covers an area of about 140,000 km^2 (54,000 mi^2) extending from onshore to deep offshore waters and is receiving sediments from the Krishna and the Godavari rivers (Figure 1). The shelf width at the Krishna mouth is very narrow and has strongly influenced the sediment dispersal in this part of the offshore basin. The Krishna delta is river dominated, with more gross sand content than the Godavari. The sand is sourced from the erosion of the peninsular gneiss and the Cuddapah basin, composed of metamorphic and metasedimentary rocks of Archean and Proterozoic ages, respectively. In contrast to the narrow marine shelf at the mouth of the Krishna River, the Godavari River forms a wave-dominated delta and an associated wider, wave-influenced shelf, although this shelf is still narrow compared to other big rivers like the Nile of Egypt or the Mississippi of North America. The Godavari River drains the Deccan volcanic country as well as the Archean Eastern Ghats; thus sequences deposited at the mouth show a mixed clay, and sand-rich character.

The Krishna and Godavari rivers delivered the largest amounts of sediment to the basin from the Cretaceous to Paleocene. This encouraged the development of growth fault activity in the basin's shelf, which helped in capturing the sediment load within the accommodation space created. A huge thickness of shale was deposited, starting from the Paleocene–Eocene, both in the basin's onshore and offshore areas.

Previous Studies

The Krishna-Godavari basin has undergone hydrocarbon exploration since the early 1980s. The shallow water areas are covered by 1980s-vintage seismic data, whereas the deeper water areas are covered by more recent seismic data. Limited outcrops of rift sedimentary rocks, as well as postrift rocks, restrict the geological understanding in the Krishna-Godavari basin's onshore part. There, workers have used the information from wells as the key data set to build the litho- and chronostratigraphy of the basin. The same approach has been taken in the offshore parts of the basin. Previous works on basin sedimentation, stratigraphy, and tectonics have been published by Sastri et al. (1973), Murty and Ramakrishna (1980), Kumar (1983), Biswas and Chandra (1992), Smith and Armentrout (1992), Mohinuddin et al. (1993), Rao (1993, 2001), Stuart and Hickman (2001), and Bastia et al. (2003, 2005). Local universities (e.g., Andhra University) have worked on delta modeling and drainage analysis to understand the sediment dispersal pattern and the morphotectonics of the basin. Gravity-magnetic studies in the onshore have been done by the National Geophysical Research Institute (NGRI) and other Indian government agencies.

During the last few years, huge amounts of seismic data have been acquired by the oil companies in the shallow and deep offshore areas of the Krishna-Godavari basin. These data allow new insights into the geological understanding of the area and help link the processes and deposits in the deep-water regions with the shallower-water areas of the basin.

Basin Tectonostratigraphy

The Krishna-Godavari basin is bounded on the west by the Archean basement rocks and toward the north and east by a major morphotectonic feature, the 85°E ridge. Toward the northeastern edge of the basin, a northeast–southwest-trending coast-parallel basement high known as the Vizag high is present, which merges with the 85°E ridge.

The east coast of India evolved as a result of rifting of eastern Gondwanaland when India separated from Antarctica during the Jurassic–Early Cretaceous. The Krishna-Godavari basin and its passive-margin setup evolved along this northeast–southwest-rifted continental margin of eastern India. This phase of rifting was superposed over an older, northwest–southeast-trending intracratonic Gondwana rift trend, the Pranhita-Godavari rift, which is a failed rift orthogonal to the younger rift

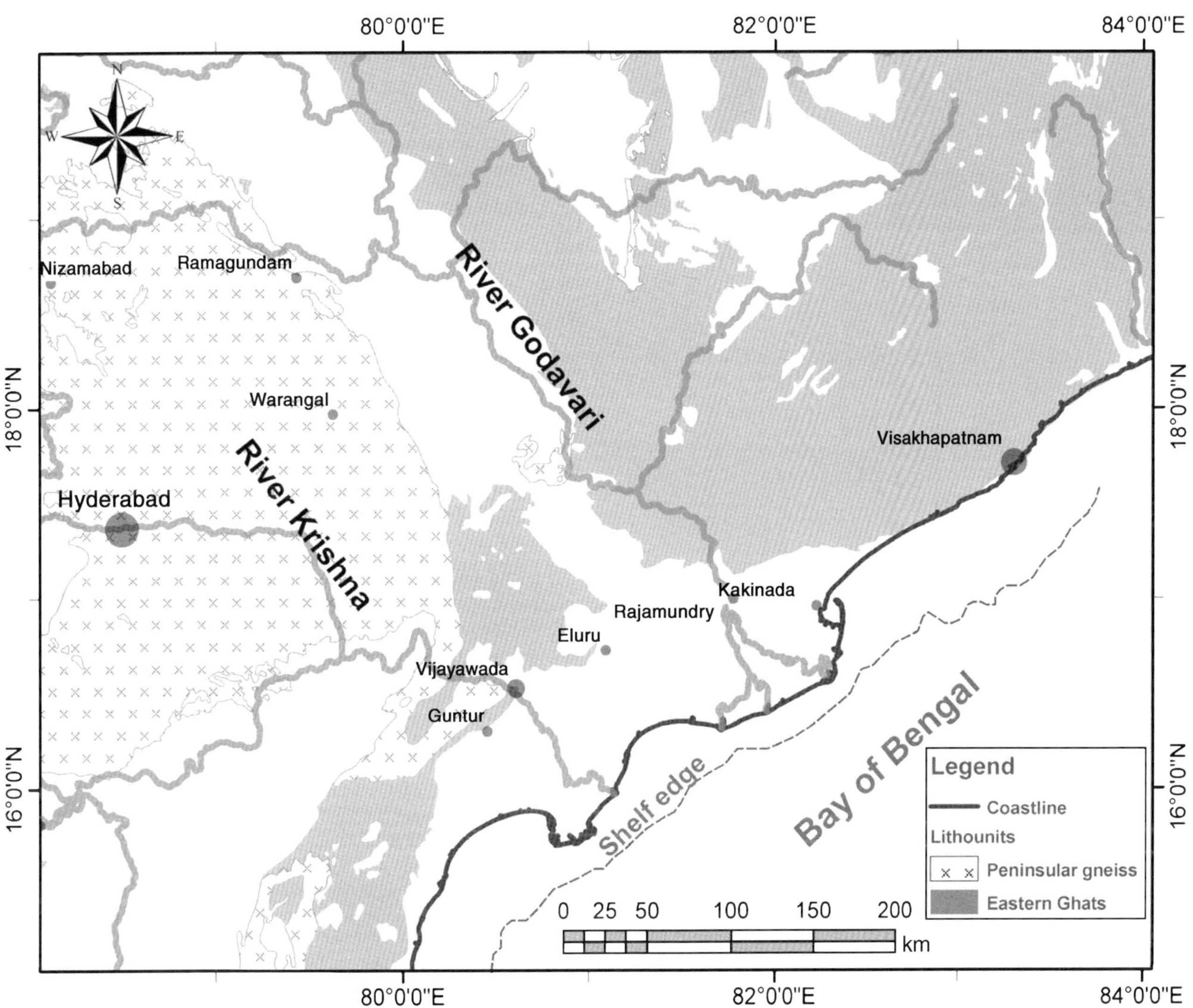

Figure 1. Map of a part of the eastern coast of India generated from Landsat Enhanced Thematic Mapper Plus mosaic (742 band combination), covering an area that includes the Krishna and Godavari deltas. Note the different configuration of the two deltas: the Krishna is a river-dominated delta with a narrower shelf width than the Godavari, which is a wave-dominated delta. The Godavari River cuts through the Eastern Ghats mobile belt (Archean) and Krishna River flows through both Eastern Ghats mobile belt and peninsular gneiss. 100 km = 62 mi.

trend (Rao, 2001). The younger rift trend comprises a series of coast-parallel northeast–southwest-trending half grabens with fluviolacustrine Jurassic–Cretaceous rift fills (Figure 2). The onset of the younger rifting started around 175 Ma (Toarcian) and produced the oldest rock dated in some of the wells drilled on the basin margin (M. Nemčok, 2007, personal communication). Around the Early Cretaceous (~132 Ma), the Indian plate rotated away from Antarctica-Australia in the northwest direction, resulting in the ultimate break up and formation of the east coast passive continental margin (Reeves and Wit, 2000; Nemčok et al., 2007).

Thermal subsidence during the early Hauterivian to late Aptian led to the differential uplift of different basement blocks and development of localized depocenters, which were inundated by marine incursion, depositing deep-water pelagic to hemipelagic claystones. Deep-water sedimentation continued during the Cenomanian, but pulses of coarse-grained clastic input in a few corridors during the Turonian and younger times were observed. Passive-margin deltaic shallow-marine sedimentation (fine- and coarse-grained clastics and limestone) occurred during the Paleocene and Eocene in the present-day, onshore part of the basin, whereas mainly thick successions of finer grained clastics were deposited in the present-day offshore part of the basin (Figure 3). Subsequent basin-margin uplift and rapid sedimentation resulted in the development

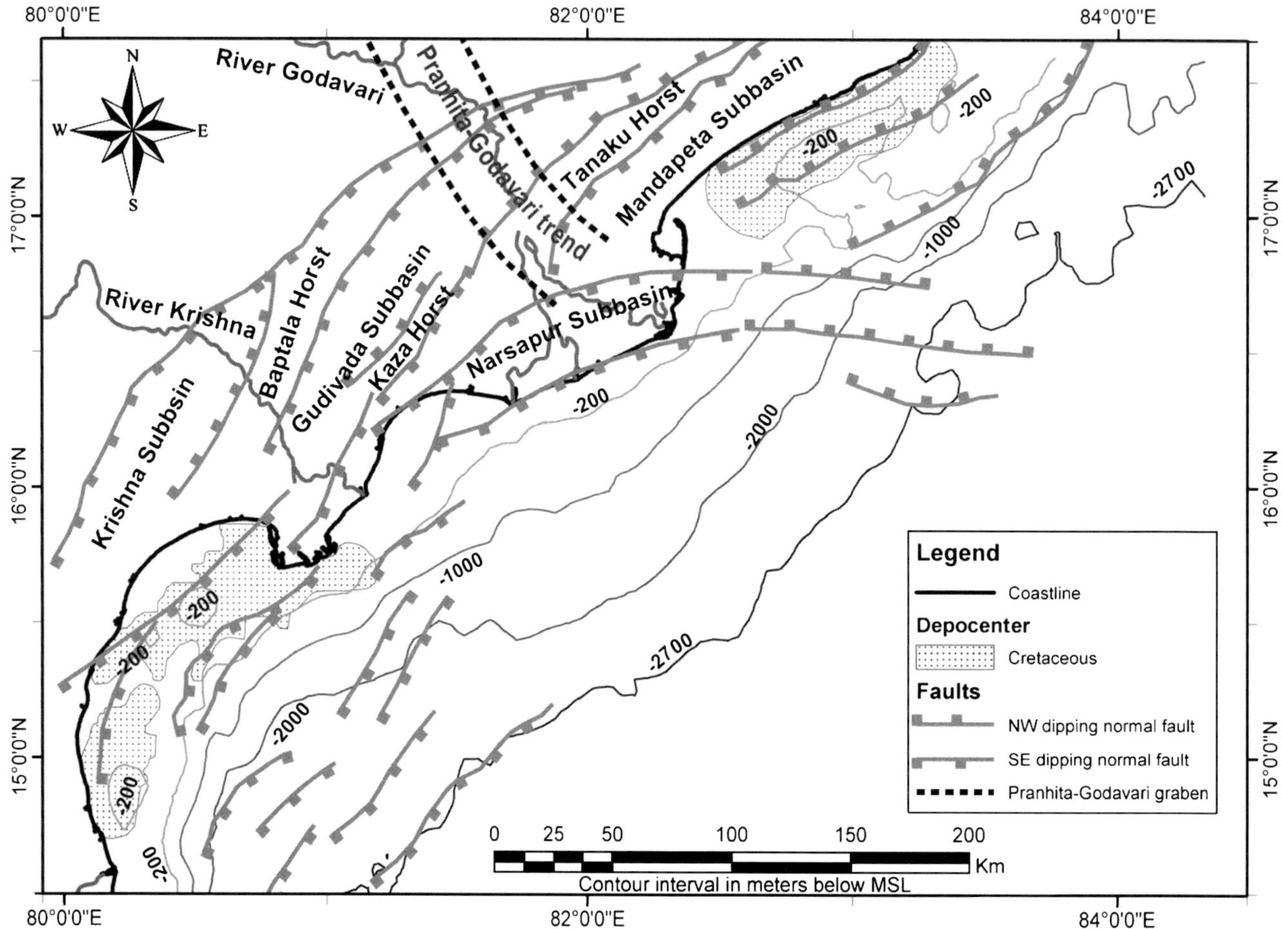

Figure 2. Onshore and offshore basement trends in the Krishna-Godavari basin showing the older Pranhita-Godavari northwest–southeast trend overlain by major basement trends of the younger rift episodes. The rift structure has been delineated in the onshore based on gravity-magnetic data (modified from Rao, 2001) and in the offshore from seismic data. MSL = mean sea level; 100 km = 62 mi; 100 m = 328 ft.

of gravity slides, creating space for the Oligocene–Miocene sedimentation in shallow-water areas. Deep-water turbidite sediments were deposited during these phases of basin-margin uplift and delta progradation. Major inputs of deep-water coarser grained clastics are seen during the Miocene and Pliocene (Rao, 2001; Raju et al., 2005a, b; Sahu, 2005).

SEDIMENT DISPERSAL PATTERN AND THIN-SKINNED TECTONICS

Understanding sediment dispersal and accumulation through time is important, because these appear to be indicators of linkages between the thin-skinned tectonic style and its localization in space and time. Several basin-scale isochore maps have been prepared and analyzed for the shelf–upper slope–lower slope region as a means of examining the sedimentation history of the basin. The gravity-driven linked systems control the most prominent Tertiary tectonic episodes in the Krishna-Godavari basin. These linked systems are a result of differential sediment loading in different parts of the basin at different times, and their occurrence is influenced by the basement architecture and the control it exerts on the basinal morphology.

Inter-Valanginian to Maastrichtian

The Upper Cretaceous top is a major decollement in the basin that dips toward the offshore at a moderate angle of 2–5°. The isochore map of the drift rock from the breakup (within the Valanginian) to the Cretaceous top (Maastrichtian) shows development of depocenters in the shallow offshore (very close to the present-day 200-m [656-ft] bathymetry) on either side of the

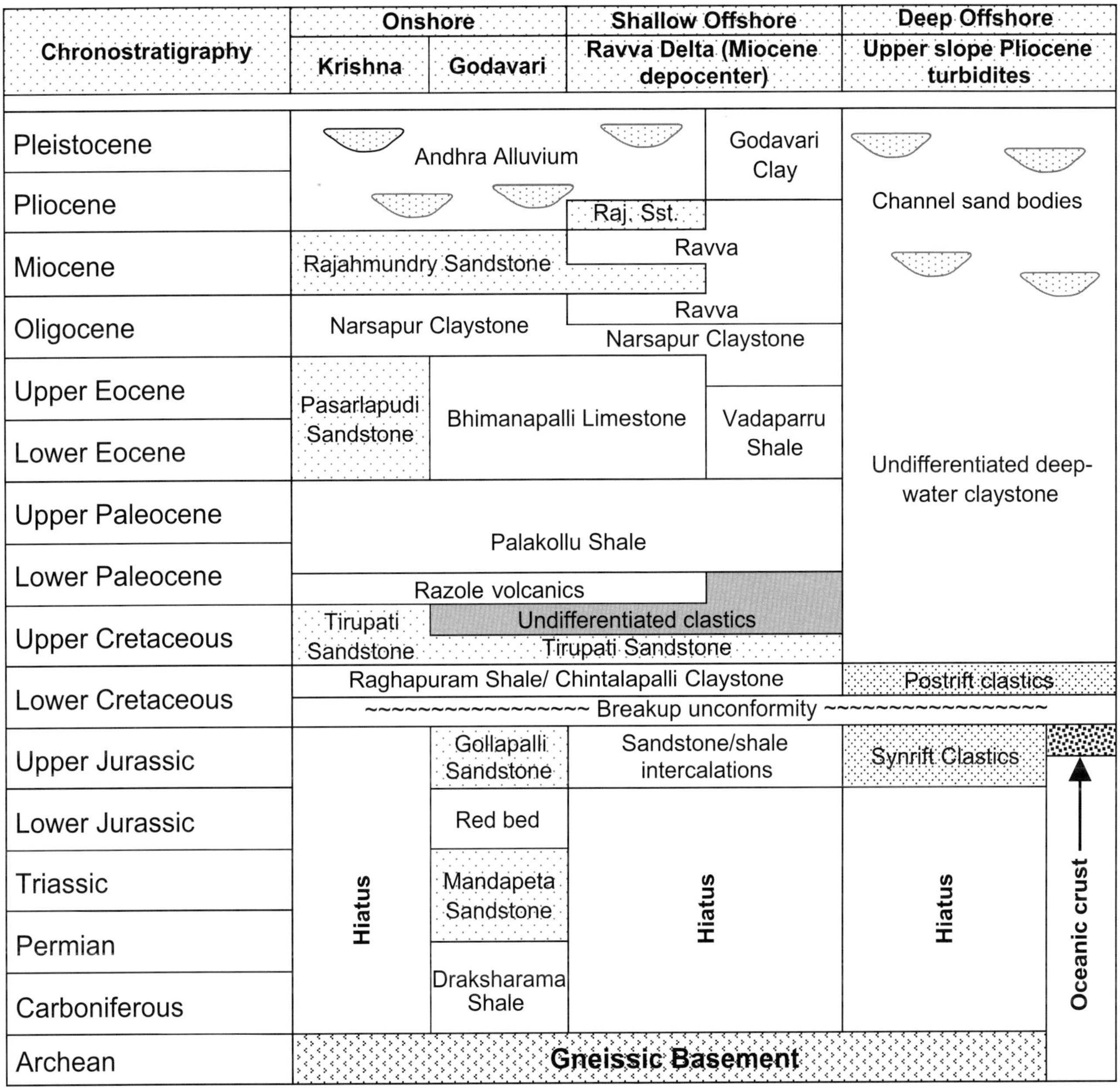

Chronostratigraphy	Onshore		Shallow Offshore	Deep Offshore	
	Krishna	Godavari	Ravva Delta (Miocene depocenter)	Upper slope Pliocene turbidites	
Pleistocene	Andhra Alluvium		Godavari Clay	Channel sand bodies	
Pliocene			Raj. Sst.		
Miocene	Rajahmundry Sandstone		Ravva		
Oligocene	Narsapur Claystone		Ravva Narsapur Claystone		
Upper Eocene	Pasarlapudi Sandstone	Bhimanapalli Limestone	Vadaparru Shale	Undifferentiated deep-water claystone	
Lower Eocene					
Upper Paleocene	Palakollu Shale				
Lower Paleocene	Razole volcanics				
Upper Cretaceous	Tirupati Sandstone	Undifferentiated clastics Tirupati Sandstone			
Lower Cretaceous	Raghapuram Shale/ Chintalapalli Claystone			Postrift clastics	
	~~ Breakup unconformity ~~~~~~~~~~~~~~~~				
Upper Jurassic	Hiatus	Gollapalli Sandstone	Sandstone/shale intercalations	Synrift Clastics	Oceanic crust
Lower Jurassic		Red bed	Hiatus	Hiatus	
Triassic		Mandapeta Sandstone			
Permian					
Carboniferous		Draksharama Shale			
Archean	Gneissic Basement				

Figure 3. Tectonostratigraphy of the Krishna-Godavari basin showing the different lithounits and facies relationships. The major lithounits have been represented on the seismic lines (profiles 1–4, Figures 8, 9, 11, 14) by their respective age correlation markers.

Krishna and Godavari River deltas and very close to the modern coast (Figure 2). A depocenter of pelagic sediments in deeper areas beyond the 2000-m (6562-ft) bathymetry contour also exists but is not shown in the map. The basin has been differentially tilted at different times within the middle and Late Cretaceous, caused by basin-margin uplifts as well as localized uplifts of the basement highs. These phases of tectonic tilting generated several small-scale gravity slides, most prevalently in the Turonian.

Paleocene to Eocene

The Paleocene to Eocene was a time of major basinal tilt, delta progradation, and deep-water finer grained clastic sedimentation. After the eruption of the Deccan volcanics, good outcrops in the catchment of the Godavari River supplied huge amounts of smectite-rich clays into the basin. The combination of tectonic instability and large quantities of smectite-rich sediments initiated a series of gravity slides and imbricate thrusts in the deep waters

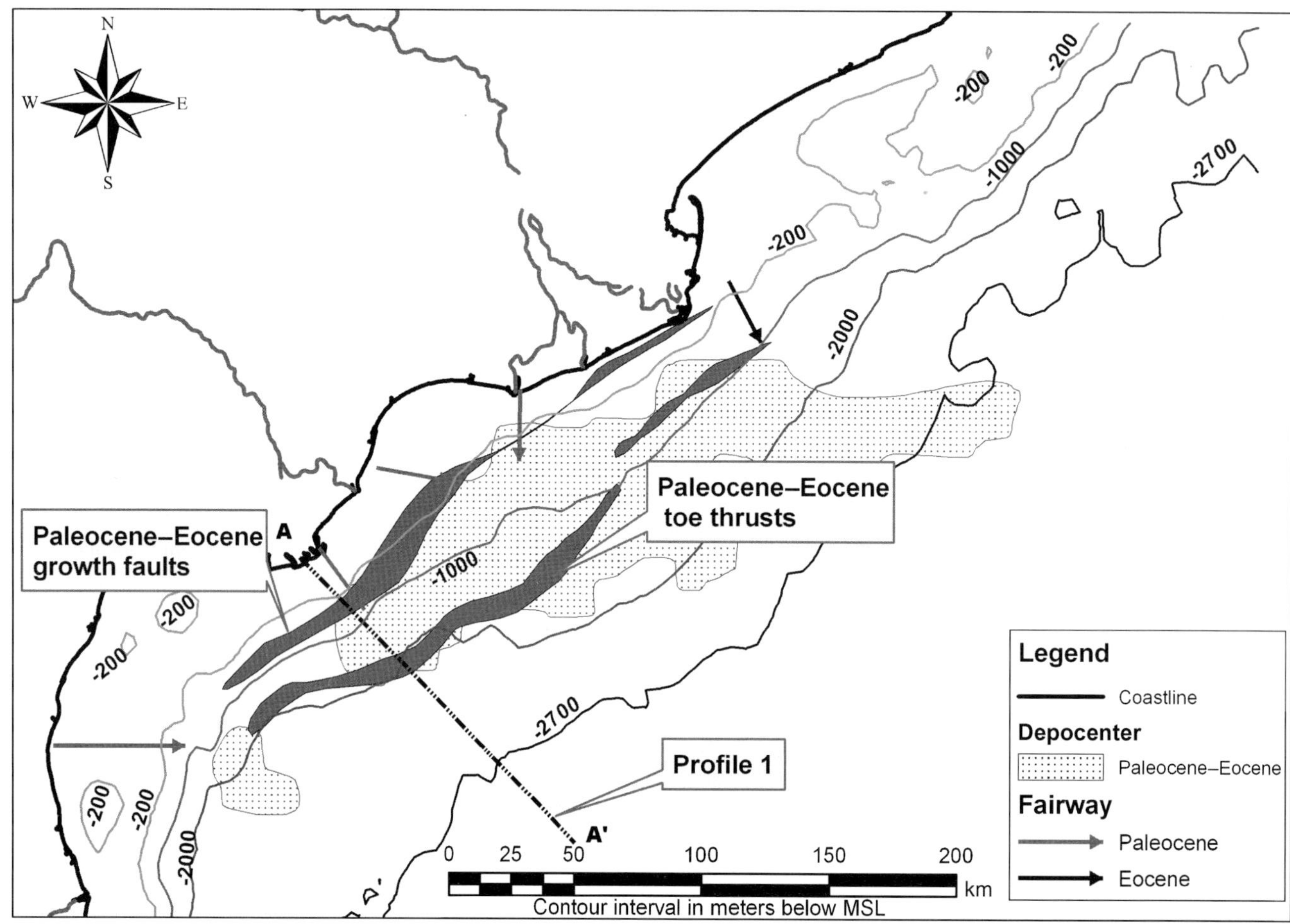

Figure 4. Map of the Paleocene and Eocene with depocenters, sediment fairways, and gravity slide and toe thrust system. Rapid loading of the north and south basin margins caused the development of this gravity-driven linked systems (see profile 1 along AA′ shown in Figure 8). The linked system is confined to a narrow corridor mostly on the present-day upper slope in the bathymetry range of 200 to 2000 m (656 to 6562 ft). MSL = mean sea level; 100 km = 62 mi; 100 m = 328 ft.

beyond the 1000-m (3281-ft) present day bathymetry. These features can be seen in the isochore map of the Paleocene to Eocene section. The gravity slides detach on top of the Cretaceous, parallel with the modern and paleocoast and are found between 2000- and 2700-m (6562- and 8858-ft) present-day bathymetry, at the convergence of the lower and upper slopes (Figure 4). The thrusts seem to have localized because of a change in the dip of the decollement, which is linked to the basin tilt episode.

Oligocene and Miocene

An isochore map of the Oligocene shows that the depocenter shifted farther south and east in the offshore during this time. This time marked a period of global lowstand of the sea level. Along the margin of the Krishna-Godavari basin, this time was also marked by limited supply of sediments coming into the basin (Figure 5).

The early Miocene marked the end of quiescence and the beginning of a major phase of basin-margin tilting. A change in sediment supply direction occurred, and depocenters shifted basinward and to the north (Figure 6). On seismic data, marine facies associated with the Godavari prodelta show successive onlaps over the Oligocene top in the deep waters, suggesting major areas of sediment bypass on the shelf and the upper slope. This onlap is overlain by middle to upper Miocene progradation of the delta and development of the Ravva growth fault system. Shale movement associated with this episode of sediment loading is evident on the prodelta parts of the Miocene Ravva delta. Deepwater toe-of-slope thrusts are evident in this middle to upper Miocene phase of deep-water sedimentation. The shale bulges occurring in association with these linked growth

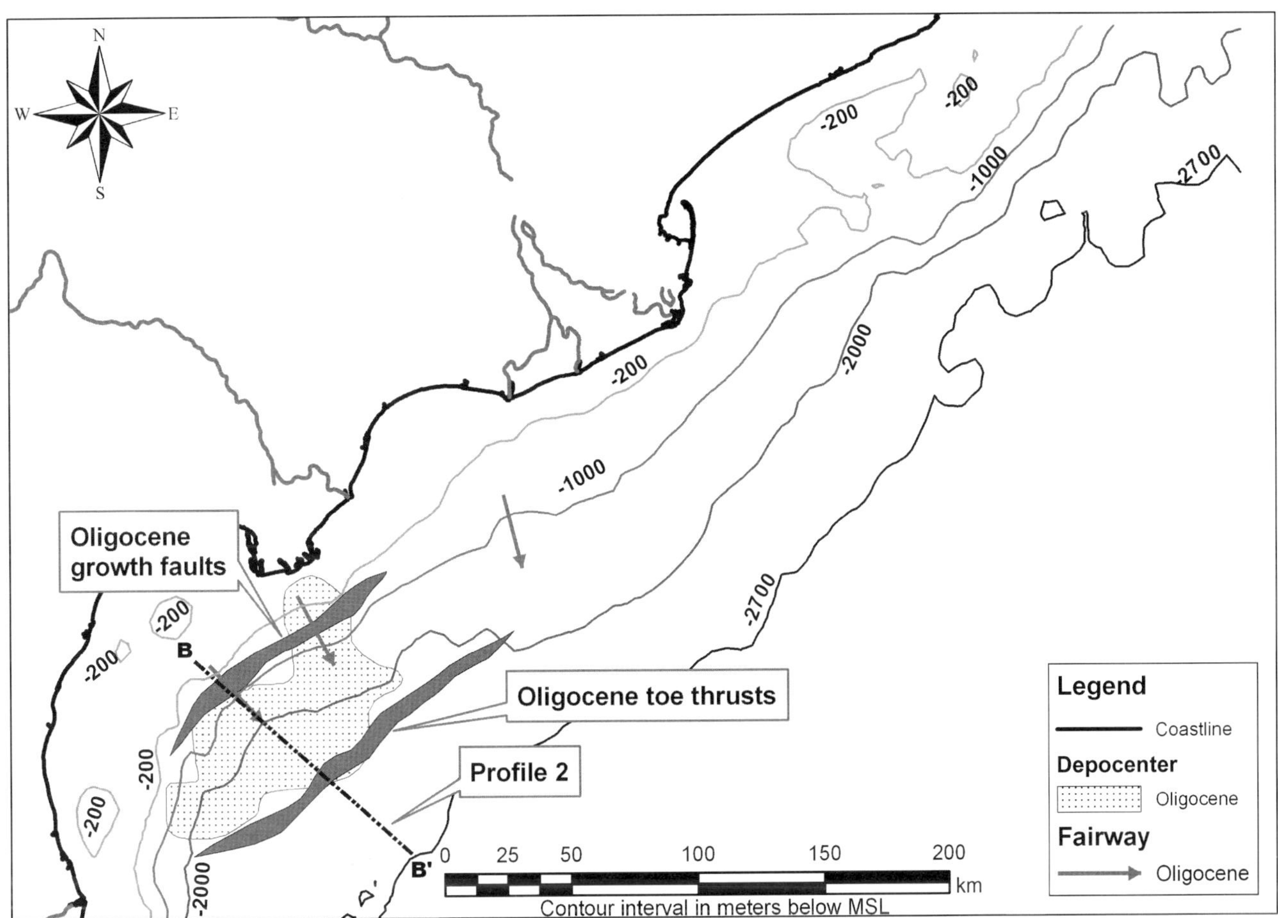

Figure 5. Map of the Oligocene showing the depocenter, Oligocene sediment fairways, and gravity slide and toe thrust system. Note the localization and shift of the depocenter toward the south. A steep basement at this location favors the development of the gravity slides and linked basinward thrust system, and consequent huge sediment loading, which further accentuates the process (see profile 2 along BB′ shown in Figure 9). MSL = mean sea level; 100 km = 62 mi; 100 m = 328 ft.

fault and toe-of-slope-thrust systems are referred to as the proximal and distal shale bulges.

Pliocene to Holocene

During the Pliocene, the proximal shale bulge continued to grow because of the shifting of depocenters onto the upper slope. In contrast, the distal shale bulge almost died out, and the Pliocene rocks pinch out toward deeper water areas. A local decollement at the Miocene level developed in response to this sediment loading on the upper slope. Continued sedimentation in the outer shelf and upper slope areas distal to the Godavari delta and shelf front helped this localized linked system to remain active to the present day. Growth fault activity is also evident during this time near the Krishna River mouth, although associated prodelta shale movement is not seen. Sediment bypass to deeper water limited any further growth on these normal faults during the Pliocene in the southern part of the basin (Figure 7).

STRUCTURAL MODELING AND BASINAL EVOLUTION

Four representative seismic profiles were selected for palinspastic modeling. Palinspastic modeling involves sequential restoration of a section to its original depositional configuration before the occurrence of any structural deformation. The goals of the modeling exercise were to understand the basin development with time, to model the pulses of shale movement caused by gravity slides and sediment loading, and to understand the basin tilt episodes and their effect on localization of thin-skinned tectonics. Profiles were selected for structural modeling with consideration of their coverage of key features of the sediment dispersal and

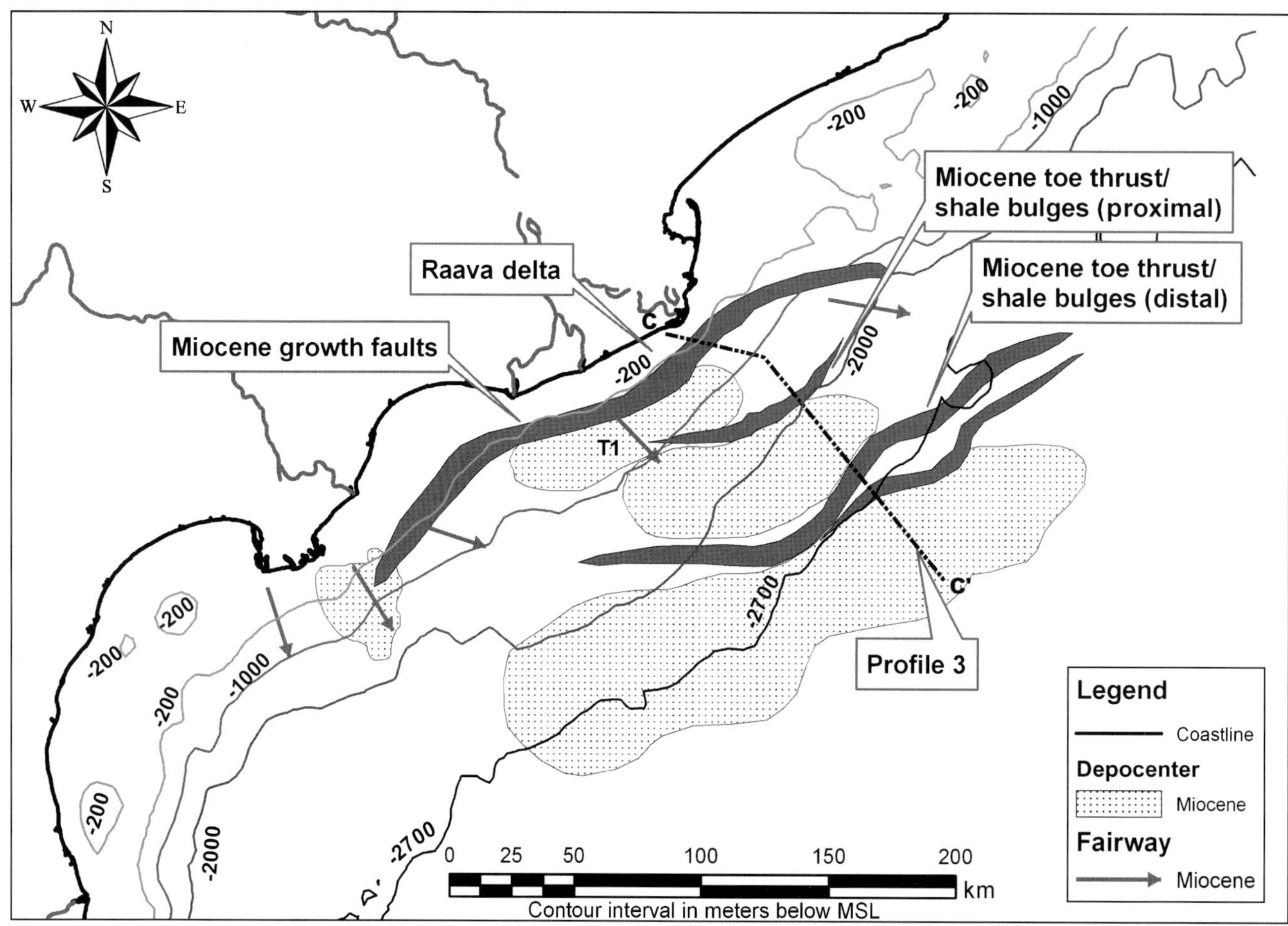

Figure 6. Map of the Miocene with depocenters, Miocene sediment fairways, and growth fault-toe thrust and shale bulge system. Note a deepening of the basin and consequent basinward shifting of depocenters. The distal shale bulge formed as a result of huge sediment loading in the Raava delta during the Miocene. The proximal shale bulge formed as a result of a basinward shift of the Raava depocenter during the Pliocene from location T1 to location T2 (shown on Figure 7) (see profile 3 along CC′ shown in Figure 11). MSL = mean sea level; 100 km = 62 mi; 100 m = 328 ft.

thin-skinned tectonics episodes. Profiles 1 and 2 are in the Krishna offshore, whereas profiles 3 and 4 are in the Godavari offshore.

The process of structural modeling is executed in three steps. Step 1 is the removal of the topmost layer and decompaction of the underlying layers to get the paleoseabed (this is essentially backstripping in time). Step 2 is the restoration of the faults and/or folds at the topmost level to get the depositional architecture of the bed (a small area was taken for extensive restoration to get detailed deformational history). Step 3 is the repetition of steps 1 and 2 for each unit until reaching the mapped basement.

The following assumptions were used during the palinspastic restoration process. (1) Plane-strain deformation was assumed; it had been determined that the basin had experienced movement by gravity gliding (Rowan et al., 2004). (2) The sectional area remained constant during the deformation. (3) Flexural isostacy was caused by large-scale loading.

The software used for palinspastic reconstruction is 2DMove from Midland Valley Inc. For decompaction of the various lithounits and the parameters for flexural isostacy, standard values were used from within 2DMove based on the values given by Sclater and Christie (1980).

Palinspastic Reconstruction: Profile 1

Depth profile 1 along section AA′ in Figure 4 shows a classic gravity fold and thrust belt (GFTB) from the Krishna basin with a detachment at the top Cretaceous level (Figure 8). The imbricate thrusts developed on the present-day toe of the slope (the slope break between

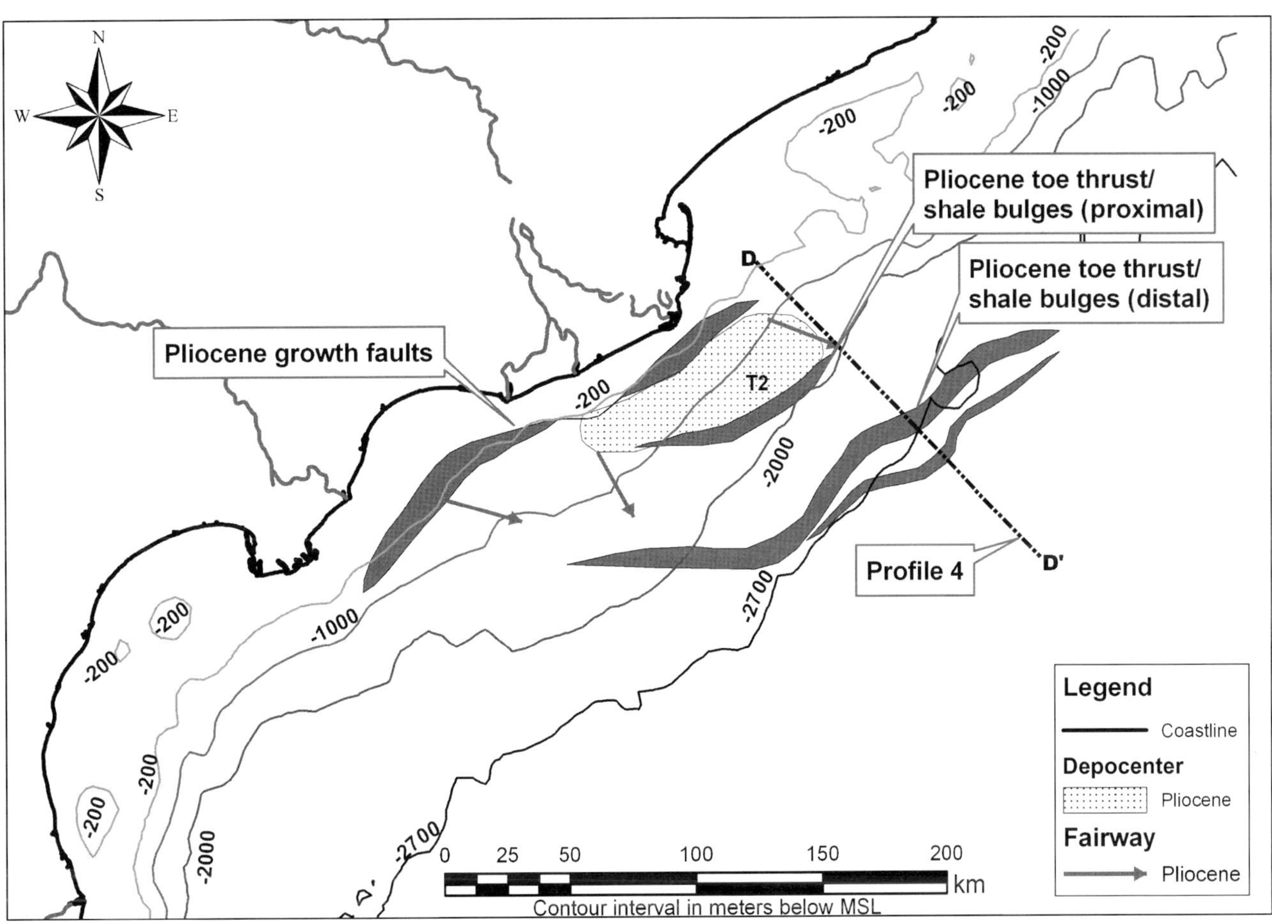

Figure 7. Map of the Pliocene with depocenters, Pliocene sediment fairways, and growth fault-toe thrust and shale bulge system. Note a shift of depocenters toward the basin from the Miocene to Pliocene (from T1 in Figure 6 to T2). The proximal shale bulge continued to grow during the Pliocene because of the basinward shift of the depocenter; the distal shale bulge almost died out at this time (ref. profile 4 along DD′ shown in Figure 14). Late Pliocene and Pleistocene loading very close to the present-day outer-shelf-slope break caused sediment bypassing to deeper water in some parts of the basin and prevented Pliocene growth across faults. MSL = mean sea level; 100 km = 62 mi; 100 m = 328 ft.

the upper and lower slope) in response to rapid deposition and growth during the Miocene. An upper Oligocene detachment may also be present, forming a shale bulge above the imbricate structures in response to the Miocene extension, possibly caused by a change of slope.

Detailed analysis of the profile led to the following observations regarding basin development. Rift architecture is not obvious because of imaging issues and if present would be at a burial of 6000–7000 m (20,000–23,00 ft). The most obvious marker picked for modeling is the top Cretaceous decollement. Thick Cretaceous rocks are common in this part of the basin with maximum thickness in the range of 1000–2000 m (3300–6600 ft). The younger mappable sequences are the Paleocene; Eocene; Oligocene; lower, middle, and upper Miocene; Pliocene top; and the sea bed. A series of Oligocene–Miocene listric growth faults detaching at the top of the Upper Cretaceous developed on the location of the present-day shelf and upper slope (faults 1–8) in response to rapid sediment loading. Maximum offset is observed along faults 3, 4, and 6; the others are minor in comparison. This extensional movement was balanced by a sequence of forward-breaking thrust sequences toward the basinal side. Forward breaking is marked by a greater dip of the landward thrusts, indicating their uplift with subsequent thrusting. After cessation of the thrust activity during the early Oligocene, the thrust crest was eroded; however, the growth faults (3, 4, and 6) remained active during the late Oligocene to early Miocene, and upper Oligocene to lower Miocene growth sequences were deposited. The extension during the late Oligocene is balanced by a bulge over the thrust complex. This bulge has an unconformable

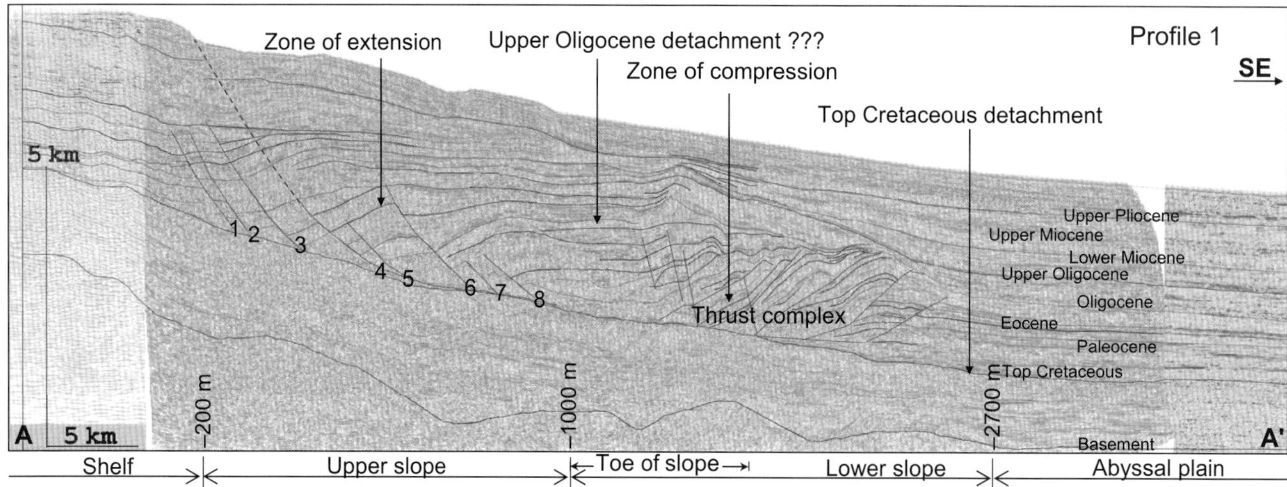

Figure 8. Depth profile showing a typical gravity fold and thrust structure from the Krishna-Godavari basin over a moderately overpressured shale decollement with a frictional detachment at the top Cretaceous. The imbricate thrusts are present toward the basinward side where the pore pressure falls sufficiently to make the frictional force greater than the shear stress along the detachment, making translation difficult. 5 km = 3 mi or 16,404 ft; −200, −1000, −2700m = intersection with bathymetric contours.

position and geometry relative to the underlying thrusts, and there may be a detachment along the lower Oligocene that is responsible for this. The bulge ceased to grow after the early Miocene. Normal sedimentation continued during the late Miocene to Holocene. Some of the listric faults (Fault 4) are possibly still active, as indicated by a fault scarp at the seabed.

In profile 1, both the updip extensional faults and the downdip imbricate thrust system detach over a shale at the top of the Cretaceous. The development of the growth faults was influenced by gravity gliding controlled by deposition of sediments. Rapid sedimentation during the Miocene caused the pore pressure to increase greatly at the depocenter, rupturing the surface to produce a growth fault system and a moderately overpressured shale decollement farther south to allow movement of the rocks along the detachment. The movement persisted till the tectonic force was less than the force of friction along the detachment because of a drop in pore pressure, giving rise to the thrust systems.

Palinspastic Reconstruction: Profile 2

Located in the same area of the Krishna-Godavari basin, but farther south in the Krishna Basin part, this depth profile along section BB′ in Figure 5 shows an unusual GFTB, where a huge amount of extension along an Oligocene–early Miocene growth fault was accommodated by repeated thrusting along a Paleocene detachment (Figure 9). Fault localization and thrusting have been correlated with a steeply dipping basement high within the basin at location A in Figure 9.

Analysis of depth profile 2 brings out the following geohistory (Figure 10). The first sediments deposited over the rifted basement were the Upper Jurassic–Lower Cretaceous synrift sediments. Breakup and drifting started in the Early Cretaceous (Valanginian). The rift-fill sediments were faulted along a basement high forming a steep (~11°) scarp and marking the initiation of the slope break at location A in Figure 9. This steep fault scarp was a key factor in the structuring events during the Oligocene–Miocene. The Albian sediments were deposited over this scarp and were slightly offset, probably as a result of instability along this fault. Until the early Oligocene, quiet basinal sedimentation occurred; the prominent markers during this span are Turonian, Campanian, Maastrichtian, Paleocene, Eocene, and lower Oligocene tops. A major listric fault developed in the area of the basement high with a detachment at the Paleocene level. The Eocene and Oligocene horizons were faulted along a thrust plane (thrust 1, Figure 10) on the basinward side. No appreciable growth is observed during this period, and the faulting is suspected to be triggered by slope failure along the basement high. Further sedimentation during the Oligocene reactivated the fault, leading to another thrust sheet (thrust 2, Figure 10) moving above the earlier one. After deposition of the upper Oligocene, further reactivation along the listric fault initiated another thrust sheet (thrust 3, Figure 10) moving above the earlier ones, forming an antiformal stack look-alike. The faulting continued up to the early Miocene, and huge growth strata of lower Miocene age

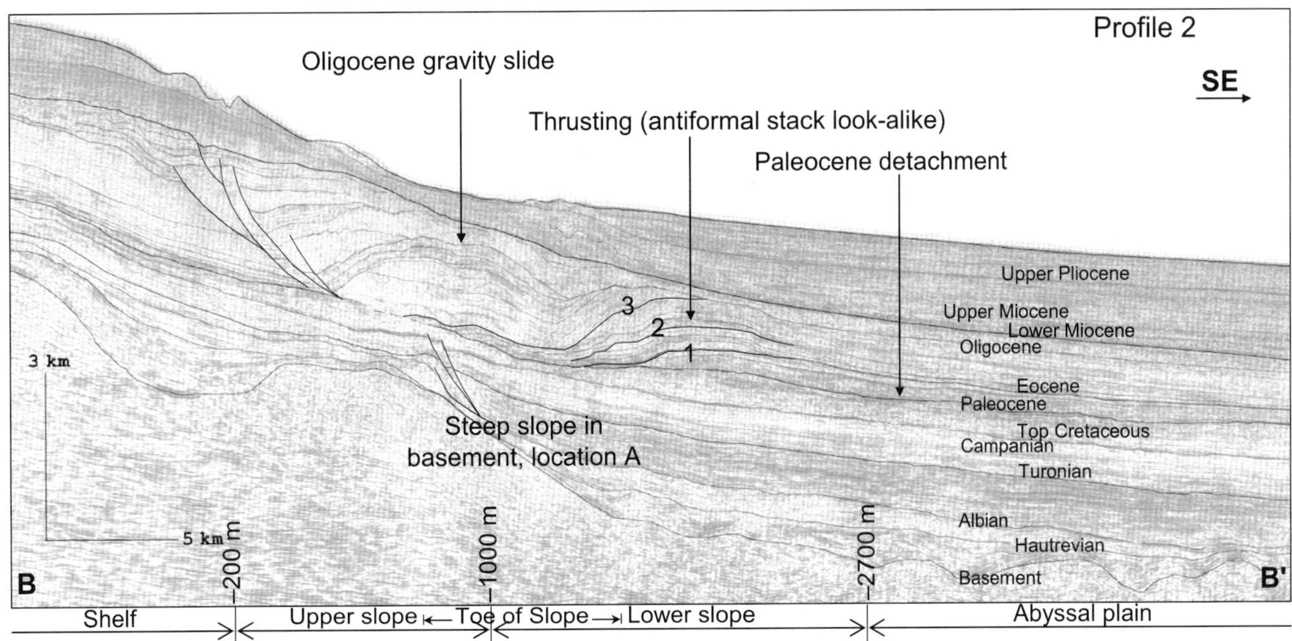

Figure 9. Depth profile showing a not so typical gravity fold and thrust structure from the Krishna-Godavari basin; a huge amount of extension along the Oligocene–early Miocene growth fault is being accommodated by thrusting along a Paleocene detachment. The detachment is moderately overpressured and has a very steep slope, allowing the development of thrusts on the basinward side. Steepening of the detachment and location of the thrusts are also correlated to the abrupt deepening of the basin beyond the basement high at location A. 5 km = 3 mi or 16,404 ft; −200, −1000, −2700 m = intersection with bathymetric contours.

were deposited into the created accommodation space. Some minor basinward migration of the normal faulting (splay faults) is observed in this section. The instability ended with a lower Miocene erosional surface truncating the thrust tip on the basinal side. Normal deposition continued during the late Miocene to the Holocene.

In depth profile 2, a steep Paleocene detachment is observed, which facilitated gravity gliding, resulting in huge amounts of extension along the Oligocene–early Miocene growth fault system on the present-day shelf margin and upper slope. Rapid sedimentation caused the decollement at the top Cretaceous to be moderately overpressured. Adequate gliding force caused basinward translation, which was maintained by the steep slope of the detachment. The decrease in steepness of the detachment, together with a fall in pore pressure in the basinal direction, resulted in an increase in frictional resistance to gliding, which was sufficient to prevent lateral translation. This led to the basinward development of thrusts.

Palinspastic Reconstruction: Profile 3

Characteristic growth faults and shale bulges are observed in depth profile 3 from the Godavari basin, along section CC′ in Figure 6, detaching at the top Cretaceous level (Figure 11). The Miocene–Pliocene growth faults controlled the development of these shale bulges. The distal bulge developed because of rapid sediment loading during the Miocene. The proximal shale bulge developed because of basinward shifting of depocenters during the late Pliocene.

After decompaction and backstripping to the basement level, the deformation history is inferred as follows (Figure 12). The rift-fill rocks of the Middle to Upper Jurassic were the oldest sedimentary rocks in this area. The Lower Cretaceous records the breakup unconformity and start of the drift phase. The basin opening and drifting occurred during this age. Until the Oligocene, there was little tectonic activity, and normal basinal sedimentation prevailed. The prominent markers within this span are the Lower and Upper Cretaceous, Eocene, and Oligocene. Listric faults (fault 1, Figure 12) with an Upper Cretaceous detachment began to develop on the landward side after deposition of the Oligocene sequence. Growth sedimentation occurred during the middle Miocene, a phase of major delta progradation and formation of the Ravva growth fault system. Synthetic faults (fault 2, Figure 12) developed toward the basinal side, sharing the extension and subsequent growth during the late Miocene. The extension was compensated by the initiation of growth at the

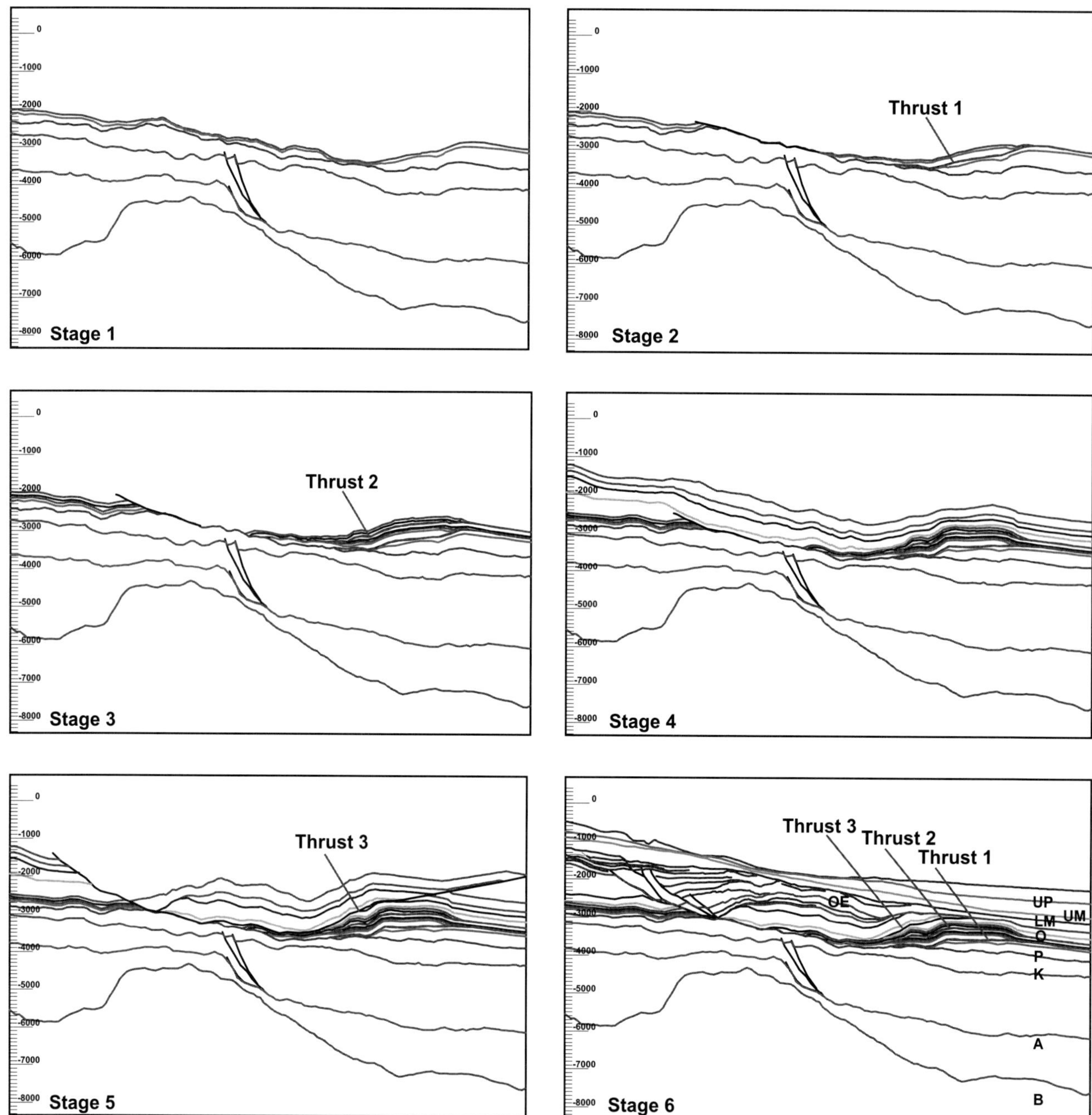

Figure 10. Profile 2 restored at different stages within the Oligocene to depict the relationship of the dip of the decollement surface and sequential development of the thrust system. The detachment is along the Paleocene horizon. Stages shown include (stage 1) unthrusted lower Oligocene seabed; (stage 2) Oligocene thrusting event 1; (stage 3) Oligocene thrusting event 2; (stage 4) deposition of Upper Oligocene beds; (stage 5) Oligocene thrusting event 3; and (stage 6) present-day seabed (UP = upper Pliocene; UM = upper Miocene; LM = lower Miocene; OE = Oligocene erosional top; O = Oligocene; P = Paleocene; K = Cretaceous; A = Albian; B = basement). 1000 m = 3280 ft.

distal shale bulge. After deposition of upper Miocene sediments, faulting moved farther basinward (fault 3, Figure 12) with a detachment in the top Cretaceous level, leading to steepening of the distal bulge. Farther basinward, movement of the listric faults during the early Pliocene (faults 3 and 4, Figure 12), caused by a basinward shift of the Ravva depocenter (from T1 to T2, Figures 6, 7), led to the development of the proximal bulge and Pliocene growth wedge along the basinward faults. The distal bulge remained inactive. A local detachment at the Miocene level developed because of loading on the upper slope, resulting in small-scale

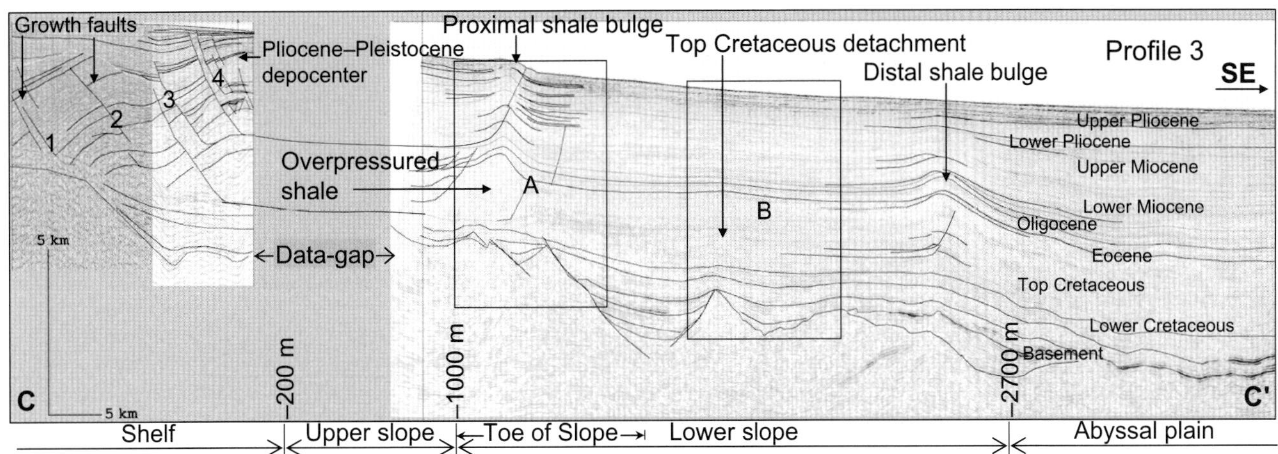

Figure 11. Depth profile showing typical shale bulges from the Krishna-Godavari basin over a highly overpressured shale decollement and detachment at the Cretaceous level. The growth across these Miocene–Pliocene faults strongly influenced the timing of development of these shale bulges. The distal bulge developed because of rapid sediment loading during the Miocene. The proximal bulge developed because of basinward shifting of the depocenter during the Pliocene. 5 km = 3 mi or 16,404 ft; −200, −1000, −2700 m = intersection with bathymetric contours.

thrusts terminating against the proximal bulge. The basinward faults (fault 4, Figure 12) continued to grow during the late Pliocene, further augmenting the proximal bulge. Continued sedimentation in the upper slope area helped this localized linked system to remain active to the present day.

In depth profile 3, the shale bulges and growth faults detached along the top Cretaceous over the Cretaceous shale unit. Rapid sediment loading during the Miocene formed the highly overpressured shale. During the early Miocene, sediment progradation along the shelf margin helped to maintain the surface slope,

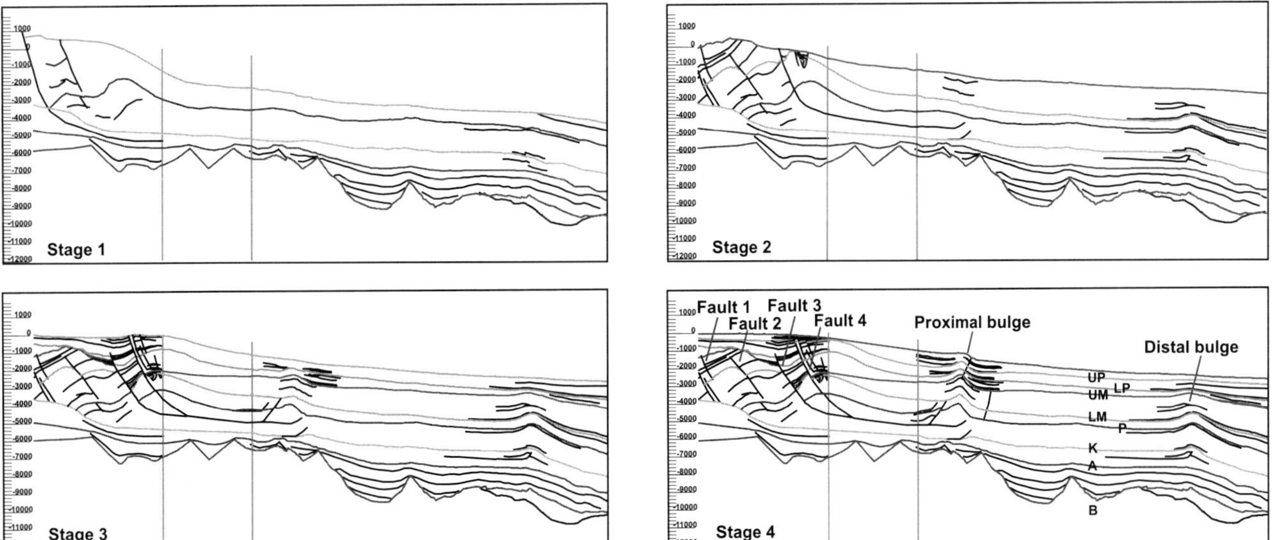

Figure 12. Profile 3 restored at the Miocene and Pliocene showing the stages of relative activity and localization of thrusting and shale movement. The detachment is along the top Cretaceous horizon and locally along the Miocene. The shale bulge at the distal end grew during the Miocene and early Pliocene but was inactive through the Pliocene to present. The shale bulge at the proximal end grew through the Pliocene, related to the growth activity along the Pliocene growth fault. Stages shown include (stage 1) unfaulted lower Miocene seabed and initiation of listric faulting at the deltaic part; (stage 2) upper Miocene faulting along landward listric faults and initiation of distal bulge; (stage 3) early Pliocene basinward shift of faults and initiation of proximal bulge; and (stage 4) present-day seabed (UP = upper Pliocene; LP = lower Pliocene; UM = upper Miocene; LM = lower Miocene; P = Paleocene; K = Cretaceous; A = Albian; B = basement). 1000 m = 3280 ft.

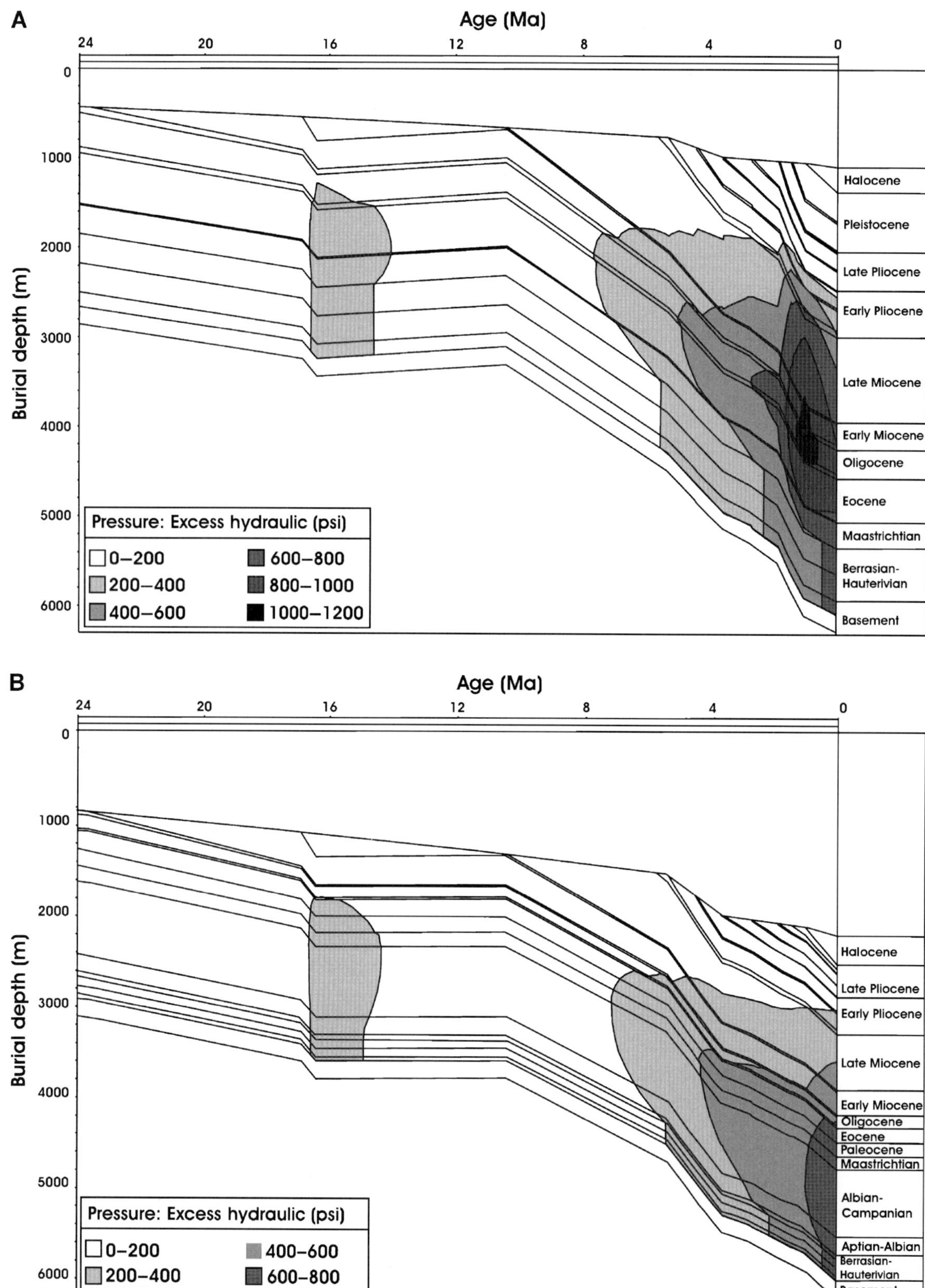

Figure 13. Burial history models with pore-pressure overlays along profile 3 shown in Figure 11. (A) Burial history model at zone A with pore-pressure overlay. Rapid sedimentation during the late Miocene to Pleistocene resulted in overpressure development in the Paleocene–Eocene deep-water shales and the formation of shale bulge, as depicted in profile 1. (B) Burial history model at zone B with pore-pressure overlay. Moderate sedimentation during the late Miocene to Pleistocene resulted in slight overpressuring in the Paleocene–Eocene shales without the formation of shale bulge. 1000 m = 3280 ft; 200 psi = 1.38 MPa.

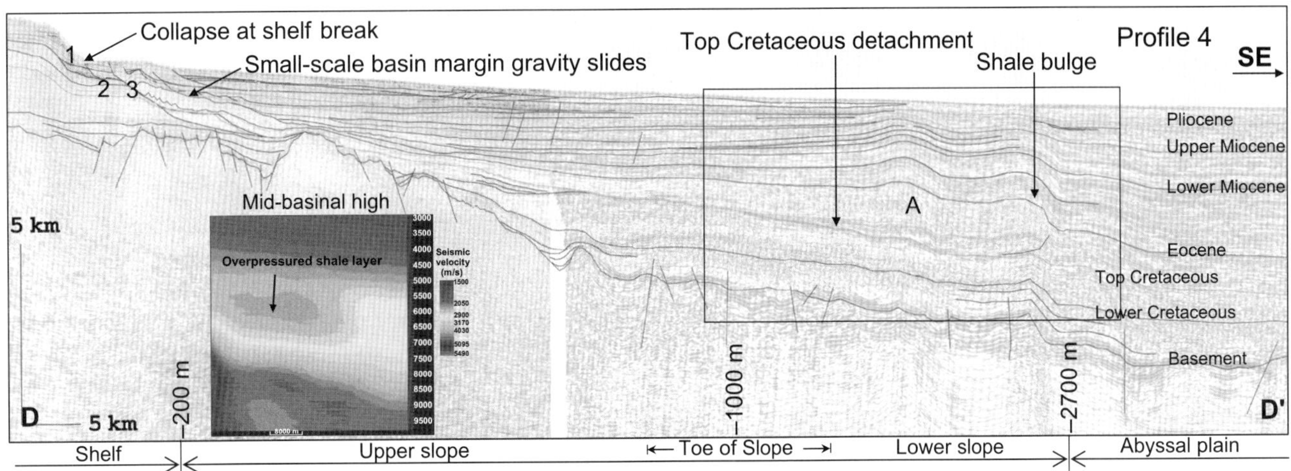

Figure 14. Depth profile showing shale bulges from the northern Krishna-Godavari basin over an overpressured shale decollement with detachment at the top Cretaceous. Tectonic disturbances during the Miocene developed small-scale gravity slides, which triggered the faulting at the shelf break. The shale bulges began to grow at the distal end with a soft linkage with the landward faults. Collapse of the shelf indicates continued instability up to the present time. This profile is slightly oblique to the tectonic transport direction, and hence the bulges do not exactly balance with the extension along the faults. The interval velocity section at zone A (inset, profile 4) shows velocity inversion at the Eocene/Paleocene level, indicating the presence of overpressured shale. Both the horizontal and vertical scales of this inset are in meters, indicating the distance and depth below sea level, respectively. The gray scale is the seismic velocity in meters per second (m/s), ranging from 1500 to 5500 (4921 to 18044 ft/s), with the white shade around 3000 m/s (9842 ft/s). This study is based on the seismic velocity only without any well control. 5 km = 3 mi or 16,404 ft; 1000 m = 3280 ft; −200, −1000, −2700 m = intersection with bathymetric contours.

providing the necessary driving force. The spread-out Miocene deposits maintained the high pore pressure along the detachment for a long distance. The fluid pressure within the decollement reduced the coefficient of friction along the detachment and initiated the translation of sediments. Farther basinward, the rate of sedimentation dropped, resulting in a normally pressured decollement and consequent piling of the sediments, forming the distal bulge. The growth of the distal bulge continued till the Pliocene. During the Pliocene, restricted depocenters led to a local bulge of the sediments at the toe of slope. This resulted in the development of the proximal bulge, perhaps initiated by a change in slope at the shelf break.

The presence and extent of overpressured shale can be established by a study of the burial history at two suitable locations along profile 3, namely zone A, lying directly over the proximal shale bulge, and zone B, lying to the south of it. The burial history plots (Figure 13A, B) show the burial history curve for the last 24 m.y., overlaid with the present-day pressure excess of the hydraulic pressure at a particular depth. The slope on the burial history curve gives an estimate of the rate of sedimentation at a particular period.

At zone A (Figure 13A), a steep burial curve slope indicating rapid sedimentation is observed during the late Miocene to Pleistocene. This rapid sedimentation rate prevented fluids from escaping from the Eocene–Paleocene decollement layer, thereby resulting in the present-day high pore pressure in this zone, mobilization of the shaley substrate, and formation of a shale bulge. The overpressure is maximum in the Oligocene–Miocene, indicating a recent buildup of overpressure.

At zone B (Figure 13B), a gradual burial curve slope during the Miocene to Pleistocene indicates a moderate sedimentation rate in the basin. Consequently, the pore fluids had some time to escape to the overlying layers, and the sediments were normally compacted. The maximum overpressure, as in Figure 13A, is not observed, and we suggest that it was present during the Cretaceous–Eocene and leaked to the present-day moderate pressure buildup. Slight overpressure is observed at this place at present in the Miocene–Pliocene; the surface acts only as a lubricant interface for lateral transfer of pressure.

Palinspastic Reconstruction: Profile 4

Farther north, in the Godavari basin, depth profile 4 along section DD′ in Figure 7 shows shale bulges over the Cretaceous detachment (Figure 14). Minor growth occurred during the early Miocene to late Pliocene along listric faults above the present-day shelf break; however, a steep basinal slope and consequent tectonic instability seem to have been the major causes for the development of the shale bulges. The collapse of the shelf at the present time indicates continued instability up to the present. Note that this particular profile is slightly oblique to the tectonic transport direction, and hence the compression at the bulges does not exactly balance with the relatively smaller extension at the shelfal region.

Detailed modeling and backstripping to the basement level give us the following development history in this part of the basin. Rifting during the Late Jurassic and breakup and drifting during the Early Cretaceous led to the basin opening and deposition of the first sediments in this area. A midbasinal high separated two subbasins to the seaward and landward sides. After deposition of the Upper Cretaceous sediments, subsidence in the basinal side created a shelf edge over the midbasinal high. Until the Eocene–Oligocene, normal sedimentation continued over the shelf and farther basinward. After the Oligocene, shelfal instability led to the formation of a series of listric faults (faults 1-2-3, Figure 14) at the shelf-slope break, making accommodation space for the Miocene deposition. The position of the faults seems to have been controlled by a steep basinal slope at that time and location. The instability may have been induced by tectonic adjustments at that time, resulting in small-scale gravity slides, triggering the faulting. The shelf edge shifted landward because of this faulting. Subsequent sediments covered up the faults, which reactivated upper deposition of the upper Miocene beds; the shale bulges began to grow at the distal end with a soft linkage with the landward faults, detaching along the top Cretaceous horizon. During the Pliocene, the activity of the faults shifted seaward (in the order faults 1, 2, 3), always maintaining a steep slope along the scarp and augmenting the bulges. The Pleistocene marked further offset along the listric faults and consequent growth of the shale bulges. The present-day seabed maintains a steep fault scarp (at fault 1), indicating continued tectonic instability at the shelf break.

In depth profile 4, a steep Cretaceous detachment surface is observed, which becomes flat toward the basinal side. Rapid sedimentation during the Miocene led to the development of an overpressured shale decollement at the top Cretaceous level. Lateral movement started as the gliding force overcame the frictional resistance, which was initiated by the rise in pore pressure within the decollement. A preexisting basement high increased the steepness of the detachment and maintained the gliding force necessary for development of the distal shale bulge; the initiation of the shale bulge may have been marked by a basement escarpment at that location. As the proximal extension basin was replenished with sediments, the tectonic force was maintained and the distal bulge continued to grow. Further extension induced additional shortening as sediment translation was restricted to the landward side of the distal bulge, leading to the formation of a proximal shale bulge. The failure along the shelf is even observed during the present time thereby indicating that the margin is presently unstable and the deformation process is still ongoing.

Seismic Velocity Indicators of Overpressure

The presence of overpressure within the decollement layer can be explained by a study of the velocities across the section. Seismic root-mean-squared (RMS) amplitude velocities were obtained by normal moveout correction of the gathers during two-dimensional processing of the seismic section. Dix inversion provided the means of converting the RMS velocities to interval velocity. A study of these interval velocities within a small zone (zone A, Figure 14) across the profile showed velocity inversion within the decollement layer (Figure 14, inset). Generally, overpressured shale formations exhibited lower P-wave interval velocities as compared to normally pressured sections at the same depth. The pore spaces within the overpressured layers were filled with fluid, which was unable to escape because of rapid sedimentation. This fluid-induced inflation of pore spaces led to an increase in porosity and thereby reduction in bulk density across the layer, resulting in decreased P-wave interval velocity (Kan and Sickling, 1993; Chopra, 2006; Sayers, 2006). Moreover, within this layer, an increase in grain-to-grain contact was also observed. Combinations of all these factors merged to cause the velocity inversion within this zone.

DISCUSSIONS

Development and Localization of Gravity-Linked Structures

Overpressure caused by excessive sedimentation rate seems to be the key factor for the formation of these gravity-driven linked structures. The development of overpressure can be correlated to the rate at which fluid escapes from the sediments. The possible cause for the generation of overpressure is rapid sedimentation over low-permeability shale units, causing disequilibrium compaction. "Disequilibrium compaction is favored as the mechanism to explain overpressure in a number of basins, including the Gulf Coast (Dickinson, 1953), Caspian Sea (Bredehoeft et al., 1988) and the North Sea (Mann and Mackenzie, 1990; Audet and McConnell, 1992)" (Osborne and Swarbrick, 1997, p. 1025). Under normal circumstances with moderate sedimentation rates, the pore fluid trapped within the deposited sediments

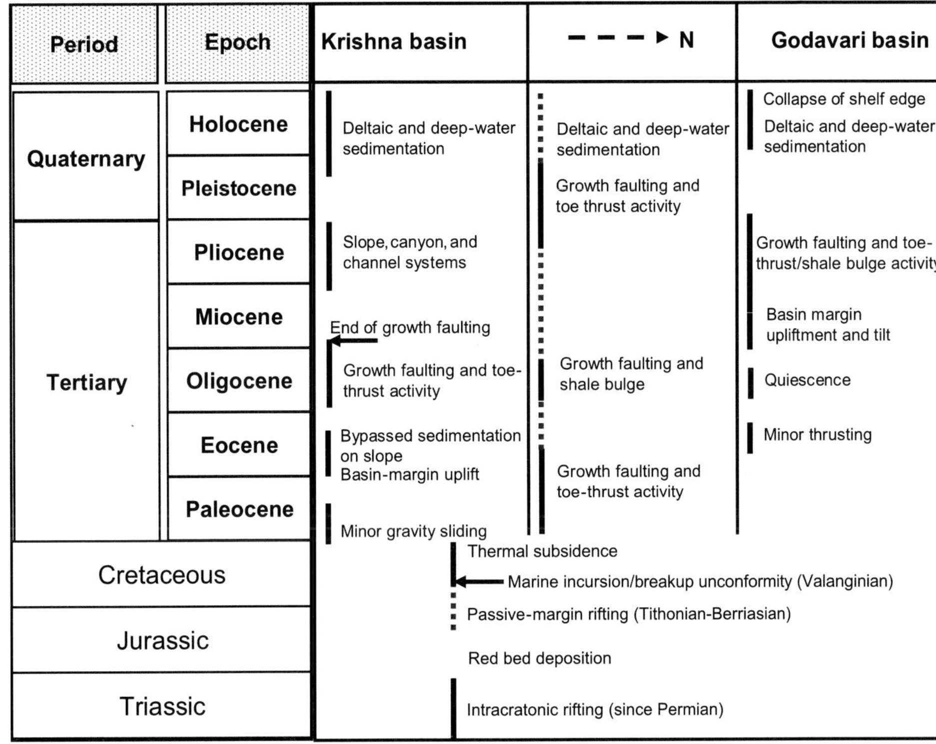

Figure 15. Tectonosedimentary events chart of the Krishna-Godavari basin, showing intracratonic rifting in the Triassic, passive-margin rifting and marine incursion during the Late Jurassic–Early Cretaceous, and the phases of growth faulting and shale bulge activity during the Paleocene–Eocene and Miocene–Pliocene.

is expelled, and the sediments are compacted, following the normal compaction trend. Once sedimentation crosses a threshold rate, the pore fluids do not have sufficient time to be expelled and, consequently, remain trapped within the sediments, building up the pressure within. The increased pressure ultimately leads to a forceful expulsion of the fluids and initiation of the growth faults responsible for these gravity-driven linked structures. In other areas, where the pore pressure is not sufficient to rupture the surface, it leads simply to detachment at the contact with the upper layer, giving a lubricant interface for smooth transmittal of stress, if not a bodily movement of the overlying rocks. Thus, logical reasoning requires that where we find overpressured shales, we would expect growth faults, diapiric shale bulges, or a cohesionless interface for transmitting tectonic stresses, in decreasing order of overpressure. A further fall in overpressure would increase the friction, leading to the formation of toe thrusts.

The slope of the detachment influences the initiation of the gravity slides because of oversteepening at the shelf break. Continuous sediment input increases the slope at the shelf, and when the slope exceeds the angle of friction, gravity gliding occurs, which continues as long as overpressure maintains the frictionless interface along the detachment. The oversteepening of the slope may also be affected by basin-margin uplifts during the early Miocene, or local uplifts, such as and subsidence, all increasing the slope of the detachment.

The localization of the shale bulges and/or toe thrusts seems to correlate with a break in gradient at the upper-lower slope junction or a sudden change caused by change in basement architecture, as most prominently observed in profile 2 (Figure 9). We observed that these contractional structures develop at locations of least gradient, and their localization is possibly a combined result of overpressure and basement structures.

Summary and Conclusion

Taking the Krishna-Godavari basin as a whole, after the rifting event during the Early Cretaceous, quiet sedimentation prevailed up to the early Oligocene, with minor disturbances. Two main events of gravity-linked faulting occurred after the early Oligocene.

The first event of gravity-linked faulting is of Oligocene age and is present in the southern part (Krishna Basin) of the area. The growth faults detach at the Upper Cretaceous–lower Paleocene level, and a huge extension is balanced by spectacular thrust faults, forming typical GFTB systems.

The second event of gravity-linked faulting is of late Miocene to Pliocene age and is mostly present in the northern part (Godavari basin) of the study area. The growth faults detach at variable depths within the Upper Cretaceous or lower Oligocene and produce shale bulges instead of thrust faults. This event is also marked by shelfal instability in the Godavari basin and seems to be mostly a slope-controlled phenomenon.

Overall, it seems that the pulse of extension climbs up from the Oligocene to Pliocene from the southern to northern Krishna-Godavari basin; intermediate areas show mixed events (Figure 15).

From the foregoing discussions, mobile shales are clearly the central factor in the gravity-driven deformation of the Krishna-Godavari basin. The shales, which overpressure during the Tertiary, act as cores for deformed toe-of-slope thrusts, as well as acting as a decollement surface for wholesale gliding of updip strata. The major factors affecting these gravity-driven linked structures in the Krishna-Godavari basin are: (1) loading rate and thickness of the sediment cover above the detachment, (2) pore pressure along the detachment layer, (3) slope of the detachment layer, (4) differential subsidence within the basin, and (5) underlying basement scarps.

Loading rate and pore pressure are key factors in the formation of GFTBs in the Krishna-Godavari basin. The triggering cause seems to be the slope of the detachment layer, with gravity gliding being the dominant mechanism of movement (Rowan et al., 2004).

ACKNOWLEDGMENTS

We are grateful to Reliance Industries Limited-Petroleum Business (Exploration and Production) in general and Achyuta Ayan Misra and Nishikanta Kundu in particular for their sincere help and cooperation during the project.

REFERENCES CITED

Audet, D. M., and J. D. C. McConnell, 1992, Establishing resolution limits for tectonic subsidence curves by forward basin modeling: Marine and Petroleum Geology, v. 11, no. 3, p. 400–411.

Bastia, R., R. J. Singh, and N. Sinha, 2003, Neogene submarine channels and fans in offshore, Godavari, east coast, India, http://www.searchanddiscovery.net/documents/abstracts/2003barcelona/extend/81881.pdf (accessed February 5, 2010).

Bastia, R., N. Sinha, V. Kolla, and R. J. Singh, 2005, Basin evolution and geomorphology of Tertiary deep-water depositional systems, Krishna-Godavari (K-G) Offshore Basin, India, http://www.searchanddiscovery.com/documents/abstracts/2005annual_calgary/abstracts/bastia.htm (accessed February 5, 2010).

Biswas, S. K., and A. Chandra, 1992, Structure and tectonic evolution of the eastern continental margin basins of India, http://www.searchanddiscovery.net/abstracts/html/1992annual/abstracts/0009.htm (accessed February 5, 2010).

Bredehoeft, J. D., R. D. Djevanshir, and K. R. Belitz, 1988, Lateral fluid flow in a compacting sand-shale sequence, south Caspian Sea: AAPG Bulletin, v. 72, no. 4, p. 416–424.

Chopra, S., 2006, Velocity determination for pore-pressure prediction: The Leading Edge, v. 25, no. 12, p. 1502–1515, doi:10.1190/1.2405336.

Dickinson, G., 1953, Geological aspects of abnormal reservoir pressures in Gulf Coast, Louisiana: AAPG Bulletin, v. 37, no. 2, p. 410–432.

Kan, T. K., and C. Sickling, 1993, Predrill geophysical method for geopressure detection and evaluation: Society of Exploration Geophysicists Technical Program Expanded Abstracts, v. 12, p. 703–706.

Kumar, S. P., 1983, Geology and hydrocarbon prospects of Krishna-Godavari and Cauvery basins: Petroleum Asia Journal, v. 8, no. 1, p. 57–65.

Mann, D. M., and A. S. Mackenzie, 1990, Prediction of pore fluid pressures in sedimentary basins: Marine and Petroleum Geology, v. 7, no. 1, p. 55–65, doi:10.1016/0264-8172(90)90056-M.

Mohinuddin, S. K., K. Satyanarayana, and G. N. Rao, 1993, Cretaceous sedimentation in the subsurface of Krishna Godavari Basin: Journal of Geological Society of India, v. 41, no. 6, p. 533–539.

Murty, K. V. S., and M. Ramakrishna, 1980, Structure and tectonics of Godavari-Krishna coastal sedimentary basins: Bulletin of Oil and Natural Gas Corporation, v. 17, no. 1, p. 147–158.

Osborne, M. J., and R. E. Swarbrick, 1997, Mechanisms for generating overpressure in sedimentary basins: A reevaluation: AAPG Bulletin, v. 81, no. 6, p. 1023–1041.

Rao, G. N., 1993, Geology and hydrocarbon prospects of east coast sedimentary basins of India with special reference to Krishna Godavari Basin: Journal Geological Society of India, v. 41, no. 5, p. 444–454.

Rao, G. N., 2001, Sedimentation, stratigraphy, and petroleum potential of Krishna-Godavari Basin, east coast of India: AAPG Bulletin, v. 85, no. 9, p. 1623–1643.

Raju, D. S. N., A. Bhandari, and P. Ramesh, 2005a, Table 27: Relative sea level fluctuations and hydrocarbon occurrences in the Cretaceous and Cenozoic in India (first version), *in* D. S. N. Raju, J. Peters, R. Shankar, and G. Kumar, eds., An overview of litho-bio-chrono-sequence stratigraphy and sea level changes of Indian sedimentary basins: Association of Petroleum Geologists, India Special Publication 1, 213 p.

Raju, D. S. N., P. Ramesh, and S. G. K. Mohan, 2005b, Table 15: Major hiatuses, stratigraphic sequences and their span in time in the Cretaceous and Cenozoic succession of India, *in* D. S. N. Raju, J. Peters, R. Shankar, and G. Kumar, eds.,

An overview of litho-bio-chrono-sequence stratigraphy and sea level changes of Indian sedimentary basins: Association of Petroleum Geologists, India Special Publication 1, 213 p.

Reeves, C., and M. D. Wit, 2000, Making ends meet in Gondwana: Retracing the transforms of the Indian Ocean and reconnecting continental shear zones: Terra Nova, v. 12, no. 6, p. 272–280.

Rowan, M. G., F. J. Peel, and B. C. Vendeville, 2004, Gravity-driven fold belts on passive margins, *in* K. R. McClay, ed., Thrust tectonics and hydrocarbon systems: AAPG Memoir 82, p. 157–182.

Sahu, J. N., 2005, Deepwater Krishna-Godavari Basin and its potential: Petromin, April issue, v. 31, p. 26–34.

Sastri, V. V., R. N. Sinha, G. Singh, and K. V. S. Murti, 1973, Stratigraphy and tectonics of sedimentary basins on east coast of peninsular India: AAPG Bulletin, v. 57, no. 4, p. 655–678.

Sayers, C. M., 2006, An introduction to velocity-based pore-pressure estimation: The Leading Edge, v. 25, no. 12, p. 1496–1500, doi:10.1190/1.2405335.

Sclater, J. G., and P. A. F. Christie, 1980, Continental stretching: An explanation of the post-mid-Cretaceous subsidence of the Central North Sea Basin: Journal of Geophysical Research, v. 85, no. B7, p. 3711–3739, doi:10.1029/JB085iB07p03711.

Smith, B. S., and J. M. Armentrout, 1992, Sequence stratigraphy of a Paleogene shelf margin, IPL Block KG-OS-IV, offshore eastern India, http://www.searchanddiscovery.net/abstracts/html/1992/international/abstracts/1126b.htm (accessed February 5, 2010).

Stuart, C. J., and R. J. Hickman, 2001, Tertiary linked extensional and compressional faults and sedimentation in the Krishna-Godavari Basin, India, http://www.searchanddiscovery.net/abstracts/html/2001/annual/abstracts/0769.htm (accessed February 5, 2010).

7

De Landro Clarke, Wanda, 2010, Geophysical aspects of shale mobilization in the deep water off the east coast of Trinidad, *in* L. Wood, ed., Shale tectonics: AAPG Memoir 93, p. 111–118.

Geophysical Aspects of Shale Mobilization in the Deep Water Off the East Coast of Trinidad

Wanda De Landro Clarke
Ministry of Energy and Energy Industries, Port-of-Spain, Trinidad and Tobago

ABSTRACT

The southeastern Caribbean is a geologically challenging region, and nowhere is this more evident than off the east coast of Trinidad. Rapid and prolific deposition in a deltaic setting and the effects of tectonic interaction between the Caribbean, North American, and South American plates serve to make this a unique area. The unpredictability that characterizes petroleum exploration in this region bears witness to this intricate geologic scenario. Similarities between this extended shelf and the Barbados accretionary prism to the north have been assumed mostly based on scant seismic coverage over this region. Analysis of newly acquired seismic data has revealed a subregion of intermittent shale mobilization, and vast sedimentary deposition exists eastward of the currently producing area. Although a great similarity to the shale belt in the region off the Darien Ridge is observed, this subregion exhibits its own distinct character. Shale flow features with associated bottom-simulating reflectors, syndepositional and postdepositional intrusives, which constrain subbasin development, and extrusive features are apparent. A distinguishing characteristic of this subregion is the association of mobilized shales with thrust faulting within the thick sequence of unconsolidated overburden, providing pathways for shale movement; this has the potential of initiating flow not only of shale but also of hydrocarbons from deeper source rocks. This may explain the presence of thermogenic hydrocarbons shown in this region and may have implications for petroleum exploration in this deep-water area.

INTRODUCTION

Hydrocarbon exploration in Trinidad and Tobago, which began as early as the 1860s, is ongoing today, with recent production rates of more than 140,000 bbl of oil per day. The region offshore of the southeastern coast of Trinidad, the Columbus Basin, has been the site of tremendous petroleum exploration and production activity, with production from several major fields, including Immortelle, Cassia, Poui, Teak, Samaan, and Red Mango. The progression of exploration has led to numerous studies chronicling the geological, geophysical, and petrophysical properties of the area that have all added to exploration success. Several key studies have led to

DOI:10.1306/13231311M933421

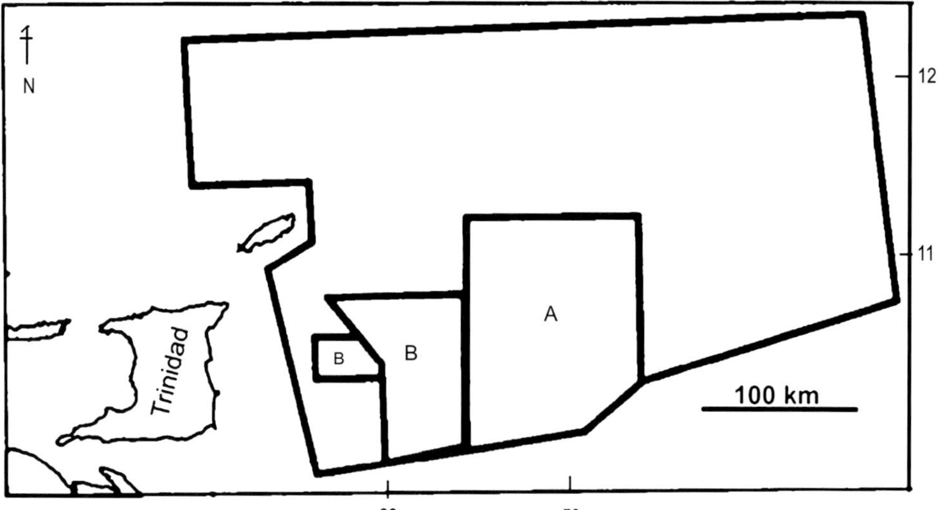

Figure 1. Map of the study area. The large polygon shows the full extent of the regional survey with area A, the location of a dense grid of seismic lines. Area B has been previously reviewed by Moscardelli et al. (2006).

a fuller understanding of the elements of the hydrocarbon system of the Columbus Basin. Gibson (1994) researched fault-zone seals in siliciclastic strata in the Columbus Basin. Baptiste et al. (2005) used seismic modeling to predict the influence of rock properties on the seismic attributes observed in the Columbus Basin. Heppard et al. (1998) investigated the widespread occurrence of abnormal pore pressures in the upper Mesozoic through Tertiary section and its effects on hydrocarbon migration. Finally, numerous field studies of petrophysical properties in the producing section have added to our understanding of proximal rock properties and seismic response (De Landro, 1985).

As exploration has matured in the most western shelfal regions of the Columbus Basin, efforts have moved eastward into more distal parts of the basin, where water depths are greater than 1000 m (3281 ft). The frontier nature of these deeper water regions of the Columbus Basin means that research studies are limited because of lack of available data (Figure 1). However, recent interest in these areas has resulted in collection of new, densely spaced two-dimensional (2-D) seismic data drop cores for geochemical analysis of fluids and sediments; and new gravity data. Academic studies have begun (Deville et al., 2003), and it is hoped that these parts of the basin will soon see a proliferation of knowledge similar to that which characterizes the more western areas.

This article uses recently acquired 2-D seismic data over the frontier, deep-water regions of offshore southeastern Trinidad to document the geophysical aspects of very prominent shale mobilization features in this deep-water environment. Such documentation and subsequent new learnings may provide significant insights into these features and their importance as traps and migration focal points in the very deep-water areas. Similar mobile shale features in more proximal areas of the basin are associated with producing fields in both Trinidad (Forest Reserve field) and Venezuela (Pedernales oil field). They occur as intrusive features (Deville et al., 2003), as shale-cored anticlines (Wood, 2000), and as fault-bounded diapirs (Wood, 2000). This article highlights the similarities and differences that exist between the shale features seen in the more proximal parts of the basin and those of the deeper waters off the east coast of Trinidad.

REGIONAL SETTING

Tectonic activity is a critical factor in defining the geology of the region. Trinidad sits near the convergence of three major tectonic plates: the east–southeastward-moving Caribbean plate; the Atlantic plate, which is subducting westward beneath the Caribbean plate; and the South American plate, which remains fairly stationary and to the south of the Caribbean plate (Figure 2). The interaction of these plates can be simplified into three major phases: (1) the implacement of the Caribbean plate during the Late Cretaceous between the separating north and south American Plates (Donnelly, 1994); (2) the downwarping of the Eastern Venezuelan basin in the Oligocene in response to the subduction of the southern edge of the Caribbean plate under northern South America (Parnaud et al., 1995), which subsequently led to the development of the Maturin and Columbus foredeep basins (Algar and Pindell, 1993); and (3) the right lateral strike-slip motion, which led to the establishment of subduction farther east and the development of the Lesser Antilles forearc complex (Pindell et al., 1998). As a consequence of this tectonic activity, fold and thrust belts, developed in the Columbus Basin

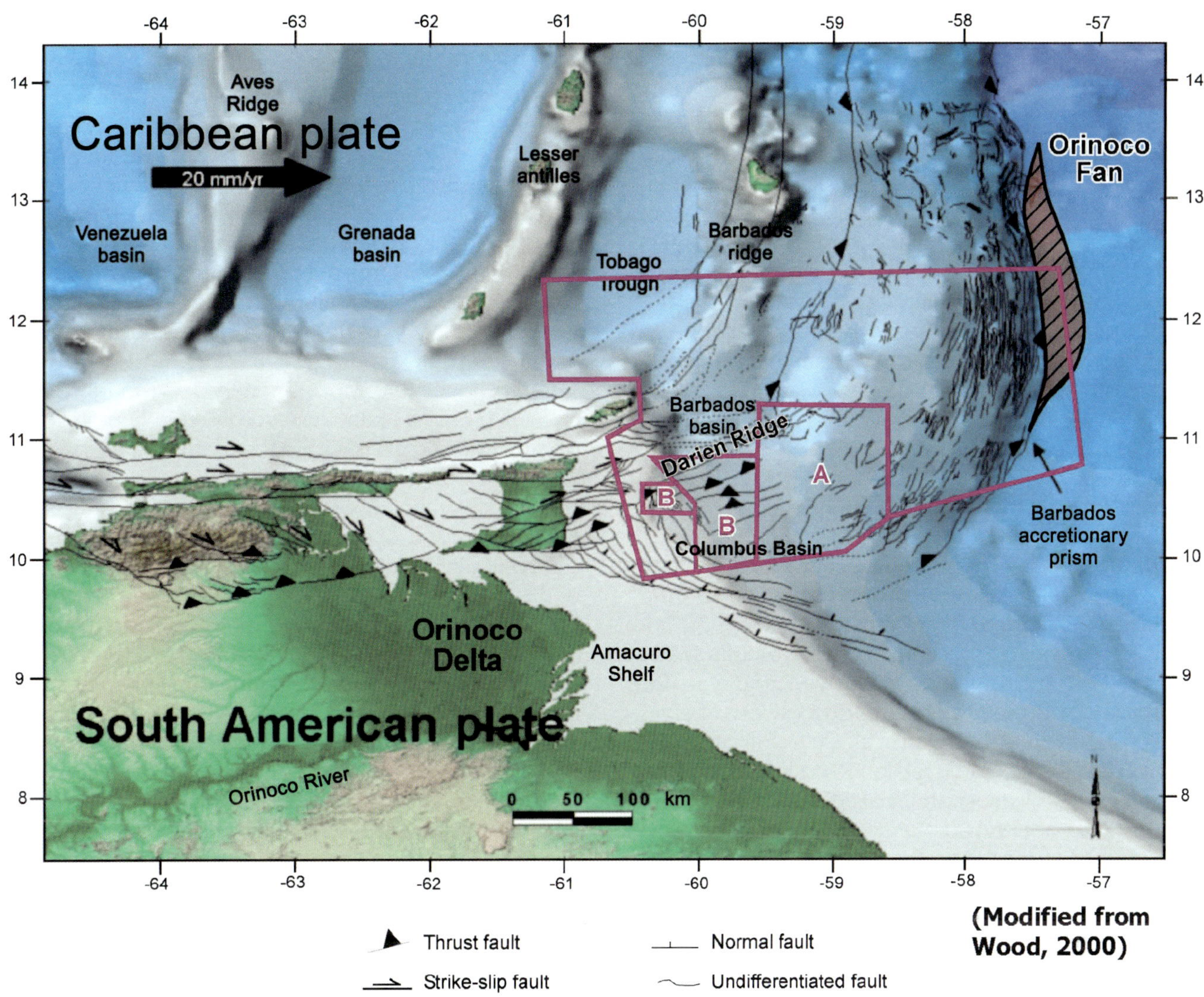

Figure 2. Regional tectonics of the southeastern Caribbean regions showing the location of the study area in offshore eastern Trinidad and Tobago. 50 km = 31 mi.

in the late Miocene to early Pliocene, created sizeable depocenters, which rapidly filled with proto-Orinoco delta sediment (Boettcher et al., 2003). Subsequent uplift during the Pliocene–Pleistocene, as evidenced by raised reefs on the island of Barbados (Chase and Bunce, 1969), re-energized sedimentary inflow and served to further complicate the stratigraphy of this region. Details of the underthrusting of the eastern margin of the Antilles by the western North Atlantic oceanic crust have been outlined as early as 1969 by Chase and Bunce (1969). Making use of low-resolution seismic reflection data, Chase and Bunce were able to delineate the strata of a forearc region as well as the deformation zone of the Barbados ridge. A further study of the area by Deville et al. (2003) relied upon both seismic and Deep Sea Drilling Project/Ocean Drilling Program data to create structure cross sections of the eastern leading edge of the Barbados prism.

Studies of the present drainage basins of the Orinoco and Amazon rivers as well as longshore currents and wave and storm activity (Warne et al., 2002) show the recent depositional setting of the Columbus Basin as a shallow-marine continuum with depositional environments ranging from beach to shoreface to inner and outer shelfal zones. These environments grade laterally into each other and are linked by various physical processes, such as the effects of wave and storm action and tidal and ocean currents, as well as fluctuations in sea level, which together produce a complex sequence of highstand, lowstand, and transgressive deposits (Rine and Ginsburg, 1985; Wood, 2000). These environments are also characterized by facies sequences that become sandier upward, coarse-grained sandstones that fine upward into mudstones, and isolated sandstone bodies, which are bounded by erosional surfaces and enclosed in marine mudstones. Mud and silt, in particular, are washed into the sea by rivers and can be transported for many kilometers in suspension. This present environment, extrapolated into the past, suggests proto-Orinoco and, to a lesser extent, proto-Amazon rivers contributing a copious supply of sediment off a once passive South American margin system, and the fluidized mud and silt transported from the basin being transformed into the mobile shale strata that now characterize this region.

REGIONAL OCCURRENCE OF MOBILE SHALES

Mobile shales have been the subject of much research (see Day-Stirrat et al., 2010). The association of mud volcanoes with surface hydrocarbon shows was the impetus for many early exploration efforts. Mobile shales occur as both intrusive and extrusive features in the study area, and they take on a wide variety of sizes and configurations (Sullivan et al., 2004; Battani et al., 2010). Deville et al. (2003) recognized a large province of active mud volcanoes and shale domes in the southern Barbados accretionary prism that the authors believed extended south into the onshore regions of Trinidad and Venezuela. Sullivan et al. (2004) furthered Deville's observations by documenting more than 160 extrusive mud features on the sea floor using three-dimensional seismic data acquired at the shelf break of the Columbus Basin. Moscardelli et al. (2006), in their study of mass transport complexes in the proximal part of the Columbus Basin (area B in Figure 2), were able to document the influence these features and shale-cored anticlines exhibited in constraining gravity-induced sediment flows in deep water. Some of these flows were more than 200 m (656 ft) thick. Mobile shale extrusions at the sea floor act to influence the extent and direction of these flows. Shale-cored and shale-intruded anticlines in this area can be massive, on the order of hundreds of meters long and oriented in an east–northeastward direction. These uplifts form intervening corridors or chutes through which gravity sediments flow downslope.

Observations from Seismic Data

The region of this study is the eastern offshore deep-water Trinidad and Tobago. Specifically, the area of seismic coverage is bounded to the northeast by the Tobago Basin and the piggyback basins of the Barbados accretionary prism and to the south by the Venezuelan shelf, and it extends eastward to the abyssal plain of the Atlantic plate margin (Figure 2). Twelve thousand five hundred line kilometers of 2-D seismic data were acquired over the study area over two separate phases of acquisition. The first phase of the acquisition consisted of regional lines on a 32 × 16-km (20 × 10-mi) grid spacing. In the second phase, a more detailed survey, 4 × 4 km (2.4 × 2.4 mi), over the southwestern area was commissioned (area A in Figure 1). The regional survey consisted of lines traversing an area north and east of Tobago and extending as far south as the Trinidad-Venezuela border, approximately 10–13°N and 57–62°W. Up to 12 s of data was recorded in hopes of seismically imaging deep within the basin.

Data quality is variable in both a northwesterly and easterly direction. The scarcity of reflectors in the seismic below 5.0 s does not allow for a high degree of confidence in the interpretation of a basement over much of this eastern basin. This lack of penetration is perhaps partly because of a lack of energy transmission through only partially consolidated layers and of poor processing.

Three seismogeomorphic regions are apparent based on seismic imaging. The first is a western region where the 2-D seismic lines show a lack of reflection continuity that is interpreted to be caused by complex normal faulting. Furthermore, extreme difficulties arise in attempts at correlation between upthrown and downthrown sides of faults, even over small distances (zone 1) (Figure 3). The second is a region (zone 3) of clearly identifiable overthrusting and piggyback basin structures characterizing the eastern edge of the study area. Oceanward, this region is dominated by horizontal seismic reflectors of the abyssal plain. Previous authors (Peter and Westbrook, 1976; Speed and Smith-Horowitz, 1995; Deville et al., 2003) noted similar seismic characters off the eastern edge of the Barbados accretionary prism in regions of abyssal plain sedimentation. The third is the middle region (zone 2) of the study area, where seismic data show angular unconformities that mark periods of changing sedimentary and tectonic activity merging into an area of shale mobilization (Figure 3).

Interpretation of Seismic Observations

The seismic evidence indicates that widespread in-situ and mobile shales underlie this extension of the Columbus Basin. Structurally synthetic, antithetic, normal, thrust, and detachment faults are featured in the upper part of the seismic section over the study area. In addition, numerous leveed channel systems are mappable from the seismic data. However, below 4 s, neither these structural nor stratigraphic features can be seen. Seismic sections over mobile shale features illustrate the

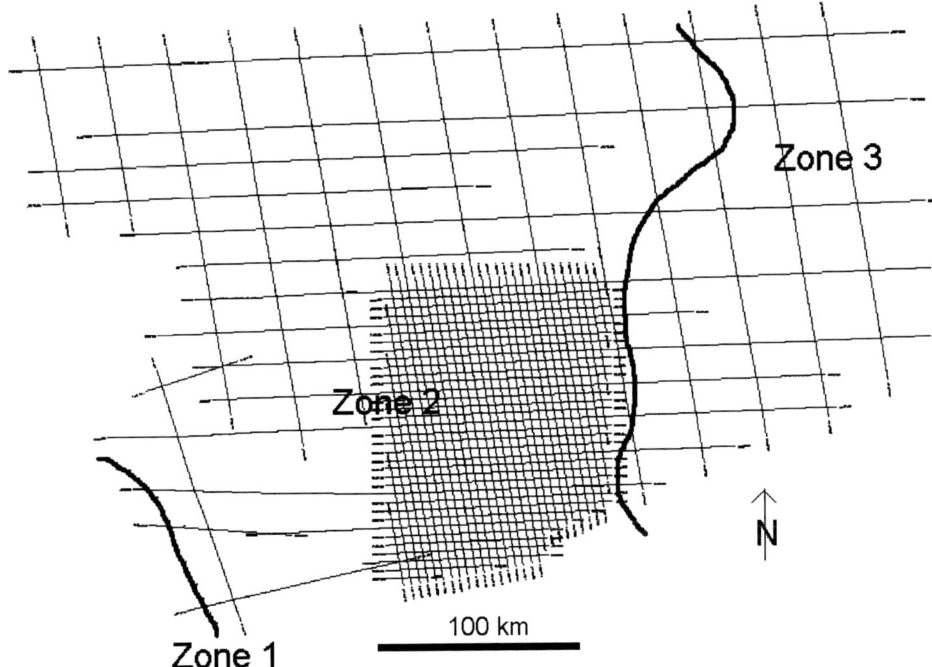

Figure 3. The study area can be divided into three zones. Zone 1 shows a lack of reflection continuity that is interpreted to be caused by complex normal faulting. Zone 2 shows angular unconformities that mark periods of changing sedimentary and tectonic activity merging into an area of shale mobilization in the southeast. Zone 3 is a region of overthrusting and piggyback basin structures; oceanward, this region is dominated by horizontal seismic reflectors of the abyssal plain.

main relationships between shale domes and surrounding strata. For many of the seismic sections, determining the depths at which shale become mobile is difficult. Several areas exist where shale can be speculated to begin to mobilize between 4.0 and 4.5 s, with shale-influenced columns as much as 2.0 s in height extending to the surface.

Several lines indicate the influence of mobile shales on deep-water sediment gravity flows. Line A (Figure 4) shows such a shale feature, acting to constrain a downslope flow that is originating from the west. The angular nature of the reflectors shows that intrusion is keeping pace with deposition. In contrast, basins to the east contain horizontal reflectors, implying that the same intrusive shale feature is predepositional in this basin, with the sediment being sourced from an area outside the plane of the section. The presence of this massive diapiric structure extending from at least 4.5 s all the way to the sea floor indicates that active shale diapirism has been occurring throughout the Pliocene–Pleistocene and illustrates the deep source of many of these muds.

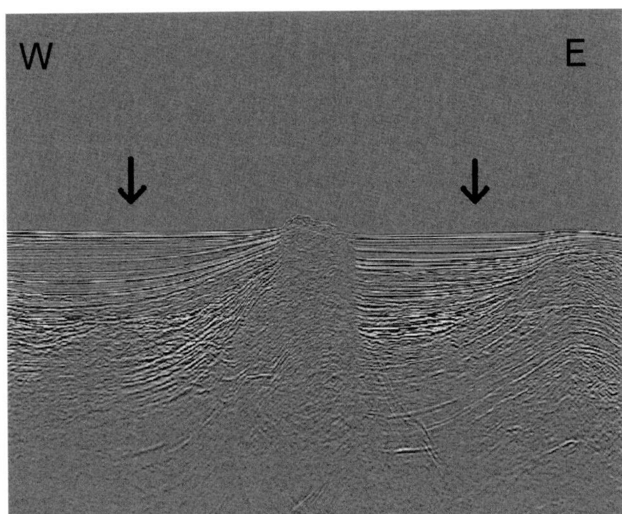

Figure 4. Line A with syndepositional and postdepositional basins; note the changing angular relationship between the strata and the shale diapir.

Mobile Shales and the Occurrence of Gas Hydrates

Gas hydrate is a crystalline solid that occurs in deep-water environments throughout the world's oceans. The volumes of gas, in particular methane, held in these crystalline structures is sufficient for methane hydrates to be viewed by many as a worthy exploration target (Kvenvolden, 1993). The association in the study area of gas hydrate with mud volcanoes was referenced by Deville et al. (2003). The presence of gas hydrate was inferred in the region of the Barbados ridge, from bottom-simulating reflectors (BSRs), and the drop cores have sampled hydrates in near-sea-floor sediments across the region (Sullivan et al., 2004). The BSRs, which mark the boundary in the shallow sea floor between frozen-methane-filled pore spaces above the BSR and free gas

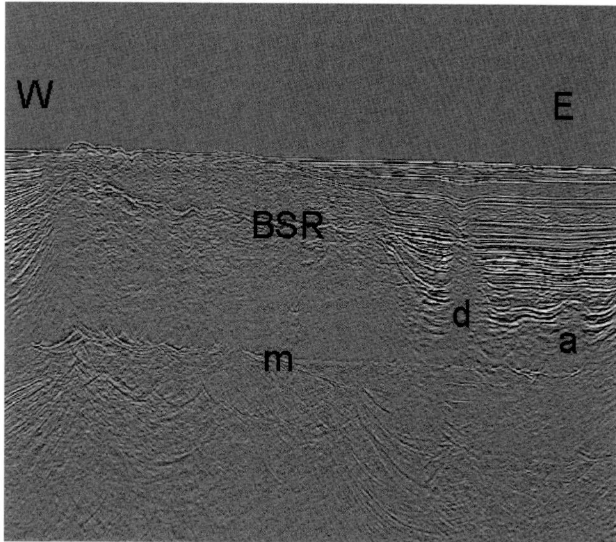

Figure 5. Line B taken from zone 2, showing bottom-simulating reflectors at an angle to the seabed distinguishable from the multiple reflection (m). A shale-cored anticline (a) and a diapir (d) are in close proximity to this zone of chaotic reflectors.

in pore spaces below the BSR, are noted to show a strong association with mud diapiric features.

Line B (Figure 5) shows an example of the BSR at 2.0 to 2.5 s and a multiple at 3.5 s. The presence of shale intrusives may have led to an interruption in the continuity of the BSR. As the shales intrude, they may alter the internal pore pressures and lead to a change in phase of the methane-water clathrate, eventually causing free gas to be released.

CONTRASTING PROXIMAL AND DISTAL SHALE MOBILIZATION

The pattern, areal extent, and degree of intrusion of mobile shale features in the distal areas of the Columbus Basin differ from those exhibited in the more proximal parts of the basin. In distal regions, features associated with mobilized shales appear large, widespread, and commonly linked to the presence of gas hydrates. The seismic data do not provide evidence of deep faults associated with these diapiric structures. These deep faults in more westward locations were cited by Boettcher et al. (2003), as providing a pivotal pathway for seepage of mud and fluids through deep strata to the sea-floor surface. In this region, mobile shale features take on a variety of forms; some are large single diapirs of more than 1 km (0.6 mi) in width and may be as deep as 2 s below the horizontal strata of the minibasins and may or may not extend to the surface (Figure 6). The association of thrust faults with the transport of shale to upper layers is notable, and several examples of the initial stages of development of diapirs in close proximity to thrusts can be seen (Figure 7a, b). These thrust faults appear to originate between 4 and 5 s. Many large massive shale domes of greater than 2 km (1.2 mi) in width are thought to be associated with thrusted folds that have breached the sea floor; remnants of the thrust structures can still be seen in some cases (Figure 7b). Additionally flexural drape and slumping at the apex of thrusted anticlines contribute to the chaotic nature of the reflections.

Evidence of regional decollements linking thrust faults is lacking. Bangs et al. (1996) mapped decollements in the northern part of the Barbados accretionary prism as strong reflectors exhibiting polarity reversal, but similar reflectors are not apparent in this data set, perhaps because these shearing planes of incompetent strata are seismically indistinguishable from the horizon reflections as a result of the poor data quality. Also, the detection of these basal detachment surfaces and their continuation into the thrust faults is reliant on the alignment of the seismic survey. Preliminary mapping of the sediment wedge between the sea floor and basement indicates a northeast trend to the major depositional

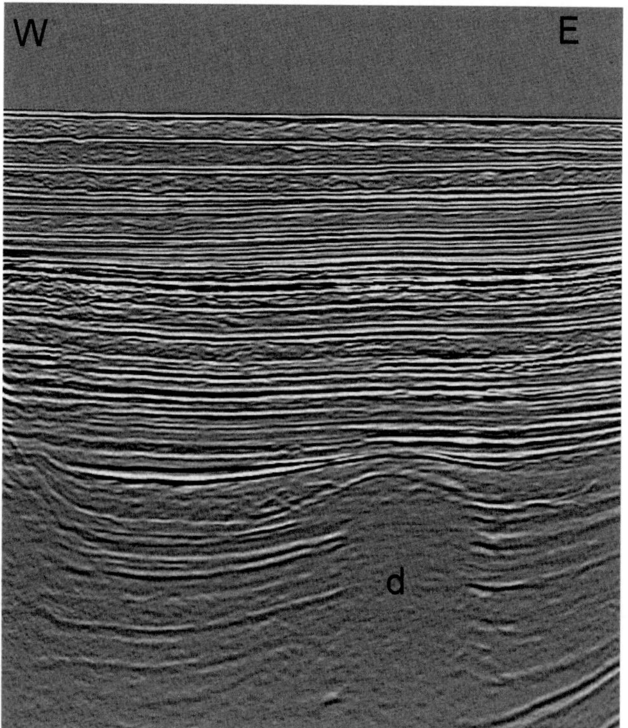

Figure 6. Diapir (d) buried by subsequent sediment deposition taken from zone 2.

(a)

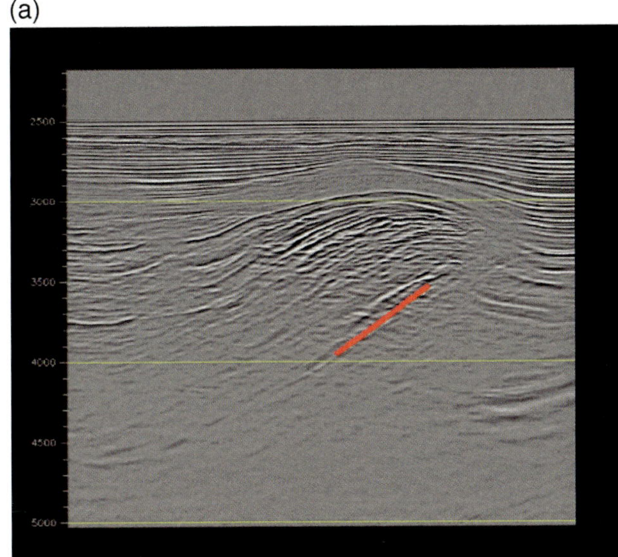

(b)

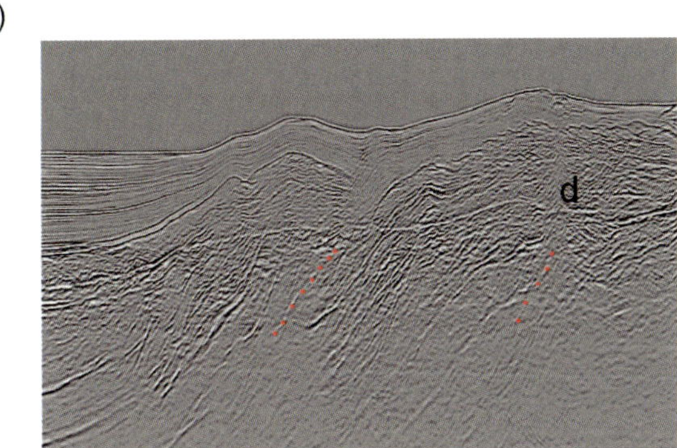

Figure 7. (a) Zone 2, low-angle thrust faulting, with an overthrusted structure being buried by subsequent deposition. (b) Zone 2, thrust fold showing relationship to shale diapir (d).

axis, whereas the seismic survey trends east to west. This alignment of the survey oblique to the strike of the main depositional trend further exaggerates problems of data capture in an area, which is already complicated by rapidly changing lithology and unconsolidated sedimentary strata.

CONCLUSIONS

Although gas has been documented in wells drilled in the deep water off the east coast of Trinidad, to date commercial quantities of liquid hydrocarbons have not been encountered. However, all the elements of the hydrocarbon system (regional rich source rock, structural traps, well-developed sandstones, and appropriate overburden thicknesses) are well documented in the offshore deep-water areas.

ACKNOWLEDGMENTS

I would like to acknowledge my indebtedness to the Trinidad Ministry of Energy and Energy Industries for permitting the use of the data for this article, and offer my grateful thanks to those who assisted me in this endeavor.

REFERENCES CITED

Algar, S. T., and J. L. Pindell, 1993, Structure and deformation history of the northern range of Trinidad and adjacent areas: Tectonics, v. 12, p. 814–829, doi:10.1029/93TC00673.

Bangs, N. L., T. H. Shipley, and G. F. Moore, 1996, Elevated fluid pressure and fault zone dilation inferred from seismic models of the northern Barbados ridge decollement: Journal of Geophysical Research, v. 101, p. 627–642, doi:10.1029/95JB02402.

Baptiste, D., R. Elliott, and S. Linn, 2005, Pushing exploration limits through seismic modeling: A case study from the Columbus Basin, Trinidad: Society of Exploration Geophysicists Expanded Abstracts, v. 24, p. 1775.

Battani, A., A. Prinzhofer, E. Deville, and C. J. Ballentine, 2010, Trinidad mud volcanoes: The origin of the gas, *in* L. Wood, ed., Shale tectonics: AAPG Memoir 93, p. 223–236.

Boettcher, S. S., L. J. Jackson, M. J. Quinn, and J. E. Neal, 2003, Lithospheric structure and supracrustal hydrocarbon systems, offshore eastern Trinidad, *in* C. Bartolini, R. T. Buffler, and J. Blickwede, eds., The circum-Gulf of Mexico and the Caribbean: Hydrocarbon habitats, basin formation and plate tectonics: AAPG Memoir 79, p. 529–544.

Chase, R. L., and E. T. Bunce, 1969, Underthrusting of the eastern margin of the Antilles by the floor of the western North Atlantic Ocean, and the origin of the Barbados Ridge: Journal of Geophysical Research, v. 74, p. 1413–1420.

Day-Stirrat, R. J., A. McDonnell, and L. J. Wood, 2010, Diagenetic and seismic concerns associated with interpretation of deeply buried 'mobile shales', *in* L. Wood, ed., Shale tectonics: AAPG Memoir 93, p. 5–27.

De Landro, W. V. C., 1985, Petrophysical study of the Samaan field shaly sands, offshore Trinidad, *in* K. Rodrigues, ed., Transaction of the 1st Geological Conference of the Geological Society of Trinidad and Tobago, p. 183–198.

Deville, E., A. Mascle, S. H. Guerlais, C. Decalf, and B. Colletta, 2003, Lateral changes of frontal accretion and mud volcanism processes in the Barbados accretionary prism and some implications, *in* C. Bartolini, R. T. Buffler, and J. Blickwede, eds., The curcum-Gulf of Mexico and the Caribbean: Hydrocarbon habitats, basin formation, and plate tectonics: AAPG Memoir 79, p. 656–674.

Donnelly, T. W., 1994, The Caribbean sea floor, *in* S. K. Donavan and T. A. Jackson, eds., Caribbean geology, an introduction: Kingston, Jamaica, UWI Publishers' Association, p. 41–64.

Gibson, R. G., 1994, Fault-zone seals in siliciclastic strata of the Columbian Basin, offshore Trinidad: AAPG Bulletin, v. 78, p. 1372–1385.

Heppard, P. D., H. S. Cander, E. B. Eggertson, 1998, Abnormal pressure and occurrence of hydrocarbons in offshore eastern Trinidad, West Indies, *in* B. E. Law, G. F. Ulmishek, and V. I. Slavin, eds., Abnormal pressures in hydrocarbon environments: AAPG Memoir 70, p. 215–246.

Kvenvolden, K. A., 1993, Gas hydrates—Geological perspective and global change: Reviews of Geophysics, v. 31, no. 2, p. 173–187, doi:10.1029/93RG00268.

Moscardelli, L., L. Wood, and P. Mann, 2006, Mass-transport complexes and associated processes in the offshore area of Trinidad and Venezuela, 2006: AAPG Bulletin, v. 90, no. 7, p. 1059–1088, doi:10.1306/02210605052.

Parnaud, F. Y., Y. Gou, J. C. Pascual, I. Truskowski, O. Galleno, H. Passalacqua, and F. Roure, 1995, Petroleum geology of the central part of the eastern Venezuela Basin, *in* A. J. Tankard, R. Saurez, and H. J. Welsnik, eds., Petroleum basins of South America: AAPG Memoir 62, p. 741–756.

Peter, G., and G. K. Westbrook, 1976, Tectonics of southwestern North Atlantic and Barbados ridge complex: AAPG Bulletin, v. 60, no. 7, p. 1078–1106.

Pindell, J. L., R. Higgins, and J. F. Dewey, 1998, Cenozoic palinspastic reconstruction, paleogeographic evolution and hydrocarbon setting of the northern margin of South America, *in* J. L. Pindell and C. Drake, eds., Paleographic evolution and non-glacial eustasy, northern South America: SEPM Special Publication 58, p. 45–85.

Rine, J. M., and R. N. Ginsburg, 1985, Depositional facies of a mud shoreface in Suriname, South America—A mud analog to sandy shallow-marine deposits: Journal of Sedimentary Petrology, v. 55, no. 5, p. 633–652.

Speed, R. C., and P. L. Smith-Horowitz, 1995, The Tobago terrane: Transactions of the 3rd Geological Conference of the Geological Society of Trinidad and Tobago and the 14th Caribbean Geological Conference, v. 2, p. 467–488.

Sullivan, S., L. J. Wood, and P. Mann, 2004, Distribution, nature and origin of mobile mud features offshore Trinidad: Program and abstracts: Gulf Coast Section SEPM 24th Annual Bob F. Perkins Research Conference, p. 36.

Warne, A. G., R. H. Meade, W. A. White, E. H. Guevara, J. Gilbeaut, R. C. Smyth, A. Aslan, and T. Tremblay, 2002, Regional controls on geomorphology, hydrology, and ecosystem integrity in the Orinoco delta, Venezuela: Geomorphology, v. 44, p. 273–307, doi:10.1016/S0169-555X (01)00179-9.

Wood, L. J., 2000, Chronostratigraphy and tectonostratigraphy of the Columbus Basin, eastern offshore Trinidad: AAPG Bulletin, v. 84, no. 12, p. 1905–1928.

8

Soto, Juan I., Fermín Fernández-Ibáñez, Asrar R. Talukder, and Pedro Martínez-García, 2010, Miocene shale tectonics in the northern Alboran Sea (western Mediterranean), *in* L. Wood, ed., Shale tectonics: AAPG Memoir 93, p. 119–144.

Miocene Shale Tectonics in the Northern Alboran Sea (Western Mediterranean)

Juan I. Soto

Departamento de Geodinámica and Instituto Andaluz de Ciencias de la Tierra (Consejo Superior de Investigaciones Científicas-University of Granada), Granada, Spain

Fermín Fernández-Ibáñez

GeoMechanics International, Perth, Western Australia, Australia

Asrar R. Talukder

Australian Resources Research Centre, Australian Commonwealth Scientific and Research Organization, Kensington, Western Australia, Australia

Pedro Martínez-García

Instituto Andaluz de Ciencias de la Tierra (Consejo Superior de Investigaciones Científicas-University of Granada), Granada, Spain

ABSTRACT

The Alboran Basin is a back-arc basin in the Mediterranean developed during the Miocene by the extensional collapse of the thick continental orogen known as the Betic-Rif arc. Collision and basin formation occurred in the Neogene as a result of oblique convergence of the Eurasian and African plates. A major, curved depocenter (with sedimentary accumulations >10 km [>6 mi]) is located to the west of this basin, containing a diapiric province with overpressured shales and mud volcanoes. This study presents a detailed reconstruction of the three-dimensional geometry of the diapirs and associated minibasins in the northern margin of this major depocenter (offshore Spain). Basin formation began in the early Miocene when rapid initial subsidence of the basin floor was accompanied by massive sedimentation and burial of fine-grained sediments. Gravity-driven tectonics and continuous basement subsidence during the Miocene led to downslope migration of mobile shales, whereas the basin margins were affected by synsedimentary extension, and associated shale-cored thrusts occurred in the basin depocenter. Extension occurred by means of low-angle normal faults coalescing with the basement surface, which represents a master detachment surface. Thin-skinned extension during the Miocene was accompanied by punctuated diapir ascent and the advance of shale sheets. Downslope

DOI:10.1306/13231312M933422

shale advance was enhanced by counterregional high-angle normal faults isolating noncylindrical minibasins in the overburden. The Alboran Basin therefore is a useful area for analysis of the structural pattern associated with shale tectonic processes and a key basin for comparing the geometries and evolution of shale with structures formed in salt basins.

INTRODUCTION

The presence of a large province with mobilized shales and shale diapirs is fairly well known in the Alboran Sea. The major sedimentary depocenter in this basin contains several sea-floor mud volcanoes showing evidence of recent activity, with mobilization and extrusion of fluid-rich sediments. The Alboran Basin developed in the Miocene, caused by the collapse of an Alpine orogen known as the Betic-Rif arc. Shale diapirism in this setting extends the worldwide catalog of shale basins with significant sedimentary accumulations. These basins occur in passive and active margins such as the Niger Delta (e.g., Cohen and McClay, 1996; Morley and Guerin, 1996; Graue, 2000; Hopper et al., 2002), the Gulf of Mexico (e.g., Bruce, 1973; Ewing, 1986), the Baram Delta in northwestern Borneo (e.g., Van Rensbergen et al., 1999; Morley et al., 2008), eastern Indonesia (e.g., Barber et al., 1986; McClay et al., 2000), southern Australia (Totterdell and Krassay, 2003), the Panama thrust belt (Breen et al., 1988), and the offshore Trinidad and Barbados accretionary wedge (e.g., Brown and Westbrook, 1988; Wood, 2000). Mud volcanoes are found in most of these examples (Milkov, 2000; Kopf, 2002), and, sometimes accompanied by shale diapirism, they also occur in compressional settings such as the Mediterranean Ridge (e.g., Ivanov et al., 1996; Robertson and Kopf, 1998; Chamot-Rooke et al., 2005), the Apennines fold and thrust belt (Martinelli and Judd, 2004), and the Caspian Sea (e.g., Planke et al., 2003; Stewart and Davies, 2006).

Shale tectonics is a relatively common term that encompasses any deformation involving overpressured, undercompacted, fine-grained sediments. Compared with the processes involved in salt tectonics (see the recent review of Hudec and Jackson, 2007), shale tectonics also includes deformation driven only by density contrasts, promoting shale upwelling and withdrawal (e.g., Andrews, 1967; Braunstein and O'Brien, 1968; Mascle et al., 1979). Several authors have noted the similarities and differences between salt and shale tectonics (e.g., Morley and Guerin, 1996; Van Rensbergen et al., 1999; Wu and Bally, 2000). Both tectonic styles are, in part at least, examples of gravity tectonics because deformation is caused by downslope migration of a low-density source layer. The most important contrasts between salt- and shale-cored diapirs are as follows: (1) in seismic reflection imaging, shale diapirs show poorly reflective fabrics with chaotic reflections, and the surrounding beds generally lose reflectivity toward diapir borders, defining a diffuse zone with dimmed reflections; (2) shale diapirs show complex geometries, with abundant pipe and finger-like diapirs, and whether allochthonous sheets can develop as in salt basins is a point of debate; and (3) differences in geometry, structural style, and diapir evolution result from the different mechanisms causing flow in the two materials. The physical properties of shale and salt are considerably different in many attributes (e.g., Weijermars et al., 1993; Jackson and Vendeville, 1994). Salt shows moderately constant physical properties, whereas shale density, viscosity, and reflectivity depend mostly on clay mineralogy, fluid content, and temperature-depth conditions (e.g., Chapman, 1974; Bird, 1984; Kopf and Behrmann, 2000; Kopf et al., 2005; Mondol et al., 2007, 2008). Another important difference between both materials is that fluid content in shale (and consequently shale properties) can change during basin evolution and over the duration of active diapirism (e.g., Tingay et al., 2007; Morley et al., 2008). Overpressured shale may induce liquefaction and fluid migration and injection in the overburden during diapirism (e.g., Van Rensbergen et al., 1999; Maltman and Brown, 2003; Morley, 2003).

We present the geometry of shale diapirs in the northern margin (offshore Spain) of the western Alboran Basin (Figure 1), reconstructing their shape in three dimensions (3-D) by means of a dense grid of two-dimensional (2-D) multichannel seismic profiles. The structure of this basin has been exceptionally well imaged by ConocoPhillips (in 2000 and 2001), who kindly provided the seismic data for this study and a selection of previous commercial seismic profiles in the northern part of the basin (Figure 2). Because the current data set is 2-D, the 3-D geometry of diapirs and structures in the overburden should be taken as a tentative reconstruction, geometrically consistent with all the available 2-D sections. The main objectives of our study are (1) to reconstruct the geometry of shale structures and associated minibasins, (2) to characterize the structural style of shale tectonics in this margin of the basin to advance the discussion on the differences and similarities with salt tectonics, and (3) to infer the subsurface shale migration pattern.

Our study represents an advance in characterizing shale tectonics processes in this basin, focusing on their evolution during the Miocene. This analysis of shale

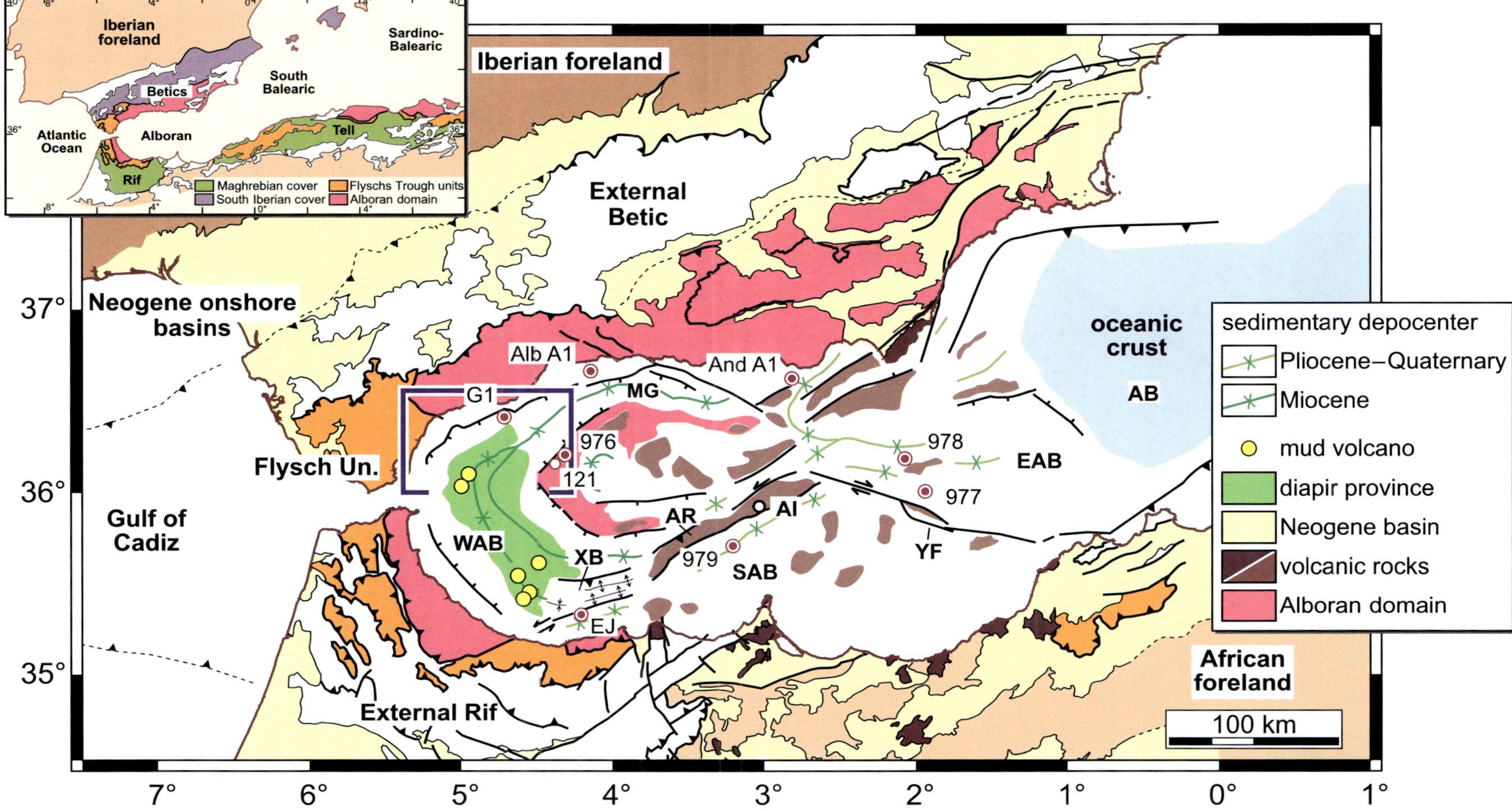

Figure 1. Geological map of the Alboran Sea and major tectonic features of the Gibraltar arc formed by the Betic and Rif mountain chains. The rectangle marks the position of the study area in the northern margin of the western Alboran Basin. The inset map shows the location of the Alboran Sea between the Betic and Rif chains in the western Mediterranean with main crustal domains (modified from Comas et al., 1999). AB = Algerian Basin; AR = Alboran ridge; AI = Alboran Island; EAB = eastern Alboran Basin; MG = Malaga graben; SAB = south Alboran Basin; WAB = western Alboran Basin; XB = Xauen bank; YF = Yusuf fault. Mud volcanoes are shown as yellow dots (Sautkin et al., 2003; Talukder, 2003; Talukder et al., 2003). Locations of Ocean Drilling Project leg 161 sites (976 to 979), Deep Sea Drilling Project site 121, and offshore commercial boreholes are also included (G1 = Andalucía-G1; Alb-A1 = Alborán-A1; And-A1 = Andalucía-A1; EJ = El-Jebha); 100 km = 62 mi.

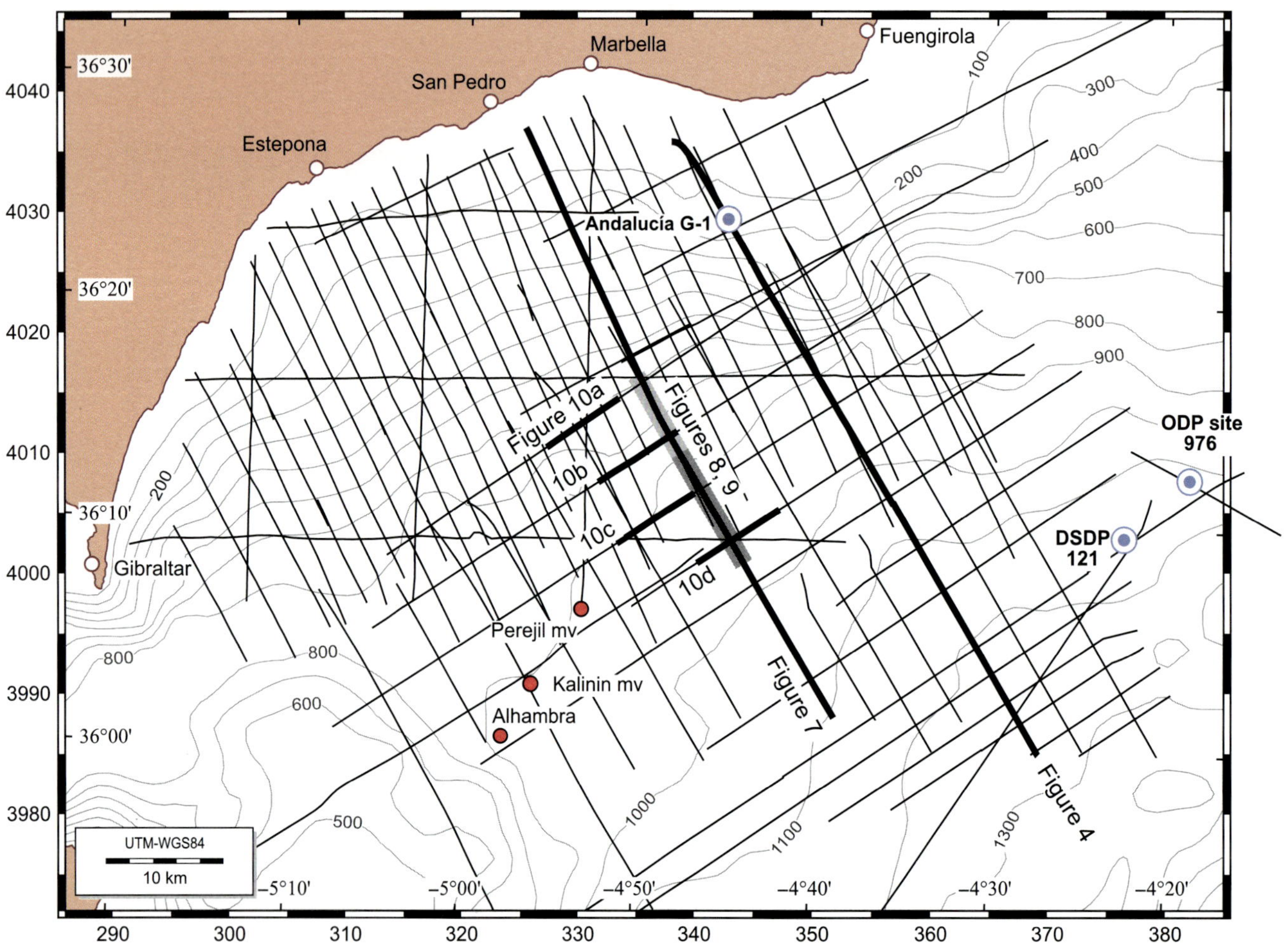

Figure 2. Location map of the complete data set of seismic lines interpreted in the northern margin of the western Alboran Basin (small rectangle in Figure 1) and bathymetry contour lines in meters. Locations of all seismic profiles in other figures, the Alhambra pock mark, and the positions of the Kalinin and Perejil mud volcanoes (mv) (see description in Sautkin et al., 2003; Talukder, 2003; Talukder et al., 2003) are shown. Locations of Ocean Drilling Project Leg 161 site 976, Deep Sea Drilling Project site 121, and the commercial borehole Andalucía-G1 are also included. 10 km = 6.2 mi; 100 m = 328 ft.

deformation and structures based on a recent, better imaged seismic data set complements previous studies done in the region, particularly Morley (1992), Pérez-Belzuz et al. (1997), Kuo et al. (2002), Mountfield et al. (2002), and Weinzapfel et al. (2003). The most recent evolution of shale diapirs and mud volcanoes (during the Pliocene and Quaternary) was depicted by Talukder et al. (2003) by means of sea-floor imaging and a selection of high-resolution seismic lines.

GEOLOGICAL SETTING

The Gibraltar arc is the westernmost Alpine orogen in the Mediterranean, consisting of the Betic (southern Iberia) and Rif (northern Africa) chains, surrounding the major extensional Alboran Sea basin (Figure 1). This region was formed during the Neogene by simultaneous westward migration of the mountain front and late orogenic extension within a general setting of the African and Eurasian plate convergence (e.g., Platt and Vissers, 1989; García-Dueñas et al., 1992; Jolivet and Faccenna, 2000). Geodetic models infer moderate, northwest–southeast contemporary convergence between these two plates in the western Mediterranean (~5.1 mm/yr [~0.2 in./yr]; e.g., Serpelloni et al., 2007).

Although the late orogenic extension is established as superimposed on an earlier collisional orogen (García-Dueñas et al., 1992; Martínez-Martínez et al., 2002; Platt et al., 2003, 2006), the causes for extension are still under debate (e.g., Platt et al., 1998). Different geodynamic models have been postulated to explain the Neogene tectonic evolution of the region, such as subduction (Royden, 1993; Lonergan and White, 1997; Jolivet and

Faccenna, 2000; Gutscher et al., 2002), lithospheric delamination (García-Dueñas et al., 1992; Seber et al., 1996; Calvert et al., 2000), and postorogenic collapse by convective removal of the lithospheric root (Platt and Vissers, 1989; Molnar and Houseman, 2004). Westward slab rollback accommodated by east–west lithospheric tearing along the southern margin of the Alboran Sea has also been proposed recently as an explanation for the strong curvature of the arc, the lithospheric slab seen in tomography, and the low north–south separation between the Eurasian and African plates throughout the Neogene (Duggen et al., 2004; Spakman and Wortel, 2004).

The Gibraltar arc encompasses the following pre-Neogene crustal domains (Figure 1): (1) the southern Iberian and Maghrebian passive continental margins, respectively corresponding to the external zones of the Betics and Rif; (2) the flysch trough units, mainly represented in the Rif and comprising rocks originally occupying troughs over oceanic or very thin continental crust; and (3) the Alboran domain, composed of a thrust stack of three nappe complexes constituting the Betic-Rif internal zones and the floor of the Alboran Sea basin.

The Alboran Sea has three main subbasins: the western Alboran Basin (WAB), the eastern Alboran Basin (EAB), and the southern Alboran Basin (SAB), separated by several ridges, seamounts, and troughs making up a complex sea-floor morphology (Figure 1). One of the most outstanding features is the Alboran ridge striking northeast–southwest and bounded by strike-slip and reverse faults (e.g., Bourgois et al., 1992; Comas et al., 1992, 1999; Fernández-Ibáñez et al., 2007). Major sedimentary accumulations are located in the western Alboran Basin (~12 km [~7 mi]), defining a curved depocenter that mimics the Gibraltar arc (Soto et al., 1996; Weinzapfel et al., 2003). This major north–south depocenter rotates to an east–west direction along the northern margin of the Alboran Sea (Alonso and Maldonado, 1992; Campillo et al., 1992; Ercilla et al., 1992; Soto et al., 1996) (Figure 1). The western Alboran Basin depocenter coincides with a major shale diapir province, approximately 1800 km^2 (~695 mi^2) (Mountfield et al., 2002; Talukder, 2003). Diapirs are formed by undercompacted shales of the lowermost sedimentary units (Comas et al., 1992, 1999; Jurado and Comas, 1992; Watts et al., 1993; Chalouan et al., 1997; Pérez-Belzuz et al., 1997; Kuo et al., 2002; Talukder, 2003; Talukder et al., 2003; Weinzapfel et al., 2003). The nature and age of the source layer are discussed in detail in the following section. In the south Alboran Basin, maximum sedimentary accumulations reach 3 km (1.8 mi), and in the eastern Alboran Basin, at the transition toward the Algerian Basin, the sedimentary thickness is up to 2–3 km (1.2–1.3 mi) (Comas et al., 1997; Booth-Rea et al., 2007; Mauffret et al., 2007).

The tectonic evolution of this basin is characterized by two major deformation episodes (e.g., Comas et al., 1992; Watts et al., 1993; Chalouan et al., 1997; Comas et al., 1999; Mauffret et al., 2007). Basin subsidence accompanied by synsedimentary faulting occurred throughout the Miocene, probably starting in the early Miocene and ceasing during the late Miocene (~9–8 Ma, Tortonian). After this stage, northwest–southeast shortening and associated strike-slip faulting began to develop in the early Pliocene. East–west folds and conjugate east-northeast–west-southwest (transtensive and right-lateral) and northwest–southeast (transpressive and left-lateral) strike-slip faults developed until the Holocene. Recent deformation caused partial inversion of older normal faults and is concentrated along major fault zones like the Alboran ridge-Xauen bank alignment in the central Alboran Sea (e.g., Bourgois et al., 1992; Chalouan et al., 1997) or the Yusuf fault zone in the transition to the Algerian Basin (e.g., Álvarez-Marrón, 1999; Domzig et al., 2006) (Figure 1). In conjunction with active faults onshore, the Alboran ridge-Xauen bank represents a major, complex, left-lateral fault zone extending from the eastern Betics to the Rif, crosscutting the Gibraltar arc (Fernández-Ibáñez et al., 2007; Fernández-Ibáñez and Soto, 2008). This fault zone limits the curved western Alboran Basin depocenter to the south (Figure 1).

STRATIGRAPHY AND SOURCE-LAYER CHARACTERISTICS

Seismic stratigraphy in the study area can be established using the Andalucía-G1 and Alborán-A1 oil wells located on the Spanish shelf of the Alboran Sea (Figures 1, 2); these have provided general chronostratigraphical information on the sedimentary infill of this basin (Comas et al., 1992; Jurado and Comas, 1992; Comas et al., 1999). Detailed biostratigraphic information is only available for the Pliocene–Quaternary sedimentary sequence, penetrated by the Ocean Drilling Program (ODP) at hole 976 (Figure 2) (de Kaenel et al., 1999; Siesser and de Kaenel, 1999). Various regional unconformities defining major seismic units are commonly accepted in the Alboran Sea. Taking the seismostratigraphy established by Comas et al. (1992) and Jurado and Comas (1992) as reference, six major units (I to VI) are commonly used in this basin (comparable to those used by Chalouan et al., 1997, in the southern margin of the basin).

The seismic horizons we have mapped through the studied region in the Alboran Sea are consistent with traditionally accepted discontinuities in the Miocene sedimentary sequence, i.e., below the conspicuous M reflector in the Mediterranean (e.g., Ryan et al., 1973; Cita and McKenzie, 1986). This reflector corresponds to an

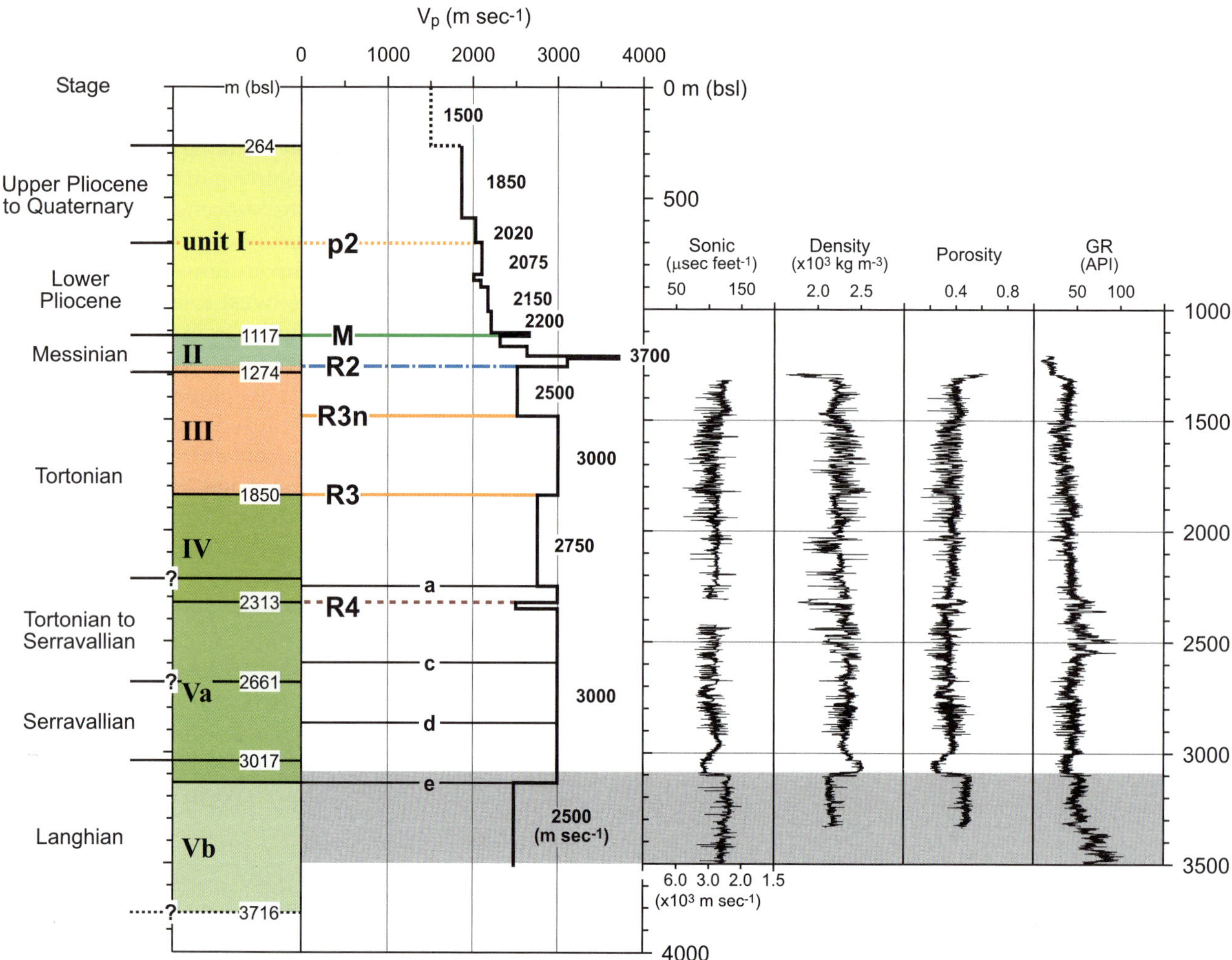

Figure 3. Sedimentary sequences sampled at commercial well Andalucía-G1 (location in Figure 2), including stages, seismic units (Roman numerals), and the interpreted seismic discontinuities. The correlation with logging characteristics is also included. The light gray band marks the sedimentary section with overpressure characteristics. Terminologies for seismic units and major reflectors are taken from Comas et al. (1999). Labels a, c, d, and e correspond to some internal discontinuities distinguished in the northern Alboran Basin. Depths are in meters below sea level (bsl). GR = gamma ray; 1 m = 3.3 ft.

erosional unconformity and constitutes the base of the Pliocene–Holocene sedimentary sequences. The discontinuities and horizons distinguished in this study are presented with a tentative stage attribution in Figure 3. Major seismic discontinuities are labeled following Comas et al. (1999) (R5 to R2 and M), including another in the upper Miocene sequence (R3n, following Talukder et al., 2003).

Unit I (Pliocene to Quaternary deposits) consists mainly of pelagic marls, muddy turbidites, hemipelagic clays, and sparse silty sand turbidites (Shipboard Scientific Party, 1996; Alonso et al., 1999). This unit has a major hiatus within the Pliocene (from 2.1 to 5.0 Ma at site 976), expressed as an angular unconformity between the lower and upper Pliocene seismic units (reflector p2, Figures 3, 4) (e.g., Alonso and Maldonado, 1992; Tandon et al., 1998; Alonso et al., 1999; Haddad et al., 2003; Talukder et al., 2003). Unit II (Messinian deposits) consists of marine and lacustrine sandy turbidites interbedded with finely laminated beds and shallow carbonate facies, with some gypsum and thin anhydrite intervals (Jurado and Comas, 1992; Iaccarino and Bossio, 1999). Rocks from unit III (Tortonian) comprise sandstone intervals, with claystone and silty clay beds, also corresponding to turbidite facies. Unit IV (Serravallian to lower Tortonian) and the upper subunit Va (upper Langhian to Serravallian) consists mainly of graded sand-silt-clay turbidites and turbiditic muds.

The lowermost unit (unit VI) overlies the basement and consists of marine deposits: olistostromes containing polymictic rocks (olistoliths and rock breccia) embedded in an undercompacted shale matrix. Unit VI has only been penetrated at the Alborán-A1 well. The top of this unit is imaged as a seismic event with high reflectivity (R5 reflector). Unit VI and the lower part of unit V (subunit Vb) are made up of clays with interbedded sandy and sandy-pebbly intervals corresponding to high-amplitude intervals in multichannel seismic profiles (Jurado and Comas, 1992). Logging data are consistent with the occurrence of undercompacted shales in unit VI (Alborán-A1 well) and also at the base of unit V (Andalucía-G1 well, Figure 3). The lowermost beds drilled in the Andalucía-G1 well (>3100 m [>10,170 ft]) show lower sonic velocity (average 2455.6 km [1525.8 mi s^{-1}]) and density (average 2176 kg m^{-3}) values and higher porosity (average 47.8%) and gamma-ray (average 60.5 API) values compared to the overlying series. The shale fraction in the lowermost unit (subunit Vb) can be computed using the normalized gamma-ray log through a linear equation (Schlumberger, 2009), resulting in a fraction of 0.14–0.32. In the same interval, the normalized gamma-ray maximum range (73 API) is consistent with a shale-dominated sequence (e.g., Vernik, 2008). Although we suggest that overpressures occurred in this layer, we have not yet computed the magnitude in this well.

The undercompacted shales of unit VI and from the base of unit V (subunit Vb) constitute the source layer for shale diapirs in the western Alboran Basin (e.g., Figure 4). The age of unit VI is still a matter of debate. Jurado and Comas (1992) first suggested a probable latest Aquitanian (?)-Burdigalian age. Sautkin et al. (2003) described extruded sediments sampled from several mud volcanoes in the region (Woodside et al., 2000; Comas et al., 2003; Talukder et al., 2003). These volcanoes root directly in the lowermost seismic unit (unit VI), and mud fluxes consist of matrix-supported breccia with assorted clasts (limestone, claystone, and sandstone) (Fernández-Soler et al., 2000). According to Sautkin et al. (2003), the matrix contains abundant lower Miocene calcareous nannofossils incorporating Upper Cretaceous and Paleocene–Eocene materials and basement-derived rock fragments. On the basis of these data, we will assume that unit VI is lower Miocene and that overpressure could also occur in the lower part of unit V (subunit Vb) with a probable Langhian age (Figure 3).

STRUCTURE

The principal features related to shale deformation in the Alboran Sea are clearly illustrated in the regional seismic line of Figure 4 and in the contour map for the shale diapir surface in Figure 5. Structural maps for the basement surface and the key regional discontinuities R4 (base of unit IV [Serravallian to lower Tortonian]), R3 (base of unit III [Tortonian]), and M (base of unit I [Pliocene–Quaternary]) are shown in the different panels of Figure 6. The basement map in Figure 6a also includes some key features of different fault surfaces, such as their tip lines, together with hanging-wall and footwall cutoff lines.

The seismic section shown in Figure 4 runs parallel to the dip direction of the basement surface (Figure 6a) and shows the evolution of shale geometries from the basin margins to the depocenter axis. Maximum sedimentary accumulation occurs around a subsidiary basement high drilled by ODP at hole 976 and is constituted by high-grade metamorphic (metapelite) rocks (Soto and Platt, 1999). This high corresponds to a horst structure, uplifted by high-angle normal faults in the Miocene and Pliocene (Comas and Soto, 1999). Taking the regional section in Figure 4 as a reference, in the following sections, we describe shale structures from basin margins to basin depocenter, following the dominant basement-dip direction that represents a probable direction for shale migration during deformation.

Peripheral Structures

Along the northern margin, shale structures show fault welds and asymmetric forms, limited by rollover and growth faults (shale rollers). The landward edge of the shale defines a complex curvilinear pinch-out (Figure 5). It lies near or is displaced by several listric faults that coalesce into the basement surface (e.g., shot points 1100–1300 in Figure 4), thus representing a clear example of a thin-skinned extension root in a basement-cover detachment surface. These fault families (peripheral faults in the sense of Rowan et al., 1999) dip congruently with the southern and southeastern basement downdip direction. In plan view, they have a curved shape (concave in the downdip direction) with an echelon pattern, showing overlapping fault segments and lateral tips (Figures 5, 6). The geometrical relationships between these faults and the updip shale pinch-out suggest that faulting extended the original distribution of the source layer, promoting basinward migration of the shale.

Faulting occurred throughout the Miocene with several extensional pulses, resuming in the upper Tortonian, i.e., post-R3 (Soto et al., 2003).

Shale Anticlines and Walls

Downdip, peripheral faults pass to a smooth shale depression that coincides with an open syncline for the Miocene sequences (e.g., between shot points 1400 and 1600 in Figure 4). The angle between fold limbs diminishes progressively up section, vanishing in the uppermost layers of unit III (Tortonian). The R2 discontinuity

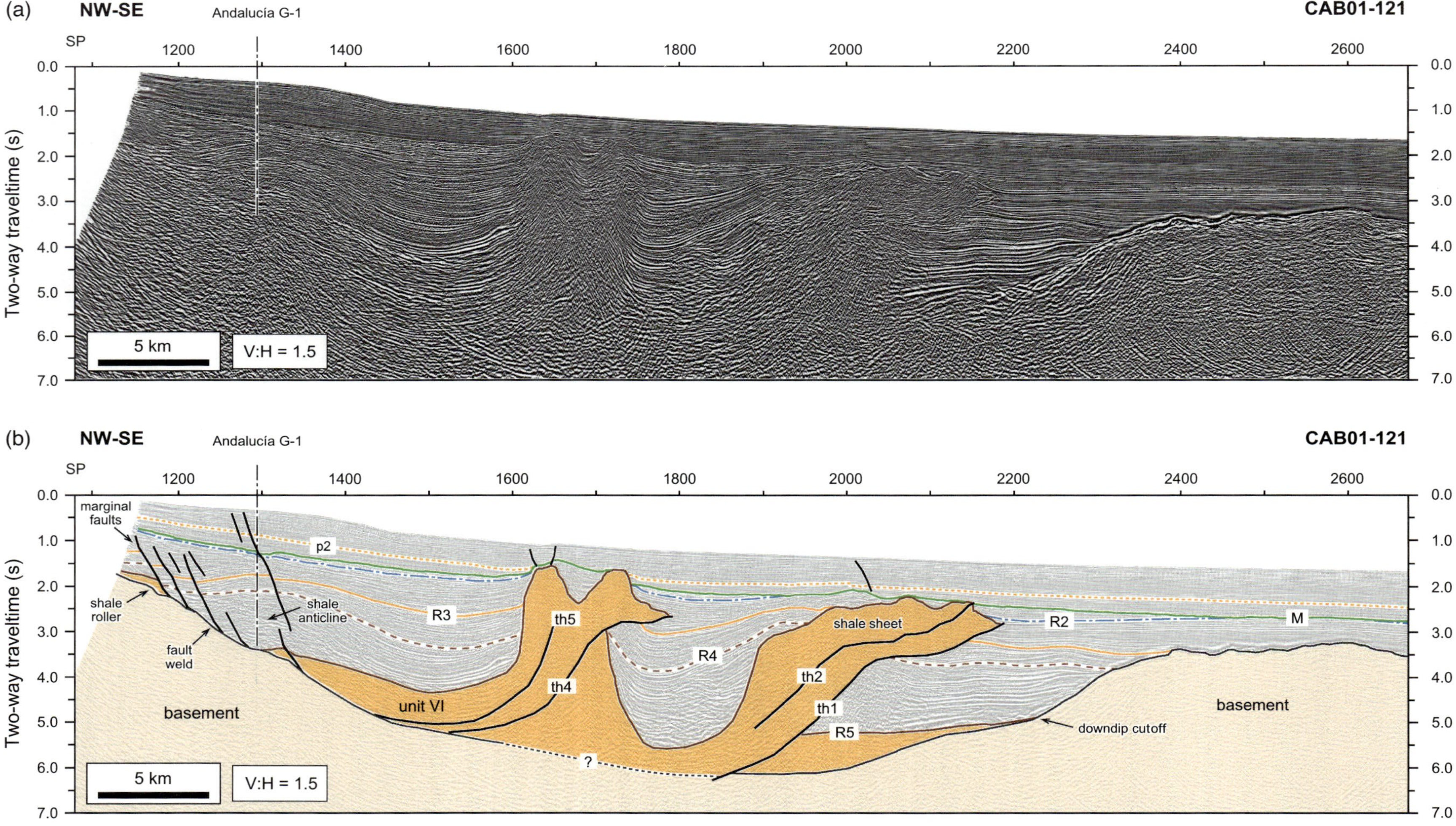

Figure 4. (a) Uninterpreted and (b) interpreted regional seismic section along the northern margin of the western Alboran Basin, showing the structural evolution of shale from basin margin to basin depocenter. Marginal structures consist of growth faults, rollover anticlines, and shale rollers, passing downdip to an open shale-cored anticline. Coinciding with the basin depocenter is a shale sheet. An open turtle-structure anticline is observed in front of the sheet and above the basement high, thus reflecting an out-of-section migration of shale. The age and color code of the major seismic discontinuities are shown in Figure 3. Interpreted shale thrusts are labeled th1–th5. Figure 2 shows the line location. Seismic data are courtesy of ConocoPhillips. V:H = ratio of vertical and horizontal scales; 1 km = 0.6 mi.

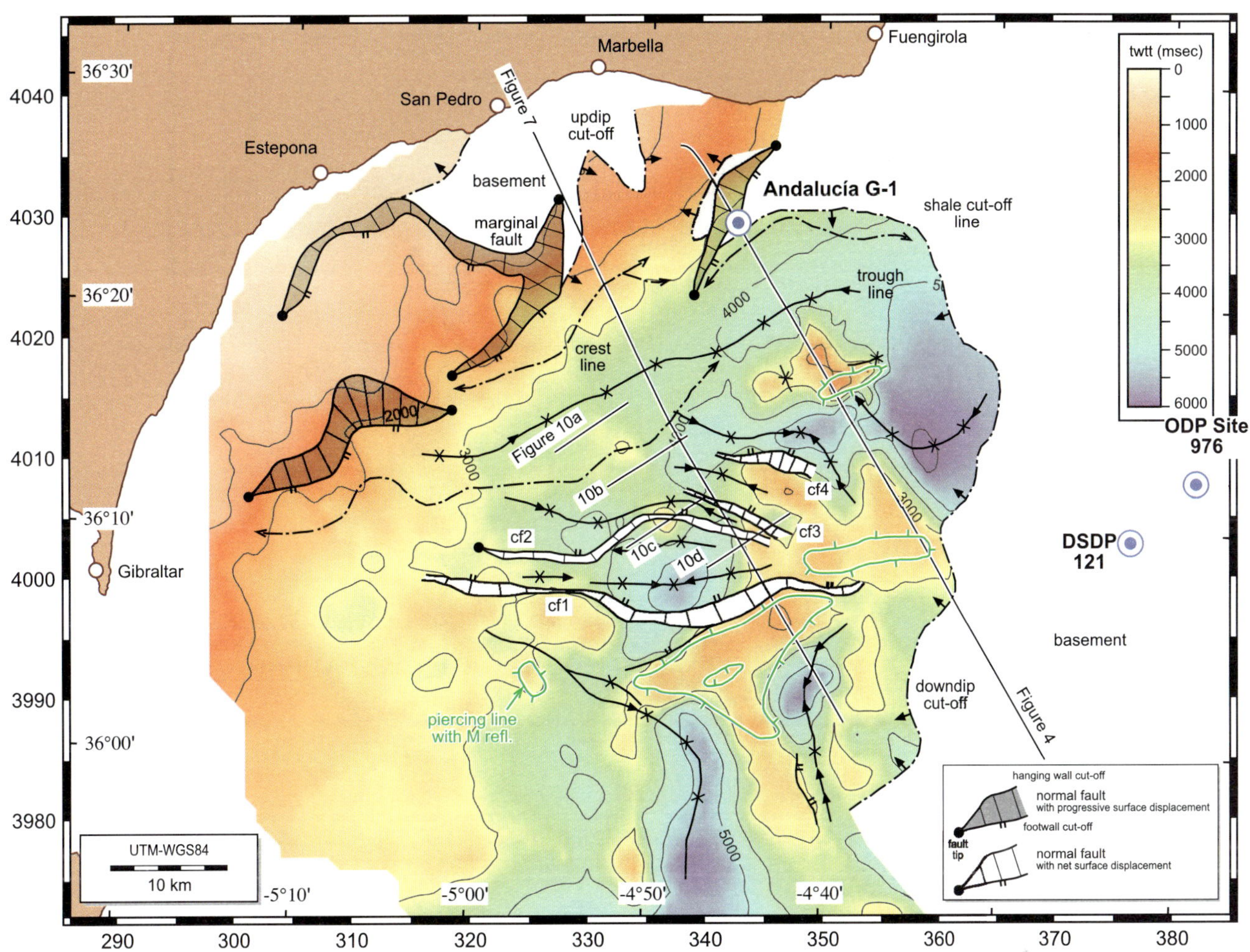

Figure 5. Structural contour map of the shale surface (R5 discontinuity) corresponding to the top surface of unit VI. The contour interval is 1 s (two-way traveltime [twtt]). The location of all seismic profiles is shown (Figures 4, 7, 10). Major normal faults associated with diapir collapse are labeled cf1–cf4. Ticks along the intersection line with the M reflector (labeled piercing line, closed green contour) point in the direction of diapir flanks where the Miocene overburden sequences are detected. 10 km = 6.2 mi; ODP = Ocean Drilling Project; DSDP = Deep Sea Drilling Project.

(base of unit II [Messinian]) and the M reflector seal this open fold. The axial trace of this structure has a curvilinear shape parallel with the horizontal direction of the basement surface (Figures 5, 6a). We have not yet analyzed the detailed folding history of this structure, but Figure 4 shows that the anticline crest migrated landward. In all, these observations suggest a synsedimentary fold accompanying downslope shale migration during the middle to late Miocene.

This structure evolves downslope forming a series of shale-cored anticlines and synclines. In the same direction, large shale walls with steep, faulted flanks are observed (Figure 7). The shape of the shale-core anticlines and peripheral sedimentary troughs is noncylindrical with the following characteristics: (1) orientation of fold axes progressively changes downdip from northeast–southwest to southeast–northwest; (2) fold axial traces are markedly curvilinear, becoming linear through time (cf. Figure 6b–c); (3) peripheral synclines evolved laterally as isolated and elliptical troughs, defining sag minibasins; and (4) shale walls show two perpendicular orientations in the deeper part of the basin, and more mature cylindrical shale plugs are developed at the intersections (Figure 5). In a basement downdip direction, shale-cored anticlines and walls become younger and longer lived structures (e.g., at shot points 1600–1700 in Figure 7).

Sedimentary sequences appear congruently folded in synforms between shale-cored anticlines and walls, with local unconformities along the flanks of the shale diapir, similar to the structures known as halokinetic sequences in salt basins (e.g., Giles and Lawton, 2002; Rowan et al.,

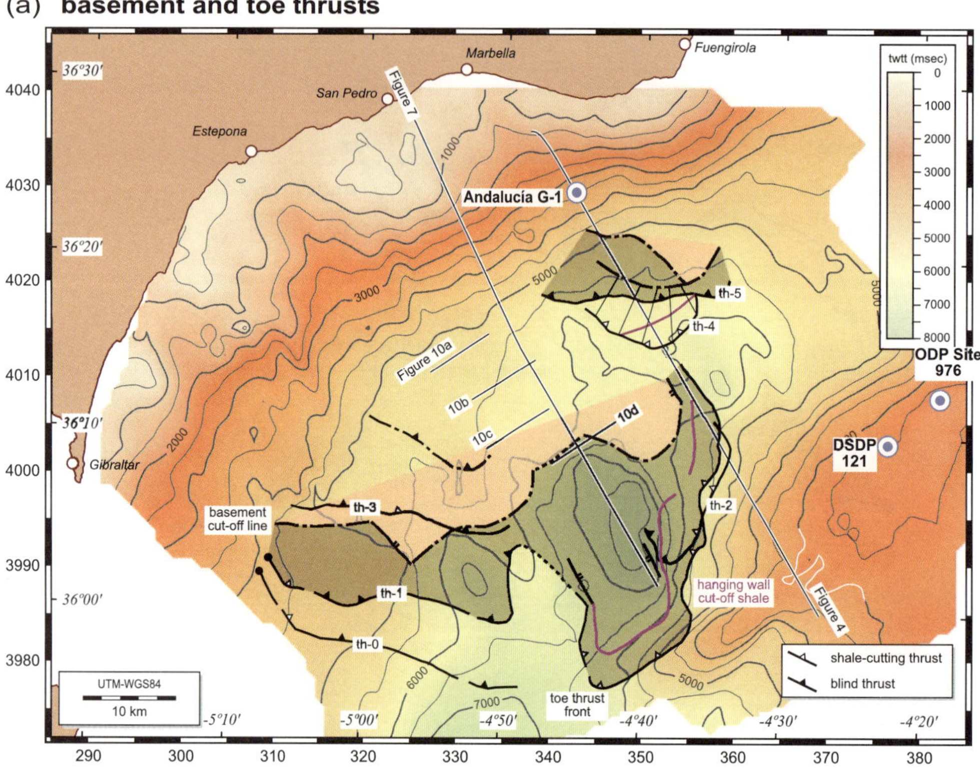

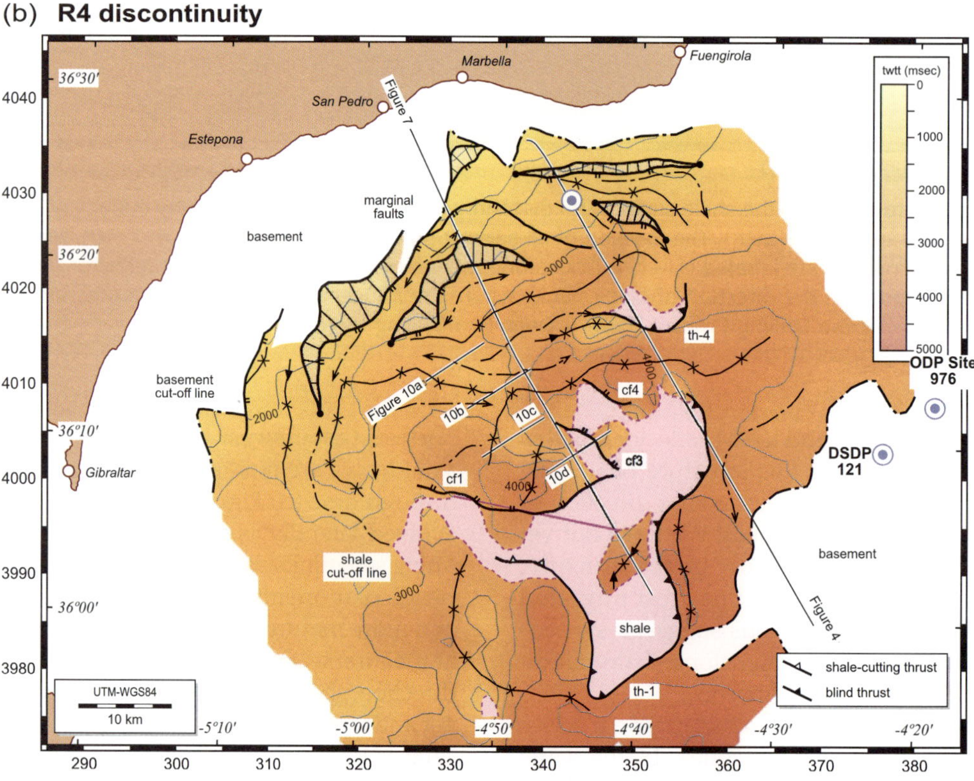

Figure 6. Structural contour maps of (a) the basement surface, (b) the R4 discontinuity, (c) the R3 discontinuity, and (d) the M reflector. The basement map includes structures of shale thrusts (th-0 to th-5). Gray areas mark the fault surface of the shale thrusts th-1 and th-5. Their downwards continuation, within the basement, are marked in light orange. In (b) and (c), cf1 to cf4 correspond to major normal faults associated with diapir collapse. Domains where the mapped surface is cut by the shale and the basement are marked in pink and white, respectively.

(c) **R3 discontinuity**

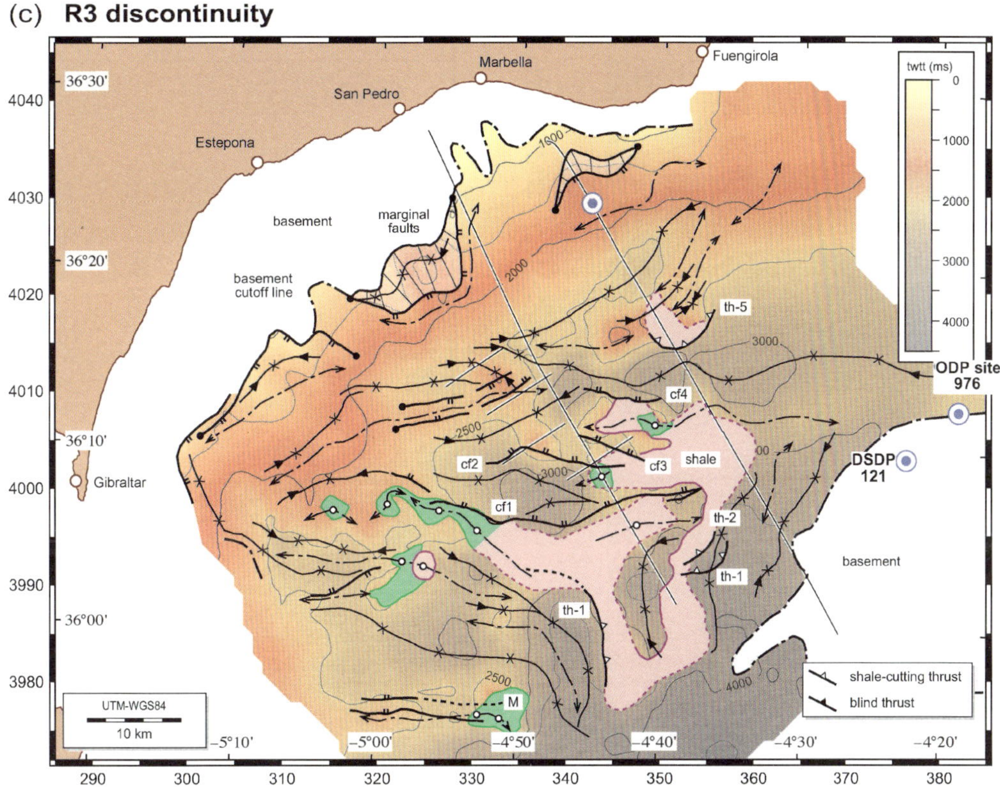

(d) **M reflector**

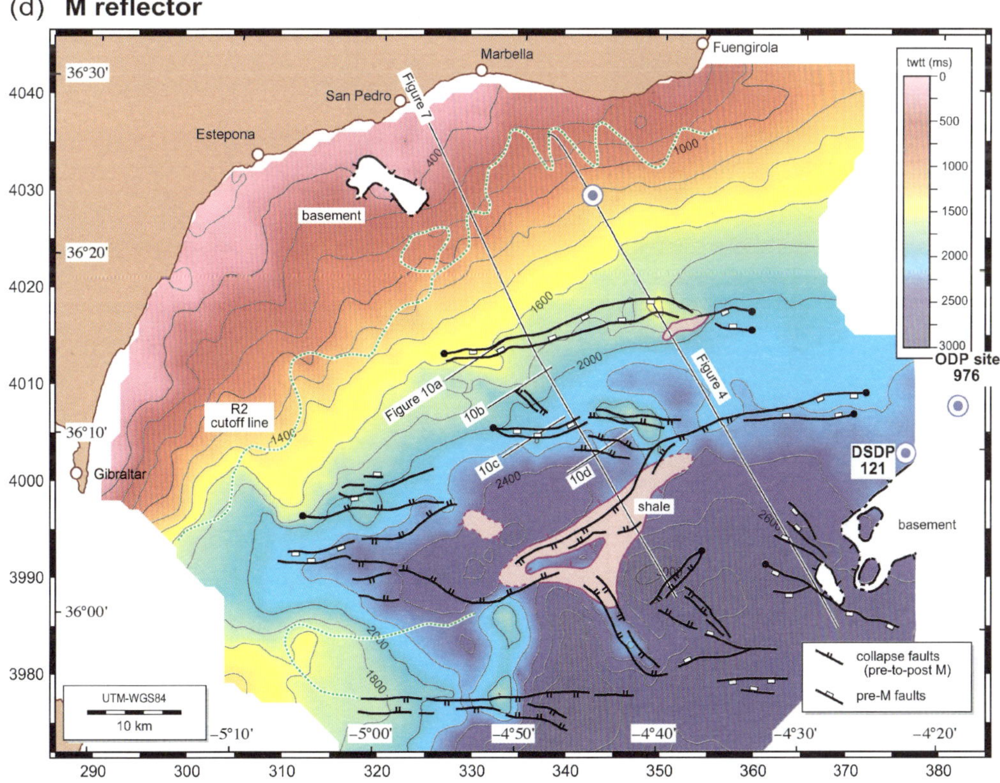

Figure 6. (cont.). Areas in green (labeled with M in panel c) represent shale culminations where the R3 discontinuity is cut by the upper, M reflector. Eroded fold culminations along anticline crest lines are indicated with open circles in (c). Contour intervals in (a) to (c) are 500 ms (two-way traveltime [twtt]) and in (d) is 200 ms (twtt). The location of all seismic profiles is shown (Figures 4, 7, 10). Other symbols are as in Figure 5. 10 km = 6.2 mi; ODP = Ocean Drilling Project; DSDP = Deep Sea Drilling Project.

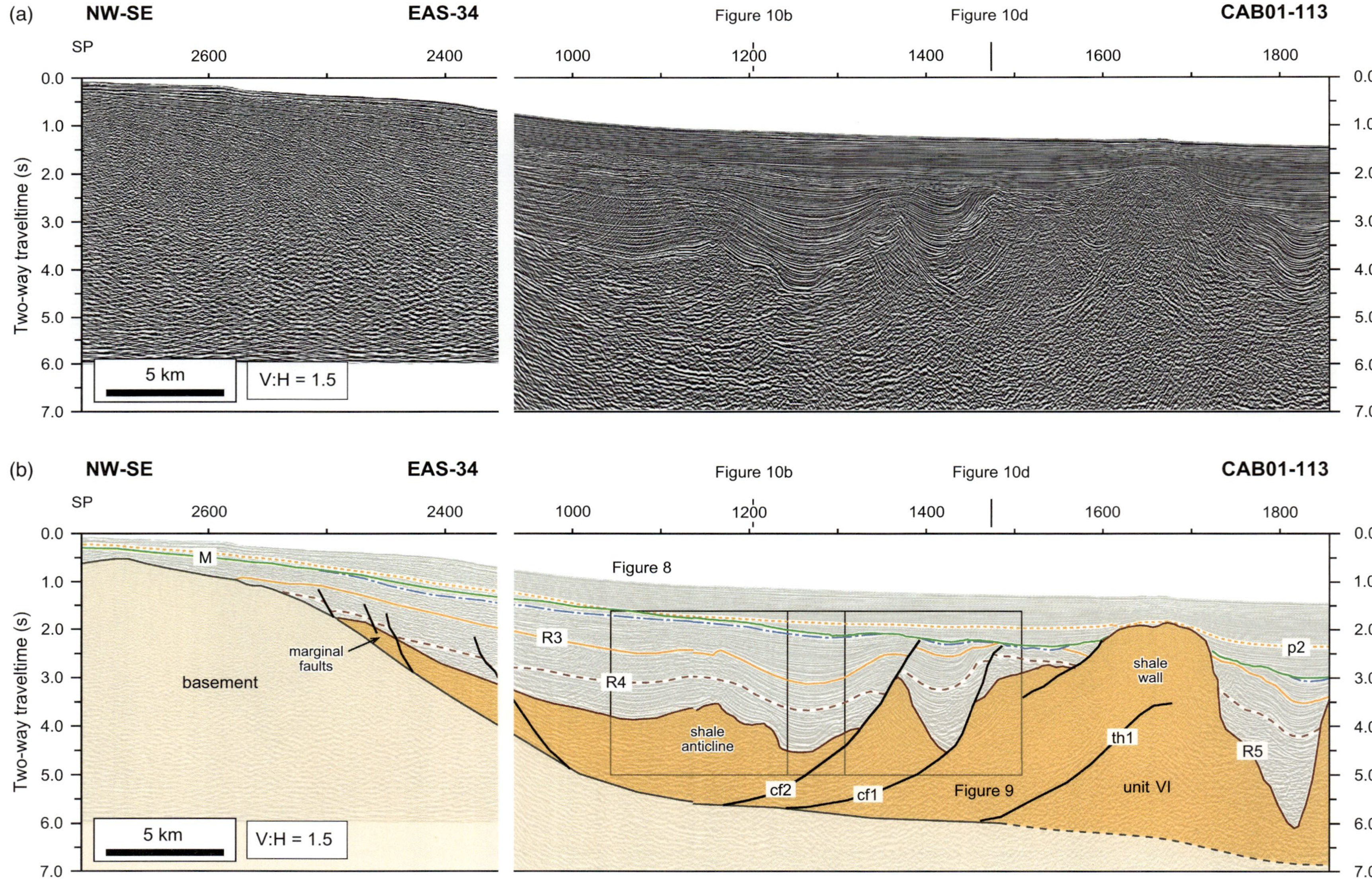

Figure 7. (a) Uninterpreted and (b) interpreted composite regional seismic section along the northern margin of the western Alboran Basin. Th-1 and cf1 to cf2 correspond, respectively, to major shale thrusts and normal faults associated with diapir collapse. Figure 2 shows the line location. Seismic data are courtesy of ConocoPhillips. V:H = ratio of vertical and horizontal scales; 5 km = 3.1 mi.

2003). The relationships between the Miocene overburden and diapirs have two extreme geometries: (1) sedimentary sequences are slightly deformed and uplifted toward diapir flanks, which show discordant piercing shapes (e.g., around shot point 1600 in Figure 4); and (2) they are progressively and congruently deformed and folded above shale-cored anticlines (e.g., around shot point 1750 in Figure 4). These geometries reflect a complex history of diapirism, with a simultaneous active and passive rise in nearby shale diapirs (Vendeville and Jackson, 1992a, b). They also reflect contrasting rates between diapir rise and sediment aggradation, which vary significantly between adjacent and nearby minibasins (e.g., Jackson et al., 1994; Talbot, 1995; Giles and Lawton, 2002). For example, average sedimentation rates during the Miocene (Langhian to late Tortonian, corresponding to units V to III, i.e., ~8.7 m.y.) in minibasins like those shown in Figure 7 range between approximately 0.3–0.4 mm yr^{-1} (0.011–0.015 in. yr^{-1}) (shot point 1250) and 0.6–0.7 mm yr^{-1} (0.023–0.027 in. yr^{-1}) (shot point 1800).

Regarding timing, these structures clearly show that diapirism becomes younger in a basinward direction. Shale-cored anticlines vanish within the lower sequences of unit III (middle? Tortonian). Shale walls pierce the M reflector and fold the upper part of unit I, above the p2 discontinuity (i.e., upper Pliocene–lower Quaternary).

Sedimentary sequences in discordant diapir flanks show examples of thickening wedges. This geometry is consistent with growth faulting, which is linked to diapir rise (reactive diapirism; Vendeville and Jackson, 1992a; Jackson and Vendeville, 1994). This geometry is observed below the R4 reflector, suggesting that diapirism started during the middle Miocene (pre-late Serravallian; Weinzapfel et al., 2003). We have not observed the typical swarms of normal faults at deep flanks of shale diapirs as observed elsewhere in association with similar salt structures (e.g., Seni and Jackson, 1983; Jackson et al., 1994; Rowan et al., 1999). The lack of such fault swarms has also been described in association with other shale diapirs (e.g., Van Rensbergen et al., 1999). The lack of these faults can occur when a low-viscosity contrast between overpressured shale and the overburden beds exists (Koyi and Skelton, 2001) or if this structural domain was absorbed by subsequent shale diapir rise.

Shale Sheets

Overhanging shale sheets occur in the basin domain with maximum sedimentary accumulations (Figures 4, 5). Shale sheets are driven by reverse and blind thrusts, and invariably maintain a connection with the source layer. They therefore constitute autochthonous sheets in the sense of Wu et al. (1990) and are precursors to the more evolved allochthonous sheets described in salt basins (Diegel et al., 1995; Schuster, 1995; Rowan et al., 1999; Hudec and Jackson, 2006). Thrust faults exhibit important ramps with moderate dip (~28–32°) and small subhorizontal flats (~2–3 km [~1.2–1.3 mi]). We estimate a maximum horizontal displacement of about 9 km (5.5 mi) for shale sheets.

Thrust cutoff lines show a marked curvilinear shape and develop progressively landward-concave shapes toward the depocenter axis (Figure 6). Remarkable changes in the orientation of thrust cutoff lines occur in this basin domain coinciding with a steep, lateral step in basement surface (Figure 6a). The resulting geometry could correspond to frontal and lateral ramps within an overthrusting shale sheet. Although seismic resolution decreases in the lowermost levels, we envisage that deep ramps in the basement surface conditioned the resultant geometry of shale sheets, also reflecting a transition from thin- to thick-skinned tectonics (e.g., Jackson and Vendeville, 1994; Vendeville et al., 1995). In detail, toe thrusts are formed by imbricate fault surfaces, and the overall geometry suggests a landward- (and upward-) faulting sequence (e.g., Figure 4). The inferred relative timing of thrusting is included for reference in the toe-thrust labels (Figure 6a).

Some open folds with progressive unconformities (below R3 and R4 reflectors) are observed in front of the blind toe thrust. These structures resemble turtle-structure anticlines, as commonly described in salt basins (e.g., Seni and Jackson, 1983; Mauduit et al., 1997). These shallow turtle-structure anticlines indicate an out-of-section migration of the shale to laterally feed the allochthonous sheet (e.g., Jackson and Cramez, 1989). This process occurred mainly during the middle Miocene (Serravallian), because turtle anticlines are more clearly seen below R4.

Crestal Faults

Shale diapir culminations show contrasting structures and geometries in the shale roof, varying from a progressive thinning and folding of the overburden (e.g., at shot points 1100–1200 in Figure 7) to discrete normal faults dipping landward (i.e., to the northwest and north; between shot points 1300 and 1500 in Figure 7). Crestal structures (following Rowan et al., 1999) in shale roofs differ also between shale-cored anticlines and shale walls; i.e., they vary following the basement downdip direction. Symmetrical grabens limited by conjugate crestal faults are commonly observed above shale-cored anticlines (Figure 8). These faults have moderate fault

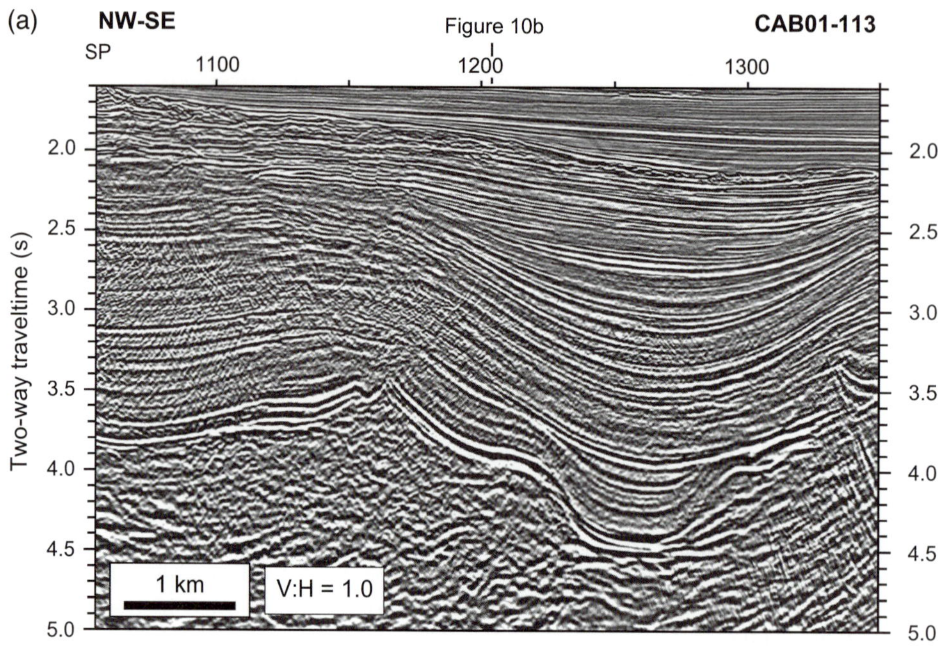

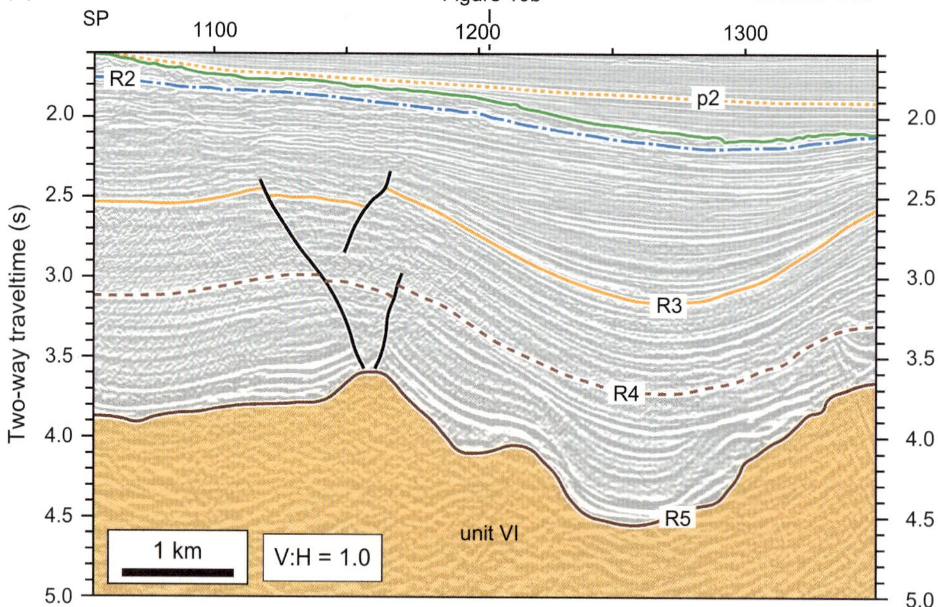

Figure 8. (a) Uninterpreted and (b) interpreted seismic profiles showing symmetric crestal faults above a shale diapir. Figures 2 and 7 show the line location. Seismic data are courtesy of ConocoPhillips. V:H = ratio of vertical and horizontal scales; 1 km = 0.6 mi.

heaves decreasing up section (horizontal offset <0.5 km [<0.3 mi]) and vanishing during the late Tortonian (below the R2 reflector). Conversely, the landward flank of diapir walls is commonly faulted, with counterregional listric faults (Figure 9). These structures have large fault throws, also decreasing up section. For example, for the seismic section shown in Figure 9b, the fault heave decreases up section, resulting in 2.5, 1, and 0.6 km (1.5, 0.6, and 0.3 mi) heaves for the shale surface and R4 and R3 reflectors, respectively.

The structural style associated with diapir culminations is illustrated in 3-D by a sequence of parallel seismic lines (Figure 10). These seismic sections show how structures evolved progressively downslope from (1) buried shale-cored anticlines with congruently folded overburden, indicating that diapir growth concluded here in the late Tortonian (Figure 10a); (2) conjugate crestal faults above shale anticlines active before the Messinian (Figure 10b); (3) passive diapir growth with a symmetric half graben active up to the early Pliocene (Figure 10c);

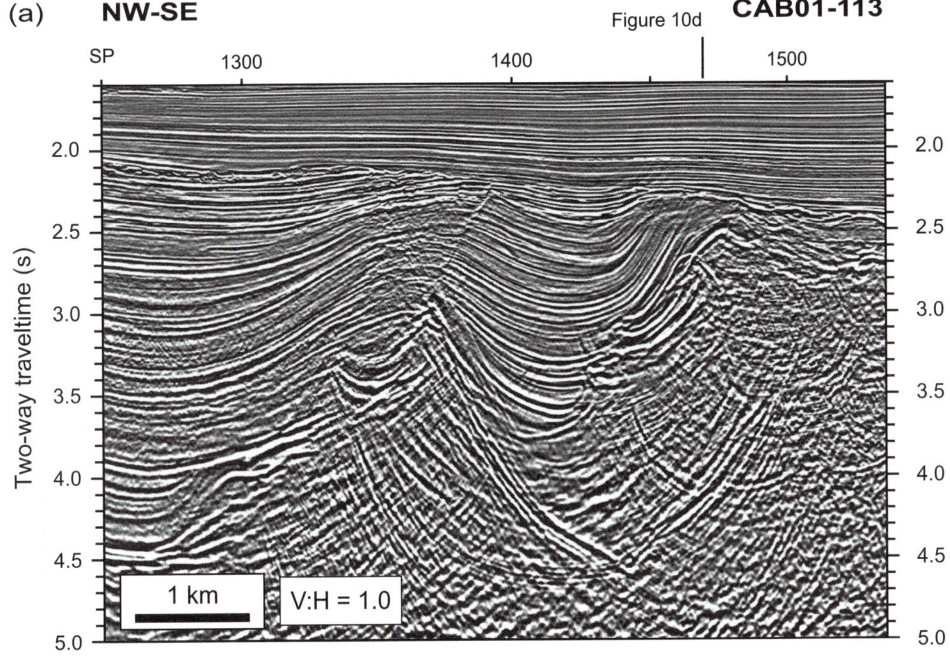

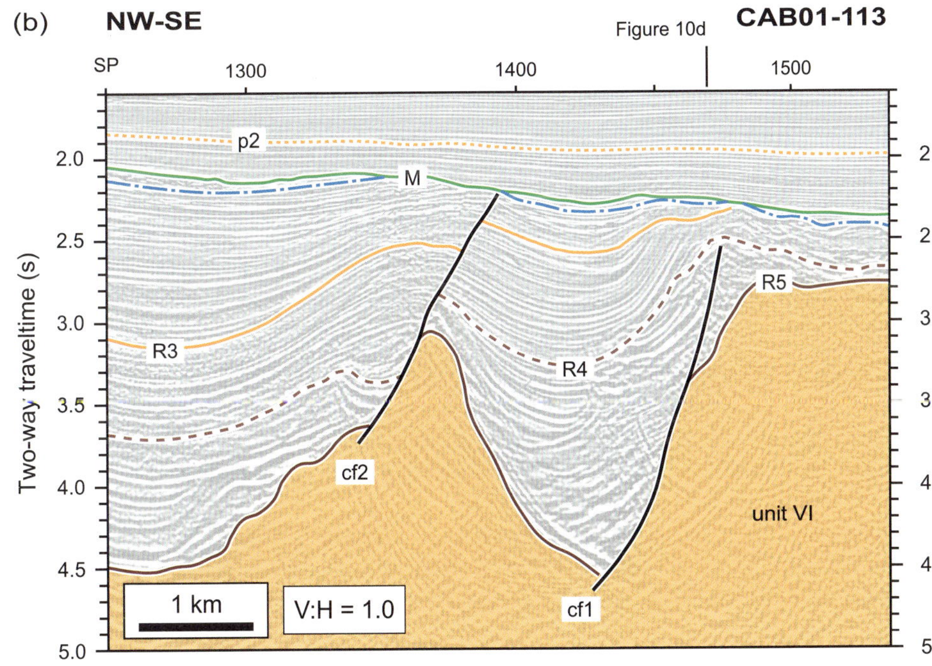

Figure 9. (a) Uninterpreted and (b) interpreted seismic profiles showing asymmetric faulting in a shale flank. Figures 2 and 7 show the line location. Seismic data are courtesy of ConocoPhillips. V:H = ratio of vertical and horizontal scales; 1 km = 0.6 mi.

and finally, (4) asymmetric collapse with passive diapir growth up to the early Quaternary (Figure 10d).

In summary, collapse faulting accompanied sedimentation and shale diapirism during the Miocene. Synsedimentary collapse began to occur in shale diapirs in at least the middle Miocene, vanishing during the late Tortonian. The contrasting structural styles depicted in Figure 10 are evidence of a downslope evolution of diapirism becoming younger.

DISCUSSION

We have focused our study on the evolution of mobile shale in the northern margin of the western Alboran Basin during the Miocene. The deformation styles described above depict a consistent tectonic history driven by lateral migration and rise of the overpressured shale. Many of the observations demonstrate that shale tectonics is controlled by basement slope and a laterally

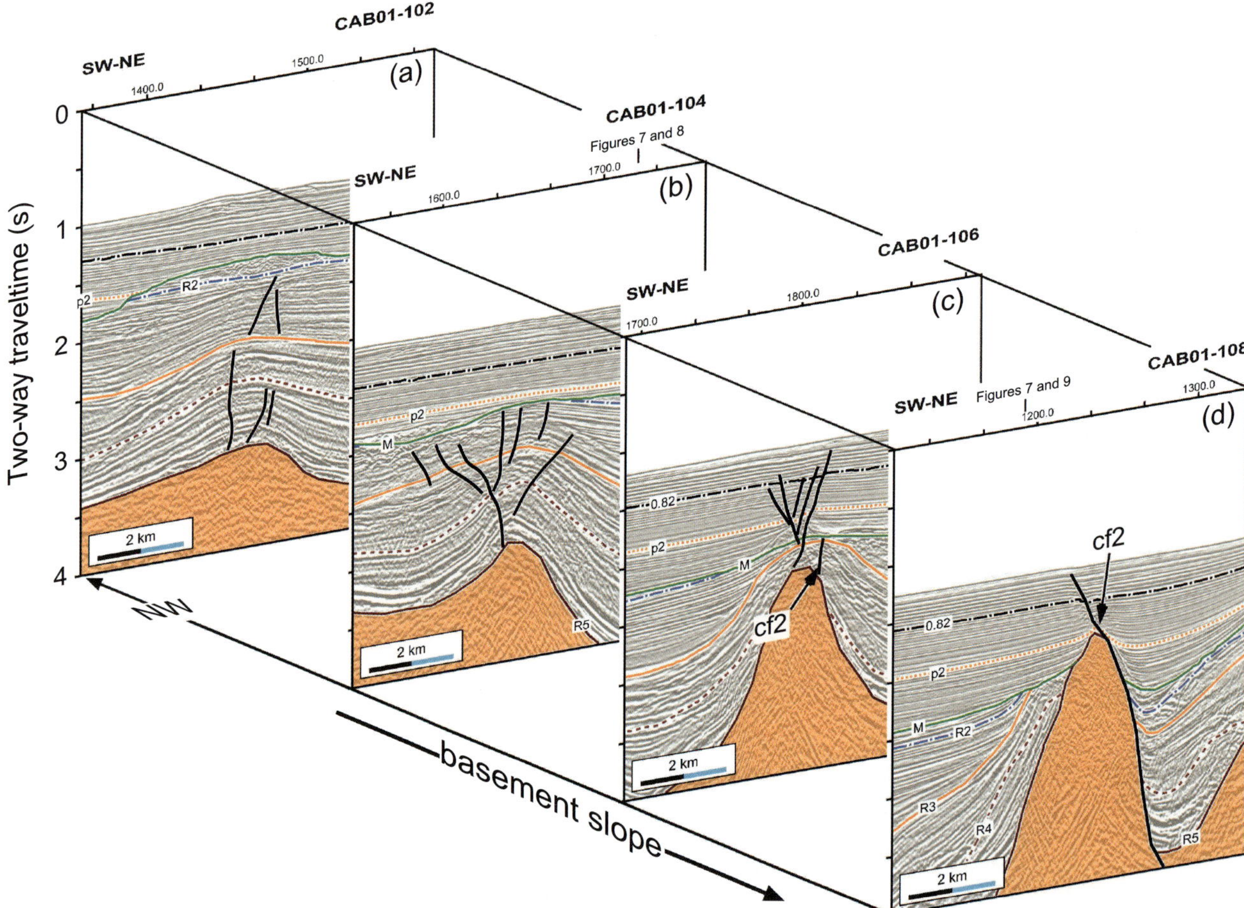

Figure 10. Perspective view of four parallel seismic profiles to illustrate the evolution in 3-D of shale collapse following the downdip direction of the basement. cf2 = major normal fault associated with diapir collapse. Figure 2 shows the line location. Seismic data are courtesy of ConocoPhillips. 2 km = 1.2 mi.

varying thickness of the overburden and the source layer. They determine an increase in maturity of shale diapirism, of associated deformational style, and of sedimentation, from margins to basin depocenter. This inference is very well established in salt basins (e.g., Trusheim, 1960; Wu et al., 1990; Jackson and Talbot, 1991; Vendeville and Jackson, 1992a, b; Jackson et al., 1994; Mauduit et al., 1997; Rowan et al., 1999; Wu and Bally, 2000; Rowan et al., 2004; Vendeville, 2005; Hudec and Jackson, 2007). We have not analyzed the burial history of the basin or the origin of the overpressure in the lowermost sedimentary sequences. In consequence, we cannot yet calculate the possible contribution of the clay transformation of smectite to illite during diagenesis (between ~70 and 180°C; Colton-Bradley, 1987; Roaldset et al., 1998) and how it could affect the overpressure history in the Alboran Basin. This dehydration transformation may have a limited contribution to the development of overpressures during diagenesis because the volume of fluid released is small (<4.0 vol.%), and the reaction is probably slow (Osborne and Swarbrick, 1997).

Downslope translation of the shale in the basement dip direction is finally blocked by a basement high (e.g., Figures 4, 5). As a result, the source layer shows an asymmetric wedge geometry with two pinch-out lines: one, the landward pinch-out with progressive thinning of the source layer above the basement-cover detachment, and another, the basinward pinch-out with maximum shale thickness cut by a counterregional normal fault (see for example Comas and Soto, 1999, their figure 7). Both pinch-outs connect to the northeast, describing a sinuous line, concave to the southwest (Figure 5). This is consistent with the regional geometry of the sedimentary infill of the western Alboran Basin in this margin, which defines a curved trough deepening to the southwest (Soto et al., 1996; Mauffret et al., 2007). The original geometry of the shale layer is markedly noncylindrical and was most likely controlled by a subsidiary basement horst. This geometry differs from the situation described in many salt basins, where a uniform increase of the salt is commonly obtained after deformation restoration (Hudec and Jackson, 2007).

The resultant geometry of the overpressured shale and the relationships with the regional discontinuities distinguished in the Miocene sedimentary sequence are illustrated by a sequence of 3-D models in Figure 11 (see also associated movies in the accompanying CD of this volume). They show in perspective the evolution of the fold and shale geometries and the piercing relationships between overburden and shale during this time interval (early Miocene to Messinian, i.e., ~15 Ma). We have also included the position of mud volcanoes for reference (Comas et al., 2003; Talukder et al., 2003).

Although some interpretations are mentioned in the section titled Structure, we now summarize some of the most important findings, centering the discussion on the following topics: (1) timing of shale tectonic history during the Miocene, (2) trigger and control factors for shale tectonics, and (3) inferences concerning subsurface shale migration.

Timing of Shale Tectonics

The different shale structures we have described show a progressive deformation history throughout the Miocene that ceased during the late Miocene (late Tortonian–early Messinian). Nevertheless, some shale diapirs rose locally during the early Pliocene (e.g., Figure 11d), forming pipelike and chimney structures feeding the mud volcanoes. Two major shale-ascent pulses exist during the Pliocene to Quaternary (Talukder, 2003; Talukder et al., 2003).

The Miocene evolution was nonetheless accompanied by a continuous instead of sporadic history of sedimentation, shale migration, and diapirism. Synsedimentary faults with associated rollover and shale rollers coalesce in the basement-cover detachment showing a continuous history of faulting in basin margins during the middle to late Miocene. The uppermost Tortonian and Messinian deposits (units III and II) seal these faults (Figure 6c, d). The most recent collapse faults, affecting Messinian and lower Pliocene series, clearly show an oblique orientation with respect to the long-lived Miocene faults (compare cf1 to cf4 faults in Figure 6b and c and faults seen in Figure 6d).

Downdip, open synclines and shale-cored anticlines formed during the same time interval as continuing sedimentation. These structures become progressively younger basinward. Development of shale-cored anticlines ended during the middle Tortonian (lower unit III), whereas shale walls pierce the upper Tortonian and Messinian (unit II) sequences (Figure 6c). Symmetric grabens above shale-cored anticlines and counterregional listric faults limiting shale walls ceased to form during the late Tortonian. Crestal faults occurred therefore during the same time interval, suggesting that diapir collapse was associated with shale rise. The structural evolution of shale sheets also reflects a congruent history of shale advance during this time interval. Toe thrusts advanced downslope during the latest Tortonian, diminishing in importance during the Messinian. This can be inferred because unit II and the uppermost beds of unit III are uplifted and overlap the toe front (Figure 4). The geometry of some of the shale sheets suggests simultaneous collapse, with the development of asymmetric half grabens affecting Messinian and lower Pliocene units. This preliminary description of shale sheets in the Alboran Sea should be extended in the future by a careful study of the geometry and timing of shale advance and collapse.

Although we have not analyzed faulting history in detail here, or any possible correlation with shale ascent pulses, Soto et al. (2003) made a first reconstruction of these processes. Analyzing fault activity in some of the marginal faults, they inferred a close link between faulting and diapir pulses during the middle–late Miocene. They reconstructed pulses of striking diapir growth in the middle Miocene and the early late Miocene (pre-R3n, see Figure 3) immediately after or before major faulting episodes.

Controlling Factors for Shale Tectonics

These observations reinforce our interpretation that basin extension and possibly basement tilting occurred from the early (post-Burdigalian?) to early late Miocene (pre-Messinian). Extension describes a centripetal pattern of stretching (to the south, southeast, and east-southeast) perpendicular to the orogenic curvature of the Gibraltar arc.

The relative age of shale structures correlates well with their position along basement dip, with more evolved structures present in basinward positions and older structures landward. An increase in thickness of the Miocene series in this direction is also noticeable (e.g., Figures 4, 7). Their general wedge shape suggests two phenomena that are probably associated. First, the high around site 976 constitutes a basement horst formed by high-angle normal faults, dipping with a northwest component (Comas and Soto, 1999). Second, the basement-cover detachment surface was active throughout the Miocene (during deposition of units V to II), and basin subsidence probably accompanied basement tilting (to the southeast). The combination of both processes caused a downslope migration of the

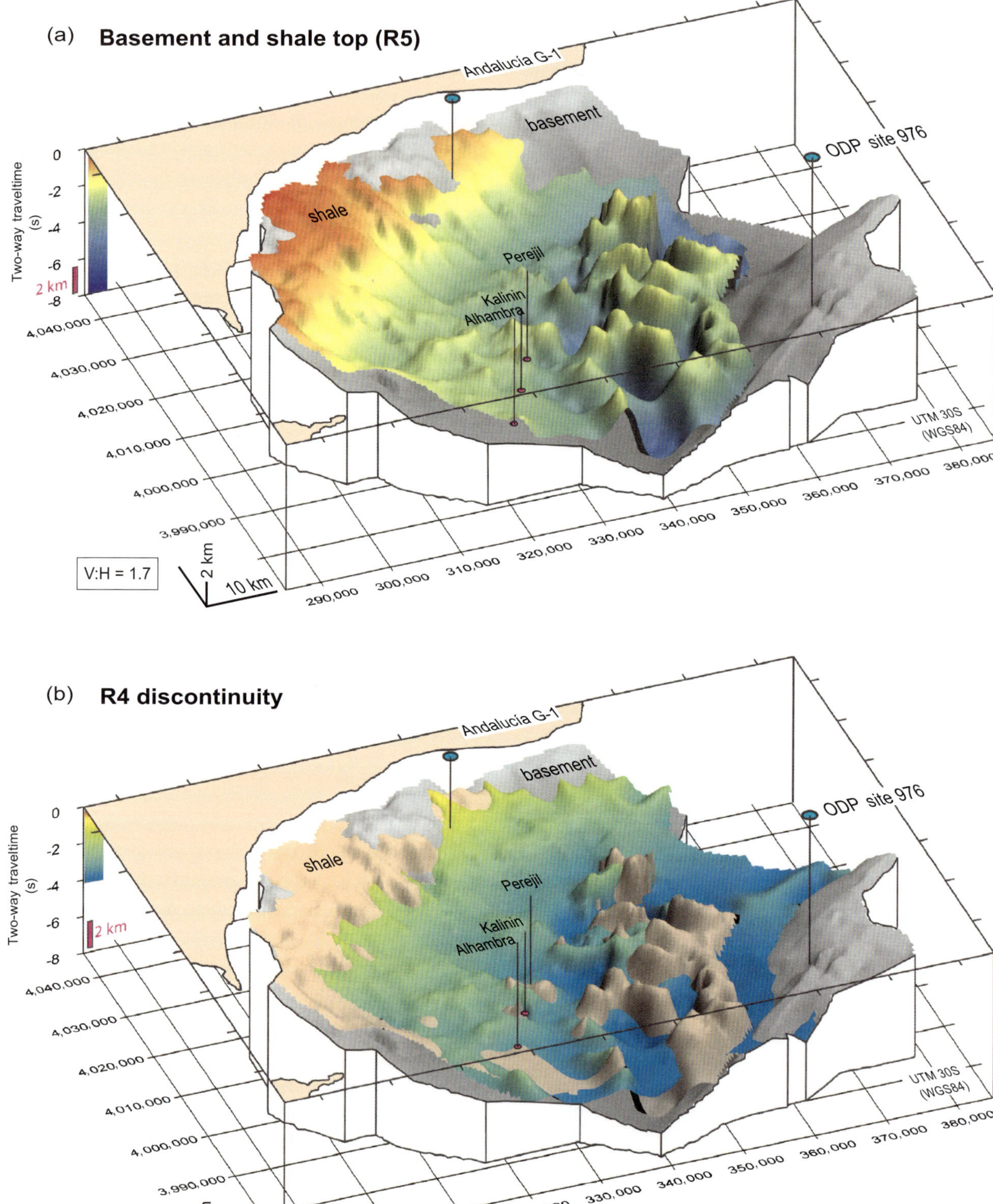

Figure 11. Three-dimensional views of (a) the basement surface, (b) the R4 discontinuity, (c) the R3 discontinuity, and (d) the M reflector. The locations of the Kalinin and Perejil mud volcanoes, together with the Alhambra pock mark, are included for reference. The vertical depth bar is calculated for the average seismic velocity of the Miocene sedimentary section (2790 m s^{-1} [9153 ft s^{-1}]). V:H = ratio of vertical and horizontal scales; 1 km = 0.6 mi; ODP = Ocean Drilling Project.

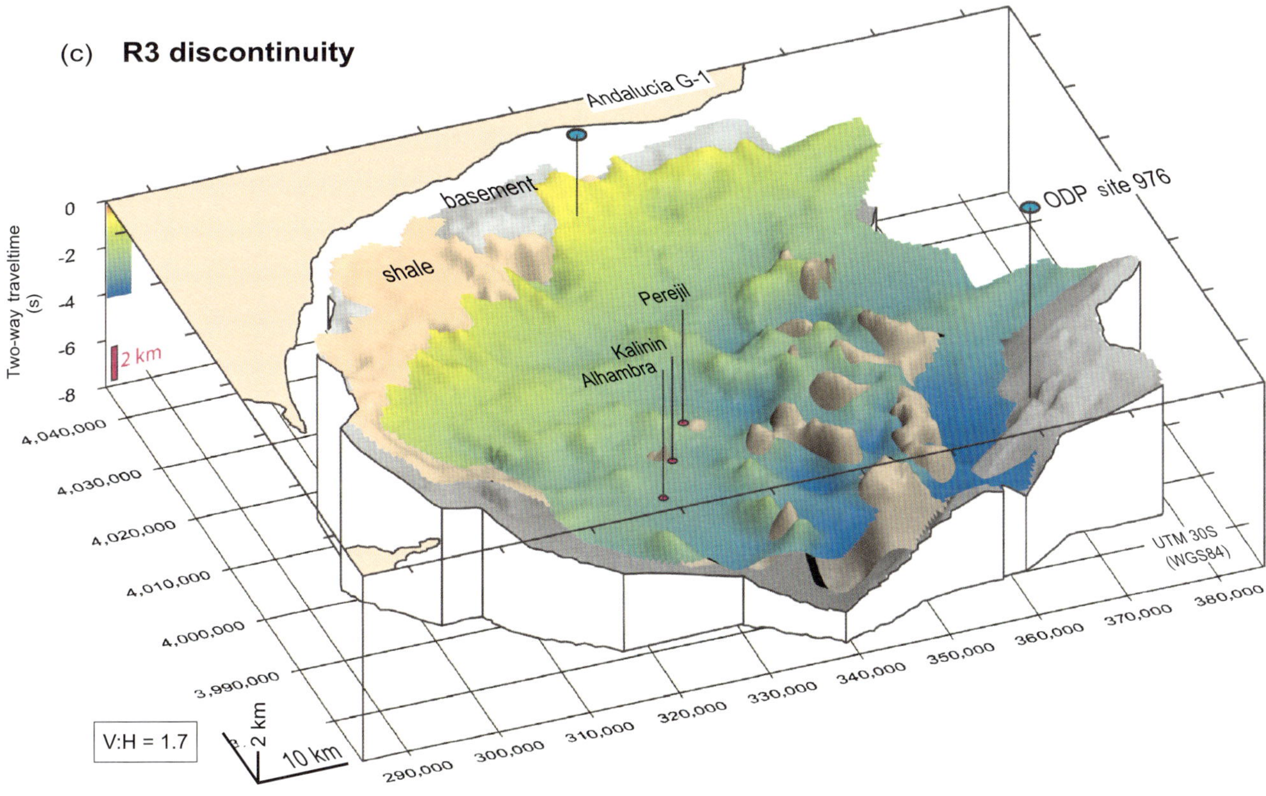

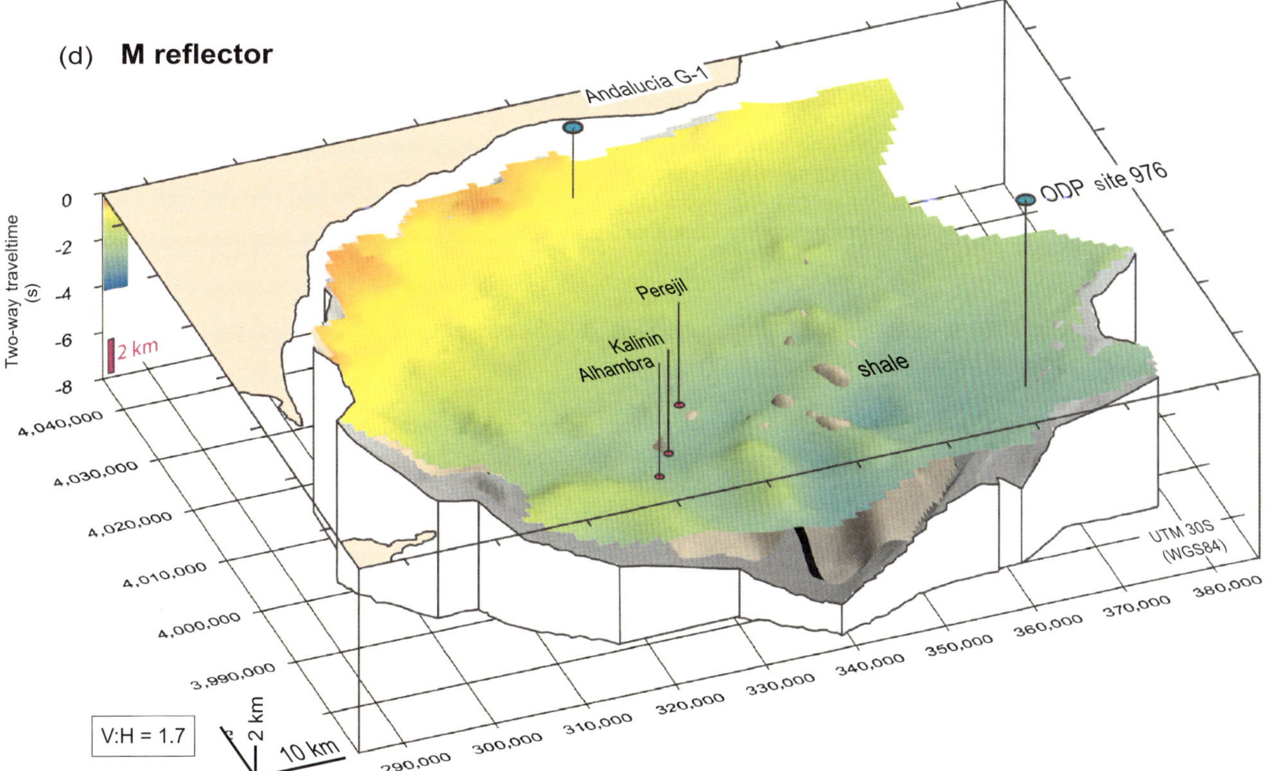

Figure 11. (cont.).

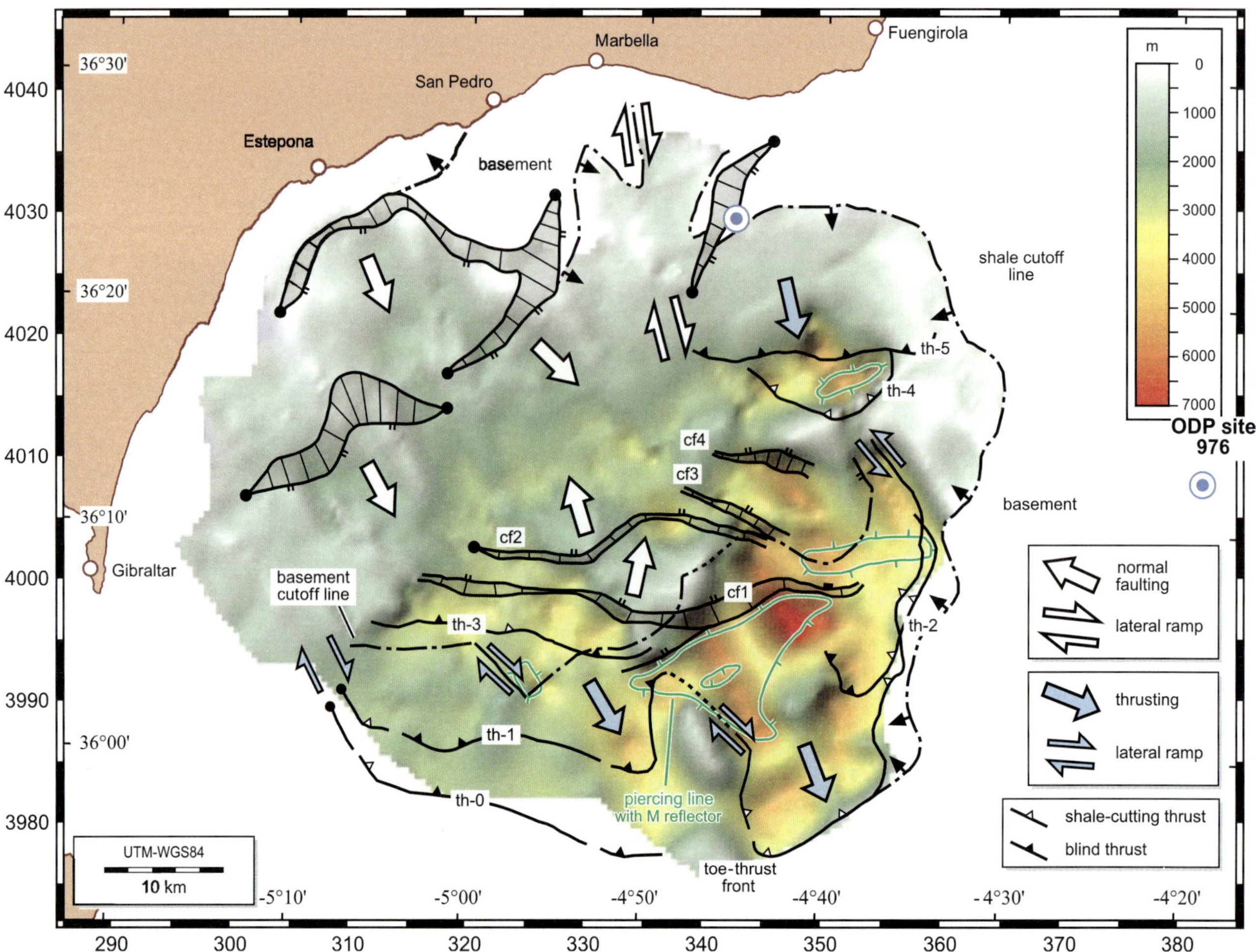

Figure 12. Thickness map of the shale, showing translation kinematics inferred for the normal faults and toe thrusts. Structures are simplified from Figures 5 and 6a. Depth conversion is made using a constant interval velocity of 2500 m s^{-1} (8202 ft s^{-1}) for the overpressured shale of unit VI. The model has a vertical exaggeration of 1.2 and is illuminated from the southeast (135° azimuth and 65° of altitude or zenith distance). 10 km = 6.2 mi; ODP = Ocean Drilling Project.

overpressured shale, resulting in marginal faulting and thickening against the site 976 high. Maximum shale accumulations around this high exceed 7 km (4.3 mi) (Figure 12). The overall tectonic history therefore consists of slope tectonics driven by both gravity and density, similar to the processes described in some passive margins with salt (e.g., Ramberg, 1963; Duval et al., 1992; Mauduit et al., 1997; Wu and Bally, 2000).

Basement geometry seems to control diapir orientation with two subperpendicular trends and also seems to control the locus of the long-lived chimney (or plug) structures. The abrupt change in toe-thrust orientation (e.g., th-1 and th-3 thrusts in Figure 6a) coincides with a lateral ramp in basement surface. This deep ramp generated a lateral thrust ramp in the overburden, which bounded the downdip advance of the shale sheet.

Other structures clearly linked to shale kinematics are counterregional normal faults bounding shale walls and sheets (cf1 to cf3 faults in Figure 5). Their geometry, timing, and position reflect shale advance synchronous with collapse. Major collapse faults are generally counterregional to the basement surface slope and likely to the direction of shale advance. This observation differs from most of the extensional fault families described above autochthonous and allochthonous salt sheets (e.g., Rowan et al., 1999), possibly reflecting features unique to shale tectonics.

Subsurface Shale Migration

As a result of these observations, we attempted to reconstruct probable migration paths of shale in the Spanish margin of the western Alboran Basin. Figure 12 presents a shale thickness map and includes the locations of major toe thrusts and main normal faults (both marginal and counterregional collapse structures). It also includes our interpretation of the possible translation kinematics of these faults. With the available seismic data set, we can only reconstruct fault kinematics by means of their general concavity or the relative position of hanging-wall and footwall cutoff lines (Figures 5, 6). Although the detailed kinematics could vary through the Miocene, we believe that this reconstruction is quite robust because fault orientation remains fairly constant during this period.

Extension direction in marginal faults is parallel with shale advance along toe thrusts (Figure 12). The inferred fault kinematics in both cases is to the southeast, parallel with the main dip direction of the basement-cover detachment surface (Figure 6a). However, the kinematics of counterregional faults describes a slip fan varying progressively from north to northeast. This pattern is also reproduced by nearby shale troughs and shale-cored anticlines (Figure 5). Counterregional extension directions are slightly oblique to the uniform orientation of marginal extension and basinward thrusting. Some of the changes in the extension direction of collapse faults coincide with oblique structures in the basement surface, interpreted as lateral ramps (e.g., at $-4°50'$ and $36°05'$). Faulting in deep, thick-skinned structures influenced fault bending in thin-skinned collapse faults, congruently switching the counterregional extension direction.

The extension direction in listric growth faults and the simultaneous parallel advance of shale driven by toe thrusts are therefore consistent with the hypothesis of gravitational gliding of shale controlled by the basement slope direction. We also propose that the bending of counterregional collapse faults is linked to the local topography of the basement-cover detachment. These faults were initiated oblique to the gliding direction, controlled by pre-existing lateral ramps in the basement surface (Gaullier et al., 1993).

CONCLUSIONS

We present the geometry of shale diapirs and associated structures in the northern margin of the western Alboran Basin, reconstructing their shape in 3-D by means of a dense grid of 2-D seismic profiles. Our study characterizes shale tectonic processes, with particular emphasis on basin evolution during the Miocene.

The overall geometry of the basin consists of a sedimentary wedge thickening from the basin margin to the basement horst at ODP site 976. The source layer for the mobile substrate comprises overpressured (undercompacted) lower Miocene shales thickening basinward. The resultant basin geometry, in conjunction with extensional deformational features and basement tilting, strongly influenced shale migration and diapirism.

Following the basement dip direction, four major structural domains are distinguished: (1) marginal growth faults with rollover and shale rollers coalescing in the basement-cover detachment surface; (2) open synclines and shale-cored anticlines with fold axes perpendicular to the basement dip direction; (3) shale walls showing two subperpendicular orientations and developing long-lived piercing chimneylike structures at intersections; and (4) autochthonous shale sheets with associated toe thrusts and a maximum horizontal displacement of approximately 9 km (~5.5 mi), coinciding with major sedimentary accumulations.

Shale tectonics during the middle–late Miocene consists of several faulting events accompanying simultaneous marginal extension and seaward shale thrusting. Shale sheet translation was driven by marginal faulting and ceased during the latest Tortonian. Shale emplacement was accompanied by asymmetric sheet collapse occurring up to the early Pliocene.

Inferred fault kinematics in marginal faults is parallel with shale advance along toe thrusts (southeastward), coinciding with the main dip direction of the basement-cover detachment surface. Kinematics of counterregional collapse faults varies systematically from north to northeast and shows local variations above oblique structures in the basement surface, interpreted as lateral ramps.

The extension direction in listric growth faults, in conjunction with the parallel and simultaneous advance of shale driven by toe thrusts, is therefore consistent with the gravitational gliding of shale, driven by marginal extension and progressive basement tilting during the middle–late Miocene.

ACKNOWLEDGMENTS

This is a contribution of the Team Consolider-Ingenio 2010 no. CSD2006-00041 (Topo-Iberia). We are sincerely grateful to Lesli Wood for her invitation to and patience with the authors. ConocoPhillips generously provided access to seismic data. This project was initially funded by this company through a research agreement promoted by R. Mountfield and Y. Chevalier. We also acknowledge the advice and support provided by A. W. Bally for this research. The manuscript benefited from the suggestions made by Y. Chevalier and the comprehensive review of T. Diggs, who are gratefully acknowledged.

REFERENCES CITED

Alonso, B., and A. Maldonado, 1992, Plio-quaternary margin growth patterns in a complex tectonic setting: Northeastern Alboran Sea: Geo-Marine Letters, v. 12, p. 137–143.

Alonso, B., G. Ercilla, F. Martínez-Ruiz, J. Baraza, and A. Galimont, 1999, Pliocene–Pleistocene sedimentary facies at site 976: Depositional history in the northwestern Alborán Sea, *in* M. C. Comas, R. Zahn, and A. Klaus, eds., Proceedings of the Ocean Drilling Program, Scientific Results, v. 161, p. 57–68.

Álvarez-Marrón, J., 1999, Pliocene to Holocene structure of the eastern Alboran Sea (western Mediterranean), *in* M. C. Comas, R. Zahn, and A. Klaus, eds., Proceedings of the Ocean Drilling Program, Scientific Results, v. 161, p. 345–355.

Andrews, J. E., 1967, Blake outer ridge: Development by gravity tectonics: Science, v. 156, p. 642–645.

Barber, A. J., S. Tjokrosapoetro, and T. R. Charlton, 1986, Mud volcanoes, shale diapirs, wrench faults and mélanges in accretionary complexes, eastern Indonesia: AAPG Bulletin, v. 50, p. 1729–1741.

Bird, P., 1984, Hydration-phase diagrams and friction of montmorillonite under laboratory and geologic conditions, with implications for shale compaction, slope stability, and strength of fault gouge: Tectonophysics, v. 107, p. 235–260.

Booth-Rea, G., C. R. Ranero, J. M. Martínez-Martínez, and I. Grevemeyer, 2007, Crustal types and tertiary tectonic evolution of the Alborán sea, western Mediterranean: Geochemistry, Geophysics, Geosystems, v. 8, doi:10.1029/2007GC001639.

Bourgois, J. A., A. Mauffret, N. A. Ammar, and N. A. Demnati, 1992, Multichannel seismic data imaging of inversion tectonics of the Alboran ridge (western Mediterranean Sea): Geo-Marine Letters, v. 12, p. 117–122.

Braunstein, J., and G. D. O'Brien, eds., 1968, Diapirism and diapirs. A symposium: AAPG Memoir 8, 444 p.

Breen, N. A., J. E. Tagudin, D. L. Reed, and E. A. Silver, 1988, Mud-cored parallel folds and possible mélange development in the north Panama thrust belt: Geology, v. 16, p. 207–210.

Brown, K., and G. K. Westbrook, 1988, Mud diapirism and subcretion in the Barbados ridge accretionary complex: The role of fluids in accretionary processes: Tectonics, v. 7, p. 613–640.

Bruce, C. H., 1973, Pressured shale and related sediment deformation: Mechanisms for development of regional contemporary faults: AAPG Bulletin, v. 57, p. 878–886.

Calvert, A., E. Sandvol, D. Seber, M. Barazangi, S. Roecker, T. Mourabit, F. Vidal, G. Alguacil, and N. Jabour, 2000, Geodynamic evolution of the lithosphere and upper-mantle beneath the Alboran Region of the western Mediterranean—Constraints from travel-time tomography: Journal of Geophysical Research, v. 105, p. 10,871–10,898.

Campillo, A. C., A. Maldonado, and A. Mauffret, 1992, Stratigraphic and tectonic evolution of the western Alboran Sea: Late Miocene to Recent: Geo-Marine Letters, v. 12, p. 165–172.

Chalouan, A., S. Rachida, A. Michard, and A. W. Bally, 1997, Neogene tectonic evolution of the southwestern Alboran Basin as inferred from seismic data off Morocco: AAPG Bulletin, v. 81, p. 1161–1184.

Chamot-Rooke, N., A. Rabaute, and C. Kreemer, 2005, Western Mediterranean Ridge mud belt correlates with active shear strain at the prism-backstop geological contact: Geology, v. 33, p. 861–864.

Chapman, R. E., 1974, Clay diapirism and overthrust faulting: Geological Society of America Bulletin, v. 85, p. 1597–1602.

Cita, M. B., and J. A. McKenzie, 1986, The terminal Miocene event, *in* K. J. Hsü, ed., Mesozoic and Cenozoic oceans: American Geophysical Union, Geodynamic Series 15, p. 123–140.

Cohen, H. A., and K. McClay, 1996, Sedimentation and shale tectonics of the northwestern Niger Delta front: Marine and Petroleum Geology, v. 13, p. 313–328.

Colton-Bradley, V. A. C., 1987, Role of pressure in smectite dehydration-effects on geopressure and smectite-to-illite transition: AAPG Bulletin, v. 71, p. 1414–1427.

Comas, M. C., and J. I. Soto, 1999, Brittle deformation in the metamorphic basement at site 976: Implications for middle Miocene extensional tectonics in the western Alboran basin: Tectonic significance, *in* M. C. Comas, R. Zahn, and A. Klaus, eds., Proceedings of the Ocean Drilling Program, Scientific Results, v. 161, p. 331–344.

Comas, M. C., V. García-Dueñas, and M. J. Jurado, 1992, Neogene tectonic evolution of the Alboran Sea from MCS data: Geo-Marine Letters, v. 12, p. 157–164.

Comas, M. C., J. J. Dañobeitia, J. Alvarez-Marrón, and J. I. Soto, 1997, Crustal reflections and structure in the Alboran Basin: Preliminary of the ESCI-Alboran survey: Revista de la Sociedad Geológica de España, v. 8, p. 529–542.

Comas, M. C., J. P. Platt, J. I. Soto, and A. B. Watts, 1999, The origin and tectonic history of the Alboran Basin: Insights from leg 161 results, *in* M. C. Comas, R. Zahn, and A. Klaus, eds., Proceedings of the Ocean Drilling Program, Scientific Results, v. 161, p. 555–579.

Comas, M. C., J. I. Soto, A. R. Talukder, and TTR-12 Leg 3 (MARSIBAL 1) Scientific Party, 2003, Discovering active mud volcanoes in the Alboran Sea (western Mediterranean), *in* M. Marani, G. Akhmanov, and A. Suzyumov, eds., Geological and biological processes at deep-sea European margins and oceanic basins: Intergovernmental Oceanographic Commission, Workshop Report 187, p. 14–16.

de Kaenel, E., W. G. Siesser, and A. Murat, 1999, Pleistocene calcareous nannofossil biostratigraphy and the western Mediterranean sapropels, sites 974 to 977 and 979, *in* M. C. Comas, R. Zahn, and A. Klaus, eds., Proceedings of the Ocean Drilling Program, Scientific Results, v. 161, p. 159–183.

Diegel, F. A., J. F. Karlo, D. C. Schuster, R. C. Shoup, and P. R. Tauvers, 1995, Cenozoic structural evolution and tectono-stratigraphic framework of the northern Gulf Coast continental margin, *in* M. P. A. Jackson, D. G. Roberts, and S. Snelson, eds., Salt tectonics: A global perspective: AAPG Memoir 65, p. 109–151.

Domzig, A., et al., 2006, Searching for the Africa-Eurasia Miocene boundary offshore western Algeria (MARADJA'03 cruise): Comptes Rendus Geoscience, v. 338, p. 80–91.

Duggen, S., K. Hoernle, P. Van den Bogaard, and C. Harris, 2004, Magmatic evolution of the Alboran region: The role of subduction in forming the western Mediterranean and causing the Messinian salinity crisis: Earth and Planetary Science Letters, v. 218, p. 91–108.

Duval, B., C. Cramez, and M. P. A. Jackson, 1992, Raft tectonics in the Kwanza basin, Angola: Marine and Petroleum Geology, v. 9, p. 389–404.

Ercilla, G., B. Alonso, and J. Baraza, 1992, Sedimentary evolution of the northwestern Alboran Sea during the Quaternary: Geo-Marine Letters, v. 12, p. 144–149.

Ewing, T. E., 1986, Structural styles of the Wilcox and Frio growth-fault trends in Texas: Constraints on geopressured reservoirs: Bureau of Economic Geology, University of Texas at Austin, Report of Investigations 154, 86 p.

Fernández-Ibáñez, F., and J. I. Soto, 2008, Rheology and crustal seismicity in the Gibraltar arc (western Mediterranean): Tectonics, v. 27, doi:10.1029/2007TC002192.

Fernández-Ibáñez, F., J. I. Soto, M. D. Zoback, and J. Morales, 2007, Present-day stress field in the Gibraltar arc: Journal of Geophysical Research, v. 112, doi:10.1029/2006JB004683.

Fernández-Soler, J., et al., 2000, Alboran Basin (diapiric province): Bottom sampling results, *in* N. H. Kenyon, M. K. Ivanov, A. M. Akhmetzhanov, and G. G. Akhmanov, eds., Multidisciplinary study of geological processes on the northeast Atlantic and western Mediterranean margins: Preliminary results of geological and geophysical investigations during the TTR-9 cruise of R/V Professor Logachev: Intergovernmental Oceanographic Commission, Technical Series 56, p. 85–91.

García-Dueñas, V., J. C. Balanyá, and J. M. Martínez-Martínez, 1992, Miocene extensional detachments in the outcropping basement of the northern Alboran Basin (Betics) and their tectonic implications: Geo-Marine Letters, v. 12, p. 88–95.

Gaullier, V., J. P. Brun, G. Guérin, and H. Lecanu, 1993, Raft tectonics: The effects of residual topography below a salt decollement: Tectonophysics, v. 228, p. 363–381.

Giles, K. A., and T. F. Lawton, 2002, Halokinetic sequence stratigraphy adjacent to the El Papalote diapir, northeastern Mexico: AAPG Bulletin, v. 86, p. 823–840.

Graue, K., 2000, Mud volcanoes in deep water Nigeria: Marine and Petroleum Geology, v. 17, p. 959–974.

Gutscher, M. A., J. Malod, J. P. Rehault, I. Contrucci, F. Klingelhoefer, V. L. Mendes, and W. Spakman, 2002, Evidence for active subduction beneath Gibraltar: Geology, v. 30, p. 1071–1074.

Haddad, G. A., Y. M. Chevalier, R. A. Mountfield, L. C. Kuo, and M. O. Strickland, 2003, Deep-water sedimentation in the west Alboran Basin since the Messinian as inferred from recent 2D seismic and swath bathymetry data: AAPG International Conference, Barcelona, Abstract Book.

Hopper, R. J., R. J. Fitzsimmons, N. Grant, and B. C. Vendeville, 2002, The role of deformation in controlling depositional patterns in the south-central Niger Delta, west Africa: Journal of Structural Geology, v. 24, p. 847–859.

Hudec, M. R., and M. P. A. Jackson, 2006, Advance of allochthonous salt sheets in passive margins and orogens: AAPG Bulletin, v. 90, p. 1535–1564.

Hudec, M. R., and M. P. A. Jackson, 2007, Terra infirma: Understanding salt tectonics: Earth-Science Reviews, v. 82, p. 1–28.

Iaccarino, S. M., and A. Bossio, 1999, Paleoenvironment of uppermost Messinian sequences in the western Mediterranean (sites 974, 975, and 978), *in* M. C. Comas, R. Zahn, and A. Klaus, eds., Proceedings of the Ocean Drilling Program, Scientific Results, v. 161, p. 529–541.

Ivanov, M. K., A. F. Limonov, and T. C. E. van Weering, 1996, Comparative characteristics of the Black Sea and Mediterranean Ridge mud volcanoes: Marine Geology, v. 132, p. 253–271.

Jackson, M. P. A, and C. Cramez, 1989, Seismic recognition of salt welds in salt tectonics regimes, *in* Gulf of Mexico salt tectonics, associated processes and exploration potential: Gulf Coast Section SEPM, 10th Annual Research Conference Program and Extended and Illustrated Abstracts, p. 66–71.

Jackson, M. P. A., and C. J. Talbot, 1991, A glossary of salt tectonics: Bureau of Economic Geology, University of Texas at Austin, Geological Circular 91-4, 44 p.

Jackson, M. P. A., and B. C. Vendeville, 1994, Regional extension as a geologic trigger for diapirism: Geological Society of America Bulletin, v. 106, p. 57–73.

Jackson, M. P. A., B. C. Vendeville, and D. D. Schultz-Ela, 1994, Structural dynamics of salt systems: Annual Review of Earth and Planetary Sciences, v. 22, p. 93–117.

Jolivet, L., and C. Faccenna, 2000, Mediterranean extension and the Africa-Eurasia collision: Tectonics, v. 19, p. 1095–1106.

Jurado, M. J., and M. C. Comas, 1992, Well log interpretation and seismic character of the Cenozoic sequence in the northern Alboran Sea: Geo-Marine Letters, v. 12, p. 129–136.

Kopf, A. J., 2002, Significance of mud volcanism: Reviews of Geophysics, v. 40, p. 1–52.

Kopf, A., and J. H. Behrmann, 2000, Extrusion dynamics of mud volcanoes on the Mediterranean Ridge accretionary complex, *in* B. Vendeville, Y. Mart, and J.-L. Vigneresse, eds., From the Arctic to the Mediterranean: Salt, shale, and igneous diapirs in and around Europe: Geological Society Special Publication 174, p. 169–204.

Kopf, A., M. Ben Clennell, and K. M. Brown, 2005, Physical properties of muds extruded from mud volcanoes: Implications for episodicity of eruptions and relationship to seismicity, *in* G. Martinelli and B. Panahi, eds., Mud volcanoes, geodynamics and seismicity: North Atlantic Treaty Organization Science Series 51: Dordrecht, Springer, p. 263–283.

Koyi, H. A., and A. Skelton, 2001, Centrifuge modeling of the evolution of low-angle detachment faults from high-angle normal faults: Journal of Structural Geology, v. 23, p. 1179–1185.

Kuo, L. C., R. A. Mountfield, Y. M. Chevalier, G. A. Haddad, A. C. Weinzapfel, and M. O. Strickland, 2002, Petroleum systems and exploration potential of the western Alboran Basin: AAPG International Conference, El Cairo, Abstract Book.

Lonergan, L., and N. White, 1997, Origin of the Betic-Rif mountain belt: Tectonics, v. 16, p. 504–522.

Maltman, A. J., and A. Brown, 2003, How sediments become mobilized, *in* P. Van Rensbergen, R. R. Hills, A. Maltman, and C. Morley, eds., Subsurface sediment mobilization: Geological Society (London) Special Publication 216, p. 9–20.

Martinelli, G., and A. Judd, 2004, Mud volcanoes of Italy: Geological Journal, v. 39, p. 49–61.

Martínez-Martínez, J. M., J. I. Soto, and J. C. Balanyá, 2002, Orthogonal folding of extensional detachments: Structure and origin of the Sierra Nevada elongated dome (Betics, SE Spain): Tectonics, v. 21, p. 3-1–3-22, doi:10.1029/2001TC001283.

Mascle, A., D. Lajatand, and G. Nely, 1979, Sediments deformation linked to subduction and to argillokinesis in the southern Barbados ridge from multichannel seismic surveys: Transactions of the 4th Latin American Geological Conference, Port-of-Spain, Trinidad and Tobago, v. 2, p. 873–882.

Mauduit, T., G. Guerin, J.-P. Brun, and H. Lecau, 1997, Raft tectonics: The effects of basal slope angle and sedimentation rate on progressive extension: Journal of Structural Geology, v. 19, p. 1219–1230.

Mauffret, A., A. Ammar, C. Gorini, and H. Jabour, 2007, The Alboran Sea (western Mediterranean) revisited with a view from the Moroccan margin: Terra Nova, v. 19, p. 195–203.

McClay, K., T. Dooley, A. Ferguson, and J. Poblet, 2000, Tectonic evolution of the Sanga Sanga block, Mahakam delta, Kalimantan, Indonesia: AAPG Bulletin, v. 84, p. 765–786.

Milkov, A. V., 2000, Worldwide distribution of submarine mud volcanoes and associated gas hydrates: Marine Geology, v. 167, p. 29–42.

Molnar, P., and G. A. Houseman, 2004, The effects of buoyant crust on the gravitational instability of thickened mantle lithosphere at zones of intracontinental convergence: Geophysical Journal International, v. 158, p. 1134–1150.

Mondol, N. H., K. Bjørlykke, J. Jahren, and K. Høeg, 2007, Experimental mechanical compaction of clay mineral aggregates— Changes in physical properties of mudstones during burial: Marine and Petroleum Geology, v. 24, p. 289–311.

Mondol, N. H., J. Jahren, K. Bjørlykke, and I. Brevik, 2008, Elastic properties of clay minerals: The Leading Edge, v. 27, p. 758–770.

Morley, C. K., 1992, Notes on Neogene basin history of the western Alboran Sea and its implications for the tectonic evolution of the Rif-Betic orogenic belt: African Journal of Earth Sciences, v. 14, p. 57–65.

Morley, C. K., 2003, Mobile shale related deformation in large deltas developed on passive and active margins, *in* P. Van Rensbergen, R. R. Hills, A. Maltman, and C. Morley, eds., Subsurface sediment mobilization: Geological Society (London) Special Publication 216, p. 335–357.

Morley, C. K., and G. Guerin, 1996, Comparison of gravity-driven deformation styles and behavior associated with mobile shales and salt: Tectonics, v. 15, p. 1154–1170.

Morley, C. K., M. Tingay, R. Hillis, and R. King, 2008, Relationship between structural style, overpressures, and modern stress, Baram Delta Province, northwest Borneo: Journal of Geophysical Research, v. 113, B09410, doi:10.1029/2007JB005324.

Mountfield, R. A., Y. M. Chevalier, G. A. Haddad, L. C. Kuo, and A. C. Weinzapfel, 2002, Petroleum systems and exploration potential of the western Alboran Basin: Marrakech International Oil and Gas Conference, Abstract Book.

Osborne, M. J., and R. E. Swarbrick, 1997, Mechanisms for generating overpressure in sedimentary basins: A reevaluation: AAPG Bulletin, v. 81, p. 1023–1041.

Pérez-Belzuz, F., B. Alonso, and G. Ercilla, 1997, History of mud diapirism and triggering mechanisms in the western Alboran Sea: Tectonophysics, v. 282, p. 399–423.

Planke, S., H. Svensen, M. Hovland, D. A. Banks, and B. Jamtveit, 2003, Mud and fluid migration in active mud volcanoes in Azerbaijan: Geo-Marine Letters, v. 23, p. 258–268.

Platt, J. P., and R. L. M. Vissers, 1989, Extensional collapse of thickened continental lithosphere: A working hypothesis for the Alboran Sea and Gibraltar arc: Geology, v. 17, p. 540–543.

Platt, J. P., J. I. Soto, M. J. Whitehouse, A. J. Hurford, and S. P. Kelley, 1998, Thermal evolution, rate of exhumation, and tectonic significance of metamorphic rocks from the floor of the Alboran extensional basin, western Mediterranean: Tectonics, v. 17, p. 671–689.

Platt, J. P., S. Allerton, A. Kirker, C. Mandeville, A. Mayfield, E. S. Platzman, and A. Rimi, 2003, The ultimate arc: Differential displacement, oroclinal bending, and vertical axis rotation in the external Betic-Rif arc: Tectonics, v. 22, p. 1017–1046.

Platt, J. P., R. Anczkiewicz, J. I. Soto, S. P. Kelley, and M. Thirlwall, 2006, Early Miocene continental subduction and rapid exhumation in the western Mediterranean: Geology, v. 34, p. 981–984.

Ramberg, H., 1963, Experimental study of gravity tectonics by means of centrifugal models: Bulletin of the Geological Institute, University of Uppsala, v. 62, p. 1–97.

Roaldset, E., H. Wei, and S. Grimstad, 1998, Smectite to illite conversion by hydrous pyrolysis, *in* D. C. Bain, ed., Clay minerals in the modern society: London, Mineralogical Society, p. 147–158.

Robertson, A. H. F., and A. Kopf, 1998, Tectonic setting and processes of mud volcanism on the Mediterranean Ridge accretionary complex: Evidence from leg 160, *in* A. H. F. Robertson, K. C. Emeis, C. Richter, and A. Camerlenghi, eds., Proceedings of the Ocean Drilling Program, Scientific Results, v. 160, p. 665–680.

Rowan, M. G., M. P. A. Jackson, and B. D. Trudgill, 1999, Salt-related fault families and fault welds in the northern Gulf of Mexico: AAPG Bulletin, v. 83, p. 1454–1484.

Rowan, M. G., T. F. Lawton, K. A. Giles, and R. A. Ratliff, 2003, Near-salt deformation in La Popa Basin, Mexico, and the northern Gulf of Mexico: A general model for passive diapirism: AAPG Bulletin, v. 87, p. 733–756.

Rowan, M. G., F. J. Peel, and B. C. Vendeville, 2004, Gravity-driven fold belts on passive margins, *in* K. R. McClay, ed., Thrust tectonics and hydrocarbon systems: AAPG Memoir 82, p. 157–182.

Royden, L. H., 1993, Evolution of retreating subduction boundaries formed during continental collision: Tectonics, v. 12, p. 629–638.

Ryan, W. B. F., J. K. Hsü, M. B. Cita, P. Dumitrica, J. Lort, W. Maync, W. D. Nesteroff, G. Pautot, H. Stauder, and F. C. Wezel, 1973, Western Alboran Basin— Site 121, *in* W. B. F. Ryan et al., eds., Initial reports of the Deep Sea Drilling Project 13, p. 43–89.

Sautkin, A., A. R. Talukder, M. C. Comas, J. I. Soto, and A. Alekseev, 2003, Mud volcanoes in the Alboran Sea: Evidence from micropaleontological and geophysical data: Marine Geology, v. 195, p. 237–261.

Schlumberger, 2009, 2009 edition log interpretation charts: Sugar land, Texas, USA, Schlumberger, 308 p.

Schuster, D. C., 1995, Deformation of allochthonous salt and evolution of related salt-structural systems, eastern Louisiana Gulf Coast, *in* M. P. A. Jackson, D. G. Roberts, and S. Snelson, eds., Salt tectonics: A global perspective: AAPG Memoir 65, p. 177–198.

Seber, D., M. Barazangi, B. A. Tadili, F. Ramdani, A. Ibenbrahim, and D. B. Sari, 1996, Three-dimensional upper mantle structure beneath the intraplate Atlas and interplate Rif Mountains of Morocco: Journal of Geophysical Research, v. 101, p. 3125–3138.

Seni, S. J., and M. P. A. Jackson, 1983, Evolution of salt structures, east Texas diapir province: Part 1. Sedimentary record of halokinesis: AAPG Bulletin, v. 67, p. 1219–1244.

Serpelloni, E., G. Vannucci, S. Pondrelli, A. Argnani, G. Casula, M. Anzidei, P. Balde, and P. Gasperini, 2007, Kinematics of the western Africa-Eurasia plate boundary from focal mechanisms and GPS data: Geophysical Journal International, v. 169, p. 1180–1200.

Shipboard Scientific Party, 1996, Site 976, *in* M. C. Comas, R. Zahn, and A. Klaus, eds., Proceedings of the Ocean Drilling Program, Initial Results, v. 161, p. 179–297.

Siesser, W. G., and E. P. de Kaenel, 1999, Neogene calcareous nannofossils: Western Mediterranean biostratigraphy and paleoclimatology, *in* M. C. Comas, R. Zahn, and A. Klaus, eds., Proceedings of the Ocean Drilling Program, Scientific Results, v. 161, p. 223–237.

Soto, J. I., and J. P. Platt, 1999, Petrological and structural evolution of high-grade metamorphic rocks from the floor of the Alboran Sea basin, western Mediterranean: Journal of Petrology, v. 40, p. 21–60.

Soto, J. I., M. C. Comas, and J. de la Linde, 1996, Espesor de sedimentos en la Cuenca de Alborán mediante una conversión sísmica corregida: Geogaceta, v. 20, p. 382–385.

Soto, J. I., M. C. Comas, and A. R. Talukder, 2003, Evolution of the mud diapirism in the Alboran Sea (western Mediterranean): AAPG International Conference and Exhibition, Extended Abstracts, p. 21–24.

Spakman, W., and M. J. R. Wortel, 2004, A tomographic view on western Mediterranean geodynamics, *in* W. Cavazza, F. Roure, W. Spakman, G. M. Stampfli, and P. Ziegler, eds., The TRANSMED atlas— The Mediterranean region from crust to mantle: Berlin, Springer, p. 31–52.

Stewart, S. A., and R. J. Davies, 2006, Structure and emplacement of mud volcano systems in the south Caspian Basin: AAPG Bulletin, v. 90, p. 771–786.

Talbot, C. J., 1995, Molding of salt diapirs by stiff overburden, *in* M. P. A Jackson, D. G. Roberts, and S. Snelson, eds., Salt tectonics: A global perspective: AAPG Memoir 65, p. 61–75.

Talukder, M. A. R., 2003, La provincia diapírica de lodo en la Cuenca Oeste del Mar de Alborán: Estructuras, génesis y evolución: Ph.D. thesis, Granada University, Granada, Spain, 251 p.

Talukder, A. R., M. C. Comas, and J. I. Soto, 2003, Pliocene to Recent mud diapirism and related mud volcanoes in the Alboran Sea (western Mediterranean), *in* P. Van Rensbergen, R. R. Hills, A. Maltman, and C. Morley, eds., Subsurface sediment mobilization: Geological Society (London) Special Publication 216, p. 443–459.

Tandon, K., J. M. Lorenzo, and J. de la Linde, 1998, Timing of rifting in the Alboran Sea basin— Correlation of borehole (ODP leg 161 and Andalucía A-1) to seismic reflection data: Implications for basin formation: Marine Geology, v. 144, p. 275–294.

Tingay, M. R. P., R. R. Hillis, R. E. Swarbrick, C. K. Morley, and A. R. Damit, 2007, ''Vertically transferred'' overpressures in Brunei: Evidence for a new mechanism for the formation of high-magnitude overpressure: Geology, v. 35, p. 1023–1026.

Totterdell, J. M., and A. A. Krassay, 2003, The role of deformation and growth faulting in the Late Cretaceous evolution of the Bight Basin, offshore southern Australia, *in* P. Van Rensbergen, R. R. Hills, A. Maltman, and C. Morley, eds., Subsurface sediment mobilization: Geological Society (London) Special Publication 216, p. 429–442.

Trusheim, F., 1960, Mechanism of salt migration in northern Germany: AAPG Bulletin, v. 44, p. 1519–1540.

Van Rensbergen, P., C. K. Morley, D. W. Ang, T. Q. Hoan, and N. T. Lam, 1999, Structural evolution of shale diapirs from reactive to mud volcanism: 3D seismic data from the Baram Delta, offshore Brunei Darussalam: Journal of the Geological Society, v. 156, p. 633–650.

Vendeville, B. C., 2005, Salt tectonics driven by sediment progradation: Part I. Mechanics and kinematics: AAPG Bulletin, v. 89, p. 1071–1079.

Vendeville, B. C., and M. P. A. Jackson, 1992a, The rise of diapirs during thin-skinned extension: Marine and Petroleum Geology, v. 9, p. 331–353.

Vendeville, B. C., and M. P. A. Jackson, 1992b, The fall of diapirs during thin-skinned extension: Marine and Petroleum Geology, v. 9, p. 354–371.

Vendeville, B. C., H. Ge, and M. P. A. Jackson, 1995, Scale models of salt tectonics during basement-involved extension: Petroleum Geoscience, v. 1, p. 179–183.

Vernik, L., 2008, Anisotropic correction of sonic logs in wells with large relative dip: Geophysics, v. 73, p. E1–E5.

Watts, A. B., J. P. Platt, and P. Buhl, 1993, Tectonic evolution of the Alboran Sea basin: Basin Research, v. 5, p. 153–177.

Weijermars, R., M. P. A. Jackson, and B. Vendeville, 1993, Rheological and tectonic modeling of salt provinces: Tectonophysics, v. 217, p. 143–174.

Weinzapfel, A. C., R. A. Mountfield, Y. M. Chevalier, L. C. Kuo, K. A. Soofi, G. A. Haddad, and M. O. Strickland, 2003, New insights into the hydrocarbon prospectivity of an undrilled mud diapir province, west Alboran Basin, Morocco-Spain: AAPG Annual Meeting, Abstract Book.

Wood, L. J., 2000, Chronostratigraphy and tectonostratigraphy of the Columbus Basin, eastern offshore Trinidad: AAPG Bulletin, v. 84, p. 1905–1928.

Woodside, J., M. Ivanov, R. Koelewijn, I. Zeldenrust, and P. Shashkin, 2000, Alboran Basin (diapiric province): Side-scan sonar data, *in* N. H. Kenyon, M. K. Ivanov, A. M. Akhmetzhanov, and G. G. Akhmanov, eds., Multidisciplinary study of geological processes on the northeast Atlantic and western Mediterranean margins: Preliminary results of geological and geophysical investigations during the TTR-9 cruise of R/V Professor Logachev: Intergovernmental Oceanographic Commission, Technical Series 56, p. 82–84.

Wu, S., and A. W. Bally, 2000, Slope tectonics— Comparisons and contrasts of structural styles of salt and shale tectonics of the northern Gulf of Mexico with shale tectonics of offshore Nigeria in Gulf of Guinea, *in* W. Mohriak and M. Talwani, eds., Atlantic rifts and continental margins: American Geophysical Union, Geophysical Monograph 115, p. 151–172.

Wu, S., A. W. Bally, and C. Cramez, 1990, Allochthonous salt, structure and stratigraphy of the north-eastern Gulf of Mexico: Part II. Structure: Marine and Petroleum Geology, v. 7, p. 336–370.

9

Wiener, Richard W., Michael G. Mann, Michael Terry Angelich, and Joseph B. Molyneux, 2010, Mobile shale in the Niger Delta: Characteristics, structure, and evolution, *in* L. Wood, ed., Shale tectonics: AAPG Memoir 93, p. 145–161.

Mobile Shale in the Niger Delta: Characteristics, Structure, and Evolution

Richard W. Wiener
ExxonMobil International Limited, Leatherhead, Surrey, United Kingdom

Michael G. Mann
ExxonMobil Exploration Co., Houston, Texas, U.S.A.

Michael Terry Angelich
ExxonMobil Exploration Co., Houston, Texas, U.S.A.

Joseph B. Molyneux
ExxonMobil Exploration Co., Houston, Texas, U.S.A.

ABSTRACT

Mobile shale exerts fundamental controls on regional structural styles and sedimentation in the Niger Delta and also impacts trap geometry, reservoir distribution, fluid flow, and seal capacity at the prospect level. In the Niger Delta, overpressured shale forms a detachment zone for normal faults, detachment folds, and fold-thrust structures in a linked extensional-contractional system. The updip extensional zone is characterized by landward-dipping and seaward-dipping normal faults superposed on older contractional detachment folds. The thick mobile shale province is characterized by high-relief, shale-cored detachment folds and minibasins. Downdip contractional structural provinces include thrusted mobile-shale, low- to moderate-relief detachment folds, and a fold-thrust belt. Mobile shale is characterized by ductile deformation style, transparent seismic facies, and anomalously low density and seismic velocity. Detachment folding and thrusting are interpreted to characterize mobile shale deformation more so than diapirism.

Cross section restorations show a balance of extension and contraction through time, a basinward progradation of extensional and contractional systems, and an evolution of the mobile shale substrate. Shale withdrawal in half grabens and minibasins is balanced downdip by tectonic thickening in cores of detachment folds.

Paleogeographic reconstructions of the shelf, slope, and basin show structural controls on paleogeography and basinward progradation. The shelf margin was localized by major extensional systems, whereas the slope-to-basin transition was controlled by the contractional structural front. Deltaic deposits formed along the shelf margin. Slope depositional systems

DOI:10.1306/13231313M933423

included minibasins and confined-channel systems influenced by paleostructures, whereas basin-floor deposits were not affected by paleostructures and hence were more weakly confined.

INTRODUCTION

Mobile shale impacts the structural style, deformation history, and depositional systems of sedimentary basins with shale-rich substrates. Mobile shale can be thought of as a zone of deep-seated overpressured mud or shale shale-rich rock that deforms in either a macroscopically ductile or a mixed brittle and ductile fashion. Seismic transparency, low seismic velocity, low density, and ductile deformation style characterize mobile shale. Structural styles associated with mobile shale include ductile detachment folds, shale domes with crestal normal faults, minibasins, and thrusted shear fault-bend and fault-propagation folds that involve a ductile substrate (Costa and Vendeville, 2002; Corredor et al., 2005). Several workers have suggested that mobile shale forms reactive, passive, and active diapirs as well (Cohen and McClay, 1996; Morley and Guerin, 1996; Van Rensberger et al., 1999; Wu and Bally, 2000; Hooper et al., 2002; Elsley and Graham, 2006). Mud volcano systems are associated with thick mobile shale but typically comprise a small part of the overall mobile-shale-related structures.

Mobile shale forms a ductile substrate to deformation in various tectonic settings, principally in passive margins but also in convergent and transform margins (Morley and Guerin, 1996; Van Rensberger et al., 1999; Rowan et al., 2004; Gipson et al., 2006). In addition, mobile-shale-related deformation is typically associated with high sedimentation rates that characterize deltas. Gravity-driven, paired extensional-contractional systems can develop in these settings where rapid burial of a shale-rich stratigraphic section forms a zone of weak, overpressured shale that serves as a detachment system. The detachment can be on individual horizons where mechanical contrast is strong or across an interval of weak shale.

Development of overpressure may facilitate deformation by weakening the rocks in a shale-rich sequence. Primary overpressure in shale may result from rapid burial of shale-rich sequences where lack of permeability will lead to entrapment of water and to high fluid pressures. Secondary overpressure may develop because of hydrocarbon maturation, especially methane generation, which results in a significant volume increase and additional pressure (Barker, 1990). Other mechanisms for secondary overpressure include superposed tectonic stresses (compression or tension). Compressive stresses may be caused by a gravitational movement downdip in a linked system and/or superposed plate-tectonic stresses in an active margin setting. Also, mineral transformations, such as the illite-smectite transition, can increase volume and pressure. Formation of overpressure can be episodic because it can be reduced by pressure release events.

Mobile substrates can deform where a weak zone that concentrates deformation exists, typically salt or overpressured shale. The mechanical behavior of overpressured shale can be modeled as a Bingham plastic with a yield strength (Albertz et al., 2006). High fluid pressures weaken shale by reducing the effective stress, which allows the shale body to overcome its yield strength (Hubbert and Rubey, 1959). The onset of overpressure at depths of 4–6 km (2.4–3.7 mi) weakens shale-rich rocks so that ductile deformation typically ensues at these depths in various mobile shale basins (Morley and Guerin, 1996; Van Rensberger et al., 1999; Rowan et al., 2004). Extreme fluid overpressures can result from gas- and fluid-charged diatremes associated with mobile-shale structures. These tend to form vertical pipes with narrow deformation haloes and lateral sills, as well as surface expulsion features (Graue, 2000; Stewart and Davies, 2006). The rheology of these features is distinct from both deep-seated mobile shale and salt and is more akin to low-viscosity magma (Vendeville, 2006).

Mobile shale is associated with prolific hydrocarbon-bearing basins, such as the Niger Delta, the Gulf of Mexico, the southern Caspian Basin, the Columbus Basin of Trinidad, the Mackenzie delta, and the Baram delta of offshore Brunei (Morley and Guerin, 1996; Van Rensberger et al., 1999; Wu and Bally, 2000; Rowan et al., 2004; Elsley and Graham, 2006; Gipson et al., 2006; Stewart and Davies, 2006). The presence of mobile shale and related deformation features affects exploration and production strategies in petroliferous basins by impacting trap definition, seal capacity, reservoir presence and quality, hydrocarbon migration, and drilling hazards. Trap definition is difficult in areas of thick mobile shale because of poor seismic imaging on the crests of structures in places. Gravity and seismic velocity analysis provide insights into trap geometry where seismic imaging is difficult. The structural evolution of mobile-shale-related features affects prospect analysis by impacting the reservoir type and distribution relative to the structural position. Overpressure in mobile shale basins also has a profound effect on fluid evolution,

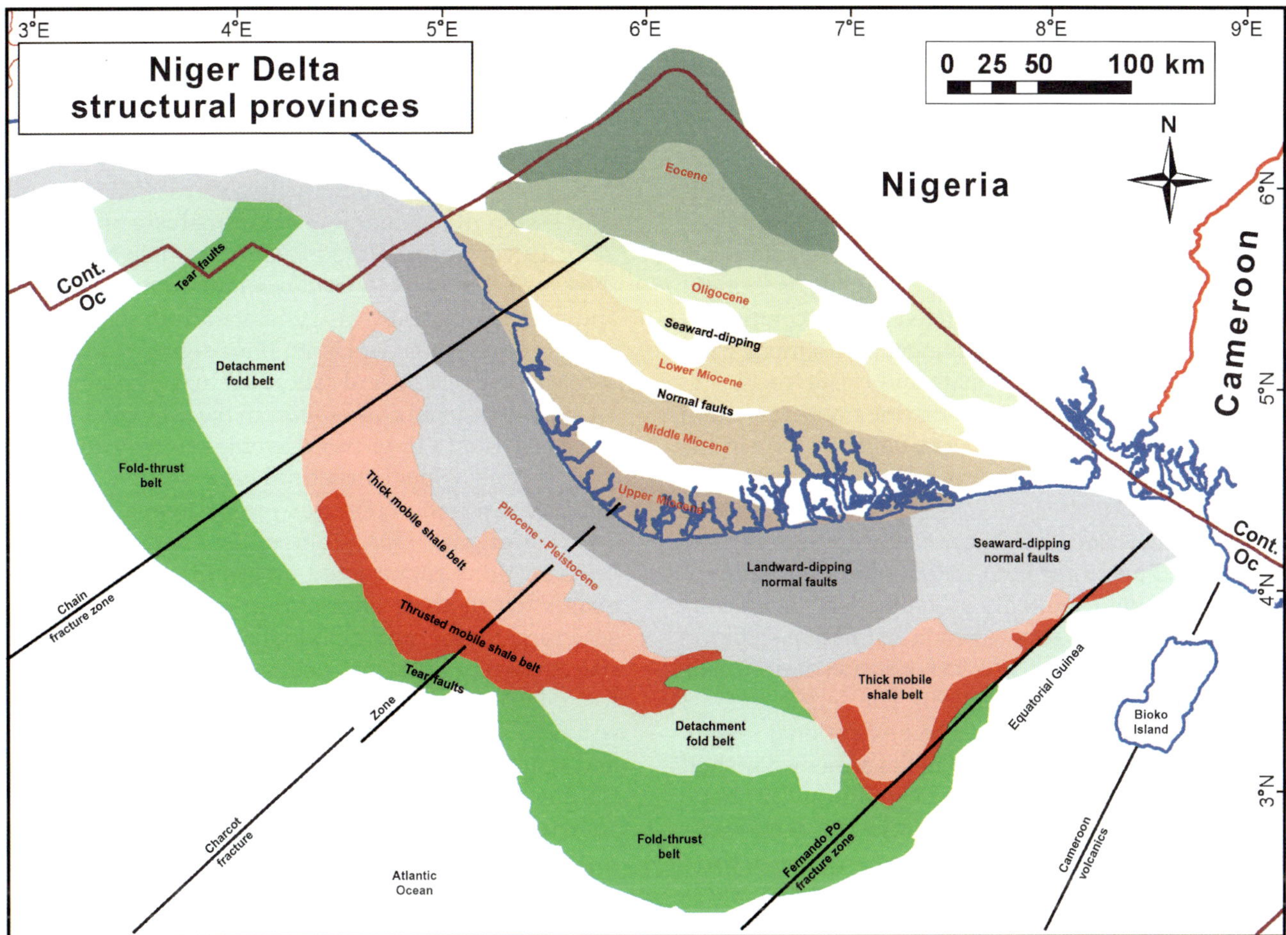

Figure 1. Regional structural provinces map of the Niger Delta. The age of detached extension is indicated in red text. Structural provinces are indicated in black text. Cont = continental crust; Oc = oceanic crust; 50 km = 31.1 mi.

pressure prediction in traps, seal capacity, and drilling hazard identification.

The Niger Delta is a major hydrocarbon-bearing province and a significant location for testing of hypotheses related to mobile shale tectonics. It is a classic example of a linked extensional-contractional system developed on a passive margin driven by gravity and sediment loading. Mobile shale forms the ductile substrate that links the updip extension and downdip contraction, and is also responsible for the formation of most of the major structural features in the delta. In the Niger Delta, the mobile substrate is a thick, overpressured, shale-rich lower Tertiary section characterized by seismic transparency, low seismic velocity, low density, and ductile deformation style. The mobile shale in a gross regional sense is a stratigraphic unit, although ductile shale layers occur at several stratigraphic levels, whose mobility can change over time depending on the migration of pore fluids.

In the Niger Delta, the rich two- and three-dimensional (2-D and 3-D) seismic, well, and gravity databases are important for formulating a regional understanding of mobile-shale-related structures and their evolution. Some of the issues regarding mobile shale tectonics that this chapter will address include (1) geometry of mobile-shale-related structures, (2) physical properties of mobile shale, (3) origins of mobile shale structures, (4) kinematics of mobile shale related structures in the Niger Delta, and (5) evolution of mobile-shale-related structures and their influence on sedimentation.

NIGER DELTA REGIONAL SETTING

The Niger Delta consists of Eocene–Holocene siliciclastic deposits up to 12 km (7 mi) thick and convex to the southwest (Figure 1) (Doust and Omatsola, 1990; Reijers et al., 1997). The oceanic basement is marked by several moderate- to high-relief Cretaceous oceanic fracture zones and associated fossil transform faults (Wu and Bally, 2000). The Charcot fracture zone basement high separates the Tertiary delta into two main lobes

(Figure 1). The Cameroon volcanic line is an Oligocene–Holocene intraplate volcanic chain that forms a southeastern buttress to progradation of the Niger Delta (Figure 1) (Burke, 1972).

The continental crust in the Niger Delta area includes Neoproterozoic granitic and low-grade metamorphic rocks overlain locally by Paleozoic and lower Mesozoic sedimentary rocks (Reijers et al., 1997). Oceanic crust, which underlies most of the delta (Figure 1), is thought to be Albian in age but is undated. An unconformity separates the oceanic and continental basement from Albian–Maastrichtian carbonate and clastic rocks. Paleocene marine shales record a marine transgression in the Benue trough. The Niger Delta began to form in the Eocene as clastics prograded over the continent-ocean boundary (Figure 1).

The Eocene–Holocene Niger Delta stratigraphy has been divided into three lithostratigraphic units that represent major environments of deposition: Benin Formation terrestrial clastics, Agbada Formation shallow-marine deltaic clastics, and Akata Formation deep-marine shales (Doust and Omatsola, 1990). The deltaic deposits formed in a shelf-margin-detached extensional system and prograded to the southwest to the present-day shelf edge. The delta assumed a convex to southwest shape as sediment built up and out over time. The deep-water Akata Formation is 2000–7000 m (6562–22,966 ft) thick (Doust and Omatsola, 1990; Reijers et al., 1997) and is the protolith of the mobile shale of the Niger Delta as referred to in this chapter. The depositional age of the mobile shale is thought to be mostly Paleocene–Eocene based on limited well data and regional seismic correlations. Other deep-water deposits include siliciclastic rocks deposited on the slope and basin floor in the Oligocene–Holocene.

DATA AND METHODS

A rich array of data was used in construction of the structural provinces map (Figure 1) and the regional cross section (Figure 2). Structural analysis used regional 2-D seismic data with a line spacing of 1–10 km (0.6–6 mi) and several proprietary and speculative 3-D seismic data sets. Wells and biostratigraphic data were used to constrain stratigraphic horizon picks and input to regional seismic lines and cross sections (Figure 3).

Gravity data were used in map-based structural analysis of mobile shale physical properties (Figure 4) and in construction of 2-D gravity and structural models (Figure 5, data courtesy of Fugro-Robertson). A gravity model was integrated with the depth model of the regional cross section using LCT 2MOD (Fugro-LCT) on gravity data from Fugro-Robertson. The gravity profile is extracted from a high-pass-filtered grid of the Bouguer gravity with the cutoff set at 30 km (19 mi). Density and depth relations in the gravity model are derived from nearby well data.

Regional seismic velocity data were used for depth conversion of seismic lines and analysis of regional velocity variation related to mobile shale properties (Figure 6). The velocity grid used for depth conversion and velocity analysis was generated from regional 2-D seismic data using GeoDepth (Paradigm). Stacking velocities were picked and calibrated to well velocities. A 3-D velocity model was built from the calibrated seismic velocities using a layer-based average velocity approach.

Two-dimensional structural cross section construction and restoration (Figure 7) were done with GeoSec 2-D (Paradigm). This is one of six regional balanced cross sections constructed across the Niger Delta to evaluate structural and stratigraphic interpretation, structural timing, and mobile shale evolution. Regional paleogeographic reconstructions of the delta were used for construction of paleobathymetry in restored cross sections. The present-day depths of the shelf margin and slope-basin transition were used in construction of the paleobathymetric profiles. Flexural-slip algorithms were used for restoration of contractional structures and inclined shear algorithms used for restoration of extensional structures. The top of the mobile shale was restored using the same algorithm as layers above the mobile shale, whereas the basal detachment, which is the base of the mobile shale, was reconstructed using area balance calculations assuming no change in the area. The cross section restoration has an implicit assumption of 2-D plane strain, which is a reasonable first-order assumption in this area because the section is perpendicular to the strike of major extensional and contractional structures. Additionally, although mobile-shale deformation is 3-D in many places, the restored section in Figure 7 is located where the mobile-shale-cored structures are linear, and plane strain is a reasonable first approximation of deformation. Decompaction and isostasy were not applied in these restorations. The impact of ignoring these corrections is mostly in small changes in the dip of restored sections and does not affect the conclusions about mobile shale kinematics.

MOBILE SHALE STRUCTURAL STYLES AND GEOMETRY IN THE NIGER DELTA

The Niger Delta is a linked, detached extensional-contractional system. Updip extension is linked to downdip contraction at depth by a ductile shear zone and by

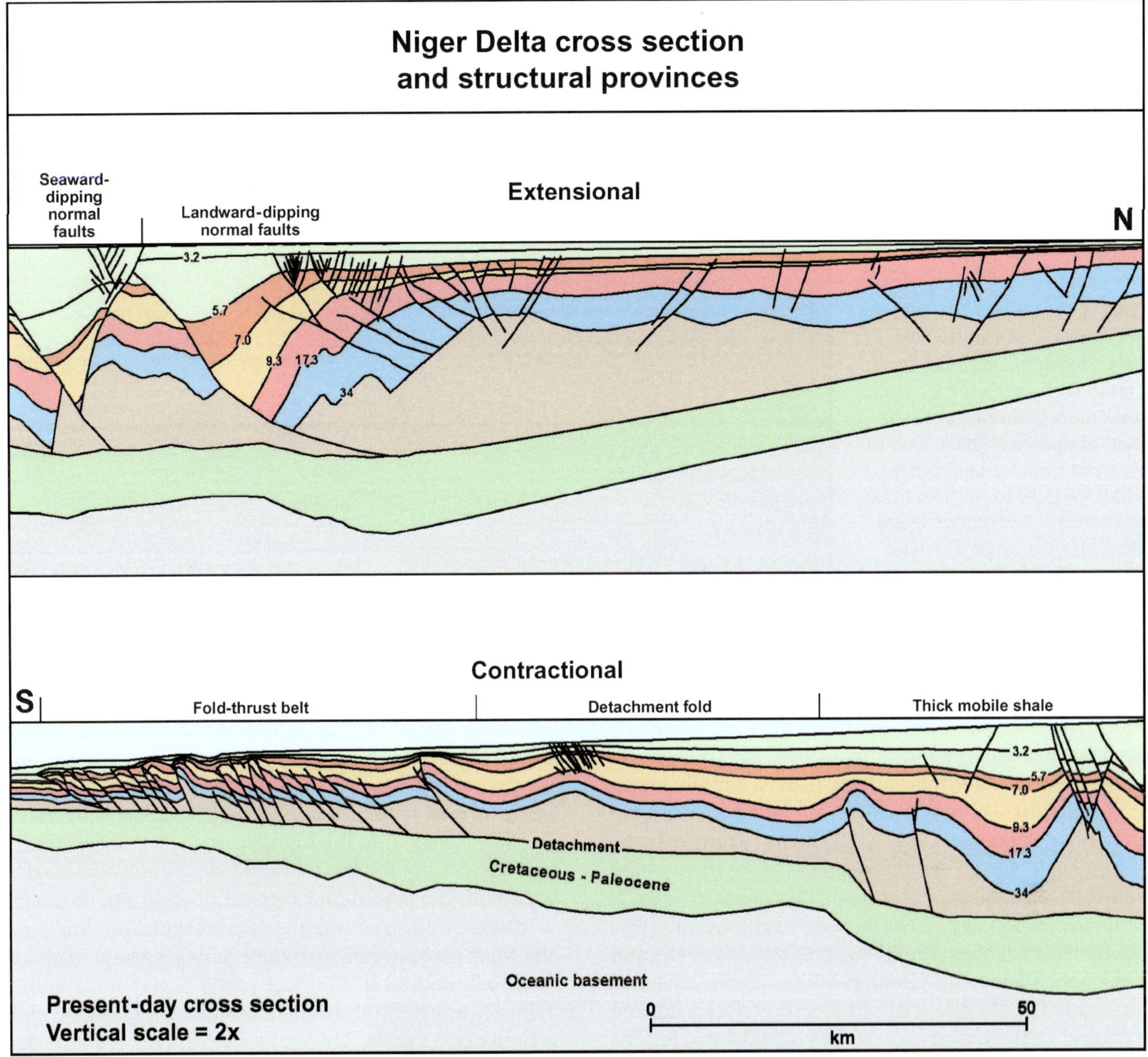

Figure 2. Regional cross section, central Niger Delta; 2× vertical exaggeration. Structural provinces are labeled. Numbers are approximate ages of the horizons in millions of years before present. 50 km = 31.1 mi.

detachment in mobile shale and linked laterally by delta-margin tear faults (Figure 1). The tear faults are localized above basement highs set up by underlying oceanic fracture zones. These tear faults, however, are detached in the mobile shale and do not penetrate to the basement.

The updip extensional province consists of landward-dipping and basinward-dipping normal faults below the area of the present-day onshore and shelf. Landward-dipping fault systems predominate in the central part of the shelf, whereas seaward-dipping faults predominate onshore and at the margins of the delta to the northwest and east. Mobile-shale-cored structures form structural highs on footwalls of normal faults in the extensional province. Most of these are interpreted to be older contractional detachment folds with younger superposed extension as indicated by the older growth packages associated with the footwall highs (Figure 2). Ductile detachment folds (Hardy and Poblet, 1994) cored by thickened mobile shale are present in all the major structural zones, including the extensional belt in footwalls, in the thick mobile shale belt, the detachment fold belt, and the outer fold and thrust belt (Figures 2, 3). Shelf-margin deltas form important hydrocarbon reservoirs in various structural fold and fault traps as well as combination stratigraphic-structural traps in the extensional province.

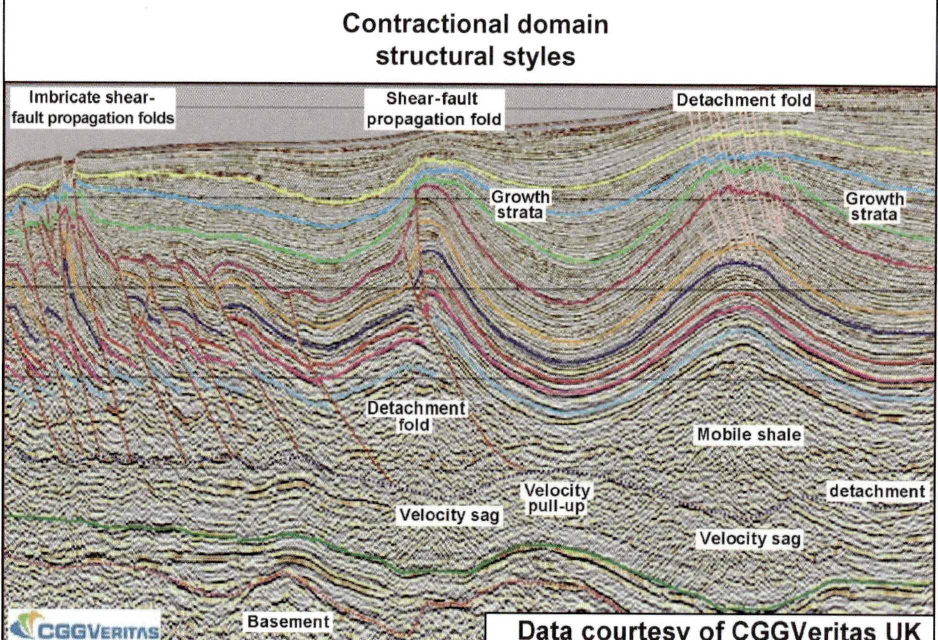

Figure 3. Seismic line across the contractional part of the Niger Delta, coincident with the regional cross section (Figure 2). Note zones of thick mobile shale beneath detachment folds, and mobile shale in shear-fault propagation folds. Also note velocity sags beneath thick zones of low-velocity mobile shale. Data are courtesy of CGG Veritas.

The thick mobile shale province (Figure 1) is a zone of high-relief mobile-shale-cored structures and shale withdrawal minibasins. The cross section (Figure 2) is located near the boundary of the thick mobile shale province and the inner fold-thrust belt and thus shows less structural relief than a typical thick mobile shale. High-relief shale-cored structures are interpreted as contractional detachment folds with a weak, overpressured mobile shale core. Previous workers have interpreted shale diapirs in the thick mobile shale province (Figure 1) (Cohen and McClay, 1996; Morley and Guerin, 1996; Van Rensberger et al., 1999; Wu and Bally, 2000; Hooper et al., 2002; Elsley and Graham, 2006). The relative importance of shale diapirism and contractional folding is treated in the Discussion section of this article. The thick mobile shale structural province also includes numerous gas escape features and mud volcanoes. These features are typically 100–1000 m (328–3281 ft) across, in contrast to the much larger shale-cored structures, which are typically several to tens of kilometers (one to several miles) across. The mobile shale province is also characterized by minibasins with thickened sections that have experienced withdrawal of the mobile shale substrate because their stratigraphic surfaces lie below regional stratigraphic levels. In places, complete shale withdrawal results in the Oligocene–Miocene sedimentary section being grounded on the detachment surface (Figure 2). This is similar to a salt weld in genesis and geometry. Similarly, the high-relief shale-cored structures are above regional stratigraphic levels, indicating an element of contraction and uplift. The thick mobile shale belt also exhibits areas of shale rollers where the highs are flanked by contemporaneous normal faults that develop on one or two of the flanks of the structural high at the same time the highs rose above regional levels. Extensional faulting of mobile-shale-cored anticlines is interpreted as resulting from oversteepening and collapse of margins of the structural highs formed by either contraction and/or some form of vertical deep-seated diapirism. Movement of mobile shale into the highs seems to be kinematically and temporally related to minibasin withdrawal associated with the lateral and vertical flow of mobile shale.

Shale-cored structures in the extensional province and the thick mobile shale province have a variety of structural orientations and aspect ratios. Some are rounded with a low aspect ratio, whereas others are elongate with a high aspect ratio. The elongate features are oriented parallel, perpendicular, and oblique to the regional transport direction as determined by the orientation of contemporaneous major normal faults and fold-thrust structures. Shale-cored anticlines perpendicular to the regional transport direction are interpreted to be mostly contractional detachment folds. Some of these features have minor associated thrusts, further indicating their contractional origin. Shale-cored anticlines that are oriented subparallel with the regional transport direction are interpreted to be oblique contractional features because of their orientation perpendicular to both major extensional faults and fold-thrust belts. Some of these features also have minor associated thrusts. In addition, en echelon geometries of secondary normal faults and folds associated with the strike-parallel shale ridges indicate a component of strike slip for these features. Most of the hydrocarbon discoveries in the thick

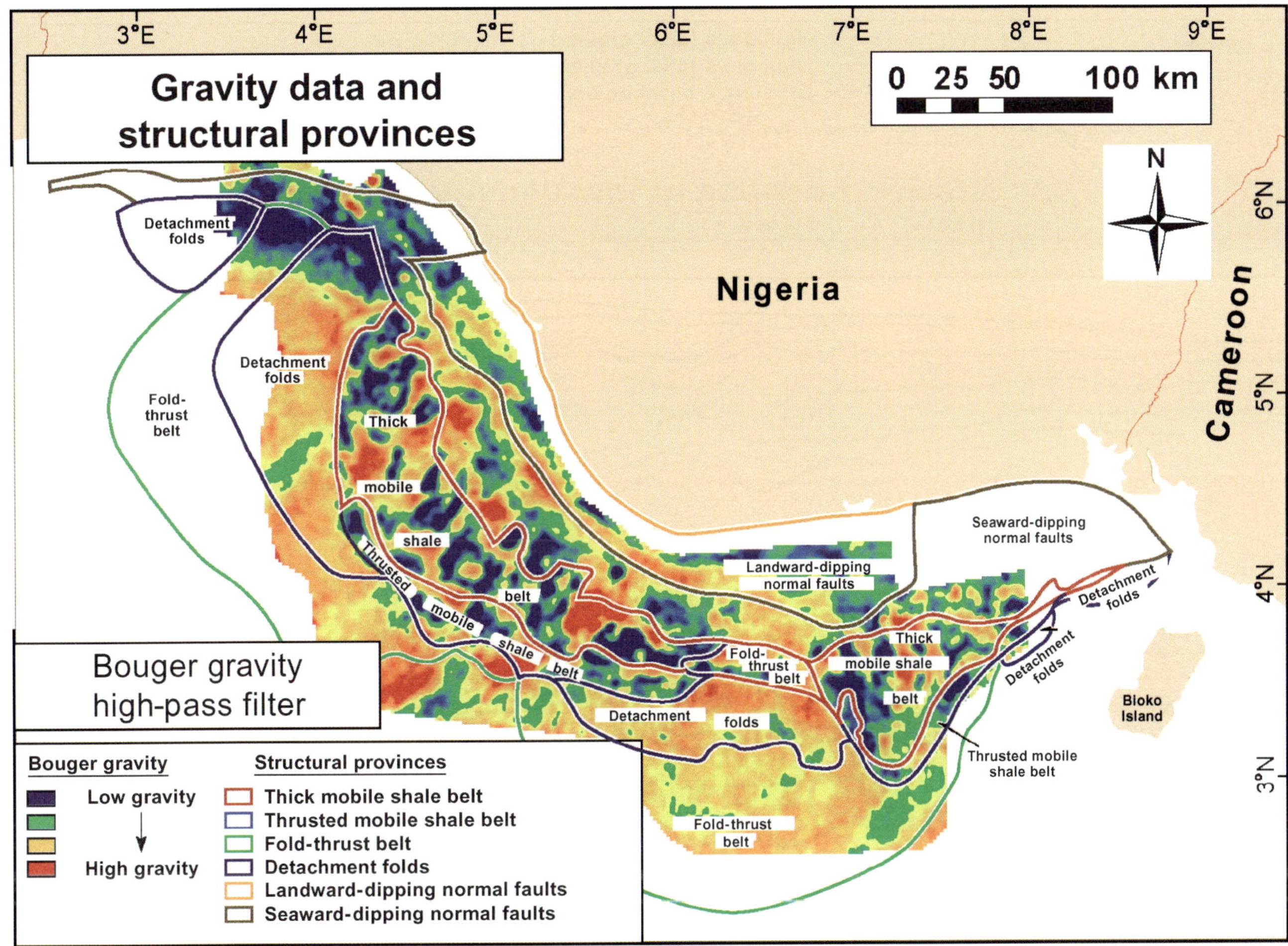

Figure 4. Gravity map of the Niger Delta; high-pass-filtered Bouguer gravity data, courtesy of Fugro-Robertson. Note the correspondence of low-gravity zones and the structural province of thick mobile shale, consistent with high degrees of overpressure in the mobile shale. 50 km = 31.1 mi.

mobile shale province have been within combination structural-stratigraphic traps on the flanks of high-relief structures.

The moderate-relief detachment fold province is an area of low contractional strain, dominated by translation on a regional detachment. Dominant structural styles include low- to moderate-relief symmetric detachment folds cored by mobile shale that may have crestal normal faults and minor thrusts (Figures 2, 3). As in the thick mobile shale province, minibasins are below regional levels and have experienced mobile shale withdrawal, whereas tectonic thickening has occurred in anticlines. These detachment folds, in association with deep-water clastic reservoirs, form important hydrocarbon discoveries in Nigerian deep water.

The outer fold-thrust belt of the Niger Delta is characterized by shear fault-bend and fault-propagation folding (Corredor et al., 2005). The fold styles include isolated structures and imbricate thrust complexes above thin mobile shale, which is the same stratigraphic horizon as the cores of updip moderate- and high-relief detachment folds described above (Figures 1–3). The mobile shale layer in the outer fold and thrust belt is responsible for the simple and pure shear fault-bend folds because shear is concentrated in this weak layer and strata immediately above the mobile shale (Corredor et al., 2005). Classic fault-bend and fault-propagation folds, including duplex and imbricate styles (Suppe, 1983; Suppe et al., 2004), are present where the mobile-shale layer is thin or not acting as a ductile layer. The outer fold and thrust belt generally propagates basinward, but in detail, variation in forward- and back-stepping imbrication patterns exists (see also Corredor et al., 2005). Exploration in the outer fold and thrust belt is immature, but some discoveries have been made in combination traps.

Structural styles in the contractional part of the Niger Delta can be interpreted in terms of critical taper wedge

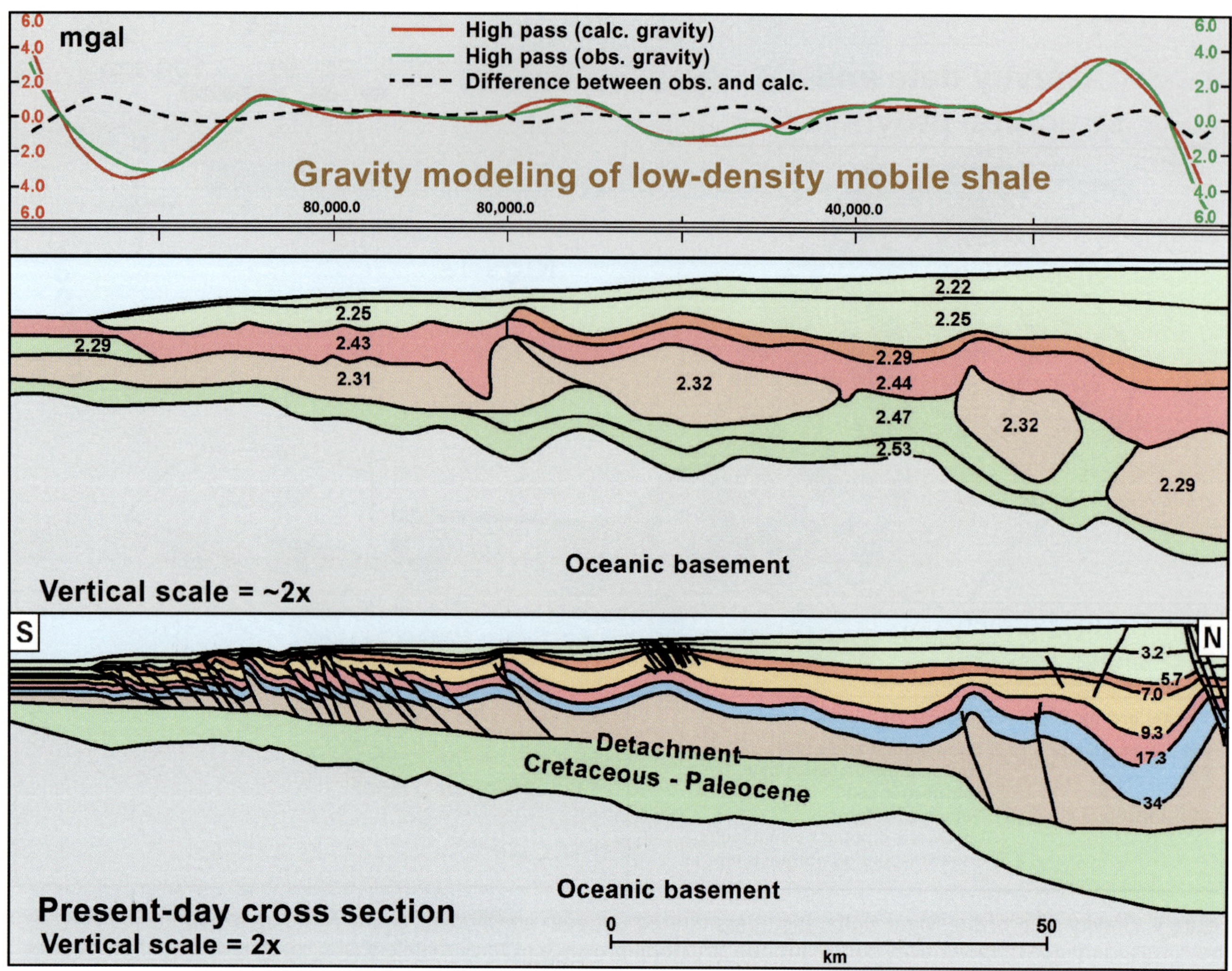

Figure 5. Two-dimensional gravity model of the cross section in Figure 2. Note the correlation between low-density zones and mobile-shale-cored structural highs. Numbers in gravity profile are density values (g/cm^3). Calc. = calculated; obs. = observed; 50 km = 31.1 mi.

mechanics (Billotti and Shaw, 2005; Corredor et al., 2005) and have implications for properties of the mobile shale detachment. The critical taper wedge angle is measured between the bathymetry and basal detachment; the strength of the detachment relates to the critical taper wedge angle, with strong detachments producing high taper wedge angles (Billotti and Shaw, 2005). Critical taper wedge angles have been measured on the regional cross section in different structural provinces (shown in 2× vertical exaggeration in Figure 2 and 1:1 in Figure 7). In the thick mobile shale belt, the critical taper wedge geometry has an internal angle of 2°. In the detachment fold belt, the critical taper wedge angle is 3°, and in the fold-thrust belt, the angle is 4°. These changes show an increasing critical taper wedge angle basinward, which implies an increasing strength of the basal detachment basinward. This is consistent with seismic evidence for thinning of mobile shale and increased faulting basinward.

MOBILE SHALE PHYSICAL PROPERTIES

Mobile shale in the Niger Delta is characterized by low density and low seismic velocity. Density analysis of the mobile shale was performed using both map-based comparison of gravity and structural elements, and 2-D gravity modeling of depth sections created from seismic interpretations. Results of these models were used to evaluate the relation of density to overpressure and structure. A map-based analysis shows a zone of low gravity and density in the thick mobile shale belt (Figure 4). Densities increase basinward in the moderate-relief detachment fold belt and the outer fold-thrust belt. These

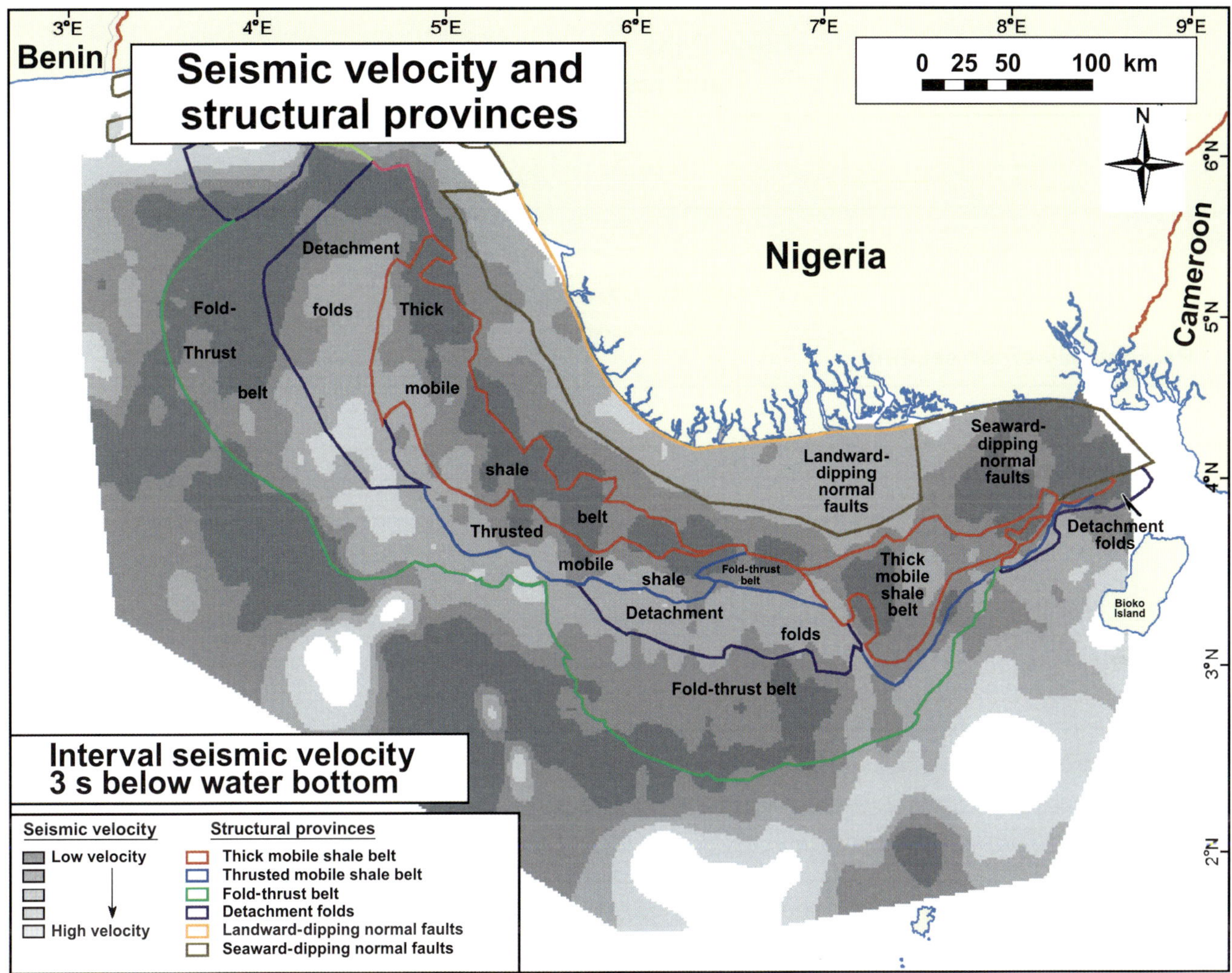

Figure 6. Seismic velocities of the Niger Delta. Contours are interval velocities at 3 s below water bottom. Note the correspondence of low-velocity zones and the structural province of thick mobile shale, consistent with high degrees of overpressure in the mobile shale (compare with Figure 4). 50 km = 31.1 mi.

changes in density are interpreted to correspond to the basinward-decreasing thickness of overpressured mobile shale, which corresponds to changes in structural style, from high-relief detachment folds to moderate-relief detachment folds to shear fault-bend folds. This represents a change from a more ductile to a more brittle structural style basinward. Individual structures in the thick mobile shale belt show gravity lows over structural highs, indicating a strong correspondence of structural highs with thick overpressured low-density shale-rich material (Figure 4).

Two-dimensional gravity modeling was conducted on several cross section depth interpretations to evaluate the density contrast necessary to match the gravity anomalies (Figure 5). Density data from nearby wells were used to populate the density model in the section above the mobile shale. The model shows that the cores of detachment folds correspond to low-density zones, which are interpreted as thick overpressured mobile shale. The density contrast between the mobile shale and surrounding rocks is 0.12–0.18 g/cm^3. Some synclines in the model require no low-density material, corresponding to thin to absent mobile shale. Two-dimensional gravity models in other areas of higher relief mobile shale structures require a greater thickness of low-density material, indicating thick zones of overpressured shale.

Zones of thick mobile shale correspond to low-seismic-velocity zones. Low seismic velocity is evident in seismic traveltime sags beneath thick mobile shale zones (Figure 3), by sonic velocities in wells that penetrate overpressured shale, and in regional velocity trends.

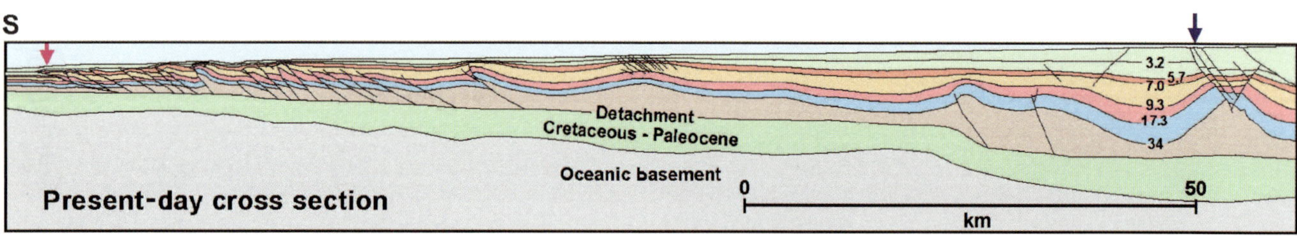

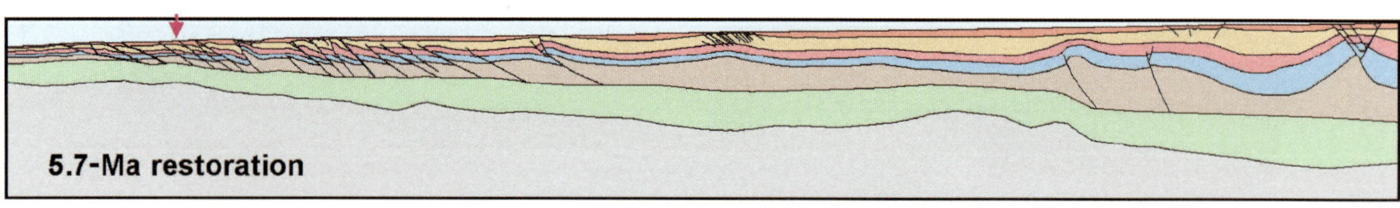

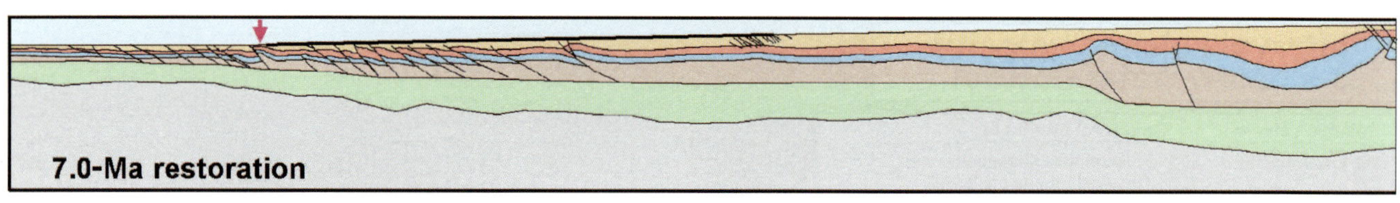

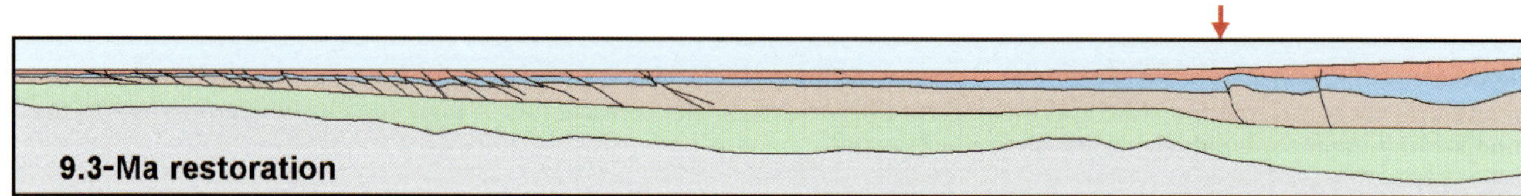

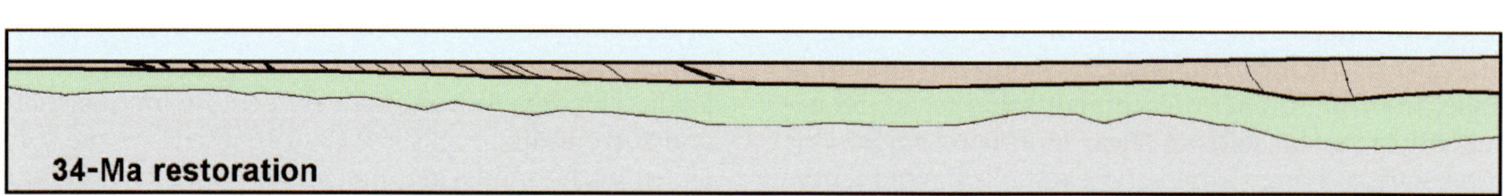

Figure 7. The 1:1 regional cross section and restoration, central Niger Delta. Blue arrows mark the downdip limit of regional extension, and red arrows mark the downdip limit of regional contraction for each restoration. Note the basinward propagation of extension and contraction through time, as well as the lateral and vertical movement of mobile shale substrate. 50 km = 31.1 mi.

A deltawide 3-D velocity model was built from stacking velocities derived from 2-D seismic data and calibrated to well data. Where seismic data are poor quality, the velocity analysis is uncertain, but velocities determined in the well-imaged section below the mobile shale help constrain the velocity model. This model shows low-velocity zones in areas of inferred thick mobile shale at time slices from 2.0 to 3.5 s below water bottom (Figure 6).

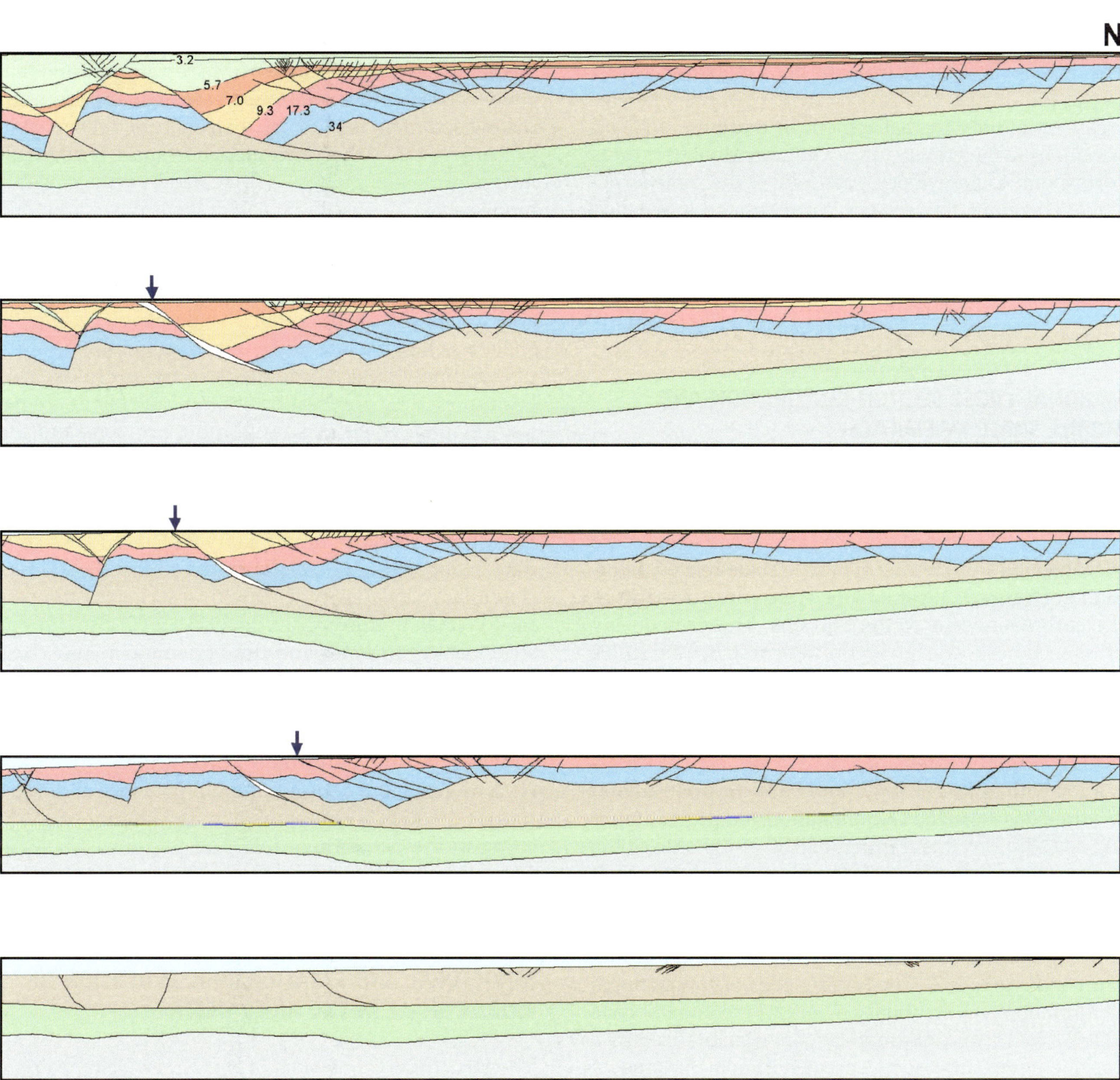

Figure 7. (cont.).

Also, the mobile shale is inferred to be thick in these areas because of a thick zone of poor seismic data quality. Figure 6 shows zones of low velocity coincident with both the structural province of thick mobile shale and on the abyssal plain downdip of the fold-thrust belt. In contrast, the detachment fold belt and fold-thrust belt are characterized by relatively high velocity. These changes in velocity are related to changes in stratigraphic levels at 3 s below water bottom (the velocity slice in Figure 6). In the thick mobile shale province and abyssal plain, the Paleocene–Eocene section is 3 s below water bottom, whereas in the detachment fold and fold-thrust provinces, the Eocene–Miocene section is present at 3 s below water bottom. This relationship indicates that the Paleocene–Eocene section is of relatively low velocity and likely shale rich and overpressured even where it is undeformed on the abyssal plain. The increase in velocity from the thick mobile shale belt to the fold-thrust belt is consistent with a similar increase in gravity in that region. In addition, detailed variations in the gravity map over individual structures correspond to low-velocity zones over the same structures. Therefore, low seismic velocities and low densities characterize the structural province of thick mobile shale with increasing velocity and density in the zones of detachment folds and fold-thrust belts.

REGIONAL CROSS SECTION RESTORATION AND MOBILE SHALE KINEMATICS

A two-dimensional cross section and restorations were constructed to evaluate the structural-stratigraphic framework of regional interpretations and to examine the kinematics and timing of mobile-shale-related deformation (Figure 7). Restorations were constructed from the early Oligocene to the Pliocene. The present-day cross section shows the distribution of structural styles from updip extensional systems to downdip contractional systems.

The restorations show basinward propagation of the leading edge of extensional and contractional deformation with time. The large landward-dipping fault on the shelf has localized sedimentation and growth faulting as sediment was trapped on the updip side of the fault until the half graben filled and spilled over the shelf margin. Younger extensional deformation is mostly on south-dipping normal faults downdip of the landward-dipping faults (see fault south of major north-dipping fault on Figures 2, 7). The leading edge of contractional deformation was stationary from early to late Miocene (23–9.3 Ma). This deformation front was localized by a basement step and also by a greater original thickness of mobile shale on the north side of this basement ramp, which sets up a zone of weakness to the north and a buttressing effect to the south. The contractional deformation front propagated southward from 9.3 Ma to the present day (Figure 7).

The restorations show an approximate balance of extension and shortening through time because the sections are pinned at the south end, and the north end is a loose line that remains within 5% of the original line length. The total amount of shortening in the restoration is 19 km (12 mi) based on the restoration of the northernmost thrust fault in the 34-Ma (early Oligocene) restoration.

In Figure 7, the mobile shale is labeled in the cross section as the interval above the regional detachment and below the 34-Ma (base Oligocene) surface. The restorations show that the original thickness of the mobile shale decreased basinward (34-Ma restoration) and that a mass transfer of mobile substrate seaward occurred during sediment loading by shelf-margin deltas in half grabens. Much of the accommodation space for the sediments in the half grabens was created by withdrawal of mobile shale. In places, the half grabens have grounded on the regional detachment and all the mobile substrate has been withdrawn. Although mobile shale was being withdrawn from beneath the shelf half grabens and minibasins in the mobile shale province, mobile shale was flowing laterally and vertically into the cores of anticlines, creating ductile detachment folds. This is supported by observations that the synclines are below regional levels of stratigraphic horizons and the anticlines are above regional levels.

Other workers in the Niger Delta have published cross section interpretations and restorations similar to Figure 7 based on high-quality 2-D and 3-D seismic data. Hooper et al. (2002) published a cross section that is similar in location and structural styles to our cross section (their figure 6). Corredor et al. (2005) also showed structural geometries and timing similar to our cross section interpretation in their figures 18 and 19. Their figure 19 is west of our Figure 7 location and involves a deeper detachment level in the outer fold belt. The shortening estimate in their cross section is 17.5 km (10.8 mi) from the detachment fold province and outer fold belt, which is comparable to the amount of shortening in the detachment fold and fold-thrust belts in our restoration (Figure 7).

STRUCTURAL AND STRATIGRAPHIC EVOLUTION OF MOBILE SHALE IN THE NIGER DELTA

The structural framework of the Niger Delta exerts a fundamental control on paleogeographic boundaries in the present and the past. Present-day bathymetry is aligned with structural boundaries and therefore can be used to predict paleogeographic boundaries in the evolution of the Niger Delta. The present-day shelf break is defined by the downdip limit of major normal faults. Shelf-margin deltas were deposited in the hanging walls of normal faults as the shelf edge prograded through time across the Niger Delta. The present-day basinward limit of contractional structures sets up a change in slope

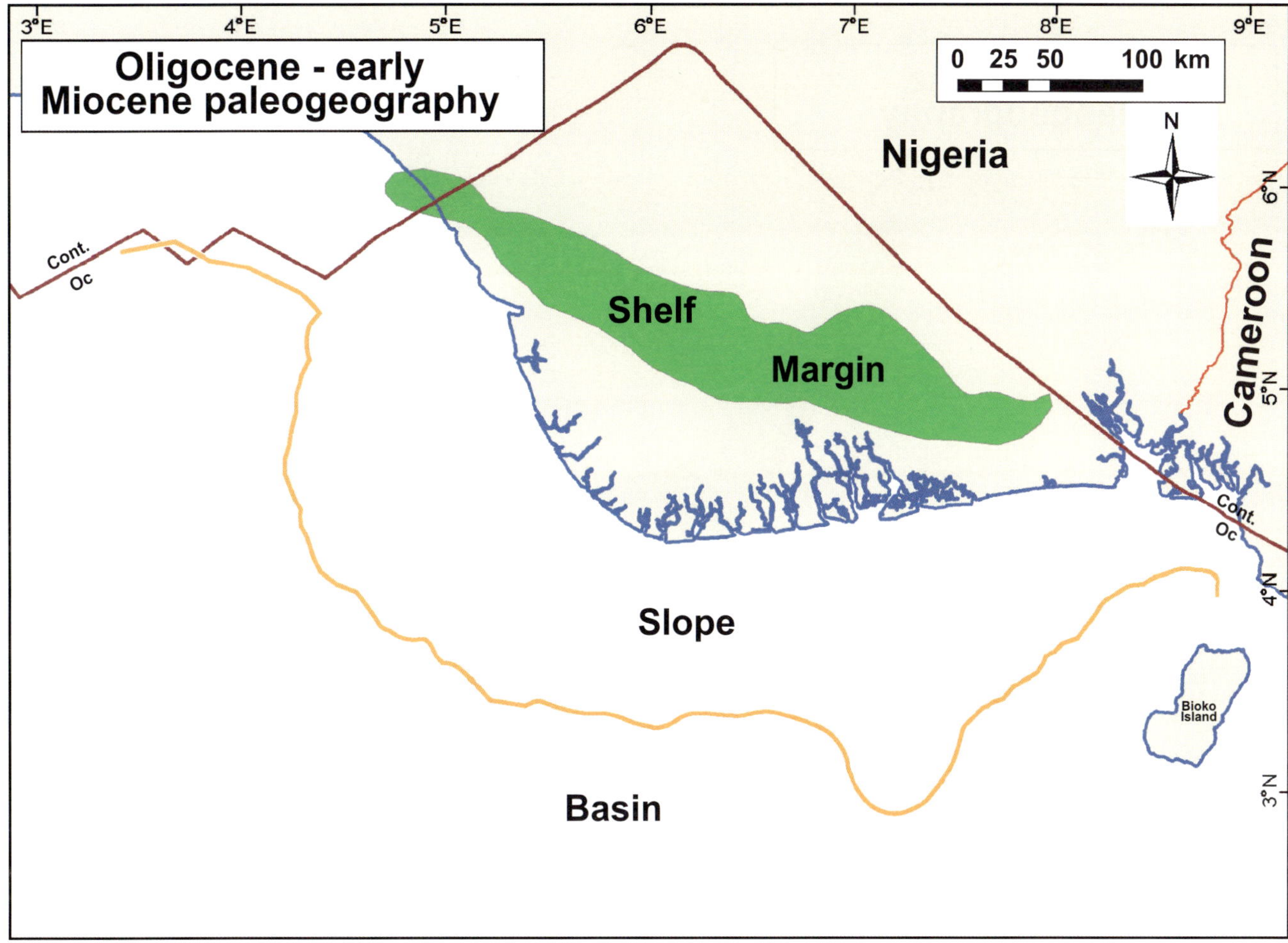

Figure 8. Oligocene–early Miocene paleogeographic reconstruction. Cont = continental crust; Oc = oceanic crust; 50 km = 31.1 mi.

that defines a structurally controlled toe of slope. Although not the actual depositional toe of slope, the steep gradient change separates more weakly confined, distributive systems downdip from more channelized systems and minibasin depositional systems on the slope.

By defining the paleostructural limits of extension and contraction through time, the paleogeography can be better defined. Structural timing maps were created across the Niger Delta based on growth timing on individual structures as observed in seismic data. These maps were used to constrain the paleoshelf margin and paleo-toe-of-slope for several time intervals. Figures 8 and 9 show the paleogeographic boundaries for the Oligocene–early Miocene and late Miocene, respectively.

In the Oligocene–early Miocene, the downdip limit of extension defines the shelf margin, which is onshore, and the outboard limit of contraction defines the structurally controlled toe of slope (Figure 8). On the slope, deposition is controlled by actively growing structures and minibasins, characterized by confined channel complexes on the flanks of structures and ponded deposition in minibasins. Downdip of the toe of slope, deposition is more distributive on the flat basin floor.

During the late Miocene (9.3 Ma), the shelf margin is dominated by large landward-dipping normal faults near the present-day shelf margin (Figure 9). These landward-dipping faults lock the position of the shelf from late Miocene to Pliocene (9.3 to 3.2 Ma, Figures 7, 9). The downdip limit of contraction controls the toe of slope. Deposition is mainly occurring in the large extensional systems and on the highly deformed slope, with thin deposits reaching to the basin floor (Figure 9). Progradation of these paleoenvironments leads to an overall regressive stacking of facies so that more proximal facies are located higher in the stratigraphic section.

The contractional front of the eastern lobe of the Niger Delta prograded rapidly basinward in the late Miocene (9.3–5.7 Ma, Figures 7–9). The large influx of sediments in the eastern lobe in the late Miocene may have perturbed the critical taper wedge angle of the delta and led to basinward progradation of the contractional front (see also Billotti and Shaw, 2005).

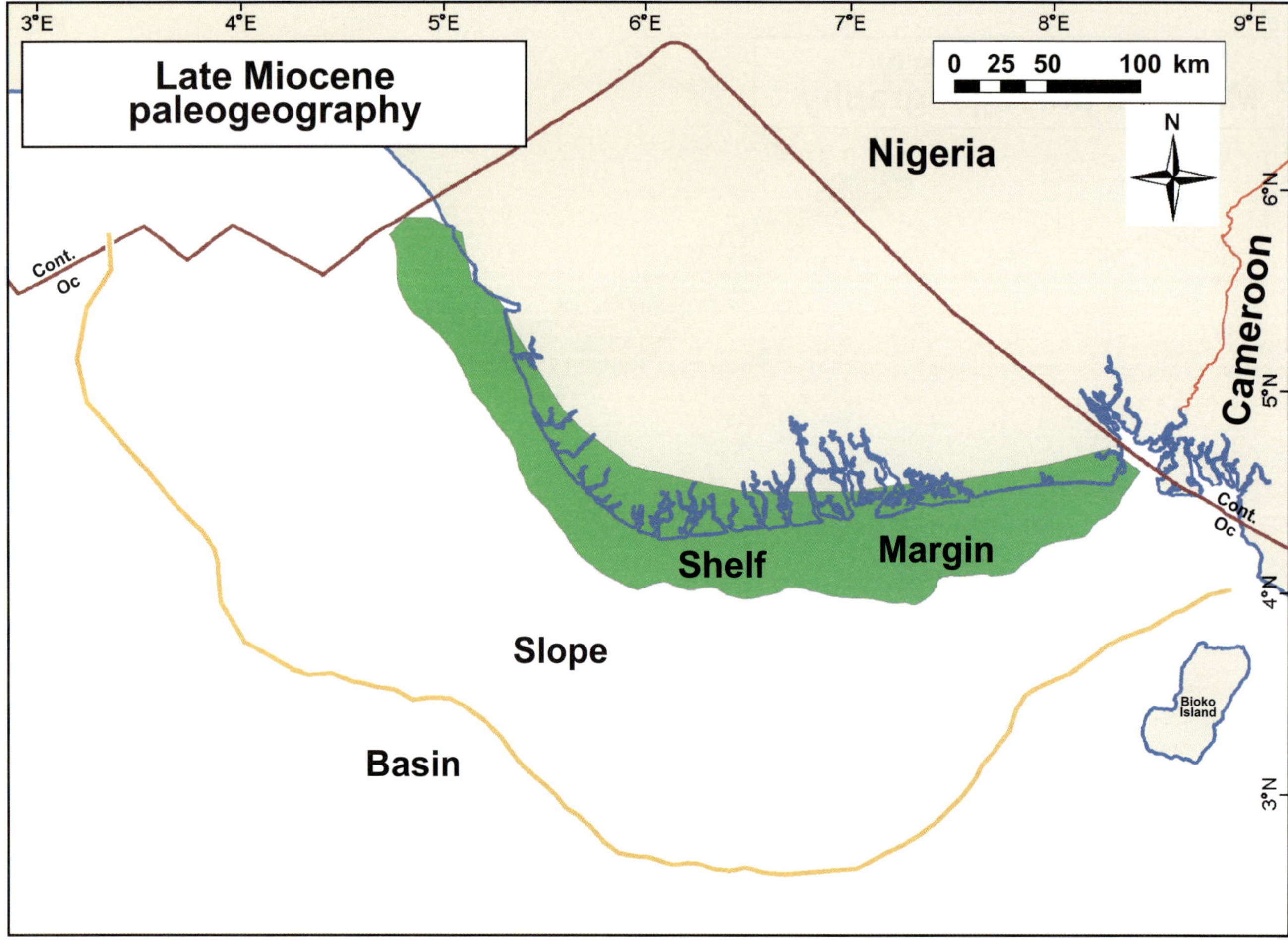

Figure 9. Late Miocene paleogeographic reconstruction. Note the progradation of the shelf margin and structurally defined toe of slope from the early (Figure 8) to late Miocene. Cont = continental crust; Oc = oceanic crust; 50 km = 31.1 mi.

DISCUSSION OF MOBILE SHALE IN THE NIGER DELTA

Mobile-shale-cored highs can be interpreted as diapirs, contractional detachment folds, diatremes, or some combination of these types. Many authors have cited shale diapirism as a major structural style in the Niger Delta (Cohen and McClay, 1996; Morley and Guerin, 1996; Van Rensberger et al., 1999; Wu and Bally, 2000; Hooper et al., 2002; Elsley and Graham, 2006). The function of shale diapirism in the sense of reactive or passive diapirs is difficult to substantiate in the Niger Delta. Diapirism as shown by the authors above is, however, a permissible interpretation because poor imaging allows multiple interpretations in much of the mobile shale province. With that caveat, we shall discuss some of the evidence against reactive or passive shale diapirism and evidence in favor of an interpretation of a structural genesis as contractional detachment folds in the mobile shale and detachment fold belts of the Niger Delta (Figure 1).

The structural styles in the contractional part of the Niger Delta are consistent with physical models of weak substrate with strong overburden undergoing contraction as shown by Costa and Vendeville (2002). The lower part of the cross section in Figure 2 shows moderate- to high-relief mobile-shale-cored anticlines on the right (north) side, comparable to the structural style in figure 1b of Costa and Vendeville (2002). Their physical model has a thick weak substrate of silicone and strong overburden of sand and has undergone contraction. The experiments vary the thickness of the weak layer and the presence of buttresses, which impact the resulting structural styles. The structural styles in the physical model include high-relief ductile detachment folds with secondary thrusts, and detachment folds and shear fault-bend folds with varying vergence, similar to structural styles in the Niger Delta (Figure 2). The left (south) end of the lower part of the cross section in Figure 2 shows a series of shear fault-bend folds in the fold and thrust belt. These structures are analogous to a physical model shown in figure 4 of Costa and Vendeville (2002). Thus, physical models provide an analog for weak substrate and strong overburden undergoing contraction as a genetic model for structures in the Niger Delta. Furthermore, the thickness of the weak substrate has an impact on the structural expression.

Additional evidence for the contractional nature of shale-cored highs is that growth wedges associated with these anticlines have the appearance of onlapping wedges formed by limb rotation and/or kink-band migration such as found in contractional detachment folds (Figure 2) (see also Shaw et al., 2005). The geometry of the detachment folds is also similar to that found in salt-cored detachment folds (Rowan et al., 2004).

The fact that crests of the anticlines are above regional stratigraphic levels indicates that there has been uplift of the cores of the folds (Figure 2). In addition, erosion of strata on the crests of folds and development of angular unconformities and onlaps attest to the uplift of these structures. Finally, the balanced cross section restoration (Figure 7) incorporates flexural slip and shortening in the thick mobile shale province. The balance of extension and contraction through time is consistent with a component of contraction in the mobile-shale province to help balance the amount of extensional strain observed updip. These observations favor a contractional origin and argue against simple vertical downbuilding and passive diapirism as mechanisms for creating the shale-cored highs.

Synclines in the mobile shale province have mobile-shale withdrawal as indicated by the absence of the mobile shale layer at the bottom of some of the synclines where layered rocks are in contact with the regional detachment as in a weld (Figure 2). Also, the stratigraphic level in the syncline is below regional, indicating withdrawal of mobile substrate.

Extension can initiate reactive diapirism by the local structural thinning of overburden load (Vendeville and Jackson, 1992). In the Niger Delta, extensional faults are associated with shale-cored highs (Van Rensberger et al., 1999; this study). However, we do not interpret these features as reactive diapirs, because extension is genetically unrelated to the formation of the mobile-shale features for the most part. In most places where normal faulting is associated with shale-cored highs, the extension is a younger event superposed on an older, genetically unrelated, contractional anticline cored by mobile shale. In other places, normal faults are crestal faults with very small displacement, unrelated to the formation of the shale-cored high, and represent a consequence of the shale deformation instead of a cause. In other places, evidence of normal faulting associated with initial stages of anticlinal growth exists. However, these fault systems are generally limited to one flank of the anticline whose other flanks typically have a contractional detachment-fold growth wedge characterized by limb rotation or kink-band migration. Therefore, no compelling evidence for extension triggering diapirism exists as has been described for salt-cored reactive diapirs (Vendeville and Jackson, 1992).

Passive diapirism can be thought of as downbuilding where the crest of the diapir does not rise and the diapir has a long history of growth near the paleo-sea bottom (Vendeville and Jackson, 1992). The function of passive diapirism in genesis of mobile-shale-related features is difficult to evaluate because of the poor seismic data near the crests of these features. Data from a few wells at the crests of these high-relief structures indicate a continuous section above the Eocene mobile shale. This suggests that these mobile-shale-cored features are not passive diapirs because they should have large amounts of missing section above them. Mechanically, passive diapirism of shale-cored bodies is unlikely because overpressure tends to bleed off near the surface, which would immobilize the shale. Also, unless highly overpressured (approaching lithostatic pressure), shale is more viscous than salt and so is less likely to be able to lift even minor amounts of overburden in the mode possible for passive salt diapirs. In places where the crests of the shale-cored highs show erosion, the erosion generally appears to be related to the development of asymmetric imbricate thrust belts as opposed to passive diapirism.

Regarding active diapirism, distinguishing a contractional detachment fold cored by overpressured shale from an active diapir undergoing contraction may be hard. One of the fundamental genetic questions about these features is whether they are detachment folds modified by faulting or diapirs modified by contraction (folding and/or thrusting). These can be hard to distinguish, but the lack of evidence of piercement above the shale features and the association of growth wedges with contractional elements such as folds and thrusts suggest a contractional origin. In addition, gravity modeling and seismic velocity analysis (see above) argue for a thick zone of overpressured shale that cores these structures. Mechanically, this would be consistent with high-relief detachment folds such as those present in salt-cored fold and thrust belts.

Diatremes or explosive mud volcano pipes are prevalent in the mobile shale province. The mud volcano pipes may contain deformation haloes, but these typically are at least an order of magnitude smaller in diameter than the large shale-cored features (100–1000 m [328–3281 ft] across versus ~5–20 km [~3–12 mi] across).

CONCLUSIONS

The Niger Delta is a classic province of mobile-shale-related deformation in a linked extensional-contractional system. Gravity and seismic velocity data indicate that a thick zone of overpressured shale exists in the middle of the delta that thins basinward. The change in thickness of the overpressured shale zone results in a change in structural style basinward from high-relief detachment folds to low-relief detachment folds to shear fault-bend fold systems. Cross section restorations show that withdrawal of mobile shale in extensional half grabens and minibasins was an important sediment accommodation mechanism, and that shale withdrawal zones were balanced downdip by inflation of thick mobile-shale-cored contractional anticlines.

Extensional and contractional fronts prograded downdip through time. The extensional front was marked by lowstand shelf-margin deltas deposited in thick sedimentary basins associated with normal faults. The contractional front was localized by a transition from thick mobile shale to thin mobile shale over a ramp in the basement. A stillstand of this front occurred from the Oligocene to late Miocene. Massive shelf-margin deposition in the late Miocene–Pliocene perturbed the critical taper wedge equilibrium profile of the delta so that there was a significant basinward progradation of the contractional front in the late Miocene.

Sedimentary paleoenvironments were associated with the structurally controlled shelf margin and contractional fronts. The active extensional system localized the position of shelf-margin deltas through time, whereas the contractional front localized the structurally defined toe of slope and the slope-to-basin transition.

ACKNOWLEDGMENTS

The authors thank ExxonMobil Exploration Company for the permission to publish and for underwriting publication costs. We also thank Fugro-Robertson for the permission to publish gravity data. We acknowledge Paradigm Inc. for the use of GeoSec software for cross section construction and balancing, as well as LCT for the software for 2-D gravity modeling, Paradigm Inc. for the GeoDepth software for 3-D velocity model building, and Schlumberbger for the use of the Geoframe software for seismic interpretation. We also acknowledge CGG Veritas United Kingdom for permission to publish their seismic data. The gravity modeling was performed by Terry Angelich, and the seismic velocity analysis and depth conversion were performed by Joe Molyneux. David Advocate of ExxonMobil Exploration and Valerie Goggin of BP Exploration (formerly with ExxonMobil Exploration) assisted in the paleogeographic reconstructions. The article benefited from reviews by Martin Jackson, Peter Bentham, Chris Shaw, Pam Darwin, Kevin Biddle, and Malcolm Wilson.

REFERENCES CITED

Albertz, M., C. Beaumont, and S. J. Ings, 2006, Role of sedimentation-controlled overpressure development in shale tectonics: Insight from numerical modeling of passive continental margins, *in* L. Wood, S. Wharton, R. Davies, F. Hossein, and C. Archie, conveners, Mobile shale basins—Genesis, evolution, and hydrocarbon systems: Proceedings of the AAPG/Geological Society of Trinidad and Tobago Hedberg Conference, Port-of-Spain, Trinidad, 4 p.

Barker, C., 1990, Calculated volume and pressure changes during the thermal cracking of oil and gas in reservoirs: AAPG Bulletin, v. 74, p. 1254–1261.

Billotti, F., and J. Shaw, 2005, Deep water Niger Delta fold and thrust belt modeled as a critical taper wedge: The influence of elevated basal fluid pressure on structural styles: AAPG Bulletin, v. 89, no. 11, p. 1475–1491, doi:10.1306/06130505002.

Burke, K., 1972, Longshore drift, submarine canyons, and submarine fans in development of Niger Delta: AAPG Bulletin, v. 56, p. 1975–1983.

Cohen, H. A., and K. McClay, 1996, Sedimentation and shale tectonics of the northwestern Niger Delta front: Marine and Petroleum Geology, v. 13, no. 3, p. 313–328, doi:10.1016/0264-8172(95)00067-4.

Corredor, F., J. H. Shaw, and F. Bilotti, 2005, Structural styles in the deep-water fold and thrust belt of the Niger Delta: AAPG Bulletin, v. 89, no. 6, p. 753–780, doi:10.1306/02170504074.

Costa, E., and B. C. Vendeville, 2002, Experimental insights on the geometry and kinematics of fold and thrust belts above weak, viscous evaporitic decollement: Journal of Structural Geology, v. 24, p. 1729–1739, doi:10.1016/S0191-8141(01)00169-9.

Doust, H., and E. Omatsola, 1990, Niger Delta in divergent/passive margin basins, *in* J. D. Edwards and P. A. Santogrossi, eds., Divergent/passive margin basins: AAPG Memoir 48, p. 201–238.

Elsley, G., and P. Graham, 2006, Mobile shale in the Mackenzie delta—The Pakota shale diapir complex, *in* L. Wood, S. Wharton, R. Davies, F. Hossein, and C. Archie, conveners, Mobile shale basins—Genesis, evolution, and hydrocarbon systems: Proceedings of the AAPG/Geological Society of Trinidad and Tobago Hedberg Conference, Port-of-Spain, Trinidad, 3 p.

Gipson, R., K. Meisling, and J. Bhajan, 2006, Tectonically-driven Plio-Pleistocene structural development of the Columbus Basin, offshore Trinidad, West Indies, *in* L. Wood, S. Wharton, R. Davies, F. Hossein, and C. Archie, conveners, Mobile shale basins—Genesis, evolution, and hydrocarbon systems: Proceedings of the AAPG/Geological Society of Trinidad and Tobago Hedberg Conference, Port-of-Spain, Trinidad, 2 p.

Graue, K., 2000, Mud volcanoes in deep water Nigeria: Marine and Petroleum Geology, v. 17, p. 959–974, doi:10.1016/S0264-8172(00)00016-7.

Hardy, S., and J. Poblet, 1994, Geometric and numerical model of progressive limb rotation in detachment folds: Geology, v. 22, p. 371–374, doi:10.1130/0091-7613(1994)022<0371:GANMOP>2.3.CO;2.

Hooper, R. J., R. J. Fitzsimmons, G. Neil, and B. C. Vendeville, 2002, The role of deformation in controlling depositional patterns in the south-central Niger Delta, west Africa: Journal of Structural Geology, v. 24, p. 847–859, doi:10.1016/S0191-8141(01)00122-5.

Hubbert, M. K., and W. Rubey, 1959, Role of fluid pressure in mechanics of overthrust faulting: Parts I and II: Geological Society of America Bulletin, v. 70, p. 115–205.

Morley, C. K., and G. Guerin, 1996, Comparison of gravity-driven deformation styles and behavior associated with mobile shales and salt: Tectonics, v. 15, no. 6, p. 1154–1170, doi:10.1029/96TC01416.

Reijers, T. J. A., S. W. Petters, and C. S. Nwajide, 1997, The Niger Delta basin, *in* R. C. Selley, ed., African basins: Sedimentary basins of the world 3: Amsterdam, Elsevier Science, p. 151–172.

Rowan, M. G., F. J. Peel, and B. C. Vendeville, 2004, Gravity-driven fold belts on passive margins, *in* K. R. McClay, ed., Thrust tectonics and hydrocarbon systems: AAPG Memoir 82, p. 157–182.

Shaw, J. H., C. D. Connors, and J. Suppe, 2005, Recognizing growth strata: Part 1. Structural interpretation methods: A1-3, *in* J. H. Shaw, C. Connors, and J. Suppe, eds., Seismic interpretation of contractional fault-related folds: AAPG Studies in Geology 53, p. 11–14.

Stewart, S., and R. J. Davies, 2006, Structure and emplacement of mud volcano systems in the south Caspian Basin: AAPG Bulletin, v. 90, no. 5, p. 771–786, doi:10.1306/11220505045.

Suppe, J., 1983, Geometry and kinematics of fault-bend folding: American Journal of Science, v. 283, p. 684–721.

Suppe, J., C. D. Connors, and Y. Zhang, 2004, Shear fault-bend folding, *in* K. R. McClay, ed., Thrust tectonics and hydrocarbon systems: AAPG Memoir 82, p. 303–323.

Van Rensberger, P., C. K. Morley, D. W. Arg, T. Q. Hoan, and N. T. Lam, 1999, Structural evolution of shale diapirs from reactive rise to mud volcanism: 3-D seismic data from the Baram Delta, offshore Brunei Darusasalam: Journal of the Geological Society (London), v. 156, p. 633–650, doi:10.1144/gsjgs.156.3.0633.

Vendeville, B. C., 2006, Similarities and differences between salt and shale tectonics, *in* L. Wood, S. Wharton, R. Davies, F. Hossein, and C. Archie, conveners, Mobile shale basins—Genesis: evolution, and hydrocarbon systems: Proceedings of the AAPG/Geological Society of Trinidad and Tobago Hedberg Conference, Port-of-Spain, Trinidad, 1 p.

Vendeville, B. C., and M. P. A. Jackson, 1992, The rise of diapirs during thin-skinned extension: Marine and Petroleum Geology, v. 9, p. 331–353, doi:10.1016/0264-8172(92)90047-I.

Wu, S., and A. W. Bally, 2000, Slope tectonics—Comparisons and contrasts of structural styles of salt and shale tectonics of the northern Gulf of Mexico with shale tectonics of offshore Nigeria in Gulf of Guinea, *in* W. Mohriak and M. Talwani, eds., Atlantic rifts and continental margins: Washington, D.C., American Geophysical Union, p. 151–172.

10

Warren, John K., Alwyn Cheung, and Ian Cartwright, 2010, Organic geochemical, isotopic, and seismic indicators of fluid flow in pressurized growth anticlines and mud volcanoes in modern deep-water slope and rise sediments of offshore Brunei Darussalam: Implications for hydrocarbon exploration in other mud- and salt-diapir provinces, *in* L. Wood, ed., Shale tectonics: AAPG Memoir 93, p. 163 – 196.

Organic Geochemical, Isotopic, and Seismic Indicators of Fluid Flow in Pressurized Growth Anticlines and Mud Volcanoes in Modern Deep-water Slope and Rise Sediments of Offshore Brunei Darussalam: Implications for Hydrocarbon Exploration in Other Mud- and Salt-diapir Provinces

John K. Warren[1]
International Graduate Program in Petroleum Geoscience, Department of Geology, Chulalongkorn University, Bangkok, Thailand

Alwyn Cheung
Petronas Malaysia, Kuala Lumpur, Malaysia

Ian Cartwright
School of Geosciences, Monash University, Melbourne, Australia

ABSTRACT

Modern slope and rise sediments of offshore Brunei contain thermogenic gases and bitumens atop or adjacent to actively growing mud-cored compressional ridges. This is particularly so where pressurized fluid and mud periodically breakout onto the sea floor as mud volcanoes or chimneys along ridge axes. Where modern sediment is actively accumulating in areas away from the ridges, the sediment contains organic signatures and gases that are biogenic (nonthermogenic) and related to the bacterial breakdown in and below the sulfate reduction zone.

Seismic signature implies that ridge-top volcanoes are regions of pressure buildup, and incipient fluid breakout is indicated by a combination of convex-upward seismic reflectors and zones of discontinuity in the bottom-simulating reflector (BSR, a seismic indicator of the base of stability of methane hydrates). Areas where the BSR weakens or disappears are likely zones where rising warm fluids have melted part or all of the hydrate layer. Following fluid pressure

[1]*Former address:* Shell Chair in Carbonate Studies, Sultan Qaboos University, Muscat, Oman.

DOI:10.1306/13231314M933424

release at the sea floor and a subsequent period of inactivity, sediment layers within and adjacent to a mud volcano collapse, sediment reflectors bend downward into the neck of the volcano, and a BSR re-establishes in the now-inactive neck.

Mud volcanoes in the continental slope and rise of Brunei occur along a series of gravity-driven compressional ridges. Seismic geometries are similar to the salt-cored compressional ridges of the slope and rise setting of the circum-Atlantic salt basins. In the case of the shale-cored ridges and mud chimneys of Brunei, however, the disrupted ridge and chimney cores are zones of sediment disruption created by the presence of pressurized gas and fluid. This disrupted ridge sediment has a different set of rheotropic properties compared to that of a salt-cored ridge. Where the disrupted zone has reached the Brunei sea floor, the mud-filled structure depressurizes and sediment flow ceases. Little or no evidence exists for widespread lateral flow of disrupted sediment across the Brunei sea floor, although evidence of debris-flow initiation exists in zones of instability created by ridge collapse. In contrast, where halokinetic salt breaks out onto the land surface or the sea floor, the halite mass continues to flow and spread across the deep-sea floor and ultimately constructs large allochthonous tongues and namakiers.

Active mud and salt diapirs create highs in the landscape and zones of adjacent withdrawal; this means that most reservoir styles and traps are similar in intervals above the source kinetic layer, whether it is composed of mud or salt. Once salt reaches the surface, it can flow over the landscape, but pressurized mud cannot, so no known argillokinetic equivalents to suballochthon traps are observed, nor argillokinetic equivalents to cap-rock traps.

INTRODUCTION

The Brunei deep offshore sea floor is characterized by an interesting association of gravity-driven growth anticlines, mud volcanoes, and mass transport complexes. The whole region is gas prone, and until now, certain regions of the sea floor were not known to contain gases with a thermogenic source. This article aims to (1) place the organic geochemical and sedimentological character of modern deep sea-floor sediments of Brunei into a seismic framework, (2) provide a better understanding of the nature and extent of thermogenic fluid flow and hydrocarbon seepage in a mud-dominated and pressurized sediment prism, (3) discuss hydrocarbon source signatures in the fluid outflow zone in sea-floor sediments of the modern slope and rise, and (4) compare characteristics of a mud-diapir compressional province to similar geometries in salt-diapir ridge provinces.

The literature dealing with occurrences of submarine mud volcanoes and diapirs has expanded greatly in the last decade, and so the same terms are given different emphasis by different workers. As a consequence, terminological disparities and overlap are observed with various and commonly overlapping usage of the terms mud volcano, shale (mud) diapir, mud mound, diatreme, etc. Usage in this article follows that of Milkov (2000), whereby a submarine mud volcano is considered as a topographically expressed sea-floor edifice from which water, brine, gas, or oil flows or erupts. Craters, mudflows with hummocky peripheries, and gryphons are the main morphological elements of a mud volcano. Mud (shale) diapirs are structures that have risen from the deep subsurface into shallow sediments and in some cases pierce the sea floor (sea-floor-piercing diapirs). Milkov's definitions encapsulate the important difference between a mud volcano and a diapir, namely all mud volcanoes (mud chimneys) are associated with mud diapirs but not vice versa. Kopf (2002) further refined the definitions by observing that a shale or mud diapir is a slowly upward-migrating mass of buoyant, clay-rich sediment that does not pierce all of its overburden, and the mud is behaving as a single-phase viscous fluid. Mud volcanoes, in contrast, are characterized by liquefied mud where the effective stress reaches zero.

Overpressured shale intervals act as detachments or as decollements facilitating the gravity-driven gliding or sliding of overlying more competent examples in many areas around the world, including offshore Brunei, the south China Sea, offshore Nigeria, Trinidad, and the southern Gulf of Mexico (Gee et al., 2007). The sea floor of offshore Brunei is characterized by the effects of mud volcanoes fed from deeper shale (mud)-cored diapiric ridges defining zones of overpressure.

Mud volcanoes in many places on Earth are commonly associated with compressional tectonics at convergent margins (Kopf, 2002, 2003). Their abundance shows a positive correlation with (1) sediment overpressuring associated with thick, rapidly deposited sediments with high clay mineral contents; (2) sediment overpressuring caused by hydrocarbon formation and/or compaction disequilibrium; (3) a structural association caused by tectonic shortening and/or earthquake activity; (4) fluid emission such as gas, brines, and water from mineral dehydration and gas hydrate dissociation; and (5) polymictic assemblages of the surrounding rock present in the ejected argillaceous matrix.

Hence, mud volcanoes are predominantly aligned around subduction zones and orogenic belts where sediments suffer tectonic stress and are incorporated into accretionary wedges that later can be imbricated and uplifted in compressional belts (Milkov, 2000; Kopf, 2002). Both pore water and organic matter in a subsiding marine sediment prism may be buried by several kilometers of sediment. Increasing overburden stress drives sediment compaction and fluid expulsion, whereas increasing temperature drives maturation of organic material and expulsion of hydrocarbon fluids. Trapped pore waters and hydrocarbons in such sediments can create considerable overpressures in muds as well as zones of anomalous porosity. As a consequence of the resulting porosity and density inversion, the mud either slowly ascends through the overburden rock (mud diapirs) or extrudes vigorously (mud volcanoes, diatremes) along focused zones of structural weakness such as faults and fractures (Stewart and Davies, 2006). Focused escape of fluids from such zones may liquefy already (partially) consolidated sedimentary rock. The admixture of small amounts of gas has been shown to cause a dramatic increase in extrusion rate, so that blocks of several meters in diameter may be mobilized during mud eruption (Clari et al., 2004).

REGIONAL GEOLOGICAL SETTING

The deep sea floor of the Brunei offshore lies northwest of the Brunei Darussalam shelf and spans an area of more than 10,000 km^2 (3861 mi^2), centered on 5°N latitude and 114°E longitude (Figure 1a). Sea-floor topography in the eastern part of the study area is the continental slope and rise of Brunei and is made up of several partially sediment-covered and eroded fold ridges (Figures 1b, 2b).

Present-day geological expression indicates a longer term geological history. Antecedent structures to the fold and thrust belt that today outcrops or immediately underlies all of the slope and rise of modern Brunei Darussalam were also developed in the deep-water equivalents of the middle Miocene–Pliocene shelf sequences (Morley, 2007a). Modern growth faults located near the present shelf-slope break are prograding over older, inactive innermost deeper water folds (Figure 1c). The average dip of the slope is between 1 and 1.5° toward the deepest region of the South China Sea, the northwest Borneo Trough. The slope is affected by an extensive train of folds, spaced between 5 and 15 km (3.1 and 9 mi) apart, that verge offshore. They appear to be mostly fault propagation folds that sole out into one or more detachments (Figure 2b).

The northwest Borneo Trough is today the site of an inactive subduction zone, where the oceanic crust of the proto-South China Sea was subducted during the early Tertiary (Morley 2007a). Heat-flow calculations, using aspects of the same organic geochemical data and cores detailed in this article, confirm Morley's conclusion of a now-inactive subduction zone underlying the Brunei sediment prism (Figure 1c) (Zielinski et al., 2007). The subduction zone became jammed in the latest early Miocene, when a thinned continental crust of the Dangerous Grounds block entered the subduction zone. The current fold and thrust belt developed mostly during the Pliocene–Holocene; therefore, the present deformation is unrelated to active subduction. Global positioning system measurements show no difference in displacement direction or rate between northwest Borneo and mainland Southeast Asia, so today, no suggestion of an active plate boundary exists (Hall and Morley, 2004). The sediments forming the fold and thrust belt are not scraped off from the subducting plate but are derived from reworking of the hinterland.

The active thrust-cored growth anticlines that outcrop or underlie much of the slope and rise act as sediment baffles and sediment catchment dams (Figure 1b). These anticlines are cored by mud diapirs. Between the growing ridges are several northeastward-trending sediment-filled depressions that parallel the shelf margin and range from 2 to 10 km (1.2 to 6 mi) wide and 20 to 60 km (12 to 37 mi) long. The folds are the result of a downslope gravitational compression, which started in the upper Miocene and continues today (Hutchison, 2005; Demyttenaere et al., 2000; Ingram et al., 2004; Morley, 2007a, b). The remainder of the study area is a relatively flat terrain, mostly covered by a mass transport complex or landslide (McGilvery and Cook, 2003; Gee et al., 2007). This flow complex represents a catastrophic event that occurred in the relatively recent past and was perhaps emplaced as a single landslide and not as a series of time-separate debris flows (Gee et al., 2007). Most of its internal structure is chaotic, but locally, it still retains coherent blocks that can be kilometers across and are made up of relatively consolidated sediment with an internal layering that was not completely destroyed during emplacement. The headwall of this mass transport complex lies updip and immediately offshore from the Baram Delta (Figure 1a) (Gee et al., 2007).

In summary, the various sediment dispersal pathways that typify the slope and rise of Brunei are made up of four interactive elements (Figures 1b, 2b) (McGilvery and Cook, 2003): (1) gravity-driven mass movement, differentiated further into short-distance transport, through

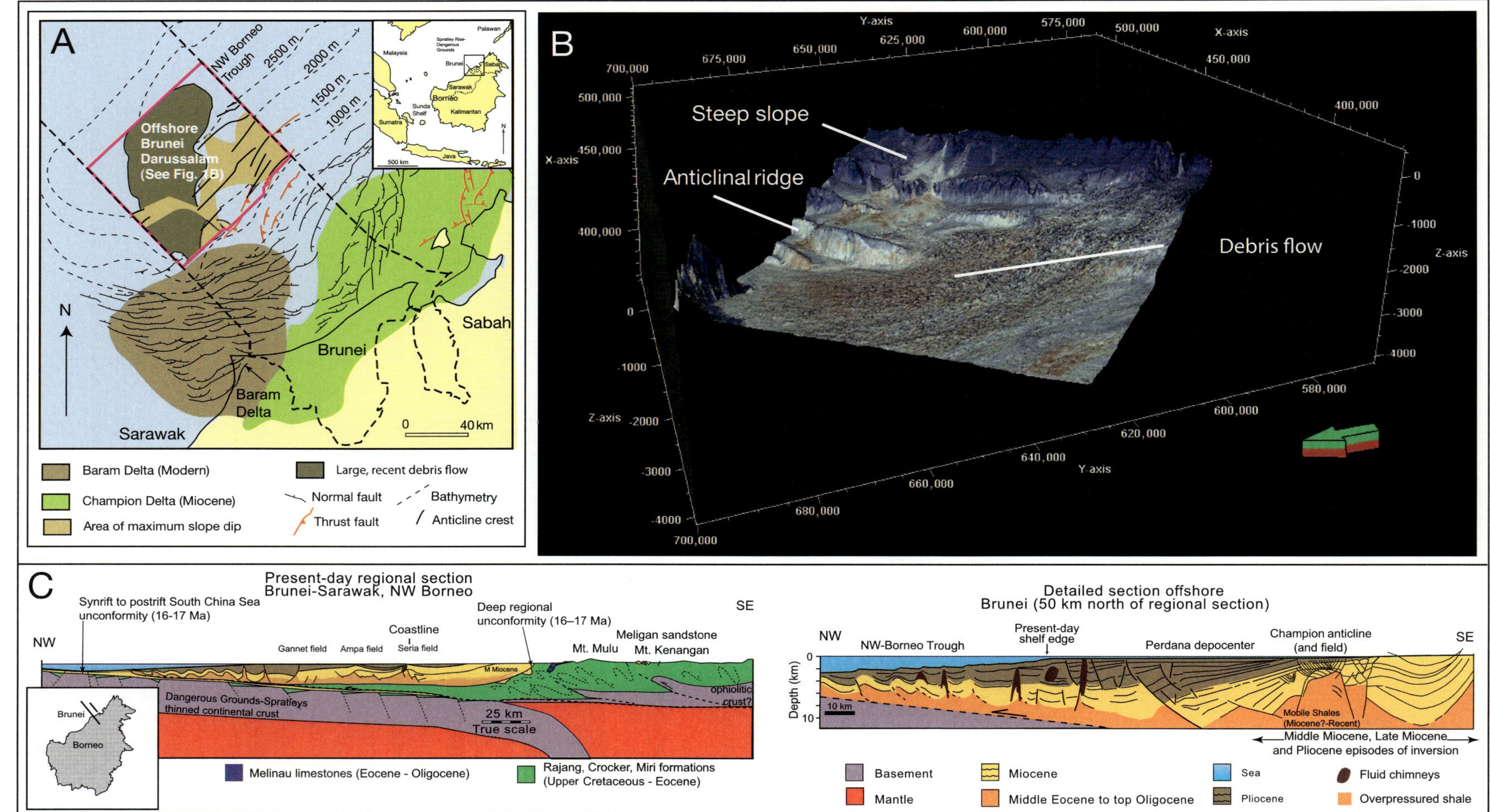

Figure 1. Regional summary for offshore Brunei Darussalam. (A) Regional sea-floor map showing Brunei-Malaysia borders (dashed line), the outline of study area (pink rectangle, centered on 5.71°N, 113.16°E), and the dominant depositional styles (after Morley, 2007a). (B) Sea-bottom map of the study area. Seismic amplitude extractions of the water bottom draped over the bathymetry of the deep-water slope and rise of Brunei showing the location of anticlinal ridges and the mass transport debris apron to the south (arbitrary z-axis scale; x and y scales UTM in m). The color range represents a range of root mean square (RMS) amplitudes. Generally, the higher the RMS amplitude, the more consolidated the sediments. Areas in brighter orange are typically active sediment flow paths (channels, downslope of deep-sea fans, and fault scarps), lighter areas indicate a more compacted mud, whereas gray indicates a relatively stable ground. Arrow points north. (C) Regional and local cross sections of offshore Brunei Darussalam showing dominant down to basin style of gravity-driven sedimentation atop an inactive subduction zone (after Hall and Morley, 2004). The pattern shows the overall progradation in a region of updip extension and growth faulting that passes downdip into compressional ridges and sediment dams and minibasins. 1 km = 0.6 mi.

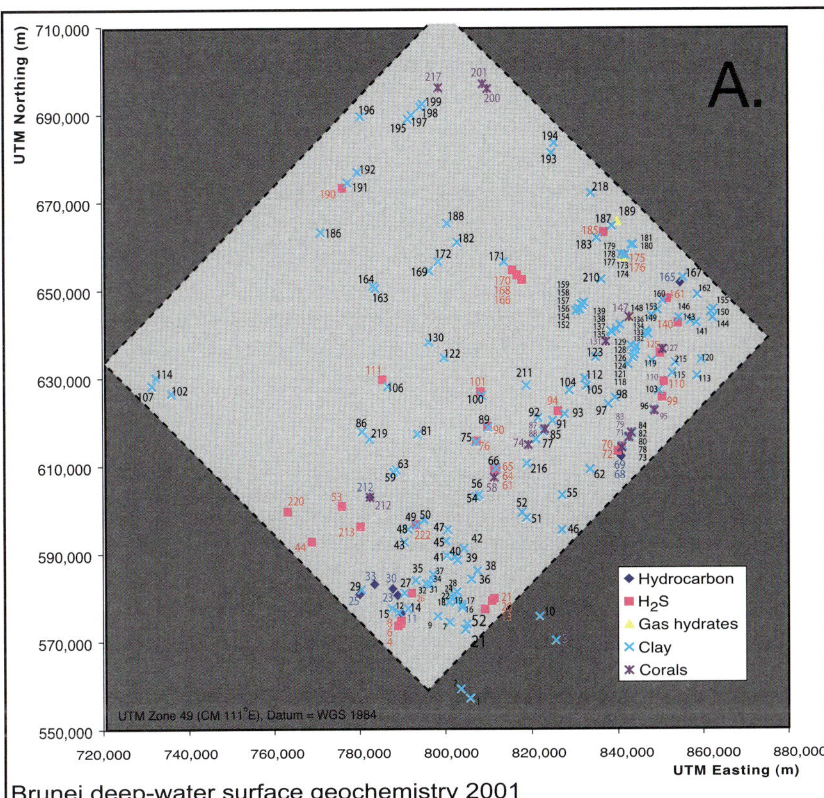

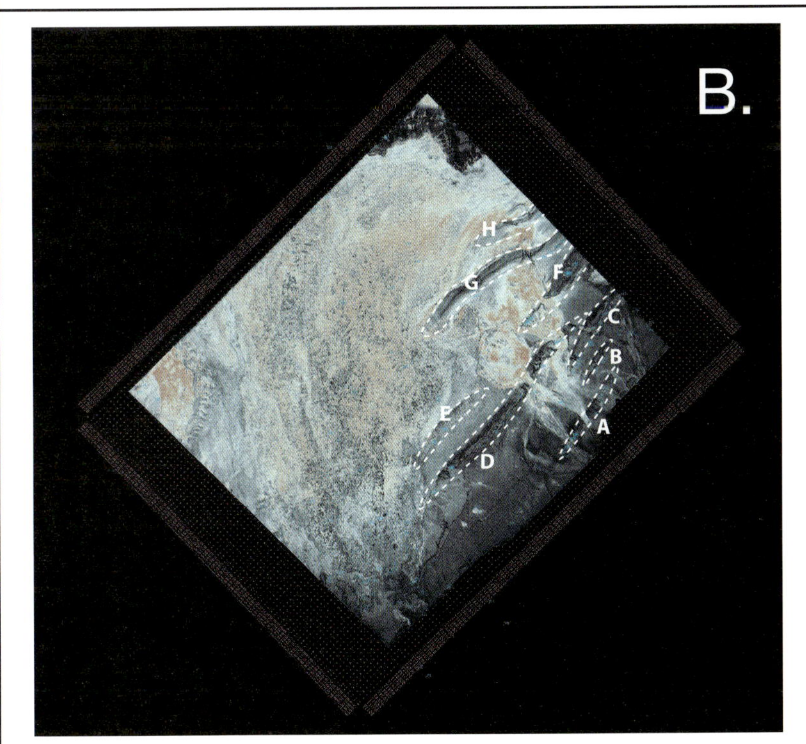

Figure 2. Brunei offshore; plan view. (A) Locations and sample numbers of the piston cores; grid orientation in the study area of Figure 1a and b. The notation in this figure is from onboard sample description sheets compiled by Geolabnor technicians in Norway. Subsequent sedimentological analysis by the authors on the same cores showed that all the collected samples are clay dominant. The term corals seems to be used nongenetically and incorrectly by the onboard crew to mean nonacquired core samples. Likewise, the onboard description of sand visible in the core barrel (not shown in this plot because it is highly misleading) was commonly wrong because most sediments described as sands were actually intervals of siderite micronodules in a clay matrix. The terms hydrocarbon and H_2S describe odors emanating from the cores at the time of collection. (B) Seismic amplitude extractions of the water bottom across the deep sea slope and rise of Brunei showing the location of anticlinal ridges (A–H) and the mass transport debris apron to the south. The color range in both images represents a range of RMS amplitudes. Generally, the higher the RMS amplitude, the more consolidated the sediments. Areas in brighter orange are typically active sediment flow paths (channels, downslope of deep-sea fans, and fault scarps), lighter areas indicate a more compacted mud, whereas gray indicates a relatively stable ground. 20,000 m = 12.4 mi.

slope creep, and long-distance transport by debris flow or landslides; (2) depressions, or submarine canyons, resulting from gravity mass wasting along the forelimbs of thrust-cored features; breaches allow the continuation of sediment transport across the thrust-cored structures; (3) point source outlets of canyons feeding deep submarine fans; and (4) sediment dispersal fairways ranging from 2 to 5 km (1.2 to 3.1 mi) in width, resembling a terrestrial braided channel in that the fairways contain multiple channels that may be straight or erosional, sinuous or leveed.

Indications of Thermogenic Seepage

The deep offshore of Brunei is a highly prospective region for hydrocarbons and has been the focus of several ongoing studies, in part funded by the Brunei government, including a regional three-dimensional (3-D) seismic study and a detailed piston coring program to test for hydrocarbon seeps. Wherever hydrocarbons are present in a sedimentary basin, they ultimately escape via leakage and are dispersed into ocean waters. Escaping hydrocarbons can be found in deep-sea-floor sediments worldwide and so, if identified by piston coring, can be used as prospectivity indicators to upgrade intervals in a prospective basin where a mature hydrocarbon source rock is present (Abrams, 2005).

Two end-member styles of mature hydrocarbon leakage patterns on the sea floor exist (O'Brien et al., 1999, 2005). One is focused leakage via conduits into narrow zones on the sea floor, such as when a fault or mud volcano breaks a seal layer and allows focused passage to the sea floor. The seal or cap to such a pattern is commonly a tight shale or salt layer, but with appropriate sea-floor conditions, the seal layer can be a clathrate (Kennett et al., 2000). The other style is more diffuse thermogenic fluid leakage through an imperfect seal over a larger area or structural culmination, so that hydrocarbons migrate to the surface over broader regions of the sea floor and do not show a highly focused outcrop trend. However, the notion of either hydrocarbon pattern residing in near-surface sediments and pointing to an underlying mature kitchen is ever balanced by the possibility that hydrocarbons and organic material in near-surface sediment originate from other organic reservoirs (Abrams, 2005). The organic signature may be derived from decomposition of recently deposited organic matter (ROM) or from biogenic methane created by methanogenic bacterial decomposition of dispersed organics in shallowly buried organic-rich muds. Therefore, distinguishing whether hydrocarbons in sea-floor sediments are residues of hydrocarbons generated from mature kerogens or from shallow and dispersed biogenic and recent sources is important.

Core Sampling and Analysis

Core recovery was attempted at 222 sites across the slope and rise of the Brunei offshore, covering an area of approximately 100 × 100 km (62 × 62 mi) (Figure 2). Of these, cores were successfully recovered from 188 sites and sent for geochemical analysis. The remaining sites were beset with collection problems (as noted in the onboard description sheets), mostly hard rock surfaces (described as corals by shipboard technicians) or very loose fine-grained sediment, which prevented recovery of any useful material fit for core description and analysis. Locations for each piston core site were pinpointed by a shipboard global positioning system. Figure 2 shows the locations and sedimentary associations of the relevant coring sites, as noted by the onboard description sheets. However, shipboard descriptions of corals and sands are incorrect, and sea-floor coral pavements are actually nonpenetrated nodular zones, whereas the sands are fine-grained sediments with numerous sand-size micronodules of diagenetic dolomite and siderite in a mud matrix (see for details the section titled Sedimentology of the Slope and Rise, Brunei Offshore Realm).

A 30-cm (12-in.) section from each recovered piston core was selected, typically from intervals free of carbonate nodules and cements, for gas analysis and chromatography. It was immediately canned, frozen, and shipped to the Geolabnor lab in Norway for analysis of organics and gases. The visible character of exposed core ends was recorded by on-board technicians during the sampling process (observations included); live oil or oil odor, presence of gas hydrates, sediment type, grain size, and color. After the removal of the 30-cm section, the remaining sections of the various piston cores were shipped to the Department of Petroleum Geoscience at the University of Brunei Darussalam for sedimentological analysis. There, cores were split, photographed, and sedimentologically described, and nodule samples were selected for stable isotope analysis.

ORGANIC GEOCHEMISTRY

Gas Analysis

Compositional analyses of frozen and canned core were done on the headspace, occluded, and adsorbed gases (Figure 3). Gas compositions in the various cores encompass a list of light hydrocarbon gases from single-carbon (methane) to seven-carbon chains (heptane and heptene), to branched, cyclic, and aromatic compounds (isobutane, cyclopentane, and benzene, respectively). Among the gases, methane was present in the highest

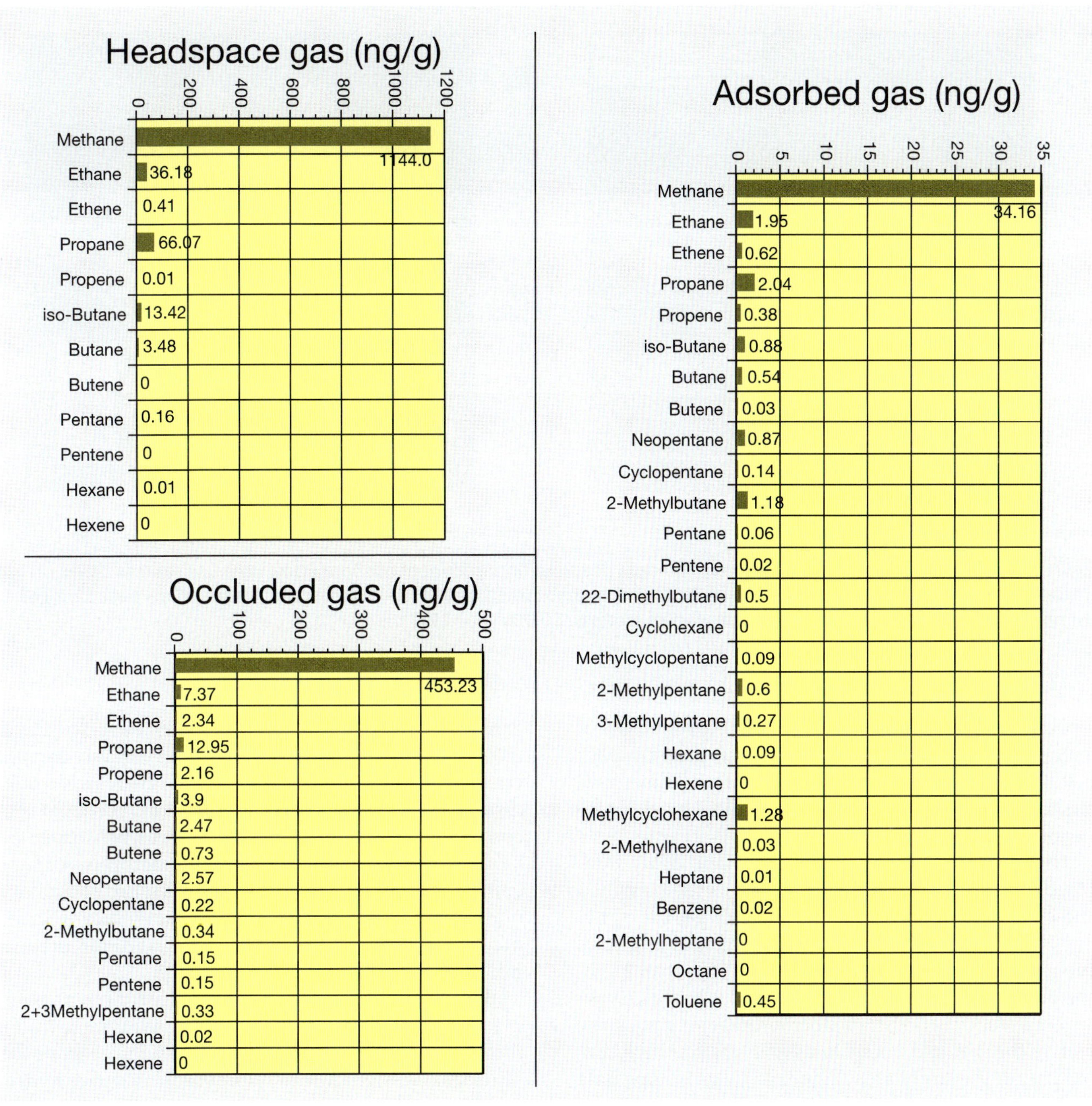

Figure 3. Average gas content (headspace, occluded, and adsorbed using gas chromatography) in the deep-water sediment of offshore Brunei based on an analysis of 188 piston cores. Headspace gas is that present in the headspace of the sample canister once the sealed, formerly frozen, canister is returned to room temperature. Occluded gas is that released from the greater than 63-μm fraction of a sample during mechanical disaggregation in a blender. Absorbed gas is the gas released from the greater than 63-μm fraction of a disaggregated sample when treated with orthophosphoric acid under vacuum.

quantities and was the most common. Methane typically shows a carbon isotope signature depleted in carbon-13 ($\delta^{13}C$ ranges from −28 to −60‰) and is the sole measured gas phase in many samples. However, methane in the cores can be either biogenic or thermogenic. The more widespread and prevalent form of methane is biogenic and comes from shallow bacterial and archaeal decomposition of ROM (Kotelnikova, 2002). The other methane source is thermocatalytic production at depth via heating of buried kerogen (Birkle et al., 2002). The resulting methane has lighter $\delta^{13}C$ values, is typically associated with other gas phases and with the release of

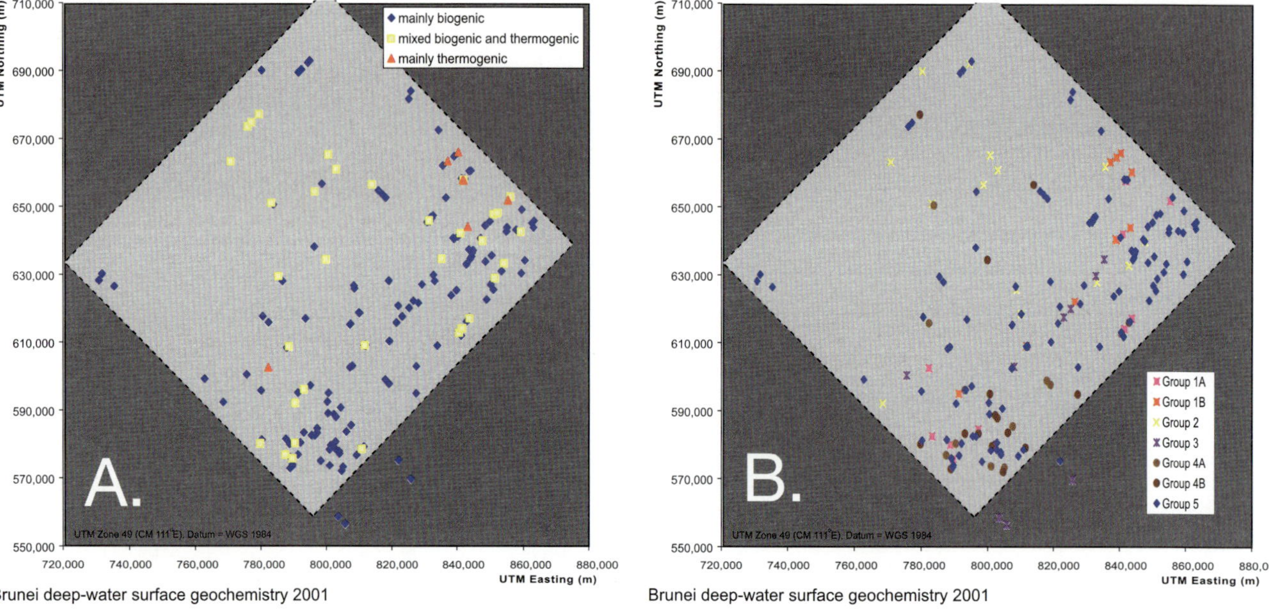

Figure 4. Plot of gas and organic associations in modern deep sea-floor sediment offshore Brunei. (A) Gas sources based on gas chromatography interpretation of headspace, occluded, and adsorbed gases. (B) Organic group associations based on gas chromatograms (see the text for the explanation of groups). 20,000 m = 12.4 mi.

cracked liquid hydrocarbons. The high proportion of methane, compared to other hydrocarbon gases, implies that much of the methane in near-surface sediments of offshore Brunei is produced from shallow biogenic instead of thermogenic processes.

The amount of nonmethane gases and other nonbiogenic associated compounds, such as the branched and cyclic compounds, and thermogenic gas (oil or hydrate and condensate associated) occurrences in several subareas are much more useful diagnostics than methane purity. Based on this analysis, broad subareas, showing consistency in occurrence of thermogenic gas, can be seen in the east and in one locality in the southern central part of the study area (Figure 4a). Gas isotope data also suggest the possibility of two populations of seepage hydrocarbons: one in the oil-window maturity upward field (0.8% vitrinite reflectance [R_o]) and the other in the condensate window upward field (1.5% R_o). Geographically, the former association is mostly restricted to the eastern part of the study area.

Gas Chromatography

The preceding gas analysis implies that piston cores from the deep sea floor of offshore Brunei contain two types of hydrocarbons: (1) thermogenic hydrocarbons and (2) hydrocarbons derived from ROM. A sample with bitumens that yields high amounts of extractable organic matter (EOM) does not necessarily have oil-related interest; the EOM may also come from variably decomposed ROM. Gas chromatograph (GC) analysis can help distinguish the two sources (Peters et al., 2005a, b). Accordingly, quantitative GC analysis was performed on bitumens extracted from the canned and frozen sediment using a hexane solvent. Results show a series of aliphatic n-alkanes and selected isoprenoids, together with remaining byproducts of unresolved complex organic mixtures (UCOM).

Based on GC signature, analyzed piston core samples are divided into five groupings, ranging from indicators of active oil seeps to inactive seeps (Figure 4b). The ideal GC signature of bitumens in an active hydrocarbon macroseep on the deep sea floor shows a combination of well-developed light and heavy molecular peaks and a well-developed hump in the undefined envelope, indicating active bacterial biodegradation (Figure 5a) (Peters et al., 2005a, b). Samples from the Brunei piston cores show various degrees of hydrocarbon biodegradation, along with fresher hydrocarbon signatures tied to numerous occurrences of ROM (Figure 5b–f, Table 1).

Subgroup 1A

Samples are generally found to have elevated amounts of EOM. Degrees of biodegradation were determined

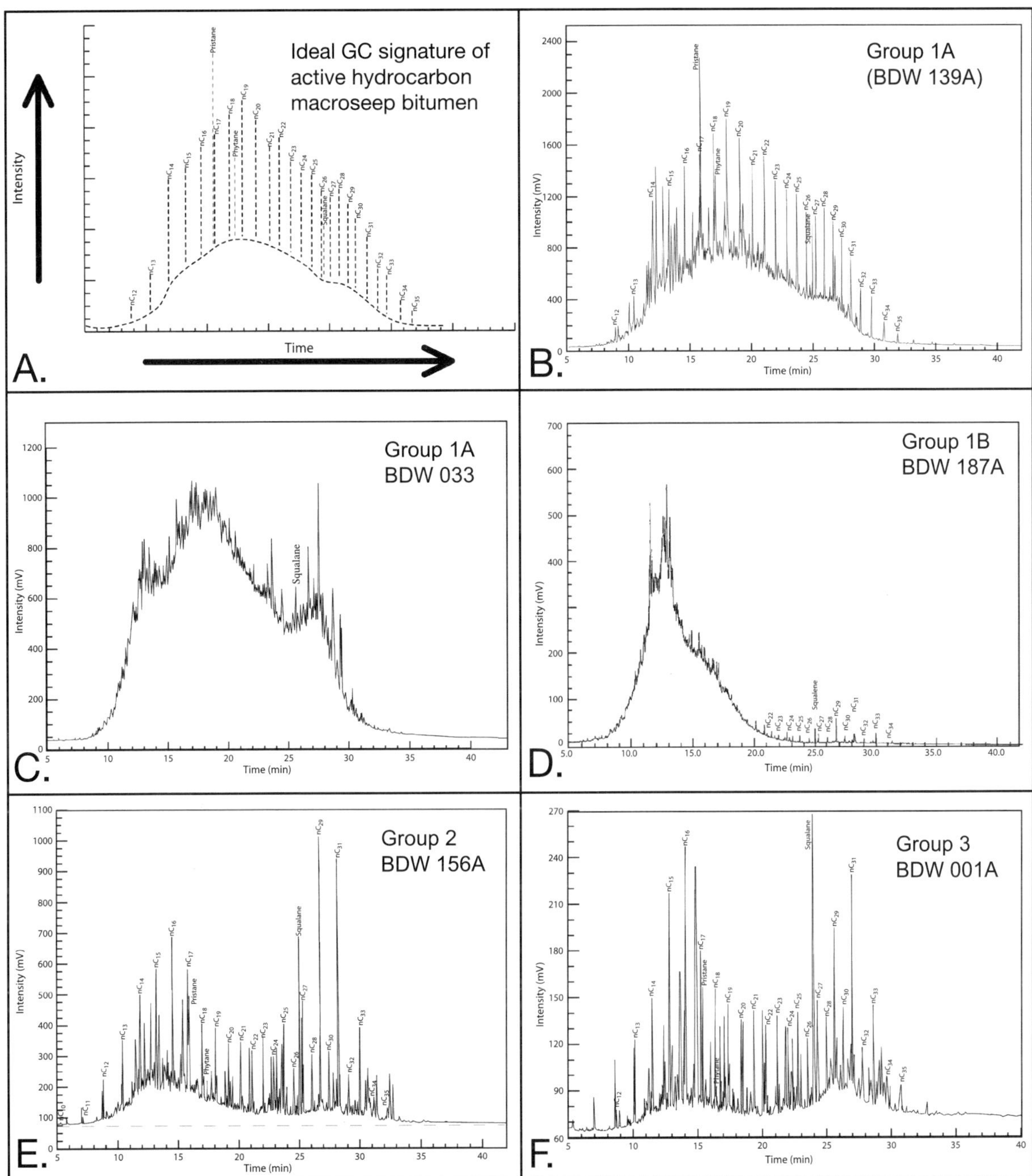

Figure 5. Typical gas chromatogram from bitumens in selected piston core samples. (A) Ideal signature of bitumen formed in a active sea-floor hydrocarbon macroseep. (B) Group 1a signature from site 139 showing a hydrocarbon macroseep signature with well-preserved short- and long-chain hydrocarbons with an increase in the intensity bulk with an overall spread from light to heavier compounds. (C) Group 1a signature from site 033 showing the shortening of peaks and a well-developed hump in the undefined envelope; both features indicate biodegradation and imply bacterial destruction of digestible hydrocarbon compounds. (D) Group 1B signature from site 187 showing a well-developed hump and loss of peak character that are typical of strongly biodegraded samples. (E) group 2 signature from site 156 showing a slight indication of biodegraded light molecular weight hydrocarbons and an increased abundance of heavier organic compounds (from recent organic matter) compared to groups 1a and 1b. (F) Group 3 signature from site 001 showing an abundance of heavy molecular weight hydrocarbons indicative of recently deposited organic matter. The sample also shows a dominant odd-carbon C_{13} to C_{21} pattern and the lack of an unresolved envelope, which would otherwise indicate biodegradation of the organic material. Chromatographic traces of group 4 and 5 samples (not illustrated) show a predominance of long-chain (heavier) hydrocarbons and are derived from recent organic matter with chromatographic signatures very similar to group 3. BDW = Brunei deep water; GC = gas chromatogram.

Table 1. Summary of gas chromatography of organic matter in piston cores from deep-water sea-floor sediment of offshore Brunei.*

	Group 1				Group 4		
	Group 1A	Group 1B	Group 2	Group 3	Group 4A	Group 4B	Group 5
Sample sites	011, 023, 033, 037, 072, 073, 082, 084, 139, 165, 173, 174, 175, 176, 212	048, 065, 094, 135, 137, 148, 180, 185, 187, 189	044, 090, 100, 105, 118, 132, 156, 164, 182, 183, 186, 188, 196, 198	001, 002, 003, 053, 056, 091, 112, 123	005, 007, 015, 027, 039, 051, 052, 219	004, 021, 022, 024, 025, 034, 035, 036, 040, 041, 046, 047, 069, 078, 122, 163, 171, 192, 221, 222	006, 008, 009, 010, 012 to 014, 016 to 020, 026, 028 to 032, 038, 042, 043, 045, 049, 050, 054, 055, 059 to 064, 066, 068, 070, 075 to 077, 080, 081, 086, 089, 092, 093, 096 to 099, 101 to 104, 106 to 111, 113 to 115, 119 to 121, 124 to 130, 133, 134, 136, 138, 140 to 146, 149 to 155, 157 to 162, 166 to 170, 177 to 179, 190, 192 to 195, 197, 199, 210, 211, 213 to 218, 220
Character and likely organic provenance of analyzed bitumen	Minor biodegraded oil (active hydrocarbon macroseep)	Biodegraded light-molecular-weight oil/condensate (likely biodegraded material is from an active hydrocarbon macroseep)	Slight indication of biodegraded light-molecular-weight hydrocarbons (possibly a slight indication of some hydrocarbon microseep activity exists)	Restricted range of light-molecular-weight and ROM hydrocarbons with low UCOM (active hydrocarbon seep source unlikely)	Large abundance of heavy-molecular-weight hydrocarbons from ROM and large UCOM (no active seep source)	Large abundance of heavy-molecular-weight hydrocarbons (ROM) and slightly less UCOM than 4A (active seep source unlikely, a part could be from relict seep)	Large abundance of hydrocarbons from ROM; no or very minor UCOM; typical modern sea-floor organics, no hydrocarbon seep activity

*See the text for the explanation. ROM = recent organic matter; UCOM = unresolved complex organic material.

by the presence of shorter, or the absence of, distinct peaks on the chromatogram when compared to GC traces of ROM (compare Figure 5b, f). Group 1A samples are interpreted as derived from relatively less biodegraded thermogenic hydrocarbons coming to the surface through focused conduits or macroseeps. This hydrocarbon style always shows variable levels of biodegradation because it was sequestered in near-surface sediments (compare Figure 5b, c).

Subgroup 1B

Samples generally have lower EOM when compared to the higher end of group 1A. They are typified by a large unresolved envelope in the GC trace, itself indicative of biodegraded hydrocarbons (Figure 5d). The unresolved envelope is typically located in the earlier times of the chromatogram curve, normally indicative of the first appearance of lighter molecular hydrocarbons. The patterns in the gas chromatograms are similar to those more biodegraded samples in subgroup 1A but show more pervasive biodegradation overprints. Group 1B associations are thought to indicate more diffuse or less active seepage (microseeps) than group 1A signatures (less overturn = more biodegradation). Thus, these condensates were derived mainly from hydrocarbon seepages and may not entirely be related to the original EOM within the sediments. The low peaks of C_{21}–C_{33} in the GC curve strengthens the interpretation of biodegraded condensates (little or no terrestrial plant waxes) (Peters et al., 2005a, b).

Group 2

Samples show low amounts of EOM and an increased abundance of the GC signature coming from higher plant material (as indicated by C_{23} peaks and higher), along with a marine signature from lighter molecular chains, C_{11}–C_{20} peaks (Peters et al., 2005a, b). The unresolved envelope is located in the same range as in group 1B, possibly indicating a contribution from biodegraded condensates (Figure 5e). However, the association of GC peaks is not conclusive for this group. The various hydrocarbons may have been oil associated or may have been derived from ROM.

Group 3

Samples show low abundances of all hydrocarbons. Likewise, the level of EOM is also low. The predominance of odd-numbered n-alkanes (in the C_{13-21} range), along with the lack of aromatic compounds, indicates that the hydrocarbon signature is mostly derived from ROM (Figure 5f) (Peters et al., 2005a, b).

Group 4

Samples contain large amounts of unresolved compound organic mixtures and an abundance of long-chain hydrocarbons from ROM. They can be assigned no more than an ambiguous genesis (Peters et al., 2005a, b).

Group 4A. Samples have UCOM of up to 40 mg/g dry sediment with a large odd-carbon predominance (C_{23-35}).

Group 4B. Samples have a UCOM of 20–40 mg/g dry sediment and possibly contain some completely biodegraded hydrocarbons; some 4B samples possibly came from earlier seeps that are no longer active.

Group 5

Samples have UCOM values less than 20 mg/g dry sediment. They show an abundance of long-chain hydrocarbons from ROM and little or no UCOM. The signature is typical of modern sea-floor organic material with no indication of hydrocarbon seep activity (Peters et al., 2005a, b).

The transition from group 3 to 5 marks an increasing abundance of heavier and more complex carbon compounds indicative of ROM. The severity of biodegradation, in contrast, decreases from 100 mg/g dry sediment to very little to none in groups 3 through 5. Worldwide, an odd-carbon preference from around the C_{23} peak upward in all ROM samples indicates the contribution of terrestrial plant waxes, independent of whether it is found in coals, lakes, or marine environments (Waples, 1985). The preference is strongest in immature, least biodegraded samples. It gradually disappears as rocks are buried, and their kerogen evolves. Thus, groups 4 and 5 GC signatures are dominantly those of modern marine organics with contributions from both marine organisms ($<C_{20}$) and terrestrial plant waxes. The average Pr/Ph ratio of 1.6 of the Brunei offshore is lower than for most coals (Waples, 1985); therefore, the terrestrial organic component was probably directly deposited in its shaly matrix without spending much time stored in a coal swamp prior to its final deposition.

In the context of groups 3 to 5 signatures, note that several piston core samples with elevated amounts of gas did not have elevated amounts of EOM. These are samples 020, 025, 048, 061, 068, 069, 096, 099, 110, 115, 135, 148, 161, and 167. All these samples have very dry gas, i.e., the gas composition is dominated by biogenic methane from biodegradation of organic material. All of these sample sites have groups 3–5 GC signatures except for sites 135 and 148, which have group 1B signatures, and 025, which has a group 2 signature.

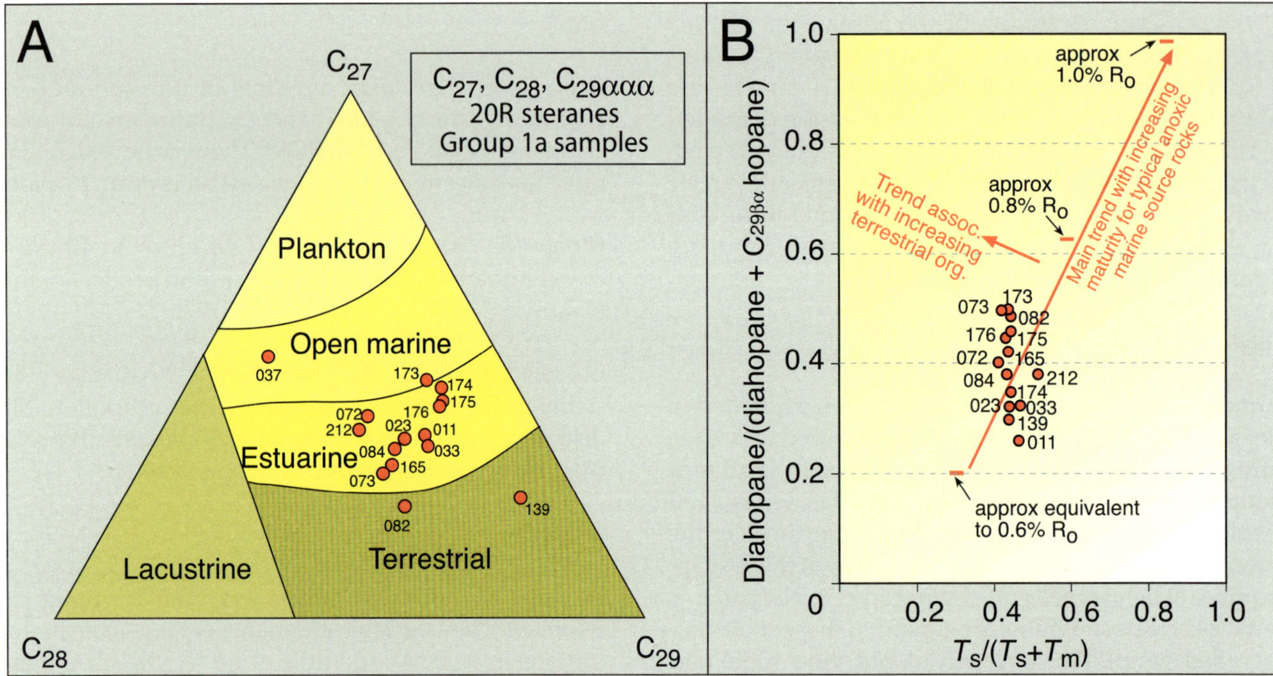

Figure 6. Source rock characteristics of group 1a organic material in piston core sediments. Sample numbers as listed, see Figure 2 for the sample locations. (A) Vitrinite reflectance (R_o) maturation equivalents based on a plot of diahopanes. (B) Source-rock typing based on C_{27} to C_{29} plots. For explanation and derivation of these commonly used standard oil-industry plots, see Peters et al. (2005, a, b). T_s = 22,29,30-trisnorneohopane; T_m = 22,29,30-trisnorhopane.

In summary, hydrocarbons within the oil window have a bulging intensity on the left side of the chromatogram in the C_{11-22} parts of a curve, which gradually flattens farther to the right in the C_{23-35} range (Figure 5a). The height of each peak is proportional to the concentration of the compound(s). Biodegradation destroys both n-alkanes and isoprenoids, leaving behind an unresolved envelope of short peaks. Groups 1 and 2 have these chromatographic patterns of partially biodegraded to heavily biodegraded seep hydrocarbons (Figure 5b–e). Group 3 is also biodegraded. However, its chromatographic pattern also shows two bulges, albeit at low intensity, in both the right and left side of the chromatogram curve (Figure 5f). So, group 3 can also be seen as an intermediate between diminishing content from hydrocarbons and increasing abundance of ROM. From group 3 to group 5, the destruction to the peaks caused by biodegradation is less common, and an obvious increase in levels of ROM in their GC curves is observed. Groups 2–5 are not considered to be directly associated with recognizable hydrocarbon seeps, and so, to relate thermogenic seepage to structure, only group 1 sites are considered in the rest of this article. That is, group 1 signatures in sea-floor sediments define sites where deep thermogenic fluids have arrived at the modern sea floor. In the context of seismic interpretation, a group 1 signature on the sea floor offers a very powerful and independent tool for testing the timing and location of pressure release and the nature of the releasing structure as interpreted from seismic data.

One of the most commonly used source provenance parameters in the oil industry is the relative abundance of the C27:C28:C29 $\alpha\alpha\alpha$ 20R steranes (Huang and Meinschein, 1979). Plots from subgroup 1A (i.e., from samples not as greatly affected by biodegradation and the effects of ROM) are plotted in Figure 6a. It implies that all the likely thermogenic seep samples, with the exception of sample 037, have an estuarine or terrestrial organic source. This estuarine or terrestrial signature is consistent with earlier source rock studies in Brunei (e.g., Curiale et al., 2000), which show that source rocks in the producing regions of Brunei tend to be from gas-prone estuarine or terrestrial (tidal-coastal embayment) shales and not from a marine-algal oil-prone precursor.

The regional plot of the GC grouping shows that the distribution of likely hydrocarbon seeps in offshore Brunei is localized to areas in the east of the survey area, along with a few sites in the south (Table 1; Figures 4b). The bitumens from the group 1 samples also indicate that most of the samples have diahopane values, indicating a vitrinite equivalence (R_o) between 0.6 and 0.8%, and so lie in the oil generation window, although we emphasize that the dominant source for Brunei hydrocarbons is undeniably gas prone (Figure 6b) (Curiale et al., 2000).

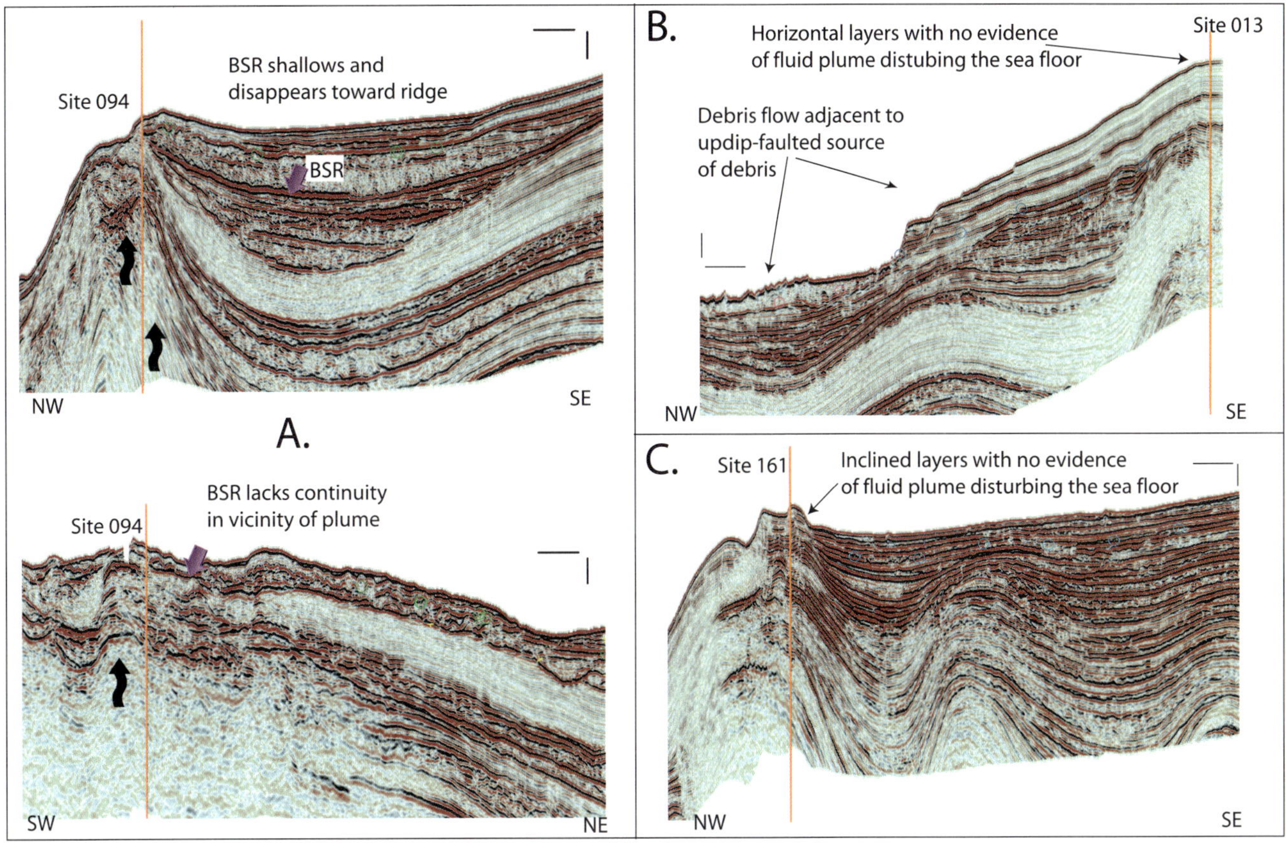

Figure 7. Seismic sections through sites of isotopic interest. (A) Site 094: the upper image is a part of inline 7276 and the lower image is a part of crossline 7862. The intersect of the two lines is shown by a red line, which is also piston core site 094. This organic group 1B site sits atop a shale-cored growth anticline with fluid plumes rising to the sea-floor surface and associated evidence of plume-related sediment collapse. The ridge is a result of ongoing gravity-driven compressional deformation, which also causes the bottom-simulating reflector (BSR) horizon to rise and break up as it approaches the sea floor. Note the disturbance and the discontinuity of the BSR caused by rising fluid plume in the vicinity of sample site 094. See Figure 9 for the location on the sea-bottom map. (B) Site 013 is part of inline 3750 where the vertical red line shows the piston core site. The sea floor at the site is not disturbed by pluming, and this is confirmed by the occurrence of group 5 organics at the site. (C) Site 161 is part of inline 10198 where the vertical red line shows the piston core site. Although mounded at the sea-floor surface, the layers underlying the sea floor at this site are not broken by fluid pluming, and this is confirmed by the occurrence of group 5 organics at the sample site. The horizontal scale bar in all images is 1 km (0.6 mi) long, and the vertical bar indicates 100 ms.

Sedimentology of the Slope and Rise, Brunei Offshore Realm

Sea-Floor Sediment

From 222 listed coring sites, 188 cores were recovered and slabbed at the University of Brunei Darussalam. Shipboard descriptions by Geolabnor technicians were useful in determining the original state of core hardness (i.e., soft, stiff, medium hard, hard), for confirming odors of H_2S and hydrocarbons in the slabbed core and for relating odors (oil versus H_2S) to the presence or absence of nodules in later sedimentological work (Figure 2a). During the sedimentological logging, special attention was given to cores described onboard as containing sand layers because these could possibly be related to features seen in seismic sea-bottom maps. As an initial guide, the onboard description was used to slab first those cores likely to contain sands. We found that onboard descriptions such as bottomed in coral and numerous notations of sand in the core barrel were incorrect. No coral fragments were ever recovered in any cores. Only carbonate-cemented nodules and almost all of the sands noted in the onboard descriptions were actually sand-size siderite micronodules; they grew in situ and were visible with a hand lens.

Only a few centimeter- to decimeter-thick siliciclastic sand intervals were recovered in any of the piston

Table 2. Sedimentary features of interest.*

Sedimentary Feature in Muds	Piston Core Location
Thin sand stringer <1 cm (<0.4 in.) thick	012, 014, 027, 032, 042, 054, 061, 064, 084, 090, 097, 098, 105, 113, 121, 122, 124, 125, 128, 129, 132, 134, 146, 157, 158, 159, 166, 172, 182, 183, 186, 188, 192, 219
Sand layer, a few centimeters (inches) thick	010, 011, 020, 025, 036, -38, 043, 052, 055, 078, 082, 084, 098, 099, 101, 105, 106, 112, 115, 119, 126, 132, 133, 135, 137, 141, 152, 156, 164, 166, 168, 169, 170, 171, 183, 186, 193, 194, 216, 218
Sand layer 10 cm (4 in.) thick	061
Calcareous nodules (white-gray)	003, 007, 011, 013, 014, 015, 019, 020, 023, 024, 025, 026, 029, 030, 032, 035, 037, 039, 040, 048, 049, 051, 052, 055, 056, 059, 061, 062, 064, 065, 068, 070, 072, 073, 078, 082, 084, 085, 086, 091, 094, 096, 099, 100, 102, 105, 110, 113, 114, 115, 118, 119, 121, 122, 123, 124, 125, 126, 128, 129, 130, 132, 134, 135, 137, 138, 139, 140, 141, 148, 149, 155, 156, 157, 158, 161, 163, 164, 165, 167, 168, 171, 182, 183, 185, 186, 187, 188, 189, 190, 192, 193, 194, 196, 198, 199, 211, 213, 215, 218, 222
Fe-stained red-orange calcareous nodules and mottles	010, 040, 050, 091, 097, 115, 134, 140, 150, 155, 166, 215
Carbonate mottles and incipient nodules	073, 099, 101, 110, 115, 123, 135, 137, 148, 160
Burrows	013, 104, 020, 023, 048, 068, 070, 096, 099, 110, 115, 135, 148, 161, 167, 183, 186, 187, 196, 198
Particulate fine-grained organic material (salt and pepper texture), trace to small pieces of wood	002, 003, 005, 006, 009, 010, 013, 024, 026, 028, 031, 032, 034, 039, 047, 055, 061, 062, 064, 085, 091, 097, 098, 099, 103, 105, 110, 111, 112, 115, 119, 121, 128, 132, 140, 143, 146, 149, 150, 153, 160, 166, 171, 188, 197, 222
Shell fragments, a few whole shells	008, 009, 011, 013, 020, 023, 025, 029, 030, 032, 033, 034, 038, 044, 048, 052, 053, 055, 062, 065, 066, 068, 070, 072, 073, 085, 086, 089, 099, 105, 115, 125, 130, 135, 137, 139, 141, 143, 144, 146, 150, 155, 163, 167, 172, 174, 175, 176, 180, 183, 185, 186, 187, 189, 190, 194, 222
Smell of hydrocarbon when first slabbed	011, 023, 025, 030, 033, 068, 212
Open void (gas bubble) texture	212
Clathrates recovered in core	173, 189

*More than one feature may occur at a single piston core location.

cores. From 188 cores, 67 contained dispersed siliciclastic sand grains, in places in centimeter-scale layers and never in beds that were meters thick (Table 2). Most of these 67 slabbed cores (with up to 3 m [10 ft] of penetration into the sea floor) that contain siliciclastic sand typically contain it in only one, or less commonly two to three, thin sand laminae in an otherwise mud core. The sand layer is typically less than 1–3 mm (0.03–0.11 in.) thick, with a few layers ranging up to 4 cm (1.5 in.) thick. The thickest recovered sand layer in all 67 sand-entraining cores was some 10 cm (4 in.)-thick at site 061 (Figure 8a, b). This layer was part of a 20-cm (8-in.)-thick layer composed of sand and wood fragments within a silty clay matrix. No meter-scale sand layer was penetrated by any of the 222 piston cores.

Mud- to sand-size organic particulates and debris were dispersed through many cores so that the organic material in core samples collected for organic analysis averages 0.9% total organic carbon (and ranges from 0.3 to 1.4%. Higher values are found nearer the shelf in the eastern and southern regions of the study area. Centimeter- to decimeter-scale fragments of wood were commonly present in a core composed only of mud and nodules, and no corresponding layer of siliciclastic sand was observed. Thus, this wood is not a lag tied to a siliciclastic bed-load transport (unlike site 061), instead, it indicates a waterlogged plant material sinking to the sea bottom.

Volumetrically, aside from the ubiquitous clays and muds, the most interesting features in the cores are the numerous intervals with calcareous mottles and nodules (Figure 8c–e). Varying combinations of nodules and micronodules are found in 107 cores. Nodule dimensions range from small micronodules (<0.5 cm [<0.1 in.] in diameter) to large elongate burrowlike nodules and fragments that can exceed the diameter of the core tube (>6 cm [>2.3 in.], Figure 8c). These nodules are diagenetic features that tend to be better developed in intervals that also show elevated carbonate content from planktonics, whole shells, and shell fragments (Figure 8f). The nodules are early diagenetic features growing at depths ranging from a few centimeters (inches) below the sea floor, down to the full 3-m (10-ft) maximum depth sampled by the deepest penetrating piston core sites.

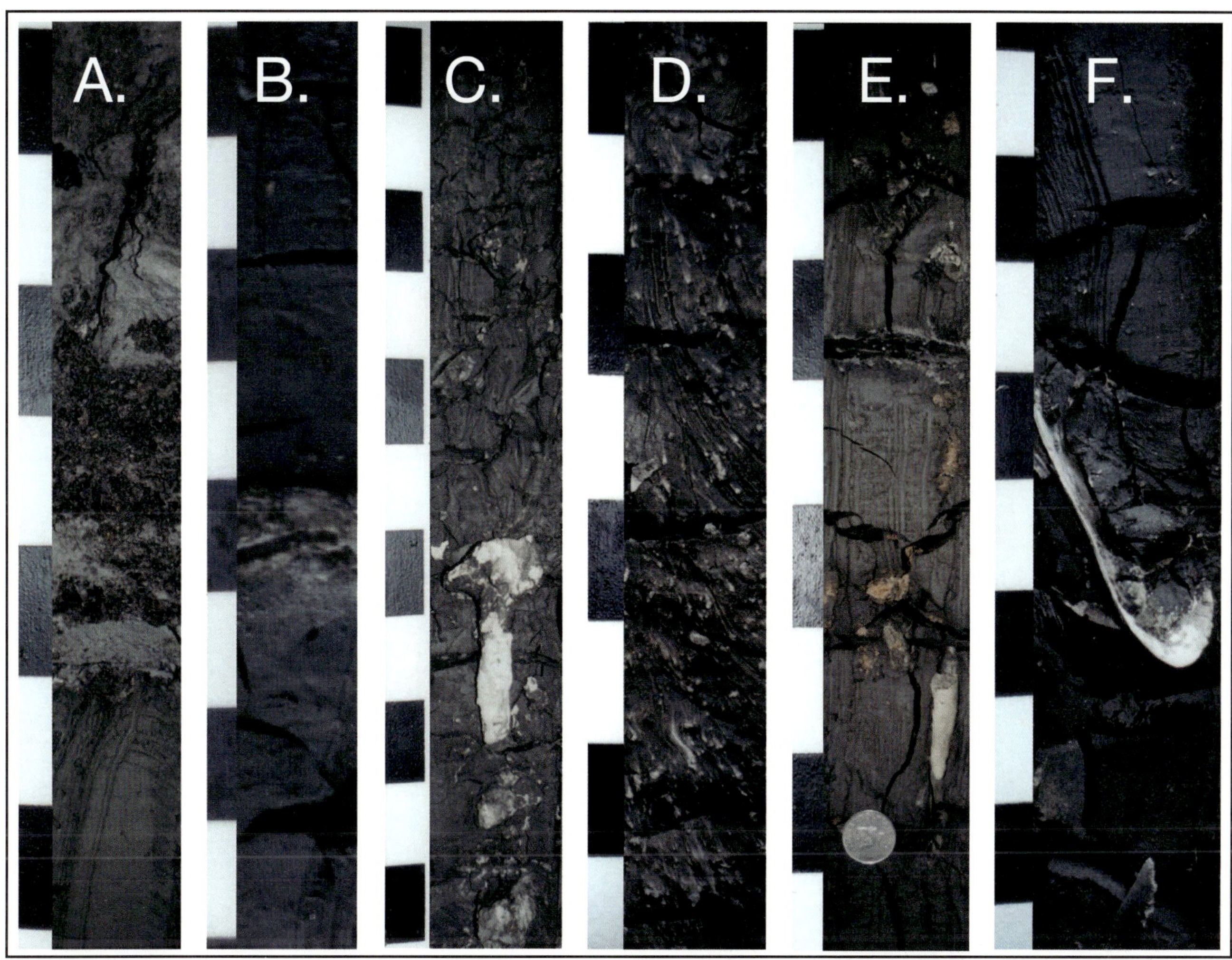

Figure 8. Sediments in selected piston cores (scale bar is shaded at 5-cm [1.9-in.] intervals). (A, B) Two thickest sand beds recovered from all 188 cores (sand is a minor component of the Brunei deep-sea floor). (A) From site 061, containing a 10-cm (4-in.)-thick fine sand with oxidized (lighter colored) clay at its base. The sand layer contains abundant fine-grained particulate organics. (B) From site 020, containing a light-colored 5-cm [1.9-in.] sand and shaley clasts. (C–F) Various styles of carbonate nodules and cemented burrows. (C) From site 187, dominated by a large 10-cm (4-in.)-long, 2-cm (0.7-in.)-wide elongate nodule probably formed by cementation of a former organic- or pellet-lined burrow. (D) From site 013, dominated by micronodules, illustrating the texture incorrectly described as sand by onboard technicians. (E) From site 115, illustrating nodules stained red by iron oxides. All the cracks are related to drying of original water-saturated muds. (F) From site 021, showing a large well-preserved bivalve shell and small micronodule flecks in the encasing mud.

An interesting association between core ends is noted in the shipboard description as having an oil or hydrocarbon odor and the presence of the calcareous nodules and shell materials in the same cores when later slabbed. That is, with the exception of one core (site 212), all core ends noted by shipboard technicians as smelling of hydrocarbons during piston core recovery (Table 2) were found later, when split, to be characterized by intervals with abundant calcareous nodules, micronodules, and shell fragments. Perhaps this reflected the need of both bacterial sulfate reducers and methane fermenters for an organic or hydrocarbon substrate. The core from site 212 retains open, millimeter-scale, elliptical void textures that indicate what appear to be gas bubbles preserved in the core. Interestingly, the core from site 212 is also the only group 1 core that comes from within the more central parts of the mass transport complex (see Figure 9 for the location and Gee et al., 2007, for deposit detail).

Worldwide, carbonate cements and nodules are commonplace in sea-floor sediments in settings that are anoxic and where bacteria and archaea break down organic matter and hydrocarbons. Sea-floor nodules typically form via varying biogenic combinations of methane oxidation and sulfate reduction (Roberts and Aharon, 1994).

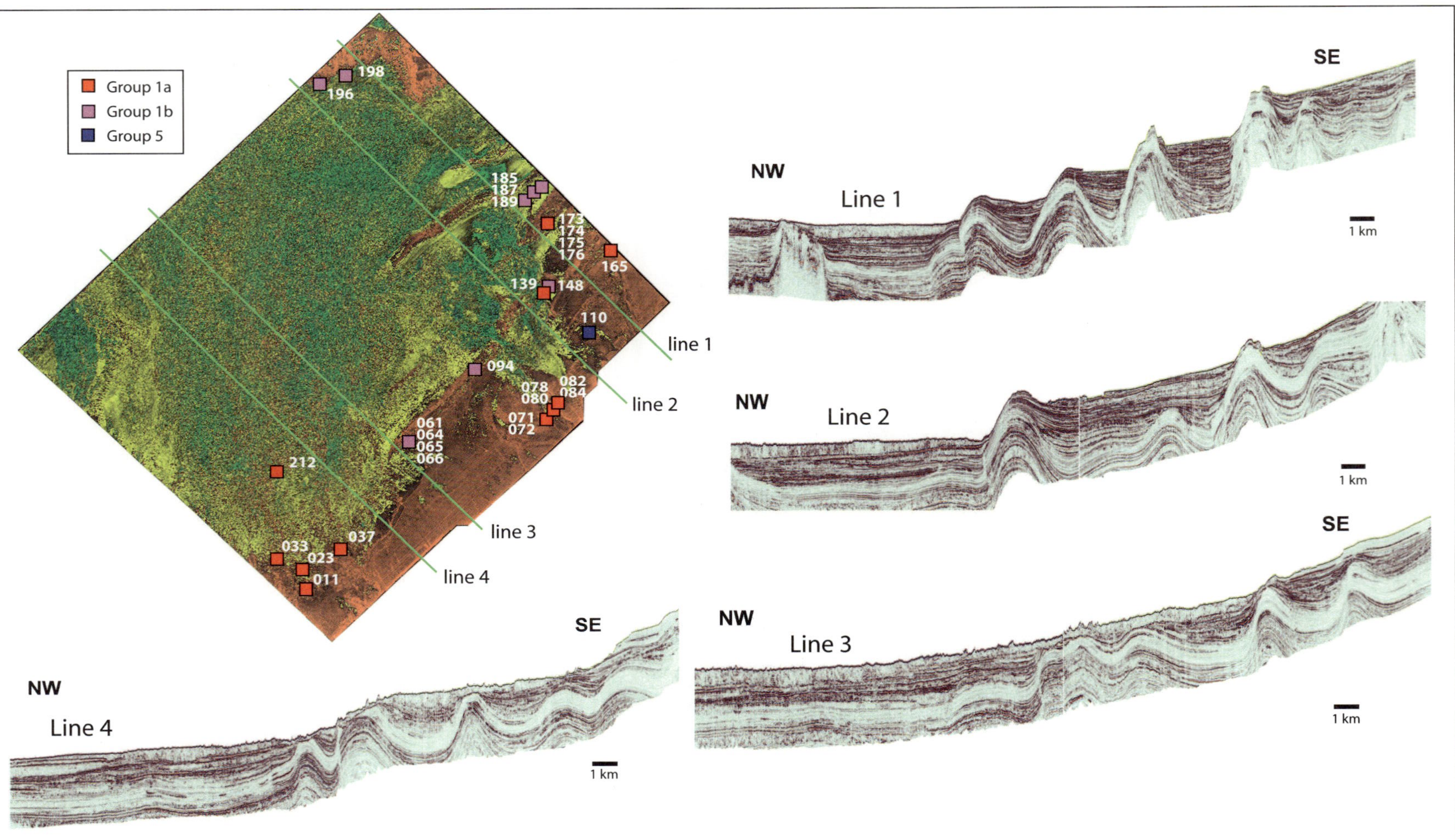

Figure 9. Sea floor and seismic character. Map view of seismic amplitude extractions of the water bottom where the color range represents a range of root mean square (RMS) amplitudes. Areas in yellow and green are typically active sediment flow paths (low RMS values); browns and grays indicate more compacted mud and a relatively stable ground. This map view also shows the position of all samples with group 1A (definite hydrocarbon seep) signatures (red squares) and selected 1B (pink squares) signatures where there are no group 1A signatures on the same feature (refer to Table 1 for details). Generally, the higher the RMS amplitude, the more consolidated the sediments. Tying the seep positions to the four sections (lines 1 – 4) shows that seeps are tied to outcropping structural culminations (except site 212, which is tied to a shallowly buried culmination with an eroded crest, as seen in line 4). The mass transport complex, shown in shades of yellow and green in map view, is a seal to thermogenic hydrocarbons (see Gee et al., 2007, for sedimentological details). The series of seismic lines 1 – 4 shows the continuation of the thrust-cored anticlinal ridges and how they plunge to the south and become increasingly covered by sediments derived from the outflow of the Baram Delta. Line 4 shows the erosional crest of one of the anticlinal ridges related to the emplacement of the mass transport complex. The scale bar in all lines is 3 km (1.8 mi), and the illustrated seismic in all lines is the uppermost 1 s of the data volume. 1 km = 0.6 mi

Bacterially mediated carbonate precipitation first occurs in early burial where methane, H_2S, organics, and bicarbonate are present in anoxic interstitial waters, for example

$$CH_4 + SO_4^{2-} \rightarrow HCO_3^- + HS^- + H_2O \quad (1)$$

The bicarbonate released during anaerobic bacterial activity increases the alkalinity and hence drives the precipitation of authigenic carbonate.

$$Ca^{2+} + 2HCO_3 \rightarrow CaCO_{3-(solid)} + CO_2 + H_2O \quad (2)$$

Bicarbonate released from dissolution of buried shell fragments can also be involved in the precipitation of sea-floor nodules. Thus, the isotopic compositions in carbonate nodules on the Brunei deep sea floor are essentially inherited from varying combinations of methane, organics, and marine bicarbonate sources.

Stable Isotopes in Nodules

Figure 10 plots $\delta^{13}C_{PDB}$ (Pee Dee belemnite) and $\delta^{18}O_{PDB}$ values for various calcareous nodules, diffusely cemented muds, and bivalve shells (as listed in Table 3). Nodules are varying combinations of siderite and dolomite. Some larger nodules are made up of coalesced smaller nodules; other nodules are transitional into lighter colored, less distinct, diffuse mottled zones of cemented mud. Isotopic compositions range from −50.8 to + 11.0‰ $\delta^{13}C_{PDB}$ and +2.3 to +6‰ $\delta^{18}O_{PDB}$. Most oxygen values cluster around +5‰ PDB, consistent with precipitation in cold marine bottom waters on a ~1000-m (~3281-ft)-deep sea floor. Carbon values plot in three distinct zones: an upper zone with heavier carbon contents ($\delta^{13}C_{PDB} > -10$‰), an intermediate carbon cluster ($\delta^{13}C_{PDB}$ of −10 to −40‰), and a group with quite depleted carbon-13 ($\delta^{13}C_{PDB} > -35$‰).

The upper cluster encompasses both nodules and marine shell material, with the shelly material clustered in the typical field for shell in deep cold sea water (indicated by the dashed rectangle in Figure 10) (Matsumoto, 1989; Warren, 2000). The typical $\delta^{18}O$ range of +4 to +6‰ for this group and the clustering of $\delta^{13}C$ around 0‰ indicate a marine carbon-bicarbonate source that was not fractionated by the activities of sulfate-reducing bacteria or methanogens.

One sample from site 094 comes from a nodule collected from the deeper part (~3-m [~10-ft] depth) of the recovered piston core has a $\delta^{13}C$ value of +11‰. This value lies well outside the normal carbon range for marine-associated bicarbonate precipitates and even farther outside values in nodules typical of sulfate reduction or methane oxidation. Some layers of ferroan dolomite and siderite nodules in Miocene reservoir sands (shoreface and tidal deposits) in the Champion oil field show similar positive carbon values (+4 to +16‰) and are the result of CO_2 produced by biogenic fermentation in a growth-fault-focused zone of methane-rich hydrocarbon seepage just below the Miocene sea floor (Warren et al., 2004). The 094 nodule is perhaps a modern example of this same set of processes. Its organic association is group 1B; thus, it is a condensate with a mixed biogenic and thermogenic source. Such heavy carbon is not uncommon a few meters below of the modern sea floor in areas where CO_2-limited methanogenic fermentation is occurring, and it is sometimes associated with methane seeps and chemosynthetic bacterial communities (Roberts and Aharon, 1994). Unfortunately, only nodules and no marine shell material was recovered from core site 094 to isotopically test this chemosynthetic hypothesis. The 094 coring site is at the apex of an anticlinal structure and near an eroded collapsed surface overlying a ramping thrust structure (Figure 7a). A homogeneous layer of weak amplitude (hydrate cemented?) and the associated bottom-simulating reflector (BSR) run updip and terminate adjacent to the core location. The suspected condensate might have risen vertically via the indicated plume pathway or have flowed laterally along the base of the low-amplitude unit to reach the surface.

All marine shells and fragments from the various cores plot within the fractionation field for a typical deep-water oxic biota (all plot in the dashed rectangle in Figure 10). Almost none of the organic matter associated with the sediment hosting bivalve shells in the various cores indicates any hydrocarbon seep association (i.e., no association with group 1A organics). The articulated shell in the core from location 011 is an exception in that it is associated with group 1A organics, yet it has a normal marine isotopic signature. This implies that healthy bivalves grew on the local sea floor in the vicinity of a hydrocarbon seep, yet the shells precipitated with a normal marine signature. Hence, no obvious evidence of a chemosynthetic signature in any of the analysed shells exists, except at site 094.

Most of the recovered nodules have carbon and oxygen isotope signatures that plot in the accepted range for sulfate reduction occurring during normal shallow-burial marine diagenesis (Matsumoto, 1989; Warren, 2000; Kotelnikova, 2002; Malone, 2002; Raiswell et al., 2002). Sulfate reducers are ubiquitous in the anoxic waters that occur in the first few millimeters to centimeters below the sea floor. These bacteria digest organics as they remove SO_4^{2-} and convert it to H_2S while increasing

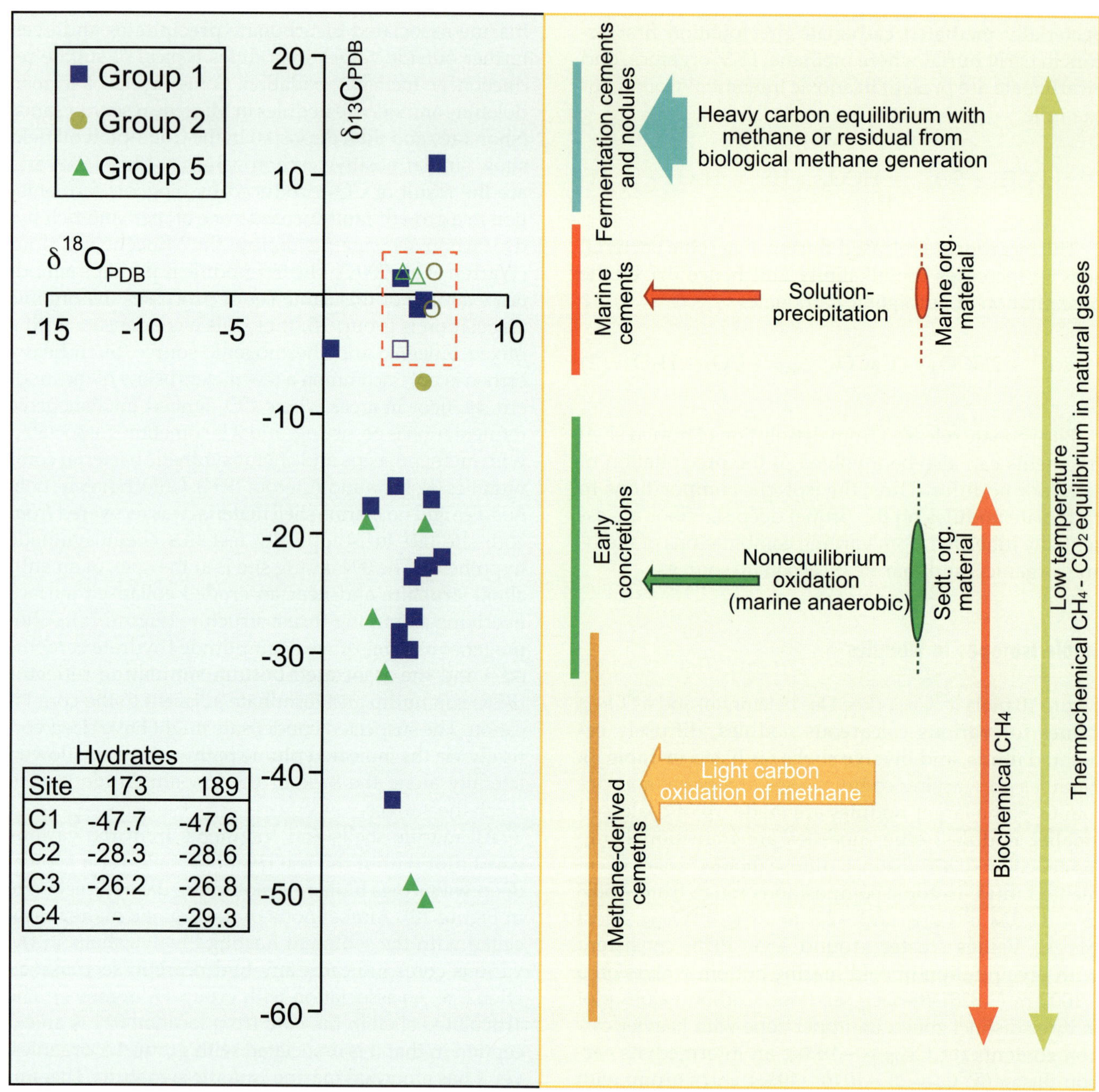

Figure 10. Carbon and oxygen stable isotope signatures from selected nodules, tied to the organic group of the host sediment. Open (unfilled) symbols indicate determinations using bivalve shell material. The dashed rectangle shows the plot field typical of oxic deep-marine shelly fauna worldwide. The inset labeled hydrates lists $\delta^{13}C_{PDB}$ values for various gases from methane hydrate samples recovered in piston cores from sites 173 and 189. Interpretive plot fields are from Matsumoto (1989,) and Warren (2000). PDB = Pee Dee belemnite.

alkalinity and so create a rising saturation state in the pore water that ultimately precipitates authigenic carbonate cement. The accepted carbon isotopic values for such carbonate precipitates in shallow sea-floor sulfate reduction lie in the range of −15 and −45‰, with oxygen values being dependent on the ambient seawater temperature.

The lower end of this carbon isotope range (>35–40‰) overlaps with isotopic signatures related to the influence of methane oxidation and thermogenic methane. This brings us to the lower cluster of carbon isotopic values in Table 3 and Figure 10 (site 013 and site 161). Depleted carbon-13 values place both samples in a possible transition range into the isotope field of carbonate

Table 3. Carbon and oxygen isotope signatures of carbonate nodules, shells, and cements listed against organic typing of host sediment.

Piston Core	Organic Group	$\delta^{13}C_{PDB}$	$\delta^{18}O_{PDB}$	Description
11	1A	1.3	4.2	Shell
11	1A	−0.5	5.5	Nodule
11	1A	−1.3	5.1	Nodule
11	1A	−4.6	0.4	Nodule
13	5	1.8	4.3	Shell
13	5	−49.3	4.9	Nodule
20	5	1.4	5.1	Shell
23	1	−17.2	5.9	Nodule
44	2	1.8	6.0	Shell
44	2	−1.3	5.9	Shell
44	2	−7.4	5.5	Nodule
61	5	−19.0	2.3	Nodule
61	5	−19.2	5.6	Micronodules (sand)
65	1B	−4.6	4.2	Shell
65	1B	−17.7	2.5	Nodule
65	1B	−22.1	6.5	Nodule
65	1B	−22.8	5.8	Nodule
65	1B	−22.8	5.6	Nodule
65	1B	−22.9	5.4	Nodule
72	1A	−27.1	5.0	Nodule
94	1B	11.0	6.1	Nodule
110	5	−27.1	2.7	Nodule
110	5	−31.6	3.4	Cemented mud
135	1B	−23.8	4.9	Nodule
148	1B	−42.4	3.8	Nodule
161	5	−50.8	5.7	Nodule
185	1B	−28.4	4.2	Nodule
185	1B	−29.8	4.6	Nodule
185	1B	−29.9	4.7	Nodule
187	1B	−29.9	4.2	Nodule
189	1B	−16.1	3.9	Nodule

cements and nodules derived from thermogenic methane, but the values lie only at the upper end of this range. Both samples can be just as readily interpreted as the result of biogenic sulfate reduction or methane oxidation.

Based on its gas composition analysis, the sediment at site 013 does not show elevated levels of methane per dry sediment weight. The presence of an obvious H_2S smell in the sediment core when slabbed, instead of the hydrocarbon odor (noted by shipboard Geolabnor technicians), and a group 5 organic association argue that the nodules at site 013 are the product of bacterial sulfate reduction. Seismic data in the vicinity of this core site likewise show a well-layered sediment with little evidence of fluids actively rising from depth (Figure 7b). For all the same reasons, the nodules at site 161 are also interpreted as byproducts of bacterial sulfate reduction acting on marine pore fluids. Site 161 does lie off the edge of a thrust-cored growth anticline and so has a higher potential than site 013 to be sourced by thermogenic waters, but the lack of any internal disturbance in seismic data beneath the coring site argues that it too is a normal sea-floor diagenetic product where nodules formed without any influence from thermogenic gas (Figure 7c).

In summary, prior to isotope analysis, we thought that the tie of hydrocarbon odor (noted by GeoLabnor shipboard technicians) to carbonate nodules logged in slabbed cores could perhaps be useful as an indicator of local chemosynthetic and seep associations. However, when various group 1–5 nodules were sampled isotopically, almost all of the resulting carbon and oxygen values are consistent with precipitation in the sulfate reduction zone, whereas the shell material shows a typical cold-water marine signature.

Thermogenic Sea-Floor Fluids (Group1 Sediment) Tied to Seismic Character and Structure

The upper sedimentary section of the northwest Borneo trough of offshore Brunei is made up of layers and chaotic units that are faulted and have collapsed and flowed around and atop discontinuous thrust-cored growth anticlines with numerous slope failures of varying extents along the flanks of various thrust anticline-controlled minibasins (Figures 1, 9). Regionally, the deep-sea floor of offshore Brunei is characterized by two depositional styles. First is an extensive, relatively flat-lying zone that defines the upper surface of a mass transport complex. It has a chaotic seismic character and extends westward and northward across the survey area to make up two thirds of the surface sedimentary section (Gee et al., 2007). Second is an undulating and sloping sea floor in the continental slope and rise with a more finely layered seismic signature (Morley, 2007b). The sea-floor ridges are the surface expression of several anticlines (A–H in Figure 1). These growth anticlines act as a catchment for point-sourced shelf sediments moving downslope along the eastern and northeastern parts of the survey area. The anticlines are in turn the shallow expression of several gravity-glide-driven toe thrusts ramping up from an underlying unit of shale that is probably equivalent to the Setap Formation (middle Miocene) in its more landward and stratigraphically lower sections (Demyttenaere et al., 2000; McGilvery and Cook, 2003). When overpressured, this shale is susceptible to focused deformation and mud diapirism that then drives the formation of the various mud-cored ridges, walls, and chimneys (Figure 9).

Thus, the stepped, slope anticline-cored features (A–H) and the various intraslope minibasins in this region are a direct consequence of gravitationally driven toe thrusts and gravity slides facilitated by decollement surfaces in the shale section (Figures 1, 9). The ridges plunge southward and underlie the upper and more landward parts of the Baram landslide (Figure 9) (Gee et al., 2007). A BSR is visible in many seismic lines and is variably present across extensive parts of the surveyed area (e.g., Laird, in press). That the BSR in the slope and rise sediments of offshore Brunei is an indicator of the base of the methane-hydrate stability zone is confirmed by clathrates recovered in piston cores 173 and 180 and the ubiquitous presence of methane in headspace gases from all piston cores. A BSR tied to the base of clathrate horizons is a common feature of slope and rise sediments worldwide (Miller et al., 1991; Grauls, 2001). Fault scarps can be seen in the south that extend toward a toe thrust in the middle and southern region of the survey area (Figure 9). When seeps identified by their organic signature (Table 1) are tied to geological features seen in seismic data, most hydrocarbon seepages in the eastern third of the study area seem to be focused by underlying geological structures (ridges and mud volcanoes). This association is not so obvious in the south, where an impervious mass-flow unit buries the ridges (Figure 9) (Gee et al., 2007).

Zones of Disturbed Mud and Diapirs and Anticlinal Ridges

At a more local scale, the various antiformal axial ridges locally encompass disturbed, overpressured, and in places diapiric mud and gas plumes, which are most obvious along faults and atop the axial core of some of the antiformal structures (Figures 11–13). Seismic data show that these disturbed and fluid-escape zones are typically accompanied by an abrupt termination of the ridge layers, commonly combined with a narrow folding of reflectors around the discontinuous core of the volcano or chimney structure (Figures 11, 12). Reflection amplitude in the middle of an anticlinal ridge or more locally in a diapiric core typically becomes weaker and more chaotic. Independent of whether or not mud has actually flowed, these weakened amplitudes and poorer reflectivities are interpreted as a seismic response to disturbed, somewhat overpressured or previously overpressured muddy and gassy cores to the ridges (e.g., Milkov, 2000).

The effects of fluid pressure buildup and its escape are clearly seen in the contrasting organic signatures in sediments above and adjacent to collapsed mud volcanoes and pockmarks in various ridges (A–H) that constitute the undulating part of the slope and rise of Brunei (e.g. Figure 11, sites 173–176) and in sea-floor sediments atop an actively growing and pressurized diapir mound that has yet to breach and collapse (Figure 11, sites 110–115). In the region of sites 173–176, the sea-floor sediments within or near the collapse cone of the formerly active mud volcano vent are characterized by group 1 (hydrocarbon seep) signatures. Deep pore fluids have breached the sea floor, whereas in the region beneath sites 110–115, the structure is still inflated and sea-floor sediments only contain the group 5 organic signature of ROM (Figure 12). Likewise, the group 1 signatures that characterize sites 065, 082, and 084 are also tied to collapse features; interestingly, both these sites are now sediment spillover sites along growth ridges that are capturing sediment on their updip sides in ponded minibasins (Figure 13a, b). The sea-floor collapse that sometimes accompanies the depressurizing of mud volcanoes may have generated a landscape low that then allowed the sediment to pass across the ridge and down into the next minibasin. Sample site 065 is not in, but in close proximity to, a pressurized mud volcano (Figure 13b). Dimitrov (2002), in a worldwide review of mud volcanoes, pointed out that gas and fluid may be emitted some distance away from the main vent via smaller secondary vents called gryphons. These gryphons commonly spew out fluid and mud that subsequently get captured in the near-bottom sediments and may explain the group 1 signature at site 065.

Fault-Breached and Eroded Anticlinal Ridges

Additional group 1 (hydrocarbon seep) sites are tied to anticlinal ridges, but these sites are not as tightly related to a single fluid conduit feature, unlike those described in the previous section. Instead, they constitute more diffuse zones of extensional normal faulting (Figure 14). Typically, they occur on the upper exposed parts of the basinward sides of growing anticlinal ridges, where ridge collapse has yet to occur and crestal sediment is yet to be redeposited as a debris flow (the debris flow lens from such a collapse is imaged in Figure 7b). Sample sites 185, 187, and 189 all contain group 1 hydrocarbons, and narrow fault-facilitated gassy fluid columns are a common occurrence at the sea floor along this part of the hosting thrust-cored ridge (Figure 14). Onlapping the landward side of this ridge is a muddy sedimentary unit sampled at site 183. There, the sea-floor muds contain only group 5 hydrocarbons, implying that this ponded sedimentary unit acts as a seal to rising deeper pore fluids. This is also a site where rates of rise were sufficiently slow to allow some of the rising fluid to be trapped as a methane hydrate, as recovered in core 189. Unlike the organics, the carbon isotope signatures in the carbonate nodules from the ridge piston cores do not show any other signature other than that of early marine concretions (Table 3; Figure 10).

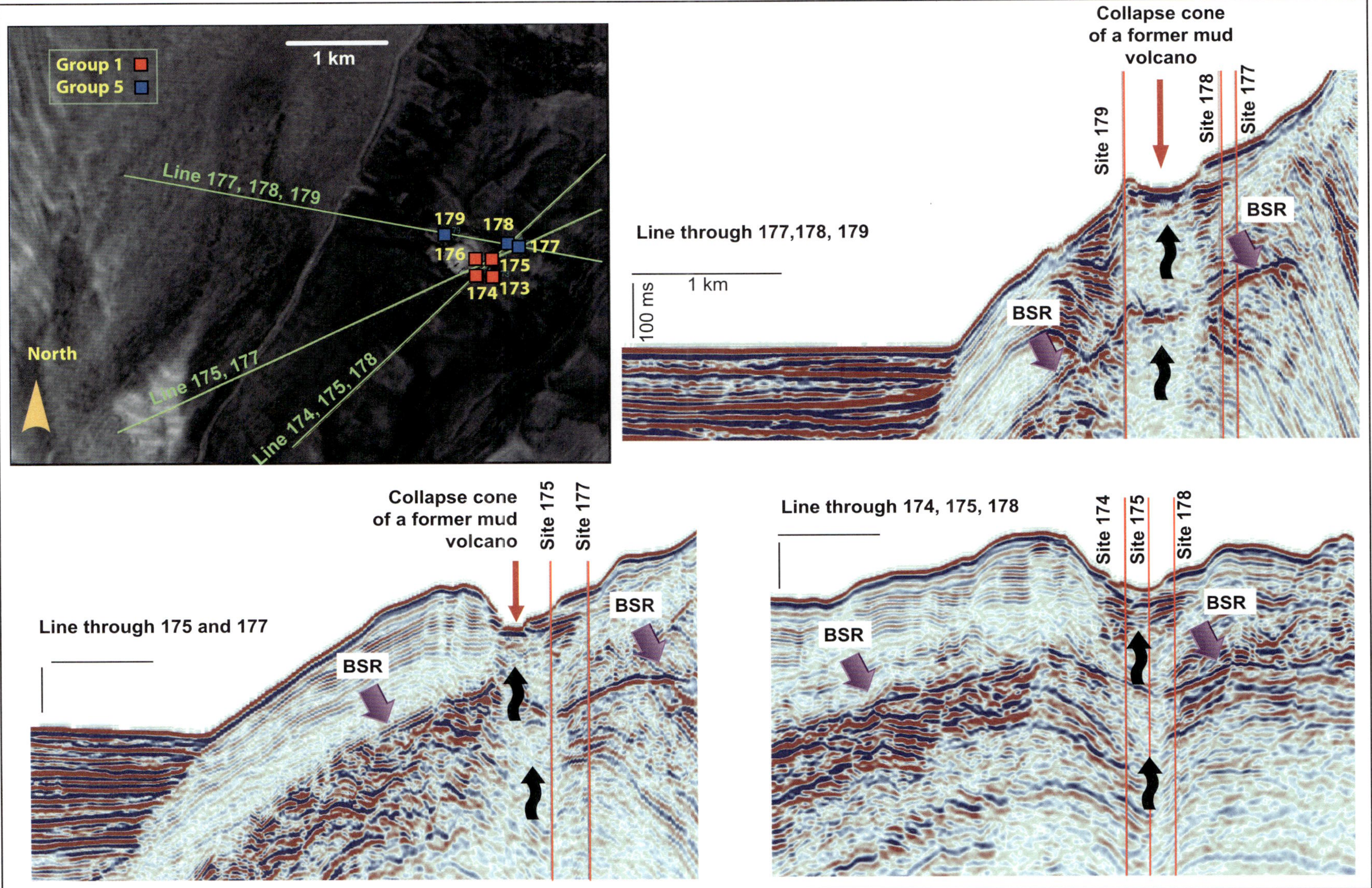

Figure 11. Group 1 organic signatures in the low with the typical depressurized signature of a now-collapsed mud volcano in the region of sites 173–179. Note the downwarping of reflectors, including the bottom-simulating reflector (BSR), in the vicinity of the volcano and the re-establishment of the BSR in the collapse cone sediments. Horizontal bars = 1 km (0.6 mi); vertical bars = 1 ms. See Figure 9 for the site locations.

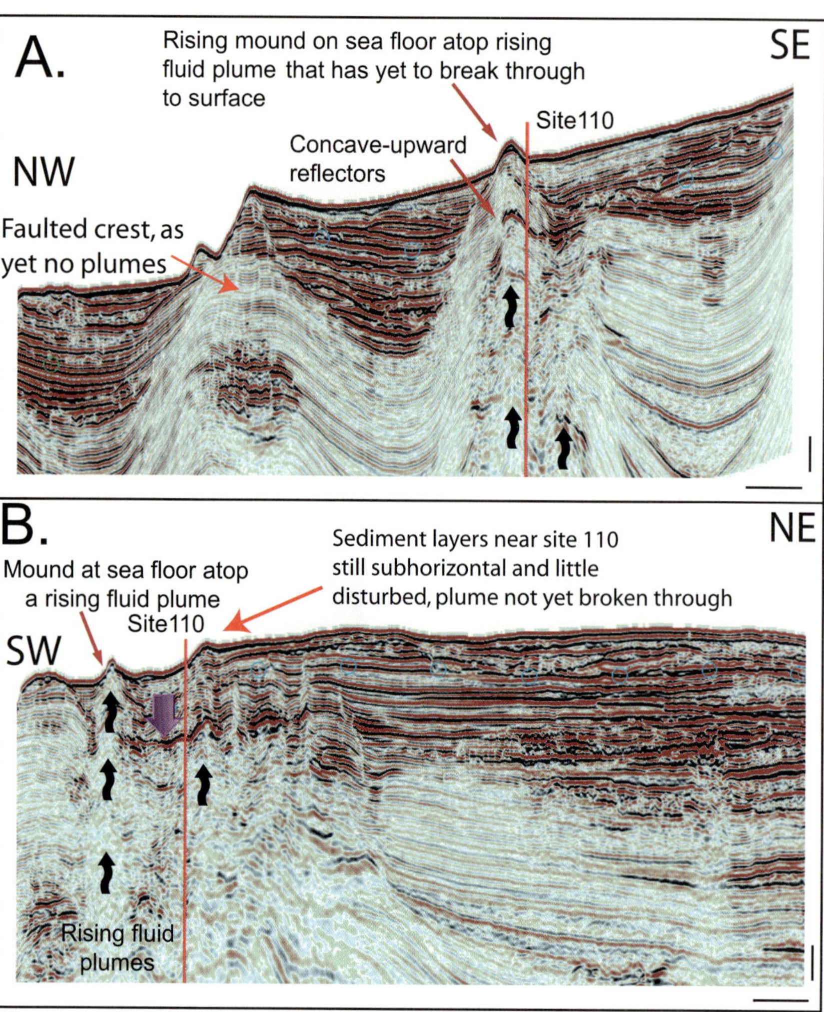

Figure 12. Incipient mud volcano. The intersection of (A) inline 9084 and (B) crossline 8858 is marked by a red line in the vicinity of piston core site 110. Sites 110–115 along this ridge all show a group 5 signature yet lie atop the thrust-cored growth anticline where rising overpressured fluids are causing the reflectors to rise into diapirlike convex geometries that can create mounds on the sea floor. The convex-upward character of reflectors that pass into the rising fluid columns and the convex-upward sea floor atop the rising plumes (no collapse cones exist) imply that the pressured fluid column is yet to breach the sea floor from below site 110 but has reached the sea floor nearby. For example, the plume that rises all the way to the sea floor in the mound downdip of site 110 and the lack of layering and a BSR in the plume feeding this mound imply that seepage of fluids onto the sea floor near site 110 occurs. This section illustrates the precursor phase and the active phase of a rising fluid plume. Horizontal bars = 1 km (0.6 mi); vertical bars = 1 ms. See Figure 9 for the site locations.

Margins of Mass-flow Deposits

Core sites 011, 023, 033, and 037 are characterized by the presence of group 1 seep hydrocarbons in the southern part of the survey area, and all are sites that occur near the upper margin of the Baram landslide (Figure 9). Cross sections reveal a covering unit characterized by weak chaotic reflections, punctuated in various places by gas columns recognizable by a vertical column of weak amplitudes, with strong amplitudes at either end (Figure 15). Major horizons are still correlatable across these rising gas columns, and a large number of the gas columns are capped with conical structures at the surface. The features may represent active or recently active mud volcano vents. This association of group 1 sediments is more enigmatic than the organic-seismic-association outline for the eastern part of the study area. This complexity reflects the emplacement of the landslide sediments over the growth anticlines and the erosion of some anticline crests as slide blocks feeding into the landslide.

Two sites are illustrated in Figure 15. Site 023 is atop the extensional crest of a now-buried growth anticline. The rising fluid plumes can be seen to be concentrated in the crestal position of the anticline (which is a closed structure) and rising within an extensively faulted and folded terrain in a situation similar to that seen at sites 185–189 (Figures 9, 14). However, this setting is complicated by the erosion of the anticlinal crest and the emplacement of the mass transport complex. In this situation, the landslide sediment retains sufficient permeability to allow thermogenic fluids to rise to the surface. The situation where group 1 organics occur at the sea floor at site 011 is even more complicated (Figure 15b). The same mass transport complex is present, and in this case, the sample site is located at the edge of a large kilometer-scale (mile-scale) slide block (olistolith) within the mass transport complex (Gee et al., 2007). The site is situated downslope of site 023, and so, two possibilities exist for the source of the group 1 signature. It could come from the olistolith block if it was derived

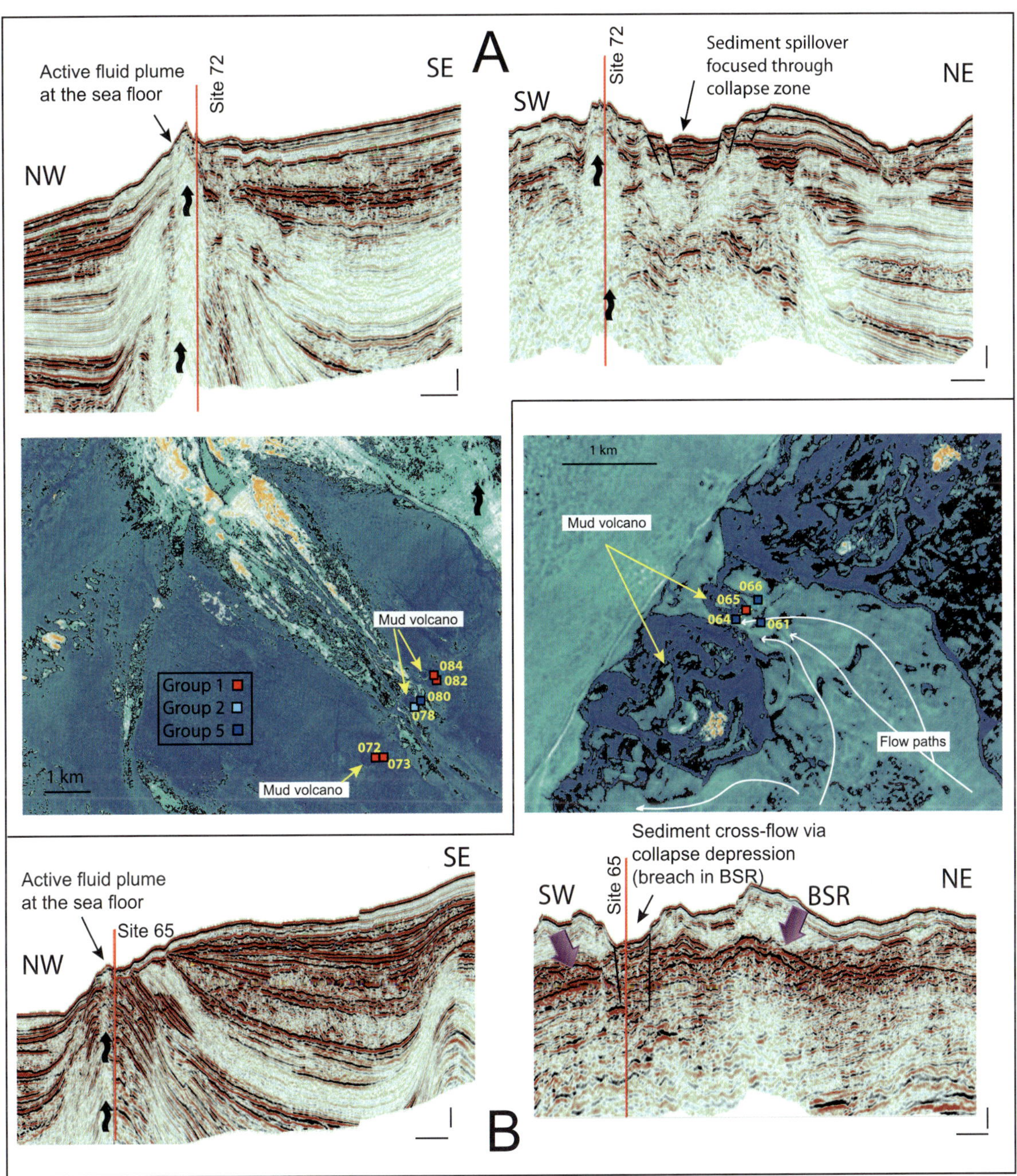

Figure 13. Mud volcanoes and the relationship to cross-ridge sediment dispersal pathways. (A) A series of active and collapsed mud volcanoes occur in the vicinity of sites 72–84 as shown in the map view. The inline shows a northwest–southeast section perpendicular to the structural strike showing location site 72 (group 1) at the edge of an active fluid plume and mud volcano. The southwest–northeast crossline through site 72 shows a depression to the northeast of the volcano where sediment is faulted and spilling down to the northwest to accumulate in the next downslope minibasin. Sediment breakout from the updip side of the ridge appears to be focused by a collapse zone located immediately northeast of the volcano. If it does not initiate sediment flow, it is highly likely that rising plumes help maintain the instability of the breach point and allow sediment to move from the updip minibasin to the downdip minibasin. (B) Site 65 is another group 1 site that the northwest–southeast inline shows located at the edge of an active fluid plume and mud volcano. Unlike site 72, this site is at a more mature stage of volcano evolution where the low created by depressurization has focused the zone of sediment crossover into the lower minibasin. Note the lack of the bottom-simulating reflector (BSR) and downwarped beds in the vicinity of the collapse. Horizontal bars = 1 km (0.6 mi); vertical bars = 1 ms. See Figure 9 for the site locations. Map views in panels A and B are seismic amplitude extractions of the water bottom where lighter colors are typically active sediment flow paths and darker colors (blues) indicate more compacted mud and relatively stable ground.

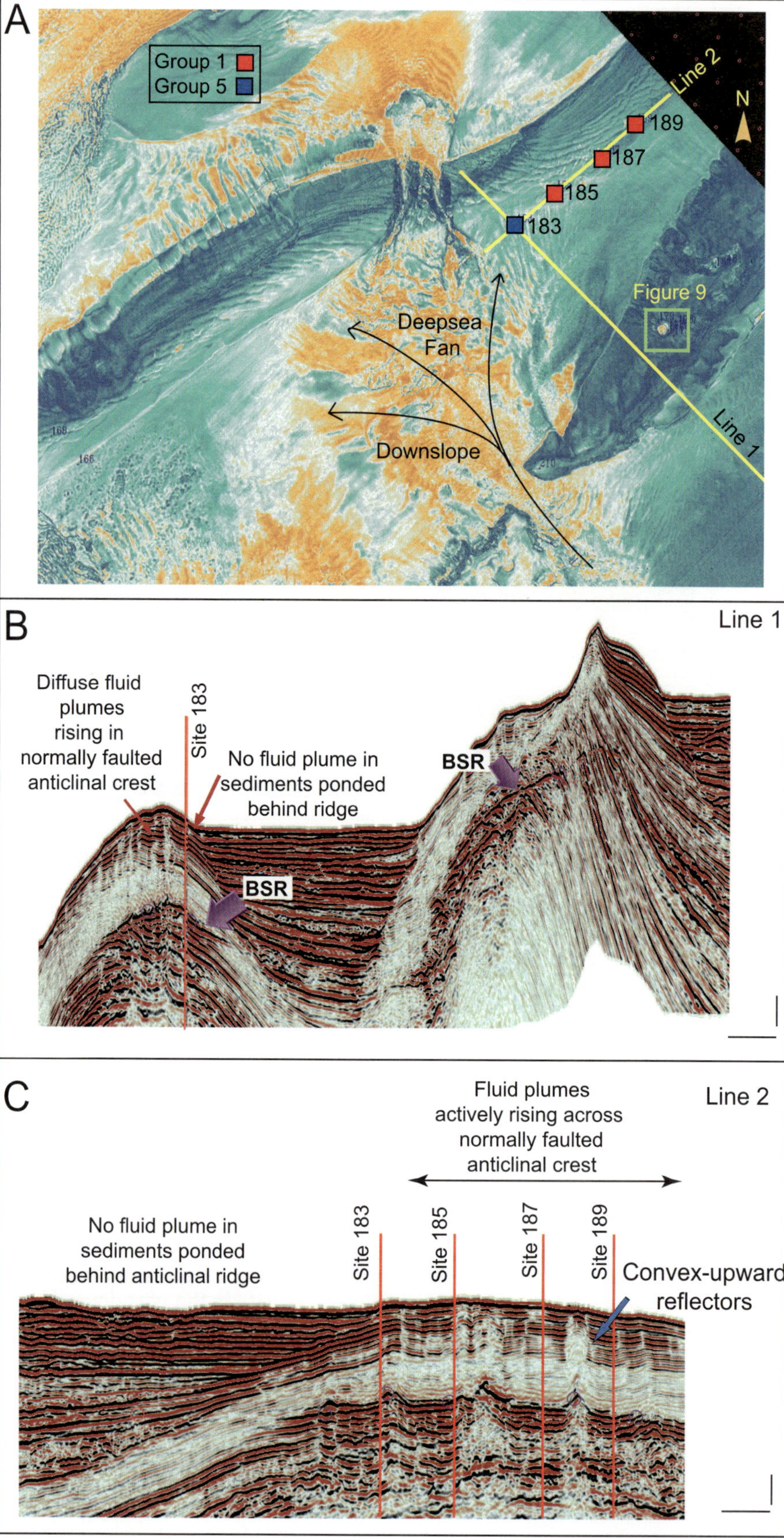

Figure 14. Diffuse zones of fluid pluming across a faulted anticlinal ridge in the vicinity of sites 184, 185, 187, and 189. (A) Inline section through site 183. (B) Crossline section through site 183. Both the inline and crossline sections show diffuse fluid plumes focused along steeply dipping extensional faults. Note the convex-upward reflectors and the truncation of the bottom-simulating reflector (BSR) implying warm fluids. Only site 183 does not have a group 1 signature, showing that the sediment ponded in the minibasin is capable of holding back the rising fluid plume. Horizontal bars = 1 km (0.6 mi); vertical bars = 1 ms. See Figure 9 for the location.

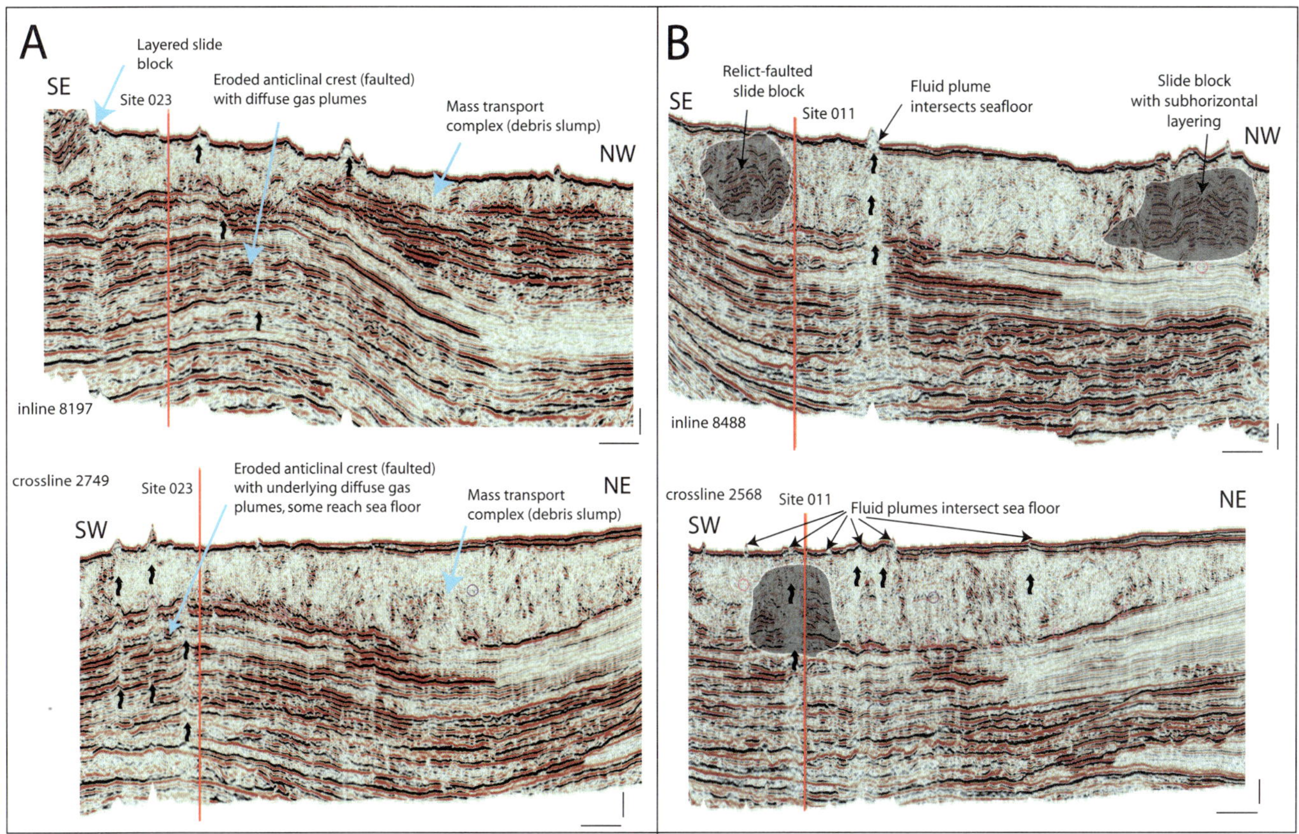

Figure 15. Interactions of the mass transport complex and faulted growth anticline crests. (A) Intersect of inline (8197) and crossline (2749) shown by the red line at sample site 023 (group 1). The site samples a diffuse zone of fluid seepage across the faulted crest of a growth anticline in a structural situation similar to sites 185–189 (see Figure 14). The anticline has some degree of closure to focus the migrating fluid toward its crest, but in this location, the crest of the anticline was partially eroded by the formation of the overlying mass transport complex. The site is at the upper end of this complex and it does not appear to seal in this region. (B) Inline (8488) and crossline (2568) intersect shown by the red line at sample site 011 (group 1). The sample site is adjacent to an obvious gas seep at the edge of a large coherent block (olistolith) in the mass transport complex. This block was possibly derived from the eroded anticlinal crest and carries a thermogenic signature farther out into the mass transport complex (see text). Horizontal bars = 1 km (0.6 mi); vertical bars = 1 ms. See Figure 9 for the location.

upslope from a hydrocarbon-charged anticlinal crest, and the obvious block faulting in the block supports this contention, or the fluid plume could be sourced from fluids rising from directly below the sample site, as indicated by some deeply sourced fluid plumes in adjacent areas (Figure 15b).

CONCLUSIONS, IMPLICATIONS, AND COMPARISONS

The Brunei database is unusual in the broad scale of its single survey 3-D seismic coverage, which is also covered by an extensive organic geochemical study of the sea floor. The area encompasses fluid escape features across several sediment-damming growth anticlines and a large landslide below the mouth of the Baram Delta. Within this broad study area, we can conclude the following:

1) Mud volcanoes, especially where associated with breached anticlinal crests, constitute the main cause of focused oil and gas seepage onto the Brunei sea floor (Figure 16a). They show a sequence of events that indicate first inflation, then breach or fluid release, then deflation of the structure. This corresponds to the transition from sea-floor mound to pockmark and to a transition from convex reflectors and loss of the BSR in the mound to concave reflectors and the re-establishment of the BSR beneath the pockmark.
2) More diffuse seepage can occur from leaky seals along the crests of anticlinal ridges, but these can

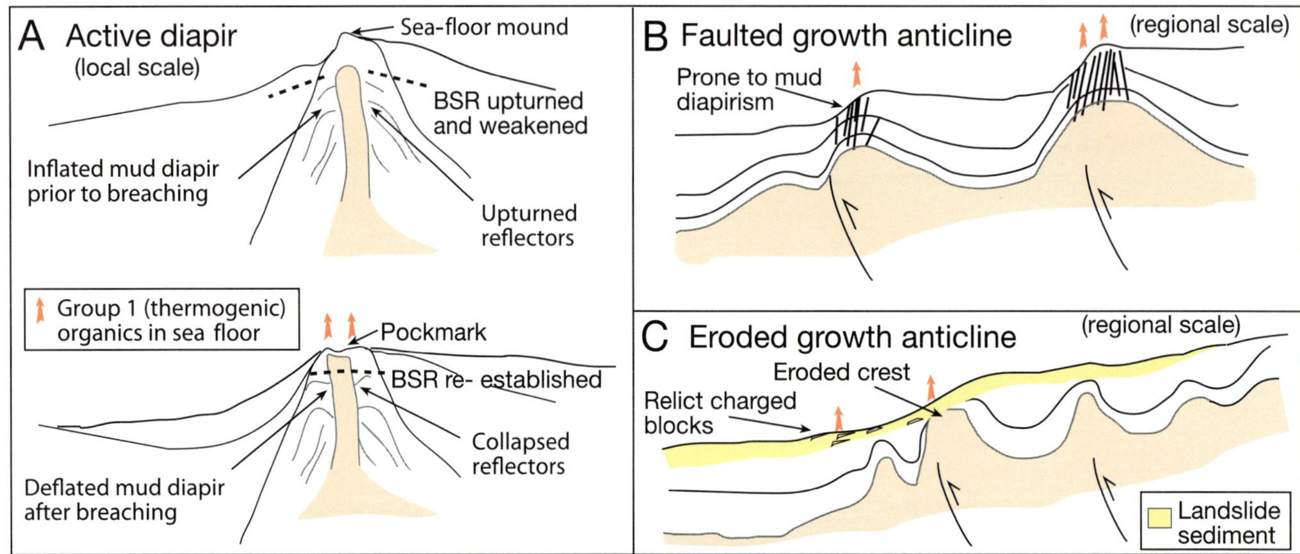

Figure 16. Relationship between the occurrence of thermogenic organic matter in sea-floor sediment, and structural and sedimentological settings. (A) Thermogenic signature indicating a breached mud diapir and found in and around pockmarks. The inflation and deflation stages show characteristic reflectors and the presence or absence of a bottom-simulating reflector (BSR) in the fluid chimney. (B) Thermogenic signature in zones of fluid leakage across the faulted extensional crests of growth anticlines, which are also regions subject to mud diapirism. (C) Thermogenic signatures related to anticlinal crests created during landslide emplacement or related to eroded relict-charged blocks within the landslide debris.

be confused with erosionally breached methane hydrates sourced from biogenic gas (Figure 16b).

3) Diffuse hydrocarbon seepage can occur along the fault-controlled edge of the huge mass flow and the disturbed landslide and sediment-creep field that defines much of the northern and western parts of offshore Brunei (Figure 16c). Thermogenic signatures in this setting are either in muds atop eroded anticlinal crests or associated with eroded and downslope crestal and collapsed fault blocks that, because of inherently low permeability, still carry a thermogenic charge.
4) Although methane seeps permeate much of the Brunei deep-sea floor, most carbonate cements and nodules found in the shallow burial environment precipitated as the result of the metabolic activities of sulfate-reducing bacteria. Also, local areas of focused seepage are observed where methanogenic fermentation dominates. Both the bacterial and archaeal communities require organics and anoxic pore waters to flourish.

These conclusions have broader implications. At the regional scale, the modern active growth anticlines of offshore Brunei are gravity driven and overlie a zone of ramping thrusts, yet the Miocene–Pliocene anticlines and the mud volcanoes of Brunei are accreting in a tectonic setting where subduction ceased in the early Miocene (Hall and Morley, 2004). This lack of active subduction beneath its muddy offshore wedge separates Brunei from many other mud-rich growth anticline systems. The latter tend to occur in active accretionary wedges atop or adjacent to zones of active subduction, as in the Mediterranean and Tethyan collision belts (Table 4). Surface slopes in such orogenically active compressional systems tend to be greater than 1–1.5° (Figure 17). The regionally quiescent tectonic setting for mud diapirism in the Brunei slope and rise is perhaps more reminiscent of argillokinesis along passive margins (Table 4). There, toe-of-slope argillokinetic delta systems constitute the lower parts of muddy outflow zones located in front of large high-sediment-load rivers. The deeper parts of the Niger and Rhone deltas are two examples of such systems, and their surface slopes tend to be relatively low, ~0.5–1.5°.

In fact, Morley (2007a) showed that sediments of offshore Brunei and Sabah are unusual when compared to other muddy submarine fold and thrust belts worldwide (Figure 17). Brunei has one of the most acute angles recorded for any submarine critical-taper wedge. Morley argued that the very low surface slopes (1–2.5°) and basal detachment dips (2° to 5°) in the Brunei offshore wedge require basal pore-fluid pressures that must be near lithostatic for the wedge to propagate at such angles. If these angles are compared to those of typical orogenic belts, Brunei plots very near fields occupied by compressional belts with a salt decollement (Figure 17). As Ford (2004) noted in her study of some of the world's orogenic belts, the presence of salt in a compressional terrain means a very low surface slope (<1°) in the slope and rise prism. Put simply, any sediment pod sitting atop a thick salt mass will tend to sink or slide and rotate atop the salt mass as it seeks gravitational equilibrium atop the fluidlike salt body (Warren, 2006).

Because of halite's inherently low strength and fluidlike character, all salt-cored compressional wedges, as plotted in Figure 17, lie below the dashed red line that indicates a 1° surface slope. Interestingly, the offshore wedges of Brunei and Barbados, two well-documented shale diapir terrains, also plot on or near this 1° line, yet both are salt free. They are, however, made up of prisms of sediment sitting atop pressurized mud terrains, with pressures in the deeply buried muds approaching lithostatic and the near-surface sediments characterized by mud chimneys and mud volcanoes.

An overburden adjustment to the fluidlike rheologic nature of both salt wedges and overpressured mud prisms occurs in a continental margin wherever zones of extremely weak sediment (salt or pressured [lithostatic] mud) lie beneath actively loading sediment pods (Table 4; Figure 17). Independent of whether the decollement sits in salt or overpressured mud, the shelf part of the continental margin is characterized by regional updip extension and growth faulting, which is balanced by gravity gliding and downdip compression of the toe-of-slope sediments. Toe-of-slope compressional anticlines in both settings tend to occur atop overlapping or stacked thrust sheets, which merge continentward into a lower decollement zone that is hosted in either salt or overpressured mud. The rising compressional ridges or growth anticlines that form above the thrust culminations act as sediment dams, thus creating minibasins between rising ridges. In terms of sediment thicknesses and sedimentation patterns in the minibasin fill, they show same-scale geometric similarities, as is seen when seismic profiles from offshore Brunei (Figure 9) are compared to profiles across salt-cored anticlinal ridge profiles across the Mississippi or Perdido fold belts in the Gulf of Mexico (e.g., Rowan, 1997, his figures 4, 6). Both settings show well-developed thrust-floored growth anticlines acting as downslope sediment barriers that pond sediment in interridge minibasins. Loads that developed in the minibasin feed back into the pulsed growth of the adjacent diapiric structures, along with the local collapse and sediment breaching of the anticlinal ridges (e.g., Rowan, 1997; Morley, 2007b).

However, key horizons generating or facilitating fluid flow within mud-cored versus salt-cored growth anticlines cannot be considered analogous (Table 4). At the scale of the interior to anticlinal growth ridges, rising pore fluids sourced from or below the decollement layer follow somewhat different escape pathways. In a mud diapir, the actual volume of deeply sourced fluid that is flowing in the diapir is tied to the distribution of overpressured muds that are periodically leaking pore fluids. What we resolve as a seismically disturbed zone in a mud-diapir province or in an overlying mud volcano or chimney is not necessarily a sediment matrix that has experienced Newtonian flow. Instead, the sediment is acoustically disturbed because it contains, or once contained, overpressured fluid on its way to the surface (Van Rensbergen and Morley, 2003). In contrast, what is seismically imaged in the compressional core of a growth anticline where a body of flowing salt is squeezed toward the surface is halokinetic salt. The salt is not a flow of gas-charged fluid, instead, the salt is a plastic mass moving through or over the adjacent sediment as a coherent impermeable whole. Halokinetic salt tends to act as a seal to gas-charged fluid, with the peripheries of the salt mass acting as foci for the flow and ultimate escape of underlying buoyant or overpressured fluids (Morley and Guerin, 1996; Warren, 2006).

Because properties and the seismic responses differ between gas-charged mud diapirs or chimneys, and salt diapirs or allochthons, the distribution of organic maturity indicators can also differ. Disturbed anticlinal cores and associated mud diapirs, volcanoes, or chimneys in offshore Brunei and elsewhere indicate sea-floor intersections with regions of overpressured mud. Within such mud diapirs, the structure itself is the source of the fluid signature on the sea floor (Figure 18a). In the case of halokinetic features, deeply circulated subsalt pore fluids tend to move laterally along the underside of the salt mass and only leak upward at the featheredge of a widespread salt allochthon sheet or across a salt weld (Figure 18b). Salt withdrawal can facilitate the rapid subsidence of muddy suprasalt sediments, and their dewatering can drive suprasalt argillokinesis. In some suprasalt situations, a combination of salt withdrawal and rapid sedimentation of water-saturated muds into the resultant suprasalt accommodation space can create instability and diapirism in the suprasalt or supraweld mud column, even if the mud itself is not overpressured. This is the case with shale diapirs of the Port Isobel fold belt, western Gulf of Mexico (Camerlo et al., 2004).

Whenever the overpressured plume of fluids, which constitutes a typical mud volcano or mud diapir, breaks out onto the sea floor, the system quickly depressurizes and collapses. Flowing mud stops flowing soon after reaching the sea floor (Figure 16a). In contrast, on reaching the sea floor, a pressurized salt mass can continue to flow and spread and can even initiate a new subhorizontal allochthon tier and a new set of sediment dams.

Table 4. Features of typical mud diapirism and salt diapirism.*

		Mud Diapir (Argillokinetic)	Salt Diapir (Halokinetic)
Tectonic setting where argillokinesis and/or halokinesis occurs	Rift to drift (passive margin (extensional) with depositional loading)	Yes, gravity-driven fold belt (especially in toe of slope). Examples: Mexican ridges and Port Isobel fold belts of western Gulf of Mexico; fold belts in the Sergipe-Alagoas and Pára-Maranhao basins of Brazil, Niger Delta fold belt; McKenzie delta fold belt, Arctic Canada; Yangtze delta, East China Sea.	Yes, gravity-driven fold belt (especially in toe of slope). Examples: slope and rise fold belts of the circum-South Atlantic salt basins and the northern Gulf of Mexico; fold belts in the Yemeni Red Sea; Nile deep-sea fan; Rhone deep-sea fan; Scotian margin, eastern Canada.
	Collisional accretionary belts, transpressional belts; both can overprint a former passive margin	Yes (forearc and backarc basins). Examples: Barbados accretionary prism; Mediterranean Sea and Tethyan belt, spanning from south of Greece over the Black and Caspian seas into Azerbaijan, the Crimea and Taman peninsulas, Iran, and Turkmenistan into the Makran coast; Borneo; South China Sea.	Yes (Especially in zones of continent-continent collision, compression can reactivate former extensional salt diapirs). Examples: Qom Basin and Zagros fold belt, Iran; Kuqa fold belt, Tarim; Betics; Carpathians; Appenines; Sierra Madre; Neuquen, Argentina.
	Intracratonic sag	No	No (unless basin inversion occurs, e.g., Amadeus Basin, Australia)
Depositional association	Marine	Typical of muddy continental slope and rise	Typical of continental slope and rise
	Lacustrine	Foreland basin	Foreland basin
Movement style and drivers		Requires overpressure. Shale can pass from ductile to fluidized, and back to ductile, numerous times as the structure oscillates between inflation (gas explosions and diapir wings) and deflation (collapse after the land surface is breached). Overpressure is caused by a combination of burial compaction, diagenesis of clay, organic maturation, and compressive tectonics. Overpressure is lost when fluid-charged mass reaches the surface, and typically only a relatively minor lateral flow component exists when the fluid escapes onto the surface. Mud-diapir provinces can form zones of extremely rapid subsidence atop salt withdrawal zones even if the mud source interval is not overpressured (Fort Isobel fold belt).	Thick beds of wet salt undergoing loading or extension can be autokinetic, almost from the time of deposition. Salt continues to move in response to extension or loading and maintains its flow until overburden touchdown occurs. Salt in compression will fold but not pierce the overburden, except under particular circumstances (see Warren, 2006, Chapter 6). When the salt reaches the surface, a strong lateral component (downslope) to flow (allochthons and namakiers) is observed, salt flow can continue via gravity spreading for many kilometers way from the vent. Overpressure is commonly focused beneath an allochthon as the salt mass is an excellent seal. Overpressure is also common in suprasalt minibasins where the feedback between loading and salt withdrawal facilitates rapid sedimentation until touchdown.

Causes of collapse of overburden or sediment immediately adjacent to the structure	Structural collapse typically occurs via venting or mud volcanism, which acts as a drain to overpressure.	Dissolution or deflation of salt feature as it flows up and ultimately out of the subsurface to land surface or sea floor. At surface, flow slows then ceases as the allochthon is cut off from the mother salt-supply layer.
Pockmarks	Indicate zones of pressure release commonly in zones of fluid escape through a zone of methane hydrate on the basin floor.	Indicate zones of salt dissolution or of escaping fluids released to the basin floor near the feather edge of salt sheet or above zones of overburden touchdown (welds).
Breccias in and around vent	Mud can carry blocks that assimilate into rising plume with diameters of several to hundreds of meters (tens to hundreds of feet). Large blocks are also captured by crestal collapse.	Salt can carry blocks with diameters of kilometers (miles). Blocks were typically originally intercalated with the mother salt or source layer.
Fault association	Shale diapirism initiated by faults that weaken the overburden. Pressurized mud then intrudes into the overlying sedimentary rock and may or may not reach the surface.	Close association between fault welds and salt welds. Much synkinematic activity is driven by feedback between loading, salt withdrawal, and creation of further accommodation space.
Fluid association	Warm subsurface fluids (typically gassy) escape along active faults or chimneys and carry mud with indicators of organic maturation. Thermal pools may or may not be present on the sea floor. Overpressuring feeds hydrocarbons to the surface (seeps associated with volcanoes). The sense of fluid flow is dominantly vertical, and no traps associated with lateral mudflow at the sea floor are observed.	Warm subsalt fluids escape at the featheredge of salt or atop active salt transecting faults or above zones of overburden touchdown. Fluids can carry mature hydrocarbons. Brine pools are present at zones of topographic closure on the basin floor. Methane seeps are commonplace. Once salt is at the sea floor, it spreads under the influence of gravity and a broad range of subsalt traps associated with lateral flow of salt are observed.
Hydrocarbon traps	Detachment folds, diapir toplap, diapir top drape or rollover rims, radial faults, tilted fault blocks, erosional truncation and associated unconformity traps, lateral minianticlines, and downbuilt anticlinal flank traps	Detachment folds, diapir toplap, diapir top drape or rollover rims, radial faults, tilted fault blocks, erosional truncation and associated unconformity traps, lateral minianticlines, and downbuilt anticlinal flank traps, suballochthon traps, and intrasalt traps

*Derived from Morley and Guerin (1996), Van Rensbergen et al. (1999), Rowan et al. (2004), Wood et al. (2004), and Warren (2006).

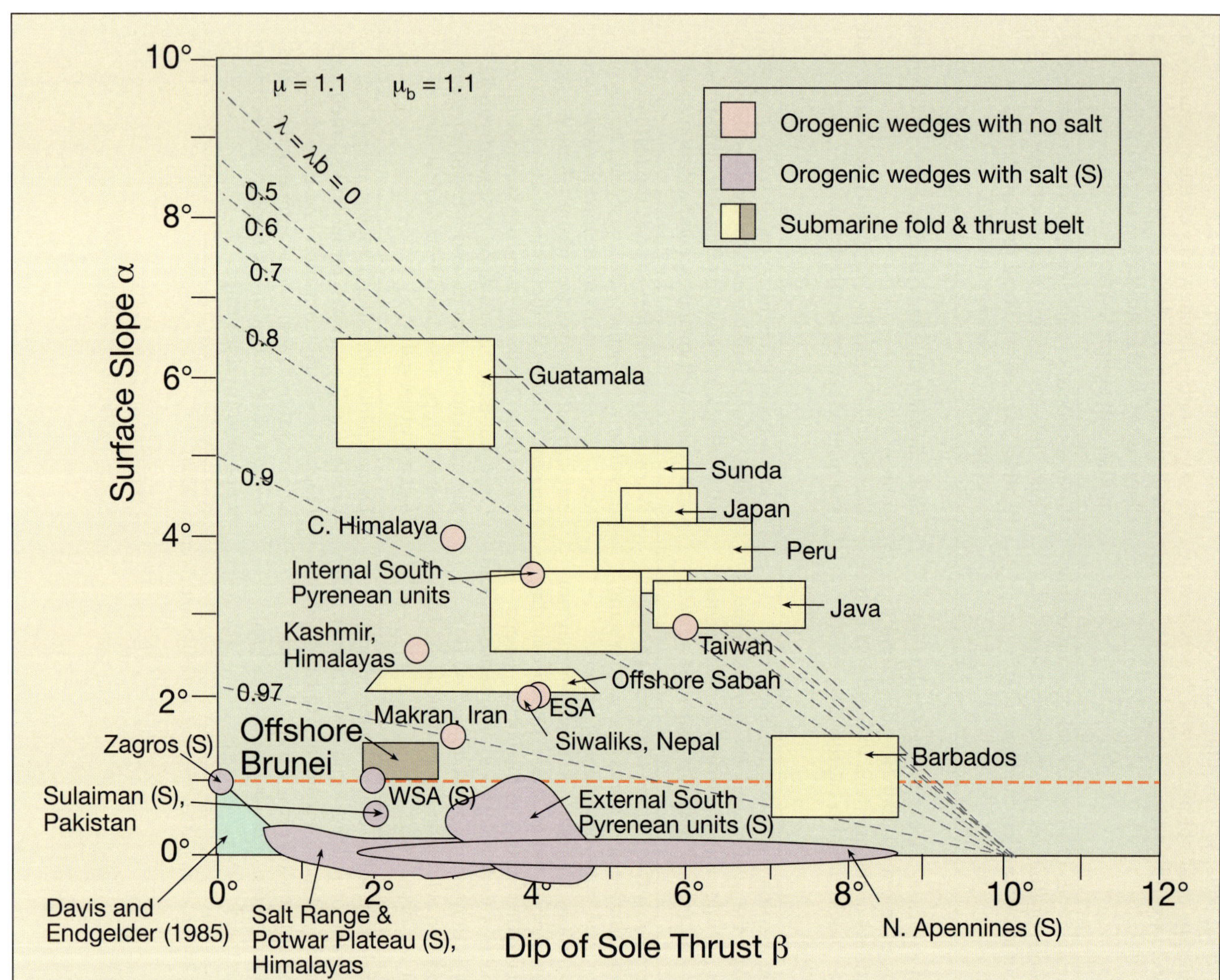

Figure 17. Pressure and wedge character showing the relationship between surface slope (α), basal detachment dip (β), and pore-fluid ratio for orogenic wedges and submarine fold and thrust belts at critical taper, assuming a Coulomb material behavior (from Ford, 2004; Morley, 2007a). Orogenic belts with salt entrained in the decollement are indicated by (S). μ = postulated coefficient of internal friction for the wedge; μ_b = coefficient of internal friction at the base of the wedge; λ_b = ratio of pore-fluid pressure to lithostatic pressure at the base of the wedge; ESA = east Swiss Alps; WSA = west Swiss Alps. The green triangle plots the region of the low basal friction wedge of Davis and Engelder (1985). As shown by Ford (2004), the region of low basal friction created by a salt decollement extends beyond the region indicated by Davis and Engelder (1985) and is indicated by the interval below the dashed red line (surface slopes <1°). Offshore Brunei and offshore Barbados lie on this boundary, although either sediment wedge entrains no salt, this is an indication of a muddy and highly pressured character in both prisms (see Morley, 2007a, for a detailed discussion of the significance of the Brunei taper).

Regionally, a multitiered allochthon style of salt in shelf progradation is more likely if the mother salt belt was deposited in a postrift, instead of a synrift, setting (Warren, 2008). Mobile salt allochthon sheets roll out over the landscape carrying along their still-accreting overburden section, much in the same way the tracks of a military tank carry the tank and its crew. In regions of mud diapirism, no same-scale equivalents to salt allochthons flowing over the sea floor are observed, nor muddy equivalents to merged salt canopies or namakiers (salt glaciers).

In argillokinetic provinces, local, typically narrow and cylindrical, subvertical chimney structures (mud volcanoes) rise from the underlying zone of overpressure (mud diapir). Small wings to the mud chimney may be visible where fluidized sediment first reached the sea floor, but there is never an extensive allochthon sheet spreading over the sea floor for kilometers (miles)

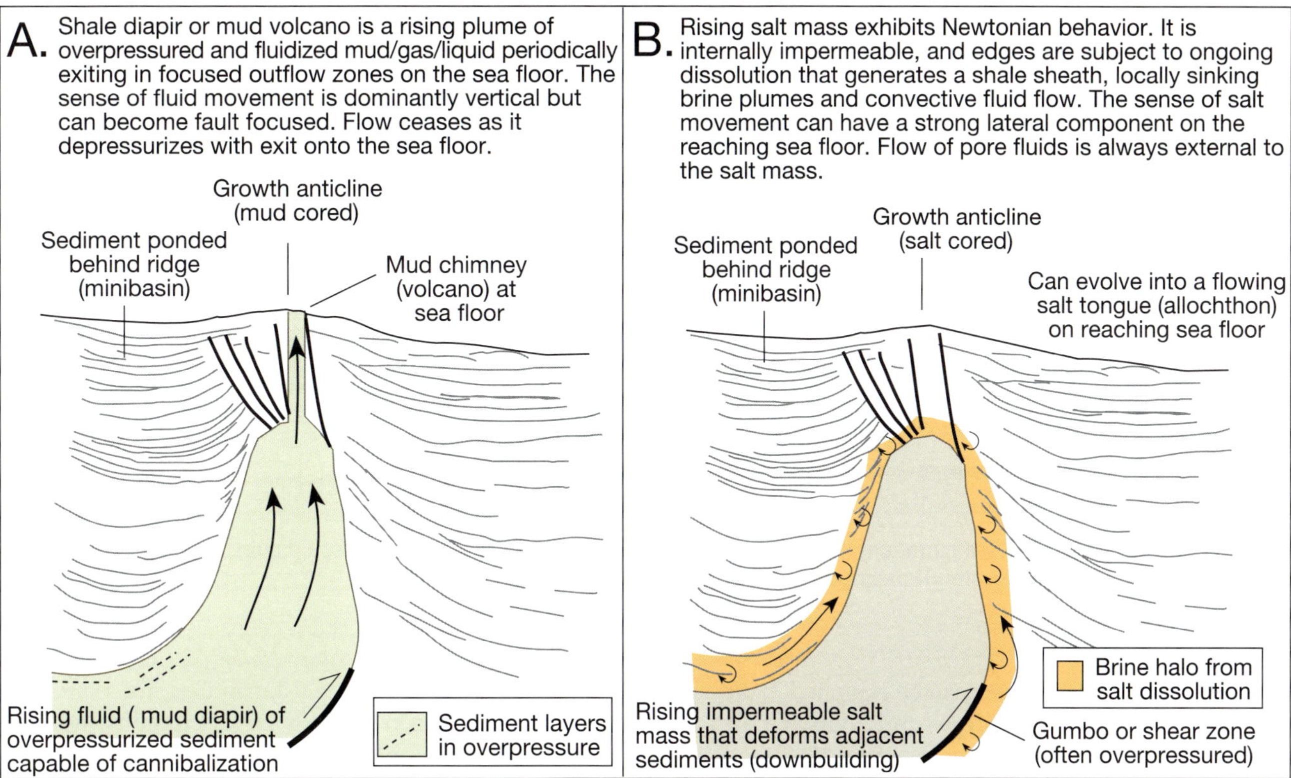

Figure 18. Comparison of signatures of mud and salt diapirism. (A) Mud diapirism indicates a zone of fluid-charged overpressured fine-grained sediment (mostly mud). It can rise via stoping subvertical mud chimneys and periodically break out onto the sea floor. The seismic signature indicates zones of active or former fluid charge. The diapir as seen in the seismic signature is not a lithology but typically a gas effect. As such, the diapir zone can expand and encapsulate or cannibalize adjacent sediments without disturbing the bed orientation. (B) Salt diapirism indicates a zone of flowing impermeable salt (halokinesis). There is both a vertical and horizontal component to this flow and a strong response to sediment loading, and, on salt reaching the sea floor, it is tied to downslope gravity spreading. The seismic signature of a salt diapir or allochthon is a direct indicator of the salt mass, which is always impermeable. Flow sets up a mobile mass of impervious salt with edges subject to dissolution (like a block of ice). This mass and its brine halo focus fluid flow and overpressuring about its edges (see Warren, 2006, chapters 6 and 8 for a detailed discussion of relevant literature).

away from the vent. Sea-floor pockmarks in a mud-diapir province define the vent and escape positions of chimneys atop overpressured fluids and commonly line up along fault zones. Regions of abundant sea-floor pockmarks in a salt-diapir province are commonly also tied to fault intersections with the sea floor but can also define the edge of the underlying salt mass or seal.

Similarities and differences in the nature of diapirism in salt and mud also lead to similarities and differences in the distribution of diapir-related hydrocarbon traps (Figure 19). Styles of reservoir distribution in sands atop salt or mud diapirs are similar (Table 4, Figure 19a). Both mud and salt diapirs tend to create local highs and ridges in the landscape, with consequent distributions of coarser grained sediments and/or reefs related to the depth of water over the high at the time the reservoir unit was deposited (e.g., Warren, 2006, chapters 6, 10). Onlap, pinch-out, rollover anticlines, growth faults, regional and counterregional faults, collapse faults, and drape traps are common to both settings (Figure 19a).

Differences in trap styles come from differences in the mechanism driving flow and in that, over the longer term, salt tends to maintain its seal integrity, whereas overpressured mud leaks. Salt can flow up and out over the sea floor for distances of several kilometers (miles). Equivalents to subsalt traps, which are tied to a strong lateral flow of salt across the landscape, are not known in argillokinetic provinces (Figure 19b). Dissolution of salt, via subsurface cross-flow of undersaturated waters, creates a residual carapace of anhydrite and limestone known as a cap rock. This too has no argillokinetic counterpart but can host low-volume hydrocarbon traps in and about a salt-diapir crest. Cap-rock traps tend to occur in relatively shallow parts of a salt-diapir

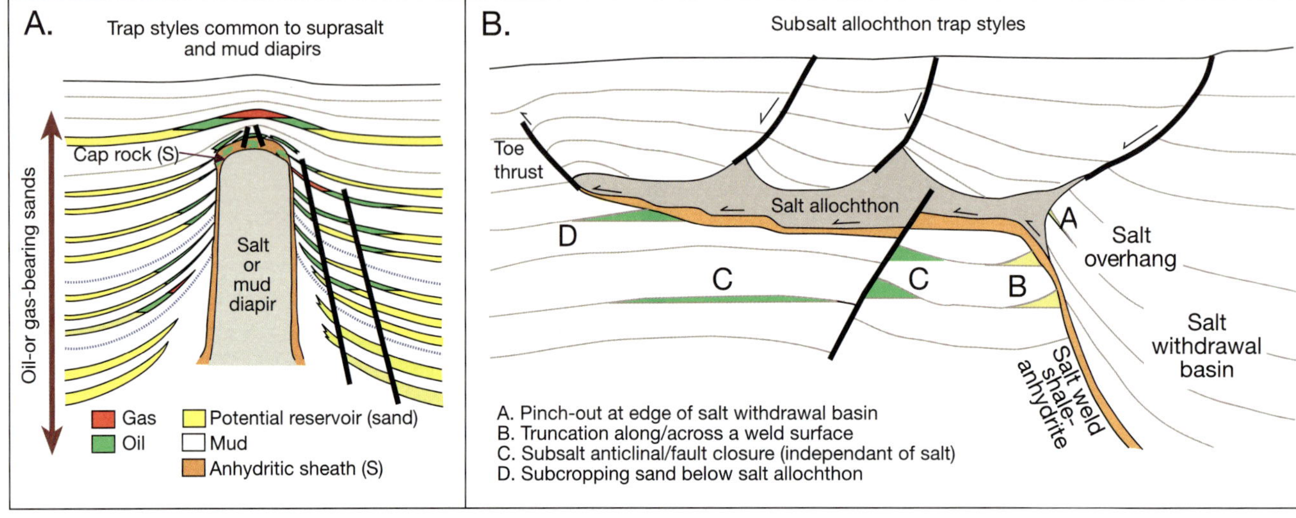

Figure 19. Hydrocarbon trap configuration. (A) Typical trap styles common to both mud and salt diapirs. (B) Subsalt traps with no documented equivalents in mud-diapir terrains.

province and are the first zones to be exploited in diapir-related oil fields of onshore Texas and Louisiana during the first 30 yr of the last century (Warren, 2006).

In summary, gas-charged faults and mud chimneys (volcanoes) lead down to zones of disturbed mud that constitute the diapiric cores of anticlinal ridges in Brunei mud diapirs and similar structures in the Niger Delta and elsewhere (e.g., Graue, 2000; Lüdmann and Wong, 2003; McGilvery and Cook, 2003). Unlike impermeable salt allochthon sheets where fluid flow occurs in deforming sediments adjacent to the salt, the site of the imaged mud diapir is the actual zone of overpressure and thermogenic (hydrocarbon-bearing) fluid flow. Thus, in a seismic line through a disturbed mud-diapir system, we are resolving the state of the fluid conduit itself (reflectors tend to bend upward when the system is pressurized and bend down or indicate collapse once the pressurized fluid has vented). In a salt allochthon province, we are resolving the edge of the salt and not the state of the fluid flow in the core of the structure. The edge of the salt mass may well act as the focus to any subsalt fluid escaping from a pressurized unit beneath the salt, but the interior of a thick flowing salt mass is impermeable.

ACKNOWLEDGMENTS

The authors thank the Brunei Petroleum unit, Total Brunei, BHP Billiton, and Shell International Deepwater for their support of this project. The senior author also thanks his former fellow faculty at Universiti Brunei Darussalam, Chris Morley, Martin Gee, Joe Lambiase, and Angus Ferguson for the lively discussions and helpful criticisms that helped shape the ideas in this article.

REFERENCES CITED

Abrams, M. A., 2005, Significance of hydrocarbon seepage relative to petroleum generation and entrapment: Marine and Petroleum Geology, v. 22, p. 457–477, doi:10.1016/j.marpetgeo.2004.08.003.

Birkle, P., J. J. Rosillo Aragon, E. Portugal, and J. L. Fong Aguilar, 2002, Evolution and origin of deep reservoir water at the Activo Luna oil field, Gulf of Mexico, Mexico: AAPG Bulletin, v. 86, p. 457–484.

Camerlo, R., D. Meyer, and R. Meltz, 2004, Shale tectonism in the northern Port Isobel fold belt, *in* P. J. Post, ed., Salt-sediment interactions and hydrocarbon prospectivity: Concepts, applications and case studies for the 21st Century: Gulf Coast Section SEPM, 24th Annual Bob F. Perkins Research Conference, (CD-ROM), p. 817–839.

Clari, P., S. Cavagna, L. Martire, and J. Hunziker, 2004, A Miocene mud volcano and its plumbing system: A chaotic complex revisited (Monferrato, NW Italy): Journal of Sedimentary Research, v. 74, p. 662–676, doi:10.1306/022504740662.

Curiale, J., W. Mueller, J. Morelos, and J. Lambiase, 2000, Brunei Darussalam: Characteristics of selected petroleums and source rocks: Organic Geochemistry, v. 31, p. 1475–1493, doi:10.1016/S0146-6380(00)00084-X.

Davis, D. M., and T. Engelder, 1985, The role of salt in fold-and-thrust belts (Canada): Tectonophysics, v. 119, p. 67–88, doi:10.1016/0040-1951(85)90033-2.

Demyttenaere, R., J. P. Tromp, A. Ibrahim, P. Allman-Ward,

and T. Meckel, 2000, Brunei deep water exploration from sea floor images and shallow seismic analogs to depositional models in a slope turbidite setting: Gulf Coast Section SEPM, 20th Annual Bob F. Perkins Annual Research Conference, p. 304–317.

Dimitrov, L. I., 2002, Mud volcanoes-the most important pathways for degassing deeply buried sediment: Earth-Science Reviews, v. 59, p. 49–76.

Ford, M., 2004, Depositional wedge tops: Interaction between low basal friction external orogenic wedges and flexural foreland basins: Basin Research, v. 16, p. 361–375, doi:10.1111/j.1365-2117.2004.00236.x.

Gee, M. J. R., H. S. Uy, J. K. Warren, C. K. Morley, and J. J. Lambiase, 2007, The Brunei slide: A giant submarine landslide on the northwest Borneo margin revealed by 3D seismic data: Marine Geology, v. 246, p. 9–23, doi:10.1016/j.margeo.2007.07.009.

Graue, K., 2000, Mud volcanoes in deep water Nigeria: Marine and Petroleum Geology, v. 17, p. 959–974, doi:10.1016/S0264-8172(00)00016-7.

Grauls, D., 2001, Gas hydrates: Importance and applications in petroleum exploration: Marine and Petroleum Geology, v. 18, p. 519–523, doi:10.1016/S0264-8172(00)00075-1.

Hall, R., and C. K. Morley, 2004, Sundaland basins, *in* P. Clift, P. Wang, W. Kuhnt, and D. E. Hayes, eds., Continent-ocean interactions within the east Asian marginal seas: American Geophysical Union Geophysical Monograph 149, p. 55–85.

Huang, W. Y., and W. G. Meinschein, 1979, Sterols as ecological indicators: Geochemica et Cosmochimica Acta, v. 43, p. 730–745.

Hutchison, C. S., 2005, Geology of Northwest Borneo: Frankfurt, Germany, Elsevier, 444 p.

Ingram, G. M., T. J. Chisholm, C. J. Grant, C. A. Hedlund, P. Stuart-Smith, and J. Teasdale, 2004, Deepwater northwest Borneo: Hydrocarbon accumulation in an active fold and thrust belt: Marine and Petroleum Geology, v. 21, p. 879–887, doi:10.1016/j.marpetgeo.2003.12.007.

Kennett, J. P., K. G. Cannariato, I. L. Hendy, and R. J. Behl, 2000, Carbon isotopic evidence for methane hydrate instability during Quaternary interstadials: Science, v. 288, p. 123–128.

Kopf, A. J., 2002, Significance of mud volcanism: Reviews of Geophysics, v. 40, no. 2, 1005, p. 2.1–2.52, doi:10.1029/2000RG000093.

Kopf, A. J., 2003, Global methane emission through mud volcanoes and its past and present impact on the Earth's climate: International Journal of Earth Sciences, v. 92, p. 806–816, doi:10.1007/s00531-003-0341-z.

Kotelnikova, S., 2002, Microbial production and oxidation of methane in deep subsurface: Earth-Science Reviews, v. 58, p. 367–395.

Laird, A., and C. K. Morley, in press, Development of gas hydrates in a deepwater anticline based on attribute analysis from 3D seismic data: Geosphere, doi:10.1130/GS00598.1.

Lüdmann, T., and H. K. Wong, 2003, Characteristics of gas hydrate occurrences associated with mud diapirism and gas escape structures in the northwestern Sea of Okhotsk: Marine Geology, v. 201, p. 269–286, doi:10.1016/S0025-3227(03)00224-X.

Malone, M. J., 2002, Variable methane fluxes in shallow marine systems over geologic time: The composition and origin of pore waters and authigenic carbonates on the New Jersey shelf: Marine Geology, v. 189, p. 175–196, doi:10.1016/S0025-3227(02)00474-7.

Matsumoto, R., 1989, Isotopically heavy oxygen-containing siderite derived from the decomposition of methane hydrate: Geology, v. 17, p. 707–710, doi:10.1130/0091-7613(1989)017<0707:IHOCSD>2.3.CO;2.

McGilvery, T. A., and D. L. Cook, 2003, The influence of local gradients on accommodation space and linked depositional elements across a stepped slope profile, offshore Brunei, *in* H. H. Roberts, N. C. Rosen, R. H. Fillon, and J. B. Anderson, eds., Shelf margin deltas and linked downslope petroleum systems: Global significance and future exploration potential: Gulf Coast Society SEPM 23rd Annual Bob F. Perkins Research Conference (CD-ROM), p. 387–420.

Milkov, A. V., 2000, Worldwide distribution of submarine mud volcanoes and associated gas hydrates: Marine Geology, v. 167, p. 29–42, doi:10.1016/S0025-3227(00)00022-0.

Miller, J. J., M. W. Lee, and R. Von Huene, 1991, An analysis of a seismic reflection from the base of a gas hydrate zone, offshore Peru: AAPG Bulletin, v. 75, p. 910–924.

Morley, C. K., 2007a, Interaction between critical wedge geometry and sediment supply in a deep-water fold belt: Geology, v. 35, p. 139–142, doi:10.1130/G22921A.1.

Morley, C. K., 2007b, Development of crestal normal faults associated with deep water fold growth: Journal of Structural Geology, v. 29, p. 1148–1163, doi:10.1016/j.jsg.2007.03.016.

Morley, C. K., and G. Guerin, 1996, Comparison of gravity-driven deformation styles and behaviour associated with mobile shales and salt: Tectonics, v. 15, p. 1154–1170.

O'Brien, G. W., M. Lisk, I. R. Duddy, J. Hamilton, P. Woods, and R. Cowley, 1999, Plate convergence, foreland development and fault reactivation: Primary controls on brine migration, thermal histories and trap breach in the Timor Sea, Australia: Marine and Petroleum Geology, v. 16, p. 533–560, doi:10.1016/S0264-8172(98)00070-1.

O'Brien, G. W., G. M. Lawrence, A. K. Williams, K. Glenn, A. G. Barrett, M. Lech, D. S. Edwards, R. Cowley, C. J. Boreham, and R. E. Summons, 2005, Yampi shelf, Browse Basin, northwest shelf, Australia: A test-bed for constraining hydrocarbon migration and seepage rates using combinations of 2D and 3D seismic data and multiple, independent remote sensing technologies: Marine and Petroleum Geology, v. 22, p. 517–549, doi:10.1016/j.marpetgeo.2004.10.027.

Peters, K. E., J. M. Moldowan, and C. C. Walters, 2005a, The biomarker guide: Biomarkers and isotopes in the environment and human history: Cambridge, UK, Cambridge University Press, v. 1, 490 p.

Peters, K. E., C. C. Walters, and J. M. Moldowan, 2005b, The biomarker guide: Biomarkers and isotopes in petroleum systems and earth history: Cambridge, UK, Cambridge University Press, v. 2, 700 p.

Raiswell, R., S. H. Bottrell, S. P. Dean, J. D. Marshall, A. Carr, and D. Hatfield, 2002, Isotopic constraints on growth conditions of multiphase calcite-pyrite-barite concretions in Carboniferous mudstones: Sedimentology, v. 49, p. 237–254, doi:10.1046/j.1365-3091.2002.00439.x.

Roberts, H. H., and P. Aharon, 1994, Hydrocarbon-derived carbonate buildups of the northern Gulf of Mexico continental slope: A review of submersible investigations: Geo-Marine Letters, v. 14, p. 135–148.

Rowan, M. G., 1997, Three-dimensional geometry and evolution of a segmented detachment fold, Mississippi Fan fold belt, Gulf of Mexico: Journal of Structural Geology, v. 19, p. 463–480, doi:10.1016/S0191-8141(96)00098-3.

Rowan, M. G., F. J. Peel, and B. C. Vendeville, 2004, Gravity-driven fold belts on passive margins, *in* K. McClay, ed., Thrust tectonics and hydrocarbon systems: AAPG Memoir 82, p. 157–182.

Stewart, S. A., and R. J. Davies, 2006, Structure and emplacement of mud volcano systems in the south Caspian Basin: AAPG Bulletin, v. 90, p. 771–786, doi:10.1306/11220505045.

van Rensbergen, P., and C. K. Morley, 2003, Re-evaluation of mobile shale occurrences on seismic sections of the Champion and Baram deltas, offshore Brunei, *in* P. van Rensbergen, R. R. Hillis, A. J. Maltman, and C. K. Morley, eds., Subsurface sediment mobilization: Geological Society (London) Special Publication 216, p. 395–409.

van Rensbergen, P., C. K. Morley, D. W. Ang, T. Q. Hoan, and N. T. Lam, 1999, Structural evolution of shale diapirs from reactive rise to mud volcanism: 3D seismic data from the Baram Delta, offshore Brunei Darussalam: Journal of the Geological Society, v. 156, p. 633–650, doi:10.1144/gsjgs.156.3.0633.

Waples, D., 1985, Geochemistry in petroleum exploration: Boston, International Human Research Development Corporation, 232 p.

Warren, J. K., 2000, Dolomite; occurrence, evolution and economically important associations: Earth-Science Reviews, v. 52, p. 1–81.

Warren, J. K., 2006, Evaporites: Sediments, resources and hydrocarbons: Berlin, Springer, 1036 p.

Warren, J. K., 2008, Salt as sediment in the Central European Basin system as seen from a deep time perspective (chapter 5.1), *in* R. Littke, ed., Dynamics of complex intracontinental basins: The central European basin system: Berlin, Elsevier, p. 249–276.

Warren, J. K., L. L. Quoc, and I. Cartwright, 2004, Ongoing dolomite cementation in the Neogene sediment prism of offshore Brunei Darussalam (abs.): Proceedings of the 32nd International Geological Congress, Florence, August 20–28, 2004.

Wood, L. J., S. Sullivan, and P. Mann, 2004, Influence of mobile shales in the creation of successful hydrocarbon basins, *in* P. J. Post, ed., Salt-sediment interactions and hydrocarbon prospectivity: Concepts, applications and case studies for the 21st Century: Gulf Coast Section SEPM, 24th Annual Bob F. Perkins Research Conference, (CD-ROM), p. 892–930.

Zielinski, G. W., M. Bjoroy, R. L. B. Zielinski, and I. L. Ferriday, 2007, Heat flow and surface hydrocarbons on the Brunei continental margin: AAPG Bulletin, v. 91, p. 1053–1080, doi:10.1306/01150706117.

11

McNeil, David H., James R. Dietrich, Dale R. Issler, Stephen E. Grasby, James Dixon, and Lavern D. Stasiuk, 2010, A new method for recognizing subsurface hydrocarbon seepage and migration using altered foraminifera from a gas chimney in the Beaufort-Mackenzie Basin, *in* L. Wood, ed., Shale tectonics: AAPG Memoir 93, p. 197–210.

A New Method for Recognizing Subsurface Hydrocarbon Seepage and Migration Using Altered Foraminifera from a Gas Chimney in the Beaufort-Mackenzie Basin

David H. McNeil, James R. Dietrich, Dale R. Issler, Stephen E. Grasby, and James Dixon
Geological Survey of Canada, Natural Resources Canada, Calgary, Alberta, Canada
Lavern D. Stasiuk
Shell Canada Limited, Calgary, Alberta, Canada

ABSTRACT

A new method for recognizing hydrocarbon seepage and migration in exploration wells is documented from the Immiugak A-06 exploration well that drilled through a hydrocarbon-related diagenetic zone (HRDZ). The HRDZ is seismically conspicuous as part of a gas chimney on a shale-cored anticline in the Tertiary of the Beaufort-Mackenzie Basin, Arctic Canada. The HRDZ contains classic diagenetic minerals, notably greigite (Fe_3S_4) and calcite with $\delta^{34}S$ and $\delta^{13}C$ values diagnostic of hydrocarbon-related, sulfate-reducing, microbial activity. The HRDZ also contains exceptionally preserved calcareous benthic foraminifera with conspicuous bitumen-filled chambers and agglutinated foraminifera with bitumen and diagenetic silica with bound particles. Silica was highly mobile within the seepage or migration system and was precipitated and dissolved extensively in the agglutinated foraminifera. Seismic profiles, resistivity anomalies, diagenetic minerals, and altered foraminifera all suggest that significant hydrocarbons migrated or seeped through sandy Oligocene and Miocene strata at the crest of a shale-cored anticline in response to late Miocene tectonism. Hydrocarbon-related diagenesis can be distinguished from standard burial diagenesis using the foraminiferal coloration index (FCI). Foraminiferal coloration within the HRDZ was controlled by silicification in a bitumen-rich environment. The FCI values in the HRDZ are much higher than predicted for normal burial and show abnormal variance caused by variable dissolution of foraminiferal silica. The FCI values from agglutinated foraminifera outside the HRDZ show a uniform linear trend increasing with depth. The extent of hydrocarbon-related diagenesis observed in foraminifera can be used to assess the relative magnitude of hydrocarbon seepage in the Beaufort-Mackenzie Basin and potentially other petroleum basins.

DOI:10.1306/13231315M933425

INTRODUCTION

Hydrocarbon vents and seeps are a common feature of continental shelves and ocean basins globally. Their sea-floor environment has been studied extensively for a variety of purposes, ranging from analysis of the unique biota they support to their significance as indicators of subsurface hydrocarbon potential. Significant seepage can also indicate a breached reservoir and increased exploration risk. In the subsurface, these seepage systems are commonly expressed as gas chimneys or diagenetically mineralized zones that can be recognized on seismic profiles as distinctive vertically oriented zones of chaotic reflectors (Cowley and O'Brien, 2000). In cores or cutting samples, they can be recognized by distinctive assemblages of diagenetic minerals (carbonates, magnetic sulfides, radioactive minerals) produced by the bacterial degradation of migrating hydrocarbons (Schumacher, 1996; Saunders et al., 1999). O'Brien and Woods (1995) referred to these systems as hydrocarbon-related diagenetic zones (HRDZs), which are commonly present within prolific hydrocarbon-producing basins (Cowley and O'Brien, 2000).

We introduce a new method for recognizing hydrocarbon seepage and migration events in the subsurface, based on evidence from the foraminiferal record in the Gulf et al. Immiugak A-06 exploration well situated in the Beaufort Sea of Arctic Canada (Figure 1). The Immiugak A-06 well penetrates part of a seismically imaged gas chimney that appears to be a major hydrocarbon seepage system extending from the sea floor of the Arctic Beaufort Sea into Miocene and Oligocene strata of the Mackenzie Bay and Kugmallit sequences (Figures 2, 3). Well-cutting samples from the Immiugak A-06 HRDZ yield a microscopic but striking record of hydrocarbons (bitumen and traces of oil) trapped within the tests (shells) of exceptionally preserved calcareous and agglutinated benthic foraminifera of Oligocene and Miocene age (Figure 4a–d). Bitumen is also conspicuously trapped within clusters of magnetic minerals dominated by greigite (Fe_3S_4) and in diagenetic calcite. Magnetic mineral associations are typical of hydrocarbon seepage systems and are considered to have been produced by anaerobic magnetotactic bacteria acting on subsurface hydrocarbons (Machel, 1996). Diagenetic carbonates are widespread in near-surface hydrocarbon seepage systems and were formed by the microbial oxidation of hydrocarbons (Schumacher, 1996). Machel and Foght (2000) also noted that sulfate-reducing bacterial degradation of oil can lead to the precipitation

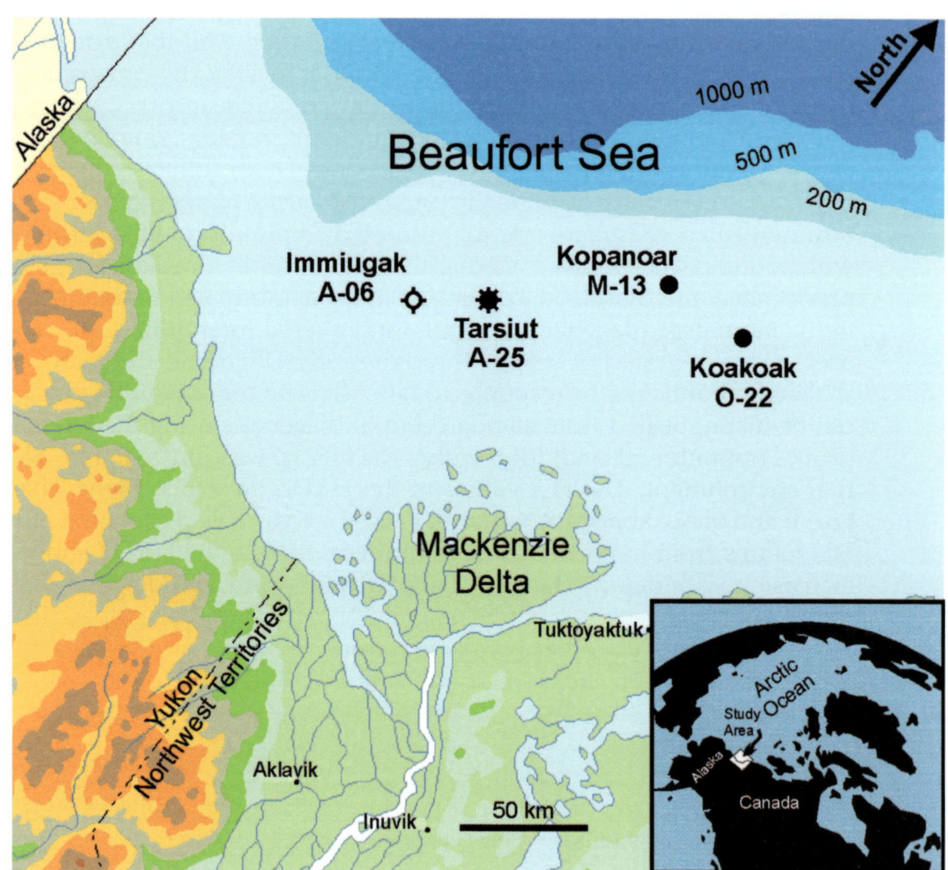

Figure 1. Location of the Immiugak A-06 exploration well and three significant hydrocarbon discovery wells in the Beaufort Sea. Contours are bathymetry. 50 km = 31 mi; 100 m = 328 ft.

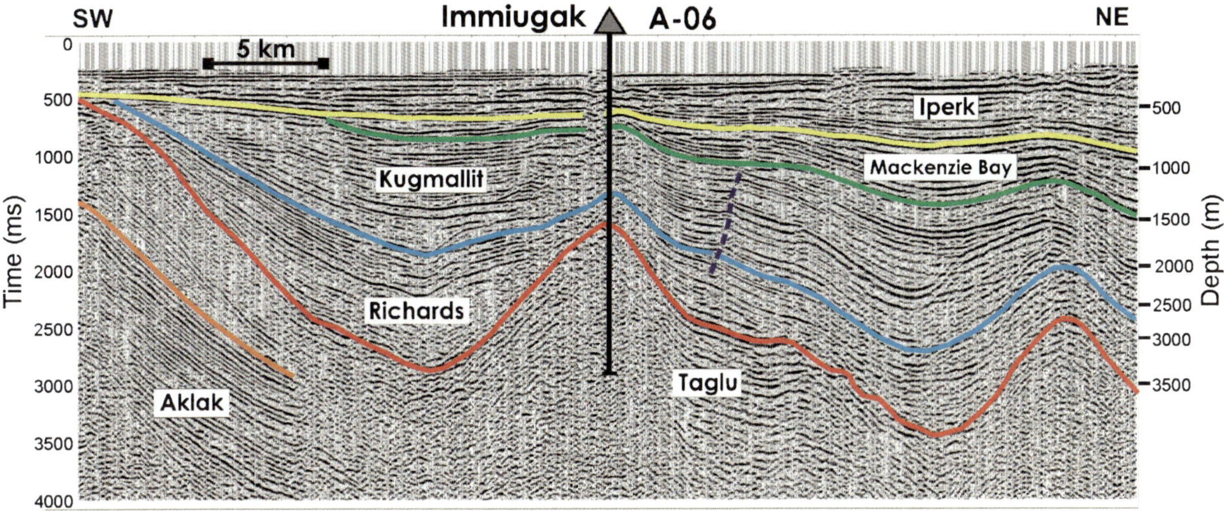

Figure 2. Regional seismic profile across the Immiugak anticline, illustrating Cenozoic sequences and reflection amplitude anomalies of the Immiugak A-06 gas chimney and hydrocarbon-related diagenetic zone. Iperk (Pliocene–Pleistocene); Mackenzie Bay (latest Oligocene–middle Miocene); Kugmallit (Oligocene); Richards (late Eocene); Taglu (early–middle Eocene) Aklak (Paleocene-earliest Eocene). 51 cm = 3.1 mi; 500 m = 1640 ft.

of calcite. They have set general depth and temperature limits for subsurface bacterial activity at 2000 m (6562 ft) and 70°C (158°F), although deeper and higher temperature exceptions occur.

Conspicuous evidence of diagenetic alteration, primarily in the form of silica precipitation and dissolution effects in agglutinated foraminifera, is also observed in the Immiugak A-06 well (Figure 4f–h). These alteration affects differ significantly from normal burial diagenetic trends, which would be characterized by progressively darker coloration and silicification in agglutinated foraminifera with increasing depths, as previously documented in the Beaufort-Mackenzie Basin by McNeil (1997b). Together, these diagenetic minerals and altered foraminiferal tests provide evidence that points toward a major hydrocarbon seepage system initiated in the late Miocene, although microseepage might have occurred earlier. Seismic profiles suggest that the system continues to the present day as a gas chimney venting into the Beaufort Sea continental shelf. This association of observations clearly suggests that microfossil analyses can record evidence of hydrocarbon seeps, migration events, and breached reservoirs. The amount of foraminiferal alteration and bitumen impregnation observed in foraminifera has potential for estimating the scale of the breaching, at least in a relative sense.

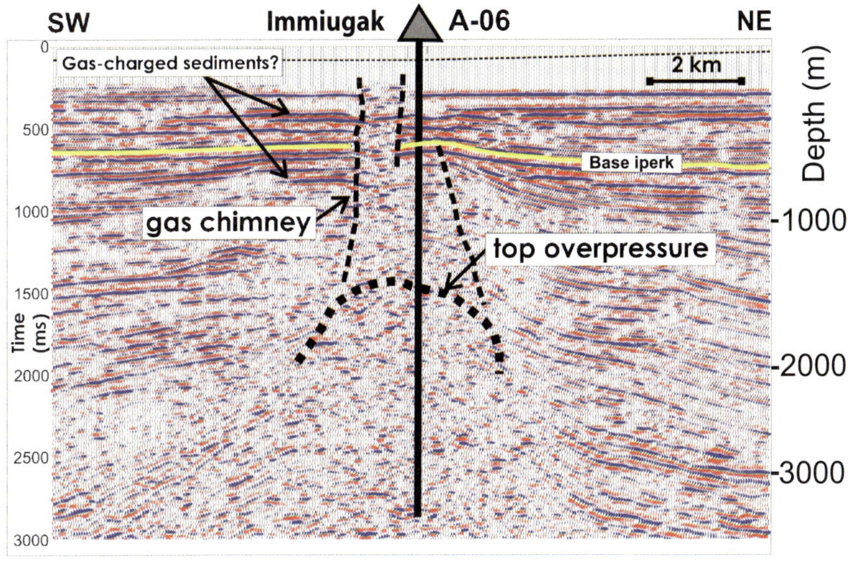

Figure 3. Seismic profile across the Immiugak A-06 well location, with the interpreted gas chimney. Reflection amplitude anomalies adjacent to the chimney may indicate gas-charged sediments. Near-sea-floor reflections in this seismic profile were muted during data processing. 2 km = 1.2 mi; 1000 m = 3281 ft.

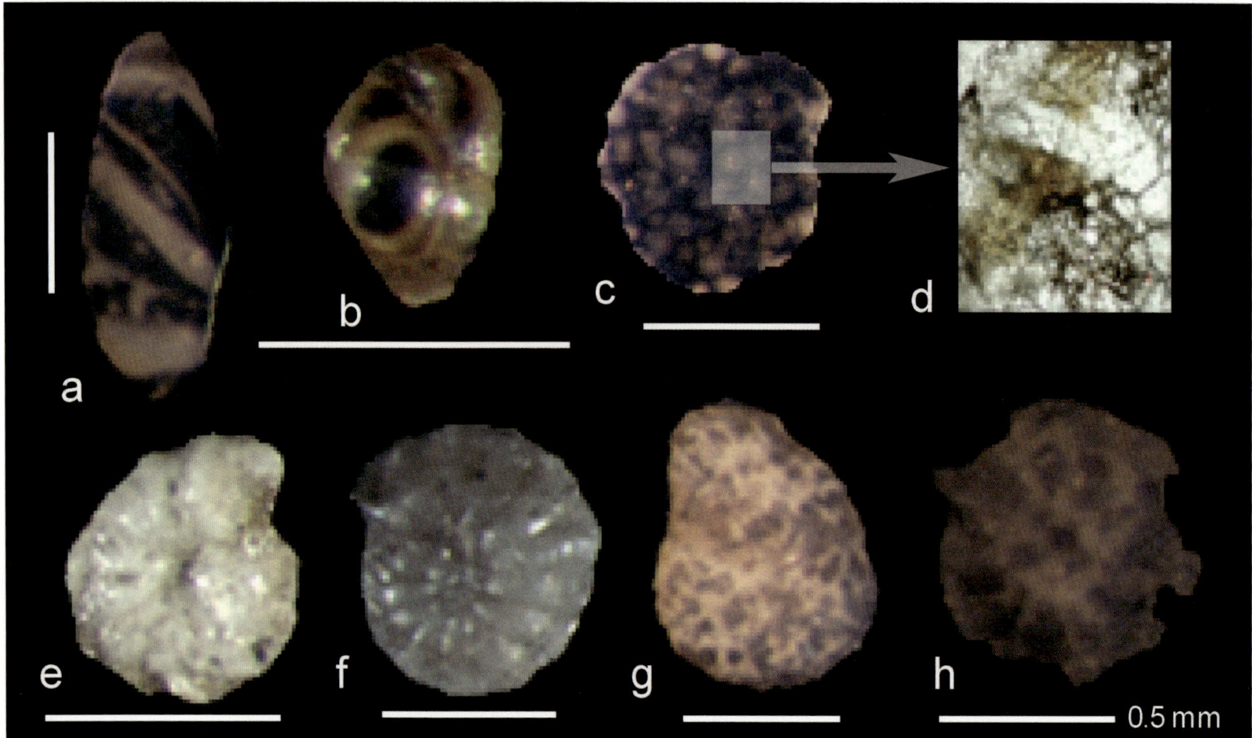

Figure 4. Microfossil illustrations from the Immiugak A-06 well. (a, b) Calcareous benthic foraminifera with bitumen-filled chambers (Geological Survey of Canada [GSC] 124229, 124230). (c) Agglutinated foraminifera with silica and bitumen interspersed (GSC 124231). (d) Thin section of agglutinated foraminifera showing quartz and bitumen (GSC 124231). (e–h) Illustrations of agglutinated foraminifera showing various stages of silicification or dissolution (GSC 124232–134235): (e) unaltered, (f) silicified, (g) combined silicification and dissolution, (h) silica alteration and dissolution within agglutinated foraminifera.

GEOLOGICAL SETTING

The Immiugak A-06 gas chimney and HRDZ are seismically imaged as zones of chaotic reflections on the crest of a large shale-cored anticline in the Tertiary fold belt of the Beaufort Sea continental shelf (Figure 2). The seismic data indicate a broad, 2-km (1.2-mi)-wide HRDZ in Oligocene and Miocene strata and a much narrower, less than 1-km (0.6-mi)-wide gas chimney in Pliocene–Pleistocene strata. The gas chimney extends vertically across a major late Miocene unconformity (base Iperk sequence) and may extend to the sea floor. The Immiugak A-06 well does not appear to have penetrated the shallowest part of the chimney (Figures 2, 3), although it penetrates the lower part of the chimney or HRDZ at depths of approximately 530 to 1200 m (1740 to 3940 ft). Like many occurrences of gas chimneys and breached hydrocarbon reservoirs, the Immiugak A-06 well is situated in an area of intersecting structural trends (Figure 5). The well penetrates a northwest–southeast-trending anticline that is part of a regional Tertiary fold belt in the western Beaufort-Mackenzie Basin. The anticline is intersected and disrupted by southwest–northeast-oriented normal (and strike-slip?) faults that form part of the Tarsiut-Amauligak fault zone (Lane and Dietrich, 1995). The structural development of the anticline is primarily associated with (syndepositional) compressional tectonics, and although no evidence of significant mud diapirism is observed here, mud diapirism has been recognized in the Beaufort shelf by Bergquist et al. (2003) and Elsley and Tieman (2010). Seismic patterns of stratal thinning over the anticline indicate that (Figure 2) structural growth occurred during deposition of the Taglu to Mackenzie Bay sequences (Eocene to middle Miocene), with the most significant phase of growth during deposition of the upper Eocene Richards sequence. Regional late Miocene tectonism and faulting may have caused breaching of the anticline and seepage of hydrocarbons.

The stratigraphic section in the Immiugak A-06 well consists of marine mudstones and coaly sandstones of the lower to middle Eocene Taglu sequence, marine mudstones, and sandstones of the upper Eocene Richards, Oligocene Kugmallit, upper most Oligocene to middle Miocene Mackenzie Bay, Pliocene to Pleistocene Iperk, and Holocene Shallow Bay sequences (Figure 6). No significant hydrocarbon accumulations were encountered in

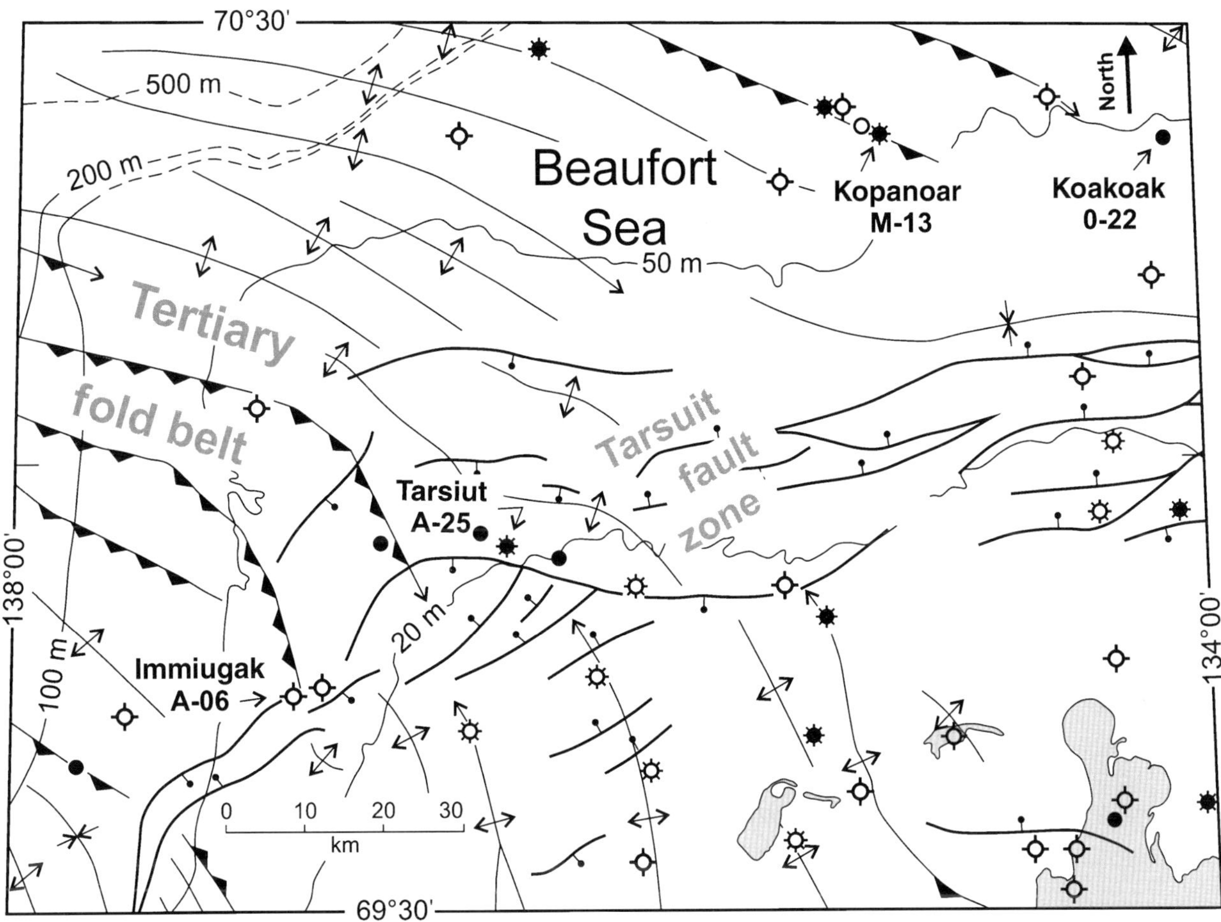

Figure 5. Tertiary structural trends for the Beaufort Sea continental shelf and location of exploration wells (structural features are from Lane and Dietrich, 1996; reprinted by permission of the Geological Survey of Canada). Contours are bathymetry. 10 km = 6.2 mi; 100 km = 328 ft.

this well. General information on the petroleum geology of other Beaufort Sea wells can be found in Morrell (1996).

METHODS

Mineral and foraminiferal assemblages from the Immiugak A-06 well were derived from bulk well-cutting samples composited into 30-m (98-ft) intervals through the entire well. The well cuttings were washed over a 75-μm mesh sieve, and the foraminifera and minerals were hand picked. All samples are curated at the Geological Survey of Canada Calgary (GSCC) and are accessible by reference to the sample number and/or microfossil type number. Color images of the foraminifera and mineral grains were photographed using a Leica DFC480 digital camera. Mineral grains from washed cutting samples were mounted in resin pellets, which were ground and polished for fluorescent examination. Fluorescence intensity images were captured using a Zeiss MPM II Universal incident light microscope (HBO 100 ultraviolet source), 03 photomultiplier, a continuous filter monochromator b (1/2 band width of 14 nm at 54 nm), and a Zeiss UMSP controller pc-based computer system. Mineral separates were identified using a Philips PW1700 x-ray diffraction (XRD) system and a Philips XL30 scanning electron microscope with an energy dispersal spectrometer (EDS) analysis at the GSCC. Isotope analyses ($\delta^{13}C$, $\delta^{18}O$, $\delta^{34}S$) were performed at the University of Calgary Isotope Laboratory. Continuous-flow isotope ratio mass spectrometry was used to analyze $\delta^{34}S$ from separated grains of greigite. The International Atomic Energy Agency (IAEA) standards were used, and the precision and accuracy of 1 sigma of (n = 10) laboratory standards were 0.3 for $\delta^{34}S$. Measurement of $\delta^{13}C$ and $\delta^{18}O$ from carbonate grains followed the classic methods in which isotopically representative CO_2 is prepared by reaction of carbonate grains with concentrated H_3PO_4. Results are relative to IAEA standards; precision and accuracy of 1 sigma of (n = 10) laboratory standards were 0.4 for $\delta^{13}C$ and $\delta^{18}O$. The foraminiferal coloration index (FCI)

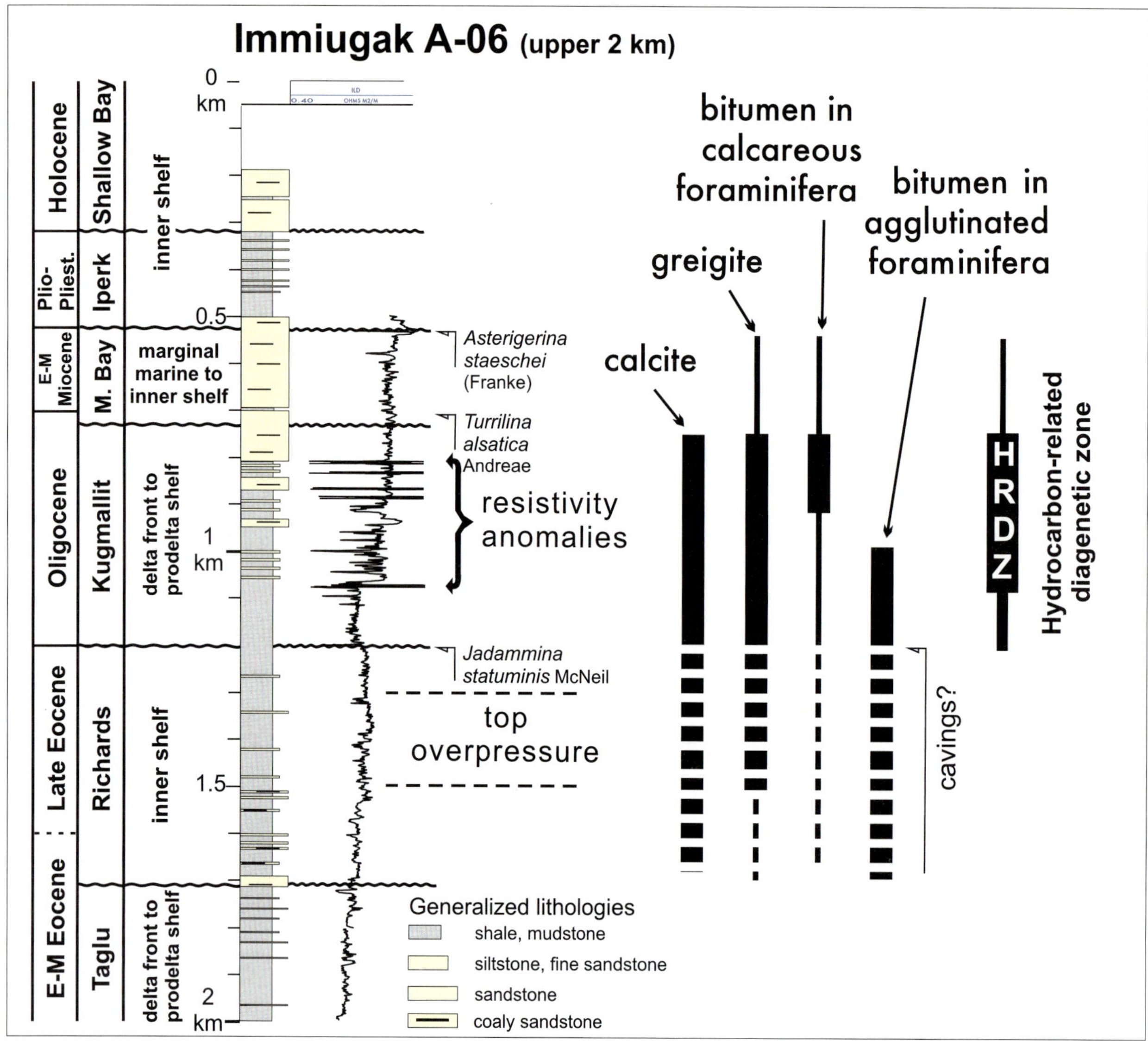

Figure 6. Overview of the upper 2 km (1.2 mi) of the Immiugak A-06 well showing stratigraphic ages, sequences, depositional facies, lithologies (estimated from gamma and sonic logs and washed well cuttings), resistivity (ILD) log, foraminiferal indices, and generalized distribution of hydrocarbon-related diagenetic indicators. Depths are below kelly bushing.

of McNeil et al. (1996) was used to assess alteration effects in foraminifera. The FCI measurements were compiled using a binocular microscope and standard color chips (Munsell Color, 1975).

BITUMEN AND ALTERATION EFFECTS IN FORAMINIFERA

Seismic profiles indicate that the Immiugak A-06 well first penetrates the HRDZ near the top of the Mackenzie Bay sequence at 530 m (1739 ft) (Figures 2, 3). This corresponds to the first downhole observations of bitumen in foraminiferal tests and the first evidence of hydrocarbon-related diagenetic minerals. Relative abundance of bitumen in foraminifera is closely associated with diagenetic mineral distributions (Figure 6). Bitumen occurs in both calcareous and agglutinated foraminifera, but the physical emplacement of bitumen is different in these two very different types of foraminifera. In calcareous benthic foraminifera, bitumen simply fills the internal chamber cavities (Figure 4a, b) presumably via the apertural opening. No evidence is observed, such as aberrant morphology and specially adapted vent biotas, to suggest that the HRDZ foraminiferal assemblages lived in a petroleum-rich environment. In agglutinated foraminifera, the bitumen is typically pervasive but always interstitial to secondary quartz cementation of the test, with the virtually complete loss of individual primary quartz grains as well as the original chamber definition (Figure 4c, d). Data from the North Sea Basin indicate that quartz cementation occurs at temperatures exceeding 70°C (158°F) (Giles et al., 1992) and does not become significant until greater than 84°C (183°F), suggesting that warm, migrating fluids produced the pronounced alteration effects in the agglutinated foraminifera. Regional studies of silicification in agglutinated foraminifera by McNeil et al. (1996) and McNeil (1997b) indicated that silicification from sedimentary burial does not occur until depths of 2 km (1.2 mi) or more and approximately 75°C (167°F) are exceeded.

Silica appears to have been very mobile in the HRDZ, with quartz crystals growing internally within the chambers of calcareous and agglutinated foraminifera. In agglutinated foraminifera, silica was precipitated and dissolved, and Figure 4e–h illustrate (e) an unaltered specimen, (f) a silicified specimen, (g) combined silicification and dissolution, and (h) silicification with significant dissolution.

The biostratigraphic succession through the Immiugak A-06 well appears to be unaffected by alteration effects in the HRDZ in that no vertical displacement or mixing of assemblages is apparent. Three representative foraminiferal indices are listed in Figure 6. Their order of occurrence is consistent with the regional biostratigraphic succession and foraminiferal zonation of McNeil (1997a).

The generalized distribution of bitumen-filled benthic foraminifera and diagenetic mineralization in Immiugak A-06 is shown in Figure 6. Calcareous benthic foraminifera from the upper Mackenzie Bay show minimal amounts of bitumen. In the lower Mackenzie Bay, however, the relative abundance of bitumen increases substantially and continues into the upper Kugmallit in a pattern closely paralleling the occurrence of magnetic minerals (Figure 6). Carbonate mineralization becomes conspicuous in washed well-cutting samples at 680 m (2231 ft) within the Mackenzie Bay and continues to 1430 m (4692 ft) (Figure 6), but the occurrences below 1075 m (3527 ft) (base of resistivity anomalies) or at the very most 1200 m (3927 ft) (base of Kugmallit sequence) are probably caved. Agglutinated foraminifera do not make their first downhole appearance until 1010 m (3314 ft), well into the Oligocene Kugmallit sequence and continue downhole through the Richards sequence. This distribution of calcareous and agglutinated foraminifera, with the Miocene strata being dominated by calcareous and the Oligocene being dominated by agglutinated, is typical for the Beaufort-Mackenzie succession and is controlled by water mass and substrate (Schröder-Adams and McNeil, 1994).

Foraminifera diagnostic of the Richards sequence occur at 1490 m (4888 ft) and show only normal burial diagenetic effects for that amount of burial (i.e., slightly darker coloration). No hydrocarbon-related diagnosis (HRD) effects were observed in the foraminifera of the Richards or Taglu sequences much deeper in the well. The main HRDZ thus is confined to the Kugmallit sequence and to a lesser extent the Mackenzie Bay sequence.

FLUORESCENCE

Bulk washed sediments, diagenetic grains, and foraminifera from the Immiugak A-06 well HRDZ were examined using fluorescent techniques to identify and characterize hydrocarbon remains. Selected micrographs shown in Figure 7 summarize the findings and document the occurrence of bitumen and oil inclusions. Fluorescing bitumen in solid and soluble states was found to occur commonly, and traces of crude oil were found as inclusions in quartz cement and calcite crystals. Specimens of agglutinated and calcareous benthic foraminifera impregnated and stained by yellow-green fluorescing bitumen are illustrated (Figure 7a, b). At 860 m (2821 ft) in the well, bitumen in calcareous benthic foraminifera may have reached a peak where 50% of specimens in a population of 234 appeared to be affected. Grains from the washed well cuttings were found to contain oil inclusions in finely grained quartz cement and in diagenetic calcite cement (Figure 7c–e). Bitumen occurs commonly as intercrystalline deposits and inclusions in very fine-grained sandstone (Figure 7c, d). Diagenetic calcite crystals have widespread bitumen staining and contain inclusions of oil (Figure 7e).

COLORATION EFFECTS: HYDROCARBON ALTERATION VERSUS THERMAL ALTERATION

Foraminifera from Immiugak A-06 have gone through two distinctly separate diagenetic regimes: (1) a normal burial diagenetic regime and (2) a hydrocarbon-related diagenetic regime. Regionally consistent burial diagenetic effects in foraminifera from the Beaufort-Mackenzie Basin have been documented by McNeil et al. (1996) and McNeil (1997b). These studies have shown that progressive burial and associated thermal maturation produce a darker coloration in agglutinated foraminifera. Coloration effects can be measured accurately by comparing the overall color of individual foraminiferal specimens to standard Munsell soil color chips and calculating an average color value for a fossil population on a scale of 0 to 10. This scheme was referred to as the FCI by McNeil et al. (1996). In addition to color changes, agglutinated

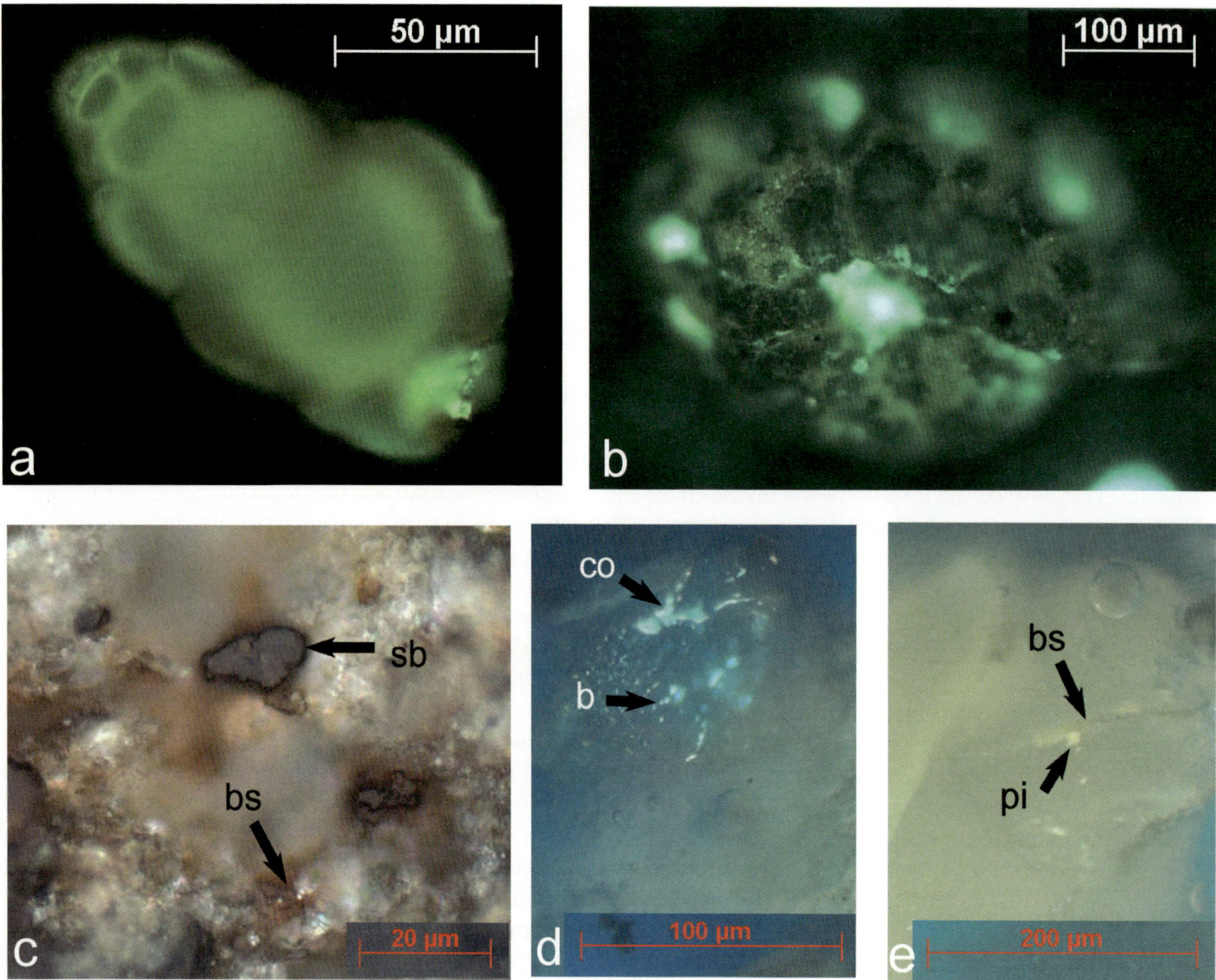

Figure 7. Fluorescent and white light photomicrographs showing hydrocarbons in foraminifera, sediments, and diagenetic grains from the HRDZ. (a) Yellow-green fluorescence and bitumen staining on calcareous benthic foraminifera (*Turrilina alsatica*). (b) Yellow-green fluorescence and bitumen staining on agglutinated foraminifera (*Haplophragmoides* sp.). (c) Very fine-grained sandstone in white light showing particles of solid bitumen (sb) and areas stained by soluble bitumen (bs). (d) Very fine-grained sandstone particle (epoxy in blue) showing crude oil (co) and bitumen (b) inclusions in quartz cement and intercrystalline bitumen. (e) Diagenetic calcite crystals (epoxy in blue) showing bitumen staining (bs) and petroleum inclusions (pi). (c–e) GSC Organic Petrology Lab sample 179/05 (sectioned and polished resin mount), Immiugak A-06, well cuttings 710–730 m (2329–2395 ft).

foraminifera in the Beaufort-Mackenzie Basin and other terrigenous clastic sedimentary basins become silicified with increasing depth and thermal maturation. These standard burial affects are shown in agglutinated foraminifera from the Immiugak A-06 well (Figure 8, red data points). The FCI values for Immiugak A-06 display a typical trend for conditions of normal hydrostatic burial (Figure 8; Table 1). At 1508 m (4947 ft), the FCI is 2.5 and at 2012 m (6601 ft), the FCI has increased to 3.5. Sediments from 1710 to 3300 m (5610 to 10,827 ft) (upper Taglu sequence) are dominantly terrestrial and do not provide any agglutinated foraminifera for FCI assessment, but marine units in the Taglu sequence below 3300 m (18,827 ft) to the well's total depth at 3800 m (12,467 ft) provide a consistent data set with values ranging from 5.5 to 5.9. A comparison to other standard thermal maturation indices at 3.3 km (2 mi) (Snowdon et al., 2004) indicates that the value of FCI 5.9 equates to a vitrinite reflectance value of 0.6 to 0.65% and a Rock-Eval T_{max} (the temperature at which the maximum amount of pyrolyzable hydrocarbons are generated) of 432–438°C. These data indicate that the Immiugak A-06 well is above the oil window, and that hydrocarbons must have migrated from depth.

A distinctly different population of agglutinated foraminifera is recovered from the HRDZ based on abnormal

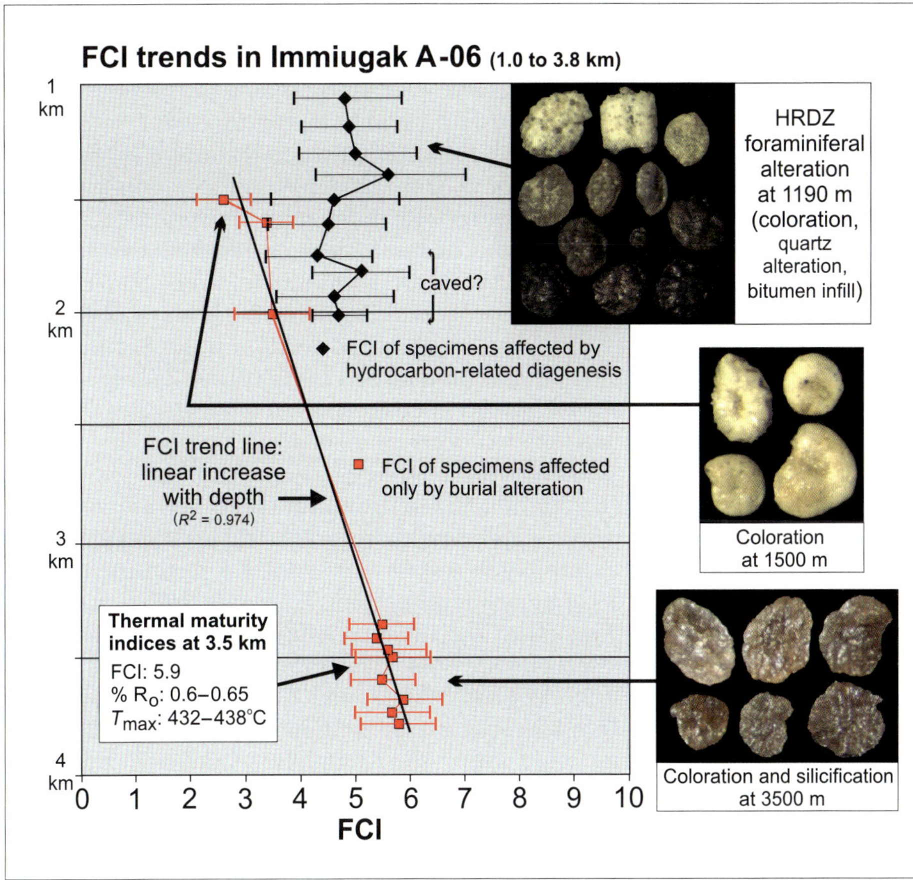

Figure 8. Foraminiferal coloration index (FCI) data from the Immiugak A-06 well with illustrations of foraminifera to show the contrasting effects between hydrocarbon-related diagenetic alteration and burial diagenesis (i.e., darkening with depth and silicification). The FCI measurements were made on agglutinated foraminifera. Data points are confined to sections of the well that are fossiliferous. The gap in data from about 2.1 to about 3.3 km (1.3 to ~2 mi) reflects dominantly terrestrial sedimentation and sediments barren of agglutinated foraminifera. Thermal maturity indices at 3.5 km (2.1 mi) are from Snowdon et al. (2004). HRDZ = hydrocarbon-related diagenetic zone.

coloration (uniformly dark to mottled and light colored), variations in silicification, impregnation by hydrocarbons, and partial dissolution of silica in agglutinated foraminifera (Figures 4g–h, 8). Using FCI as a means to quantify this anomalous coloration (data listed in Table 1), the HRDZ altered population plots on a distinctly separate trend line as indicated in Figure 8. An abnormally wide range of color variation is shown by agglutinated foraminifera from samples within the HRDZ, caused by varying degrees of silicification, dissolution, and bitumen impregnation. See Figure 8 for an example at 1190 m (3904 ft) of the variations in foraminiferal coloration and preservation within the HRDZ.

HYDROCARBON-RELATED DIAGENETIC ZONE MINERALS AND ISOTOPES

Figure 6 shows the generalized distribution of diagenetic calcite and sulfides from the Immiugak A-06 HRDZ. Calcite (Figure 9c, d) recovered from washed cutting samples occurs as crystal clusters (EDS analysis indicates $CaCO_3$). Magnetic sulfides occur as irregular rounded masses (Figure 9a, b). The XRD analysis indicates that the dominant magnetic mineral is greigite (Fe_3S_4). This mineral assemblage is typical of the diagenetic products formed by bacterial degradation of petroleum. Schumacher (1996) summarized the process as an aerobic or anaerobic degradation of petroleum, producing carbon dioxide mixing with water to form bicarbonate, in turn bonding with calcium to form carbonate. The carbon isotopic signature of the diagenetic carbonate is inherited from the hydrocarbons involved (typically −20 to −32‰ for crude oil and −90‰ for methane, according to Schumacher, 1996). The $\delta^{13}C$ values from diagenetic carbonate in our study range from −28.1 to −30.5‰ except for one anomalous sample (Table 2), thus generally concurring with predicted values for oil. For comparison, primary shelly calcite values range from 1.2 to 3.5‰, excluding one anomalous sample.

Aerobic or anaerobic bacterial degradation of hydrocarbons also produces hydrogen sulfide gas that, in the presence of iron (from sediment or meteoric water), combines to form pyrite and other sulfides, including magnetic minerals (magnetite, pyrrhotite, and greigite) known to be characteristic of HRDZs (Schumacher, 1996). A magnetic extraction of grains from washed sediment, followed by XRD analysis of the extracted iron sulfide grains from the Immiugak A-06 well, indicates that the magnetotactic mineral greigite (Fe_3S_4) occurs throughout

Table 1. Foraminiferal Coloration Index (FCI) data.*

Depth (m)	Burial Coloration (FCI ± S.D.)	HRDZ Coloration (FCI ± S.D.)**
1070	—	4.8 ± 1.0
1190	—	4.9 ± 0.9
1310	—	5.0 ± 1.1
1400	—	5.6 ± 1.4
1508	2.6 ± 0.5	4.6 ± 1.2
1616	3.4 ± 0.5	4.5 ± 1.1
1748	—	4.3 ± 1.0
1820	—	5.1 ± 0.9
1928	—	4.6 ± 1.1
2012	3.5 ± 0.7	4.7 ± 0.5
No data available from 2012 to 3356 m (6601 to 11,010 ft) (barren of foraminifera)		
3356	5.5 ± 0.6	—
3416	5.4 ± 0.6	—
3464	5.6 ± 0.7	—
3500	5.7 ± 0.7	—
3596	5.5 ± 0.6	—
3680	5.9 ± 0.7	—
3740	5.7 ± 0.7	—
3788	5.8 ± 0.7	—

*Values are derived using the technique outlined by McNeil et al. (1996).
**Hydrocarbon-related diagenetic zone (HRDZ) data below 1500 m (4921 ft) are probably caved.

the HRDZ. The $\delta^{34}S$ values for sulfur in HRDZ greigite should be inherited from the original degraded hydrocarbons. The isotopic signature of the hydrocarbon is unknown, but a comparison can be made between $\delta^{34}S$ in regionally abundant early diagenetic pyrite rods and greigite. A significant difference was found where early diagenetic pyrite has lower isotope values (average −12.15‰) than HRDZ greigite (−2.08‰), indicating very different sulfur sources for the two iron sulfides (i.e., sediment-pore-water sulfate versus oil sulfate).

Oxygen isotopic values from primary shelly calcite and HRDZ calcite indicate a clear fractionation toward lighter $\delta^{18}O$ values in HRDZ calcite, with the average decreasing from 2.9 to 0.9‰. The lower $\delta^{18}O$ values could reflect meteoric influence or lower fractionation associated with higher temperatures from warm formation waters migrating upward with hydrocarbons. Further work, e.g., fluid inclusion analysis, will be required to resolve the factors responsible for the isotope fractionation.

SIGNIFICANCE OF FORAMINIFERAL ALTERATION IN THE HRDZ

Despite the fact that seepage systems are a common feature of hydrocarbon basins and continental margins globally, and the fact that marked diagenetic processes are associated with seepage, there has been little or no recognition of these systems impacting the microfossil record. Our study indicates that evidence is conspicuously present in the foraminiferal record and provides useful information on the extent of hydrocarbon migration or seepage, its timing, and its pathway. In combination with the standard techniques, e.g., seismic, isotopic, mineralogical, or geochemical, a clearer picture of hydrocarbon migration can be obtained.

Seismic profiles indicate that the Immiugak A-06 gas chimney is a pronounced feature in the Beaufort continental margin and that hydrocarbon leakage was of apparent high intensity and on a large scale. Observations of hydrocarbon-related evidence in the form of abundant calcite, greigite, anomalous isotopic records, and foraminiferal alteration and bitumen are consistent with a significant hydrocarbon leakage system.

Immiugak A-06 is a dry well that penetrated apparently breached reservoirs. For comparative purposes, three other wells that contain significant hydrocarbon reservoirs have also been examined. These are Tarsiut A-25, Kopanoar M-13, and Koakoak O-22 (Figure 1). The reservoirs are oil dominated and are all of Oligocene age within the Kugmallit sequence in three distinct facies; prodelta shelf sands (Tarsiut A-25), slope sands (Koakoak O-22), and deep-water turbidite sands (Kopanoar M-13). The foraminiferal record is continuous through marine mudstones above and below reservoirs in Tarsiut A-25, and no evidence of HRD was observed, suggesting that the reservoirs have been consistently sealed. In Kopanoar M-13, a major oil discovery (up to 6048 bbl of oil/day in drillstem tests), only minor evidence of HRD has been observed, and that was in the form of secondary quartz crystals filling empty spaces within foraminifera. As in Tarsiut A-25, the lack of foraminiferal alteration suggests that the reservoirs have been contained by good seals. Hydrocarbon leakage at Kopanoar M-13, however, is likely going on at present, as a mud volcano is actively venting on the sea floor at the Kopanoar structure (Blasco et al., 2006). The closest comparison to Immiugak A-06 that has been observed from the Beaufort Sea area is in the Koakoak O-22 well, which encountered oil in turbidite sands of the Kugmallit sequence at 3308 m (10,853 ft). Agglutinated foraminifera in Koakoak O-22 show clear evidence of HRD in the form of silica precipitation and dissolution in a zone from 3015 m (9892 ft) to approximately 3200 m (10,499 ft), i.e., just above the hydrocarbon occurrence. No evidence of hydrocarbon-related diagenetic calcite was observed in Koakoak O-22 well cuttings, but magnetic minerals occur commonly from approximately 3000 to 3325 m (9842 to 10,909 ft). The evidence at Koakoak O-22 suggests that minor hydrocarbon leakage has occurred. More detailed studies on

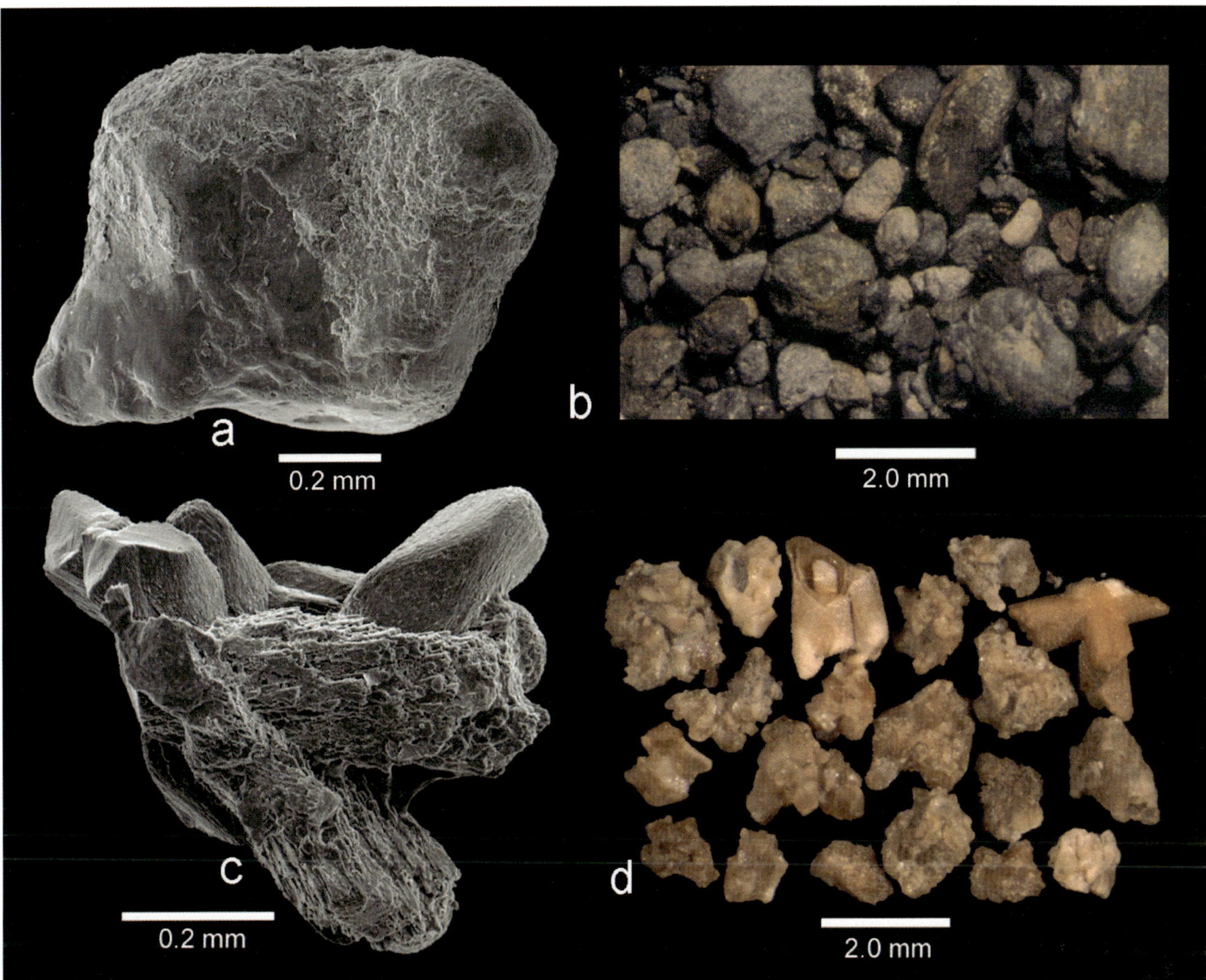

Figure 9. Diagenetic minerals from the Immiugak A-06 hydrocarbon-related diagenetic zone. (a) Biogenic greigite mass around a quartz grain (1160 m [3806 ft]). (b) Selected grains of magnetic minerals from a well-cutting sample (1160 m [3806 ft]). (c) Diagenetic calcite crystals (710 m [2329 ft]). (d) Selected grains of diagenetic calcite from a well-cutting sample (710 m [2329 ft]).

these and other wells in the Beaufort-Mackenzie Basin will be required to better understand the relationships between the various forms of observed hydrocarbon-related diagenesis and the overall hydrocarbon history in particular wells.

It appears from preliminary observations in the Immiugak A-06 well and other wells in the Beaufort-Mackenzie Basin that the diagenetic effects can be used to gauge the geographic distribution and possibly the relative magnitude of hydrocarbon seepage. Lesser effects would logically suggest minor amounts of hydrocarbon seepage and more favorable hydrocarbon prospectivity. Because foraminifera are routinely collected from exploration wells, they provide a readily available source of information during drilling and possibly in a predictive sense from regional analysis of preexisting data derived from foraminiferal collections combined with observations on the washed sediments generated from standard processing of well samples. Foraminifera that are not in pristine shape should not be discounted or ignored as poorly preserved specimens but should be examined closely for the diagenetic history they might reveal.

HYDROCARBON-RELATED DIAGENETIC HISTORY

At Immiugak A-06, very little remains of the hydrocarbons that must have been trapped or that must have passed through the large anticlinal structure on which the well was drilled. The section that appears to be most affected by hydrocarbon diagenesis occurs within the

Table 2. Isotope data ($\delta^{13}C$, $\delta^{18}O$, $\delta^{34}S$).

Depth (m)	$\delta^{34}S$		$\delta^{13}C$		$\delta^{18}O$	
	Pyrite	Greigite	Shelly Calcite	HRDZ Calcite	Shelly Calcite	HRDZ Calcite
560	—	—	3.5	—	4.1	—
710	−30.0	0.4	—	−28.1	—	1.7
860	—	—	1.2	—	4.4	—
890	−19.9	0.1	—	−28.1	—	1.6
1130	—	—	—	−1.5	—	3.1
1160	−7.0	−16.5	*	—	1.2	—
1400	2.5	−4.3	1.7	−29.5	1.9	−0.6
1556	−11.2	−3.2	—	−30.5	—	−0.1
1628	−7.3	11.0	—	−29.8	—	−0.1
Average	−12.2	−2.1	2.1	−24.6	2.9	0.9

*An anomalous $\delta^{13}C$ value of −39.2‰ for shelly calcite at 1160 m (3806 ft) could reflect fractionation in a methane-rich cold-seep environment but has been segregated here because it departs greatly from the average. Samples from below 1160 m (3806 ft) are probably caved but are nonetheless representative of hydrocarbon-related diagenesis.

Oligocene Kugmallit sequence as indicated by (1) resistivity anomalies on the well log, (2) the relative abundance of diagenetic calcite and greigite, and (3) the obvious bitumen in calcareous and agglutinated foraminifera. Diagenetic calcite that probably formed at fairly shallow depths (<2 km [<1.2 mi]) and low temperatures (<70°C [<158°F]) was recovered only from the Kugmallit. Foraminifera and shelly calcite from the Kugmallit and the Mackenzie Bay sequences show no obvious evidence of having existed in a methane- or hydrocarbon-rich environment, except possibly for one sample of shelly carbonate at 1160 m (3806 ft) that has a very negative, very light $\delta^{13}C$ value and could represent fractionation in a methane-rich cold-seep environment (Table 2). The apparent lack of surface effects suggests that a major vent did not exist during the Oligocene to middle Miocene (i.e., during deposition of the Kugmallit and Mackenzie Bay sequences). Hydrocarbons were probably generated during or prior to the late Miocene tectonic compressional movement in the basin and migrated up structure into the Kugmallit sequence where hydrocarbon diagenesis became significant as indicated by abundant evidence of microbial sulfate-reducing activity in the form of greigite and calcite with very negative $\delta^{13}C$. Porous sandstones of the Kugmallit sequence (Figure 6) would have provided the physical and geochemical environment for migrating hydrocarbon and CO^2-charged fluids to interact with formational and/or meteoric fluids. It is not known if hydrocarbons were trapped within the Kugmallit sequence, as they are in the other three wells studied (Tarsiut A-25, Kopanoar M-13, and Koakoak O-22) or migrated through the sandy facies of the Mackenzie Bay sequence at the crest of the structure during the late Miocene. To some extent, the occurrence of diagenetic calcite in the Kugmallit sequence is comparable to carbonate-cemented zones on the crests of major structures in the Gridgealpa field of the Carnarvon Basin on the northwestern Australian shelf (Schulz-Rojahn et al., 1998). There, calcite-cemented zones were developed synchronously with migration of deep-sourced hydrocarbons charging shallow reservoirs (1500–1800 m [4921–5905 ft]) near the structural crests. The common occurrence of bitumen in the Kugmallit and Mackenzie Bay sequences suggests an oil charge during the late Miocene (Figure 6), followed by gas charging in sediments deposited post-Miocene (Figure 2).

Agglutinated foraminifera are present only in the lower part of the Kugmallit sequence (sandstone and mudstone) in Immiugak A-06, and they show extensive affects of hydrocarbon-related diagenesis (Figures 4, 7, 8). The effects are most apparent in the precipitation and dissolution of silica within the foraminiferal test and the pervasive distribution of bitumen (including traces of oil) in the diagenetically altered test. The distribution of bitumen indicates that bitumen was present before and/or during silicification. The solid silicification of the agglutinated foraminifera is similar to the temperature-controlled silicification, which occurs in agglutinated foraminifera of the Beaufort-Mackenzie Basin with increased burial depth (McNeil et al., 1996), suggesting that warm fluids might have been involved in the HRD effects observed. Variable states of silica precipitation and dissolution probably attest to changing temperatures and fluid compositions and pressures during diagenesis. However, details of the diagenetic processes responsible for this mixing of bitumen and silica have not been established and will require analysis of fluid inclusions, mineral relationships, and isotopes.

An abrupt change in diagenetic mineralization and a marked reduction in bitumen impregnation are recognized in the Mackenzie Bay sequence (Figure 6). This probably represents a change in the rock or fluid properties. The record of HRD effects in sediments deposited post-Miocene cannot be evaluated in Immiugak A-06 because the well did not penetrate the upper section of the chimney (Figure 3). Seismic profiles, however, indicate that a hydrocarbon pathway existed during the late Cenozoic and that gas is probably currently migrating to the surface, perhaps escaping from gas-charged sediments in the Kugmallit and Iperk sequences (Figure 3).

CONCLUSIONS

Hydrocarbon-related alteration of agglutinated and calcareous foraminifera can provide evidence of hydrocarbon seepage and migration in the geological record. Analysis of standard microfossil samples from exploration wells, in combination with washed sediment residues from microfossil processing, can be used to identify wells affected by seepage and assess the relative extent of hydrocarbon seepage. This potential has been documented from Oligocene and Miocene foraminifera and associated diagenetic mineral assemblages from four exploration wells on the continental shelf of the Beaufort Sea that exhibit a representative range of hydrocarbon seepage; from major seepage to little or no seepage. The four wells include Immiugak A-06, Koakoak O-22, Kopanoar M-13, and Tarsiut A-25, in order of decreasing scale of seepage effects, and conversely increased reservoir seal integrity. Our study focused primarily on the Immiugak A-06 well because evidence of alteration was most pronounced in this well. Its record of exceptionally preserved foraminifera filled by bitumen is very conspicuous and is an ideal starting point to study hydrocarbon-related alteration effects. The Immiugak A-06 well also exhibits classic gas chimney features such as discontinuous vertically stacked seismic reflections, pronounced anomalies on resistivity logs, and a suite of diagenetic minerals, notably greigite and calcite, with isotopic signatures consistent with hydrocarbon-related diagenesis.

Coloration of agglutinated foraminifera in the HRDZ is markedly affected by bitumen impregnation, bitumen staining, and differential precipitation and dissolution of silica. These effects alter the standard, darkening-with-depth, color profile that can be measured by the FCI of McNeil et al. (1996) and are used to estimate thermal maturity trends. The FCI analysis provides an easy semiquantitative method of differentiating thermal maturity trends from hydrocarbon-related diagenetic trends because the foraminiferal populations can be separated visually based on their distinctive and disparate appearances.

Much further work is needed on the analysis of the Immiugak A-06 well, including mineral speciation, paragenetic sequencing, hydrocarbon characterization, and fluid inclusion studies to gain more insight into its hydrocarbon history. The main goal of this current research, however, was to establish that the fossil foraminiferal record has potential for providing useful information on hydrocarbon seepage and migration in exploration wells, particularly on continental margins with a history of rapid sedimentation and dynamic shale tectonism.

ACKNOWLEDGMENTS

The authors express their appreciation to the SEPM (Society for Sedimentary Geology) for awarding an earlier version of this work with an Honorable Mention for Best Poster at the 2005 AAPG Annual Convention. We thank the reviewers of this article, Dan Jarvie and Ron Waszczak, for their constructive comments and to Lesli Wood for encouragement to participate in the AAPG/Geological Society of Trinidad and Tobago (GSTT) Hedberg Conference "Mobile Shale Basins—Genesis, Evolution and Hydrocarbon Systems" and this AAPG Memoir "Shale Tectonics." Mark Obermajer (GSC Calgary) completed a preliminary review of the manuscript and provided many useful suggestions, including a title that did not contain more than 25 words. Microfossil samples were processed at GSC Calgary by Denise Then and picked by Judy Labadi. Krista Letourneau collected the magnetic separations from washed well cuttings and conducted the foraminiferal coloration determinations. The XRD and EDS analyses and scanning electron microscope images were conducted by Jenny Wong (GSC Calgary). Isotope analyses were supplied by Steve Taylor, University of Calgary Isotope Laboratory. This is Geological Survey of Canada contribution no. 20070325.

REFERENCES CITED

Bergquist, C. L., P. P. Graham, D. H. Johnston, and K. R. Rawlinson, 2003, Canada's Mackenzie delta: Fresh look at an emerging basin: Oil & Gas Journal, v. 101 (November 3), no. 42, p. 42–46.

Blasco, S. M., K. A. Blasco, J. M. Shearer, D. Poley, and B. R. Pelletier, 2006, Mud volcanoes, diapirs, pingos and relict topographic features on the Canadian Beaufort Shelf (abs.):

Atlantic Geoscience Society 32nd Colloquium and Annual Meeting, Program with Abstracts, p. 20–21.

Cowley, R., and G. W. O'Brien, 2000, Identification and interpretation of leaking hydrocarbons using seismic data: A comparative montage of examples from the major fields in Australia's North West Shelf and Gippsland Basin: Australian Petroleum Production and Exploration Association Journal, p. 121–150.

Elsley, G. R., and H. Tieman, 2010, A comparison of prestack depth and prestack time imaging of the Paktoa complex, Canadian Beaufort-Mackenzie Basin, *in* L. Wood, ed., Shale tectonics: AAPG Memoir 93, p. 79–90.

Giles, M. R., S. Stevenson, S. V. Martin, S. J. C. Cannon, P. J. Hamilton, J. D. Marshall, and G. M. Samways, 1992, The reservoir properties and diagenesis of the Brent Group: A regional perspective, *in* A. C. Morton, R. S. Haszeldine, M. R. Giles, and S. Brown, eds., Geology of the Brent Group: Geological Society (London) Special Publication 61, p. 289–327.

Lane, L. S., and J. R. Dietrich, 1995, Tertiary structural evolution of the Beaufort Sea—Mackenzie delta region, Arctic Canada: Bulletin of Canadian Petroleum Geology, v. 43, no. 3, p. 293–314.

Lane, L. S., and J. R. Dietrich, 1996, Geology map, *in* J. Dixon, ed., Geological atlas of the Beaufort-Mackenzie area: Geological Survey of Canada Miscellaneous Report 59, p. 11.

Machel, H. G., 1996, Magnetic contrasts as a result of hydrocarbon seepage and migration, *in* D. Schumacher and M. A. Abrams, eds., Hydrocarbon migration and its near-surface expression: AAPG Memoir 66, p. 99–109.

Machel, H. G., and J. Foght, 2000, Products and depth limits of microbial activity in petroliferous subsurface settings, *in* R. E. Riding and S. M. Awramik, eds., Microbial sediments: Berlin, Springer-Verlag, p. 105–120.

McNeil, D. H., 1997a, New foraminifera from the Upper Cretaceous and Cenozoic of the Beaufort-Mackenzie Basin of Arctic Canada: Cushman Foundation for Foraminiferal Research Special Publication 35, 95 p.

McNeil, D. H., 1997b, Diagenetic regimes and the foraminiferal record in the Beaufort-Mackenzie Basin and adjacent cratonic areas: Annales Societatis Geologorum Poloniae, v. 67, p. 274–286.

McNeil, D. H., D. R. Issler, and L. R. Snowdon, 1996, Colour alteration, thermal maturity, and burial diagenesis in fossil foraminifers: Geological Survey of Canada Bulletin, v. 499, 34 p.

Morrell, G. R., 1996, Oil and gas discoveries, *in* J. Dixon, ed., Geological atlas of the Beaufort-Mackenzie area: Geological Survey of Canada Miscellaneous Report 59, p. 146–165.

Munsell Color, 1975, Munsell soil color charts: Baltimore, Maryland, Macbeth Division of Kollmorgen Corporation, 16 p.

O'Brien, G. W., and E. P. Woods, 1995, Hydrocarbon-related diagenetic zones (HRDZs) in the Vulcan sub-basin, Timor Sea: Recognition and exploration implications: Australian Petroleum Exploration Association Journal, v. 35, p. 220–252.

Saunders, D. F., K. R. Burson, and C. K. Thompson, 1999, Model for hydrocarbon microseepage and related near-surface alterations: AAPG Bulletin, v. 83, no. 1, p. 170–185.

Schröder-Adams, C. J., and D. H. McNeil, 1994, Oligocene to Miocene agglutinated foraminifers in deltaic and deep-water facies of the Beaufort-Mackenzie Basin: Geological Survey of Canada Bulletin, v. 477, 67 p.

Schulz-Rojahn, J., S. Ryan-Grigor, and A. Anderson, 1998, Structural controls on seismic-scale carbonate cementation in hydrocarbon-bearing Jurassic fluvial and marine sandstones from Australia: A comparison, *in* S. Morad, ed., Carbonate cementation in sandstones, distribution patterns and geochemical evolution: International Association of Sedimentologists Special Publication 26, p. 327–362.

Schumacher, D., 1996, Hydrocarbon-induced alteration of soils and sediments, *in* D. Schumacher and M. A. Abrams, eds., Hydrocarbon migration and its near-surface expression: AAPG Memoir 66, p. 71–89.

Snowdon, L. R., L. D. Stasiuk, R. Robinson, J. Dixon, J. Dietrich, and D. H. McNeil, 2004, Organic geochemistry and organic petrology of a potential source rock of early Eocene age in the Beaufort-Mackenzie Basin: Organic Geochemistry, v. 35, p. 1039–1052, doi:10.1016/j.orggeochem.2004.04.006.

12

Delisle, Georg, Manfred Teschner, Eckhard Faber, Behrouz Panahi, Ibrahim Guliev, and Chingiz Aliev, 2010, First approach in quantifying fluctuating gas emissions of methane and radon from mud volcanoes in Azerbaijan, *in* L. Wood, ed., Mobile shale basins: AAPG Memoir 93, p. 211 – 224.

First Approach in Quantifying Fluctuating Gas Emissions of Methane and Radon from Mud Volcanoes in Azerbaijan

Georg Delisle, Manfred Teschner, and Eckhard Faber
Bundesanstalt für Geowissenschaften und Rohstoffe, Hannover, Germany
Behrouz Panahi, Ibrahim Guliev, and Chingiz Aliev
Geology Institute of Azerbaijan National Academy of Sciences, Baku, Azerbaijan

ABSTRACT

We report the results of a first attempt to monitor the gas emission from two distinct vents of two mud volcanoes in Azerbaijan, representing in both cases a fraction of the total amount of gas emitted by these structures. One monitoring station was placed on a salse lake in the crater area of the Dashgil mud volcano. It recorded methane flux from December 2003 to March 2006 and additionally radon emission in 2004. The second monitoring station was placed in October 2004 on the Perikushkul mud volcano. Both vents show significant variations in gas output with time. Month-long phases of near-constant gas output and periods characterized by large-scale daily variations were both observed. We proposed that fluctuations in gas output are connected with short-term variations in vent geometry (Dashgil) or with periodic buildups of methane at depth and sudden discharges (Perikushkul). Methane emission from the vent at Dashgil shows a well-developed periodicity of 24.05 hr, whereas no periodicity at Perikushkul was observed. During the observational period, both vents exhibited slowly decreasing gas output with time.

INTRODUCTION

More than 500 mud volcanoes in Azerbaijan pose a major geohazard. The country experiences several eruptions annually, some of which may be of a violent nature associated with the emission of a high-rising, burning gas flame and/or with the extrusion of mud on the order of greater than 10^5 m^3 (35 million ft^3) (Aliev et al., 2002). Many of these mud volcanoes are located in the vicinity of technical installations (oil and gas fields or pipelines) or even within the limits of villages or cities (e.g., Baku). As in the case of igneous volcanoes, developing techniques to predict imminent mud volcano eruptions would be highly desirable. Although a long history of scientific observations of the activity of mud volcanoes in Azerbaijan exists, we are still lacking an in-depth understanding of the processes responsible for the pressure buildup of fluids and mud at depth as

DOI:10.1306/13231316M933426

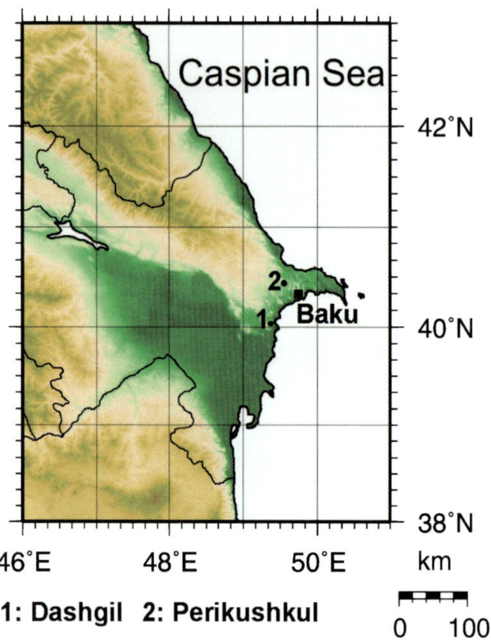

Figure 1. Location of the Dashgil and Perikushkul mud volcanoes, where gas emission patterns from distinct vents have been under investigation by ANAS-BGR since December 2003. 100 km = 62 mi.

well as the processes active during the ascension of the fluids and/or mud along pipes or fractures. To improve our understanding of these internal processes, in December 2003 the Azerbaijan National Academy of Sciences (ANAS) and the Bundesanstalt für Geowissenschaften und Rohstoffe (BGR) initiated a systematic long-term investigation on gas emissions (methane, carbon dioxide, and radon) from mud volcanoes. The locations of both monitoring sites are shown in Figure 1.

PREVIOUS WORK

Locations, dates, and details of previous eruptions of all mud volcanoes of Azerbaijan have been cataloged (Aliev et al., 2002) for the period of 1810–2001. The dominant gas component emitted from these volcanoes is methane (84 to 99% by volume) with traces of heavier hydrocarbons, carbon dioxide, and some noble gases (Aliev et al., 2002). However, only limited data are available on the specific activity patterns of mud volcanoes in Azerbaijan (e.g., Guliev, 1992) and very little is known about possible changes in mud volcano activity prior to major eruptions. Mud volcanoes have long been known to exist in a so-called quiet stage most of the time, followed by violent eruptions (Jakubov et al., 1971; Kopf, 2003; Milkov et al., 2003; Etiope and Milkov, 2004), which typically last from a fraction of an hour to several days. The amounts of gas emitted in quiet as well as violent stages are thought to be roughly comparable (Milkov et al., 2003).

For decades, researchers have searched for the triggering mechanisms of violent eruptions. Two principal ideas have been proposed: (1) the influence of the Earth's tides on the geometry of subsurface conduits (variations in fracture width) and (2) the correlation between seismicity and eruption events.

Huseynov and Guliyev (2004) provided a summary of observations, which suggests a close correlation between the tidal forces of the moon-sun system, as well as seismicity, with mud volcano eruption events. Sixty percent of mud volcano eruptions have been reported to occur during a new or full moon, that is, when the deformational forces exerted by the moon-sun system are strongest. A strong correlation between the Sun's activity (11-yr solar cycle) and mud volcano eruptions has been reported (Mekhtiev and Khalilov, 1988), although no specific mechanism linking these two processes has been clearly established.

An in-depth investigation of the correlation between seismicity and mud volcanism was undertaken by Guliev and Panahi (2004). They showed that earthquakes in the areas of high mud volcano concentrations occur commonly within the sedimentary pile with magnitudes (m) of less than 4–4.5 and concluded that intensive mud volcanism and thick sedimentary deposits closely relate to seismicity with low to moderate magnitudes.

Mellors et al. (2007) examined the potential triggering relationship between large earthquakes and mud volcano eruptions in Azerbaijan. Their findings suggest a pronounced temporal relationship between earthquakes of Mercalli seismic intensities greater than 6 and mud volcano eruptions within a radius of 100 km (62 mi) and a weak correlation between heightened numbers of eruptions within 1 yr after an earthquake occurrence.

An extensive survey to estimate the total discharge of greenhouse gases by mud volcanoes of Azerbaijan was provided by Etiope et al. (2004). Their approach was based on spot measurements of gas discharge from gas vents of major mud volcanoes and from soil microseepage near quiescent mud volcanoes.

To date, the questions, to what extent gas emissions fluctuate, and if characteristic gas emission patterns allow us to come to additional conclusions concerning processes controlling emission and eruption, remain unanswered.

SITES

The first monitoring site was stationed in a salse (lake) of the Dashgil mud volcano (Figure 2). This volcano is located about 2 km (1.2 mi) to the west of the village Alat

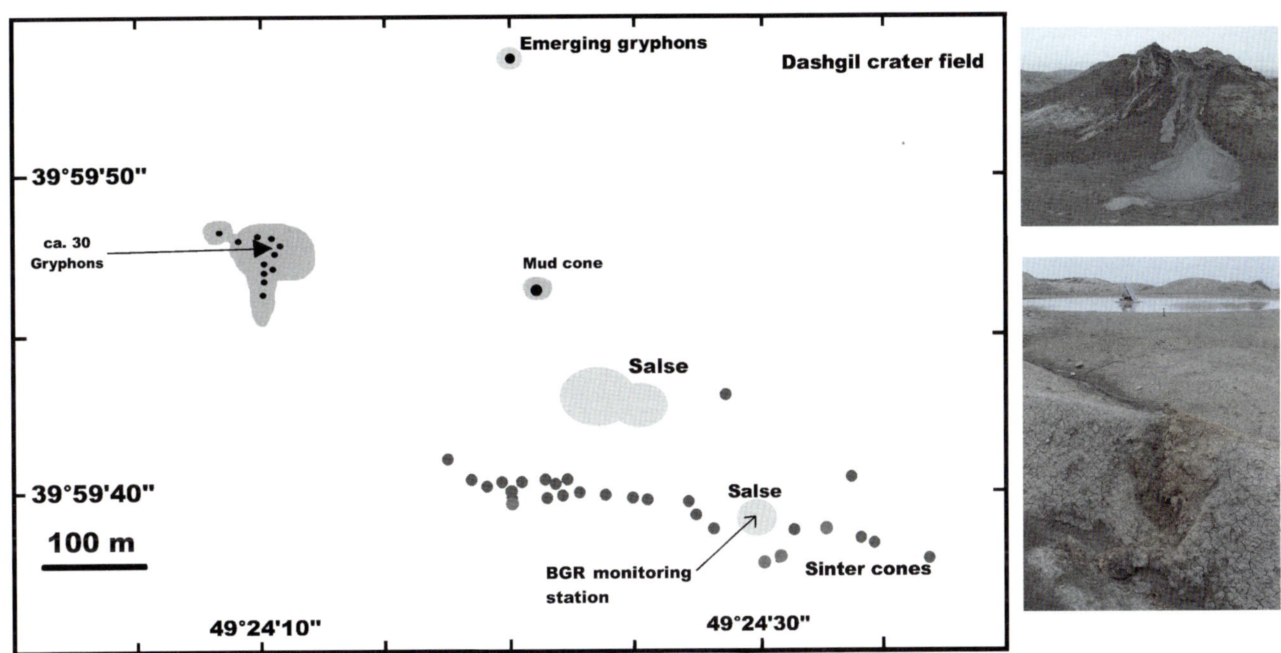

Figure 2. Map of the crater area of the Dashgil mud volcano (adapted from Planke et al., 2003). Gas is emitted from various gryphons and two salses (lakes). The last explosive eruption was associated with burning gas escaping from a fissure, which is delineated now by baked sediments (black dots) and sinter cones. The top right shows the typical form of a gryphon on Dashgil; the bottom right shows baked sediments near the small salse (background). 100 m = 328 ft; BGR = Bundesanstalt für Geowissenschaften und Rohstoffe.

near the western coast of the Caspian Sea (Figure 1). The summit rises 262 m (859 ft) above sea level. The known major eruptions on Dashgil occurred in the years 1882, 1902, 1908, 1926, and 1958 (Jakubov et al., 1971).

Descriptions of the site, in particular of the summit area, can be found in Hovland (1997) and Planke et al. (2003). A map contained in the latter article shows three different locations where gas emission is or has occurred (Figure 2).

In the southern sector, an east–west-oriented field of active gryphons produces a low-level outflow of mud of low viscosity. Toward the northern edge of the caldera, two lakes (salses) exist. In both, gas bubbles rise from the lake bottom. Water depth measurements with

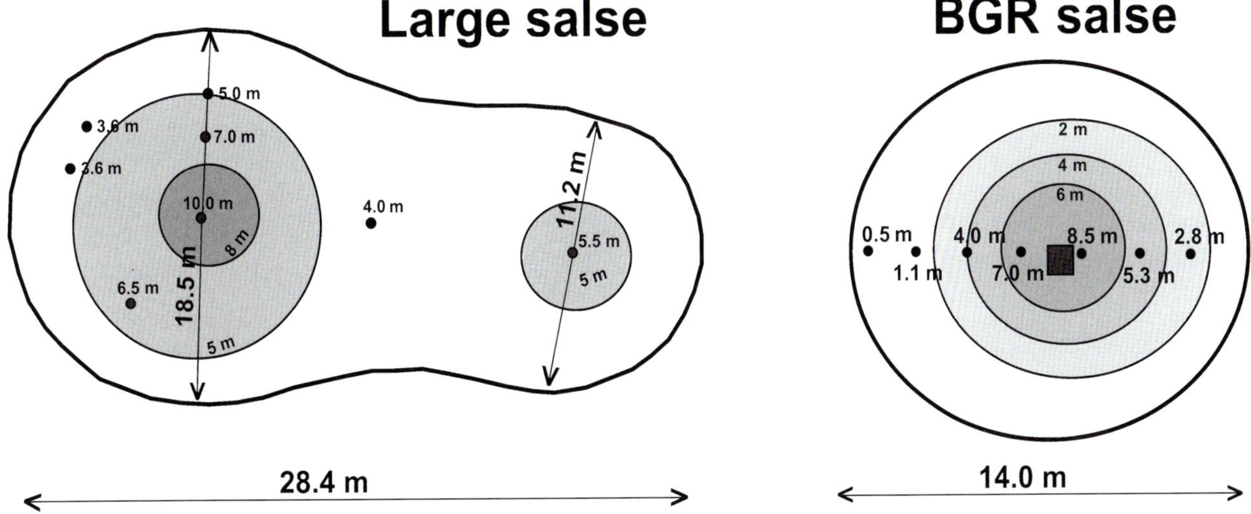

Figure 3. Outline and inferred water depths of the two salses (lakes) on Dashgil. The Bundesanstalt für Geowissenschaften und Rohstoffe (BGR) monitoring station (black rectangle) floats on the smaller salse (right). 1 m = 3.2 ft.

(a)

(b)

Figure 4. (a) Gas flux monitoring station mounted on a 1 × 1 m (3 × 3 ft) platform, floating on a salse (lake) of the Dashgil mud volcano and (b) view over the gas collection site on the Perikushkul mud volcano. Poles are about 2 m (6 ft) high.

a weight attached to a taper were conducted on both salses (Figure 3). The area of most vigorous gas bubbling at the lake surface coincides roughly with the location of the maximum lake depth. Monitoring since 2003 has shown that gas seep positions slightly shift or jump, open up, or fade away with time around the point of maximum water depth. Finally, the presence of altered country rocks with a strong baked appearance along a north–south-running line points to a previous emission event from an ignited gas plume.

Our first monitoring station (Figure 4a) was placed on December 13, 2003, on the smaller of the salses of Dashgil because this location is separate from the places visited most commonly by tourists. This station has been in operation with two interruptions caused by technical failures (for periods, see Figure 5a–c) until March

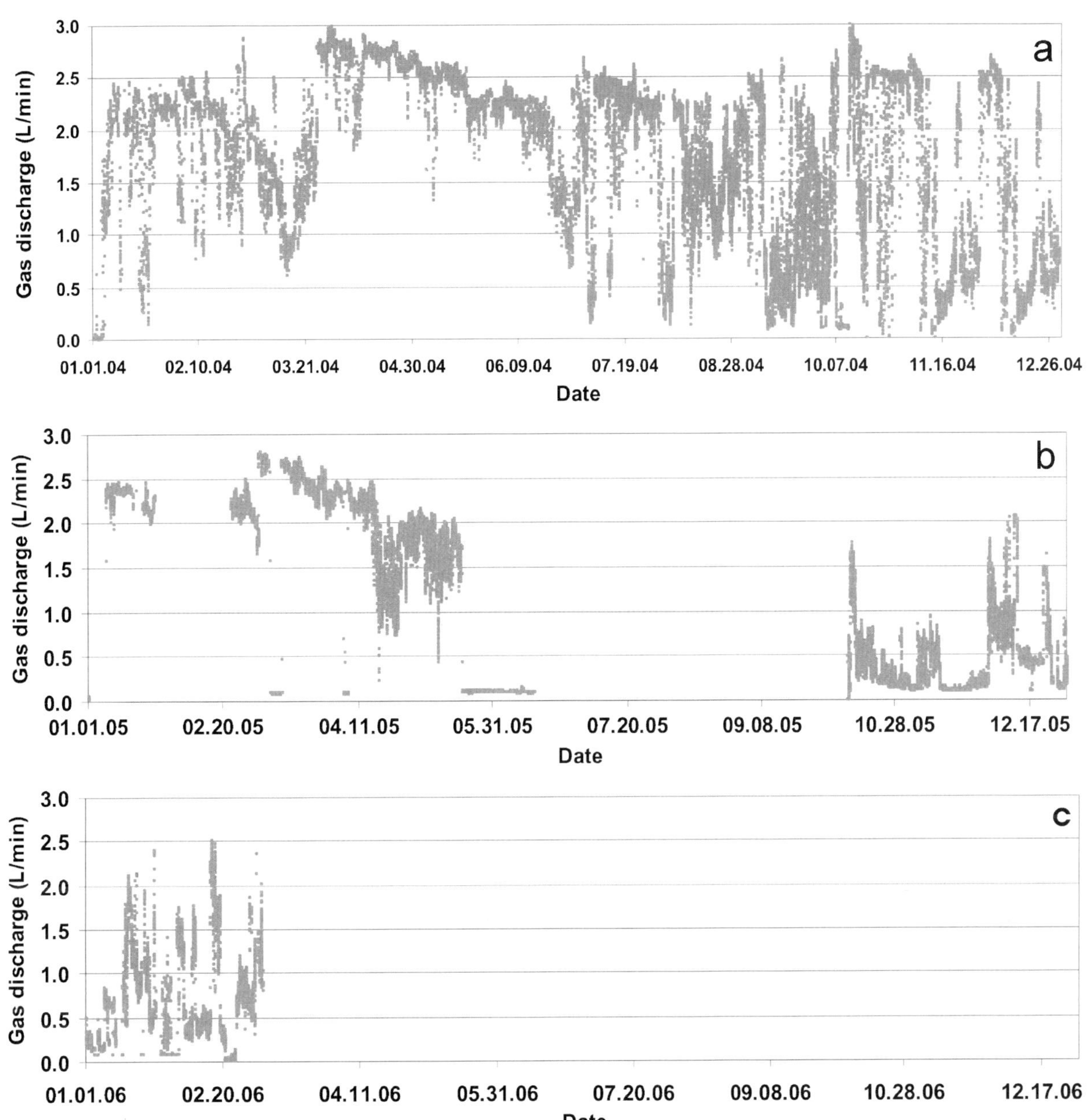

Figure 5. (a) Recorded gas discharge at the Bundesanstalt für Geowissenschaften und Rohstoffe (BGR) monitoring station floating in a fixed position on the smaller salse of the Dashgil mud volcano. (b) Continuation of record from February to May 2005. (c) Continuation of record from October 2005 to March 2006.

2006. In addition, data transmission was periodically interrupted for short time intervals caused by failures of the local power grid (mostly at night).

A gas sample collected near the larger of the two salses was analyzed (Table 1).

The sample consists mainly of methane with minor contributions of carbon dioxide, nitrogen, and oxygen. Heavier hydrocarbons are very low in concentration, especially compared to methane. Unsaturated C_3 and C_4 compounds were not found.

The carbon and hydrogen isotope data of methane indicate a thermal origin from a marine source rock, which is in the maturity range of the oil window. This is in agreement with the hydrocarbon situation in the area as oil and gas are locally produced and oil seeps are visible in place. However, neglecting the isotope data, the high methane content (or the low concentration of the higher hydrocarbons) is unusual for a gas generated from a mature oil source rock and could indicate a bacterial methane formation. We thought, however,

Table 1. Composition and isotopic signature of gas collected in December 2003 from a small vent next to the large salse on Dashgil.

N_2 (%)	O_2 (%)	CO_2 (%)	Methane (%)	Ethane (ppm)	Propane (ppm)	i-Butane (ppm)	n-Butane (ppm)	i-Pentane (ppm)	n-Pentane (ppm)	C_1 /ΣC_{1-5}	C_1/ $(C_2 + C_3)$	δ^{13}-CH_4 PDB	δ^{13}-CO_2 PDB	δD-CH_4 Standard Mean Ocean Water (SMOW)
2.81	0.81	2.93	93.40	45.60	18.6	4.9	5.6	2.3	1.6	0.9992	1455	−46.2	−17	−199.5

that bacterial oxidation reduced the concentration of predominantly higher hydrocarbons but did not change the methane content and methane isotopes considerably. Oxidation is likely because precipitation controls the water level in the salses and consequently introduces water into the mud channels. Thus, meteoric water can provide oxygen for hydrocarbon oxidation into the rising mud.

The second monitoring station was placed in October 2004 on the Perikushkul mud volcano, located about 120 km (74 mi) west of Baku (Figure 1) at an altitude of about 250 m (820 ft) above sea level. Low-activity gas venting occurs at this site within an area of about 20 × 150 m (66 × 492 ft) from about 10 gryphons and a few centimeter-wide water pools with gas emission. The principal idea of the chosen setup was to collect gas from a representative number of openings. The gases emitted from the edifices of two small gyphons and of four pools were collected and piped to a monitoring station buried and hidden in the ground (Figure 4b).

INSTRUMENTATION TO MEASURE GAS EMISSION

The monitoring station on the Dashgil mud volcano is mounted on a float (Figure 4a). The position of the float was fixed exactly over the point where gas bubbles were observed in December 2003 to rise to the surface. The underside of the float serves as a catching device for the gas bubbles. The collected gas is forced through a flow meter and then passes methane and radon sniffers. The gas flow meter (instrument type: series D-5100 manufactured by Manger und Wittmann) can measure a maximum flow of 5 L/min with an error specified as ±3% (of full scale). The radon sensor is based on a semiconductor and detects the alpha decay of ^{222}Rn at 5.4 MeV and ^{218}Po at 6.0 MeV. All data, which were collected in a 15-s interval, were transferred for storage via telemetry to a host computer in the nearby village Alat. In addition, all data were stacked over a period of 10 min, and the resulting value was stored (for safety reasons) on site in a data logger.

In addition to the gas measurements, air pressure was recorded at the same intervals of measurement. The temperature of the lake water at 1- and 3-m (3- and 10-ft) depth was recorded in 2004 by handylogs (temperature sensor, memory, and battery integrated into one device), which were attached to a wire hanging down from the float.

Gas bubbling eventually emerged on the water surface beyond the float and slowly increased throughout the observation period. In October 2005, this observation caused us to submerge a 2-m (6-ft)-high tent with a 4 m^2 (43 ft^2) large, quadratic, open bottom in the salse. It features a narrow opening on the top to channel the gas flow. The tent was centered over what was considered to be the main gas emission point near the point of deepest water depth. Nevertheless, some gas bubbling continued to occur sidewise, which points to a branch-type configuration of the near-surface section of the vent.

The monitoring device at the Perikushkul site is housed in a buried aluminum box, in which a flow meter for gas flux measurement (same type as above) is mounted. The emitted gas is collected in PVC cylinders, which were pressed into the ground over gas seeps. The collected gas flows from the cylinders via plastic pipes to a buried master pipe and into the box, passes the flow meter, and is disposed of through an exhaust pipe.

RESULTS

Methane Emission on Dashgil

After monitoring the situation at one vent at the Dashgil site from December 2003 to early March 2006, we can present a (intermittently interrupted) record of the variation in gas flux (Figure 5a–c). The visually most striking features we observe are the following.

- Phases of nearly stable gas output (e.g., March 2004) interchange with phases of rapid gas flux variations (e.g., September 2004).
- Sudden changes in the level of gas output are observed (e.g., March or September 2004).
- Peak gas production near 3.0 L/min is frequently followed by a persistent and almost linear slowdown in the gas flux before an entirely different gas emission pattern is established (e.g., end of June 2004 or mid-April of 2005).
- Repeatedly, the gas output tends to be reduced or to stop almost completely for one to several days before resumption of production at previous levels (e.g., early January 2004 or end of June 2004).

The long-term average gas output from the investigated vent, observed from early 2004 to May 2005, is near 2.5 L/min with occasional peak output near 3.0 L/min (Figure 5a). (The registered gas flux reduction at the end of December 2004 to mid-January 2005 is the result of interference by an unknown person who untied the float and shot a hole into one of the two solar panels. The latter defect resulted in a reduced power supply to the batteries, which led to episodic power shutdown during the months of January to March, visible in Figure 5b.) A technical failure occurred on May 17, 2005. The gas emission pattern observed after completed repairs in October 2005 differs significantly from earlier periods. The overall output was reduced to a level of less than 1.0 L/min, in general (Figure 5c). The fact that gas output reached at least occasionally the typical peak output of 2.5 L/min argues against the malfunction of the instrument. The float was again vandalized in February 2006, which prohibited any further search for a potential technical failure.

The dominant gas component is methane. The installed methane sniffer records an almost invariant methane concentration in the emitted gas phase.

We have noted invariably daily variations in gas discharge (Figure 6a–b). The daily variability may range from anywhere between a few percentage points (Figure 6a) to several 100% (Figure 6b). A one-day record typically shows one prolonged period with minimum as well as a maximum discharge. Small-scale, short-time perturbations are superimposed.

The integration of the output of methane for the period between January 2004 and mid-May 2005 (Figure 7a) leads to mean monthly output values of about 74 m^3 (2613 ft^3) and a total output of 887.5 m^3 (31,341.7 ft^3) at this one site. A substantial reduction in monthly gas output to about 23 m^3 (812 ft^3) was measured during the winter season of 2005/2006 (Figure 7b).

Radon Emission on Dashgil

The output in radon, indirectly measured by counts of radioactivity, is in general at a very low level. We observed infrequent bursts of radon emission (Figure 8a), which are entirely unrelated to variations of the methane emission suggesting that the source areas of both gases are unrelated.

Our measurements allow for a high time resolution (1 measurement per 10 s) of the radon gas bursts. Figure 8b demonstrates the example of a double burst without evidence of a previous increased radon emission. The first bursts seem to decay in exponential fashion, whereas later bursts show both an exponential increase in radon emission and exponential decay thereafter.

Frequency Analysis

We have applied a frequency analysis on the available gas flux data. The analysis showed that peaks in gas output occur with a well-defined and single periodicity of 24.05 hr. We have applied the same analysis to our recorded data on atmospheric pressure as measured on the site and found two well-defined frequencies at 12 and 24 hr (Figure 9).

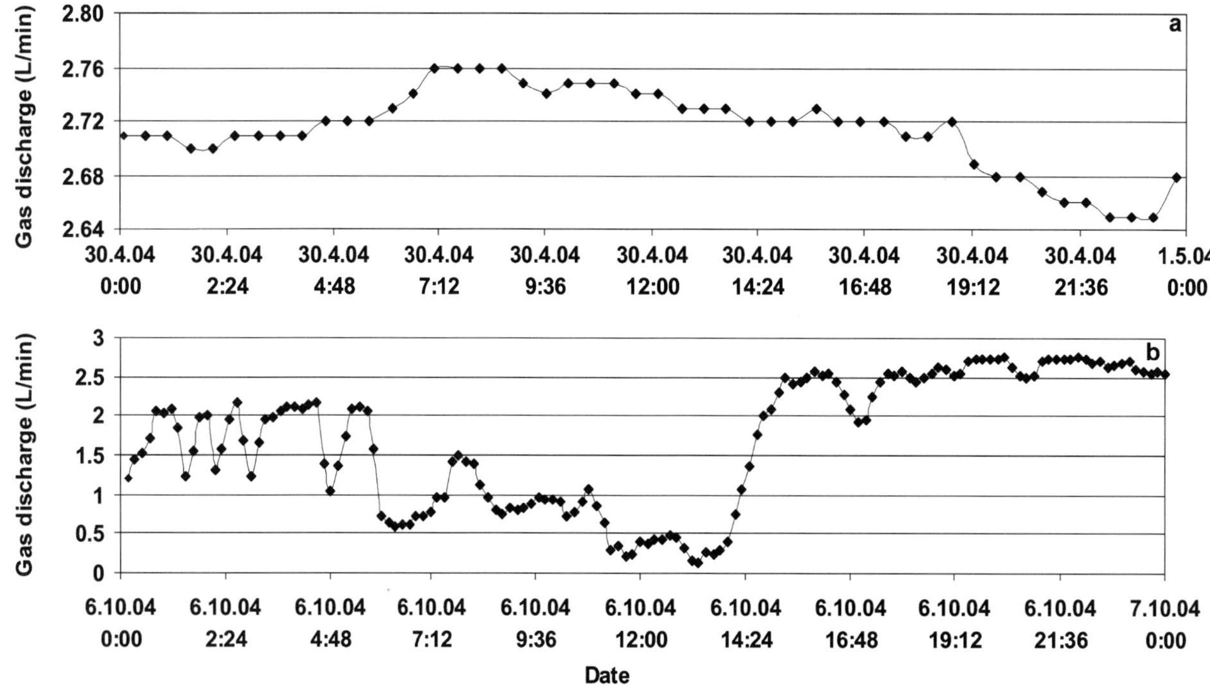

Figure 6. Examples for the daily variation of gas emission for (a) a period of near-constant gas discharge and (b) a period of strong fluctuations.

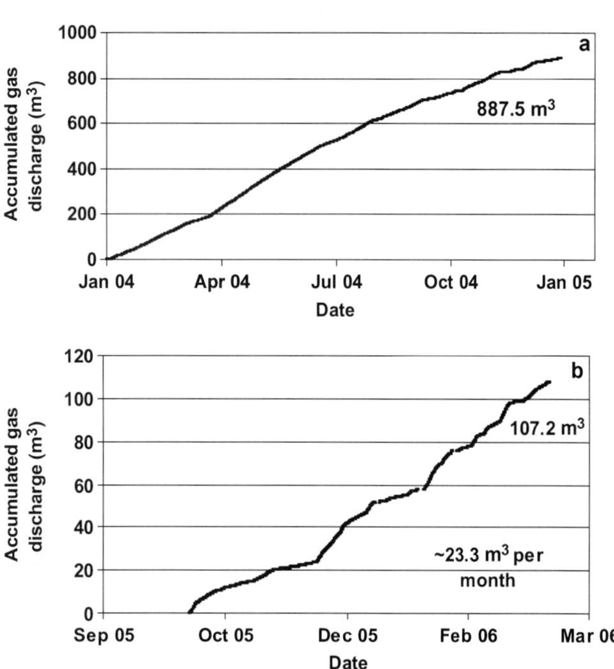

Figure 7. (a) Accumulated gas discharge through the Dashgil mud volcano monitoring system in 2004. The average discharge is about 74 m^3 (2613 ft^3) of methane per month. (b) Accumulated gas discharge through the Dashgil mud volcano monitoring system during October 2005 to March 2006. In comparison to 2004, the average discharge is reduced to about 23.3 m^3 (822.8 ft^3) of methane per month.

Our data on the lake temperatures show an effective damping of the variations of the air temperatures with water depth. The temperature fluctuation of the surface waters is already noticeably reduced at a water depth of 3 m (10 ft). The water temperature record closely follows the regional weather pattern and cannot be correlated in any way with the monitored gas flux (Figure 10).

Methane Emission on Perikushkul

Figure 11a–c shows the gas flux measured by our second monitoring station, located on the Perikushkul mud volcano since October 2004. The gas flux data from October 2004 to June 2005 are in character entirely different from the data from the Dashgil mud volcano. The Perikushkul data show a fairly constant gas flow of 0.5 L/min from October until the end of November 2004, followed by a gas burst on December 1, peaking at 4 L/min and diminished on December 10. A second gas burst, decaying in strength within 18 hr, was observed on December 13. Thereafter, the gas flux remained at a low level of between 0.2 and 0.3 L/min. Starting in April 2005, we observed rather regular short-term gas bursts (lasting for about 5 hr) on the order of every second to third day (Figure 11b). We were unable to measure the peak output during these gas burst events because the range of our gas flux recorder is limited to record a maximum gas flux of 5 L/min.

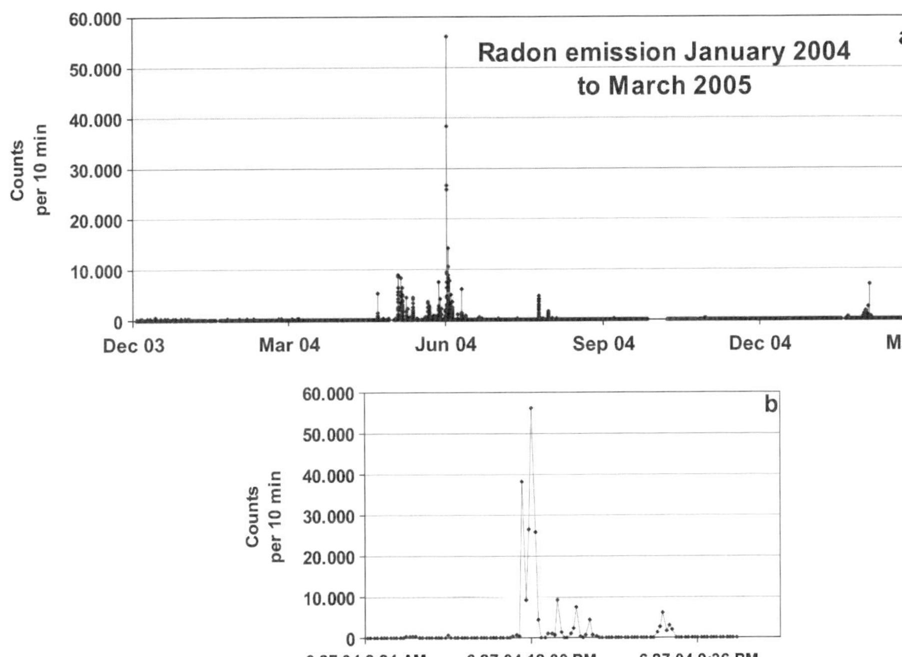

Figure 8. (a) Radon output as recorded by the Bundesanstalt für Geowissenschaften und Rohstoffe (BGR) monitoring station on Dashgil. (b) Detail of one radon burst as recorded on the end of June 2004.

Frequency analyses performed on the entire gas flux data set, as well as on parts of it, from the Perikushkul site yielded unequivocally a complete lack of any periodicity in the gas flux. This result is in striking contrast to the data from the Dashgil site.

DISCUSSION AND OPEN QUESTIONS

We had assumed at the beginning of our monitoring effort that the locations of the gas emission points at the bottom of the salse remain fixed within narrow boundaries. However, evidence for gradual shifts of the gas emission points was subsequently observed. Our visual impression is that the migration of the gas emission points along the lake bottom or along the ground is a comparatively slow process in comparison to the most observed variability patterns. Therefore, we consider the influence of the migration effect on the gas flux data to be modest, in particular in relation to the higher frequency gas flux phenomena. However, we clearly recognize that the design of future monitoring stations will have to be modified to accommodate the phenomenon of shifting gas emission points. The shortcomings identified during the operation of our first prototype monitoring station will be corrected in future operations starting in 2007.

Gas emission points at the Perikushkul site were observed to migrate with time as well. Some of them moved just outside the area covered by the PVC cylinders. The temporal background gas flux reduction is most likely caused by this development. In addition, the progressive filling of the cylinders with ascending mud, a process that tends to build up additional pressure, which has to be overcome by the ascending gas, impedes the gas collection effort. This is the key drawback of this particular setup, which cannot be overcome easily (this problem will be acerbated, if one attempts to monitor gryphons, which we have not yet done).

Observed infrequent complete blockage of gas output within the conduits of the mud volcanoes is commonly followed by slightly enhanced production implying a valve-type behavior of the ascension channel. No great overpressures at depth are apparently required to reopen this valve after shut-in. The observable slow but persistent reduction of gas output over time (i.e., April to mid-June 2004 or March to April 2005) after an initial reopening points to a gradual slow internal (diffusive?) process in the conduits toward a new blockage. Any model on the conduit configuration must also be able to explain why the particular geometry can sustain more than weeks of large-scale and systematic fluctuations of gas output as observed in September 2004.

The observed gas outputs at Dashgil (and Perikushkul) seem to drop slowly with time (see in particular Figure 7a, b). This is possibly one of the few trends that can be observed at many mud volcanoes. A case in point is the Lokbatan site (a mud volcano near the southern city limits of Baku), which experienced a massive gas eruption and outpouring of mud in October 2001 (Aliev et al., 2002). In 2003, the authors observed during a visit weak gas outputs from several conduits. During the

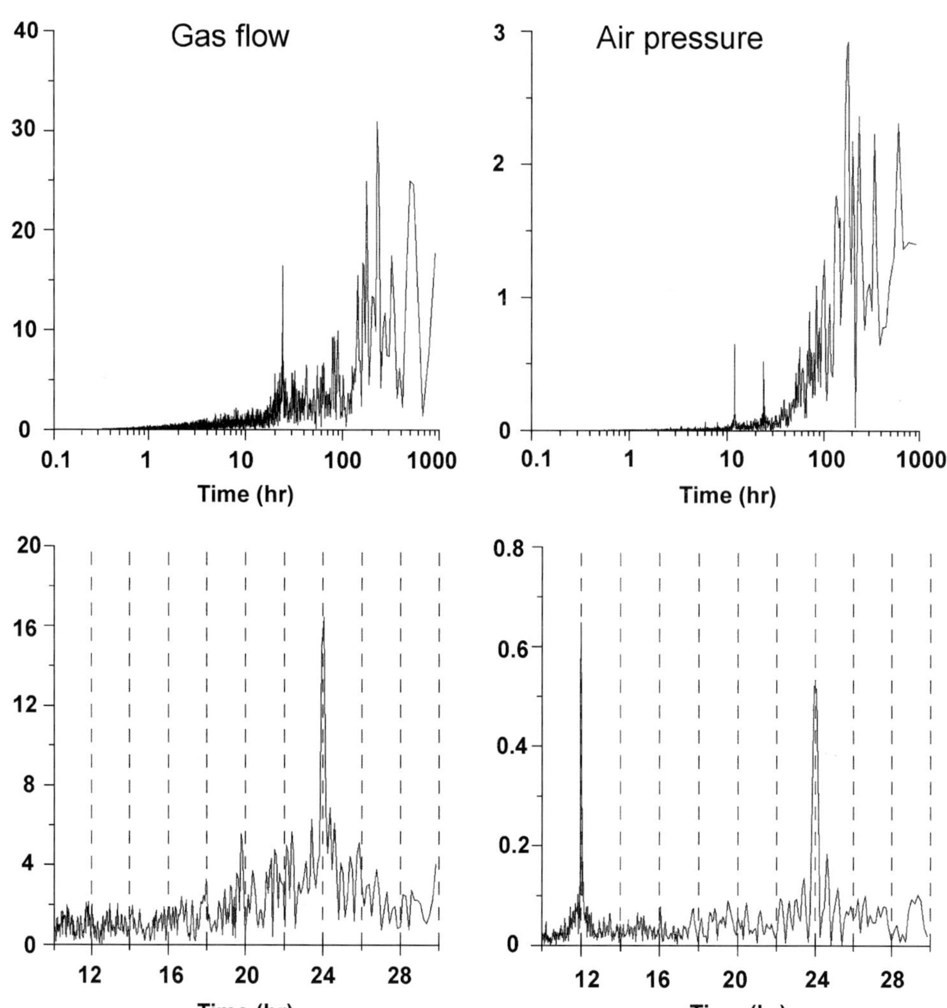

Figure 9. Frequency analysis of air pressure recorded by the Bundesanstalt für Geowissenschaften und Rohstoffe (BGR) monitoring station on Dashgil shows well-developed peak frequencies at 12 and 24 hr (atmospheric tides). The same analysis applied to gas flow data shows only one peak at 24 hr. Gas flux is therefore unrelated to air pressure variations.

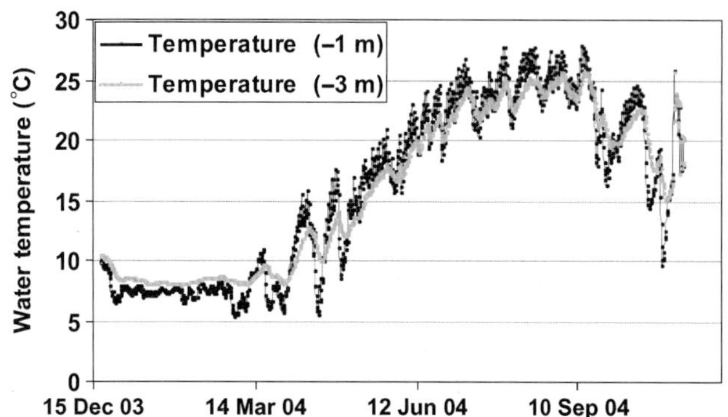

Figure 10. Water temperature record from the Dashgil salse cannot be correlated with periodicities in gas output. 1 m = 3.2 ft.

next visit in 2005, no gas emission was observed. It is not unreasonable to assume that we currently observe at all three sites parts of a cyclic behavior consisting of violent gas eruption; posteruption gas flow at low and, with time, decreasing intensity; and finally, complete blockage of gas flow from depth prior to a new eruption.

This general concept and, in particular, the available gas flux records from the Dashgil and Perikushkul sites suggest that one-time spot measurements of the gas output rate might well differ from the true average value (see Figure 6a–b). An independent study on gas emission from a mud pool in Taiwan by Yang et al. (2006), which identified short-time methane flux variations by a factor of at least 3, confirms this general conclusion.

The observed sharply developed periodicity of 24.05 hr is the more surprising because there are repeated references in the literature on the close dependence of mud

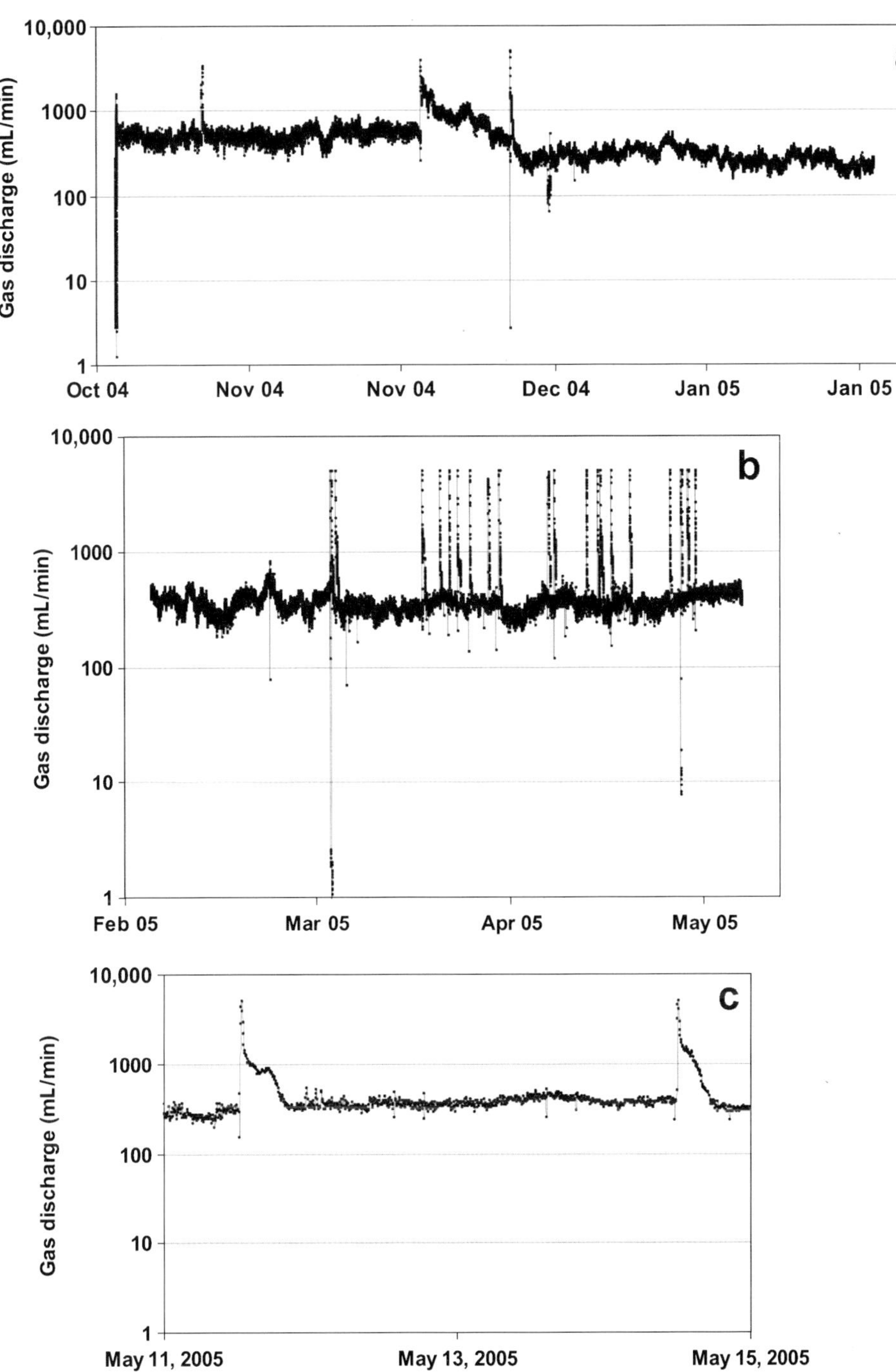

Figure 11. (a) Gas discharge on Perikushkul in October 2004 to the end of January 2005 showed one eruption event in early December and three gas bursts (top). (b) Frequent gas bursts started to occur in early April 2005 (center). (c) Each gas burst is followed by exponential decay of gas flux (bottom).

volcanism activity to Earth tide activity, which should result in an additional periodicity of close to 12 hr as well. However, other researchers (Facchini et al., 1995) found at other locations as well a dominant 24-hr periodicity in peak gas emission, without being able to identify the true cause. Earth tides occur in three modes; the lateral, tesseral, and zonal mode. Only the tesseral mode is associated with an approximately 24-hr periodicity. We point to the possibility that this mode for some reason exerts the dominant influence on the geometry and thereby on the physical properties of the vent at the Dashgil site.

The frequency analysis of the recorded atmospheric pressure shows two very clearly developed periodicities at 12 and 24 hr (atmospheric tides). The complete lack of a 12-hr periodicity in gas discharge offers strong evidence against any causality between atmospheric tides and gas discharge. Temperatures recorded in the salse suggest sinusoidal annual temperature fluctuations with an amplitude of about 19°C (34°F) at a water depth of 3 m (10 ft). At deeper water levels, the effect should be less. No causal connection or correlation between changes in water temperature and changes in gas emission rates throughout the year is evident.

The underlying cause of the observed variations in gas emission rates must be sought at deeper levels. If we assume that Earth tides affect the width of the conduits and let them fluctuate in tune with them, then we should expect that, via the associated pressure fluctuations and changes in gas solubility, the amount of free gas in the conduit should oscillate and, in consequence, should vary at the emission point. This mechanism can, however, explain only 24-hr variations. The longer-term changes, in particular the repeatedly observed moderate decline of gas output from a peak value to lower levels within weeks, strongly suggests a mechanism of slowly progressing blockage of the conduit and buildup of pressure at depth, as already discussed previously.

Apker Feyzullayev (2005, personal communication) reported that gases emitted from the gryphons on Dashgil show different isotopic compositions, which points to the existence of distinct conduits from the source depth to the surface. If this result can be further substantiated, there is an extremely complex network of conduits within the Dashgil mud volcano.

The characteristics of gas flux at Perikushkul differ strongly with respect to the Dashgil pattern (valve-type behavior). From October to the end of November 2004 as well from mid-December 2004 to the end of March 2005, we observed an almost monotonous gas flux at Perikushkul. A change in gas flux levels occurred on December 10 when the average gas flux rate dropped from 0.5 to 0.3 L/min (Figure 11a).

In the period from October 2004 to March 2005 only three gas burst events were observed (Figure 11a). Afterward, gas bursts occurred rather regularly (Figure 11b). As shown in Figure 11c, these gas bursts occur with an extremely sharp onset, followed by exponential decay of gas flow and frequently by a second low-level gas burst (Figure 11c). In stark contrast to the Dashgil site, these observations at Perikushkul point to a very limited influence of conduit geometry or permeability of the ascending mud on the gas flux. The only effect, which most likely can be attributed to changing conduit geometry, is the sudden drop in gas flux on December 10, possibly induced by the two previous gas bursts. At Perikushkul, the physical properties of the gas reservoir at depth appear to dominate the gas discharge pattern. The series of gas bursts starting in April 2005 points to an enhanced gas supply to the gas reservoir at depth. However, the background gas flux remains at a very low level of about 0.3 L/min. Alternatively, we have discussed the possibility that the gas bursts are an artifact of our experimental setup, but we were unable to identify a clear cause and effect relationship.

The comparison of the Dashgil and the Perikushkul data yields few similarities. Apparently, each mud volcano exhibits its own characteristic gas flux pattern, which may be interpreted as evidence that local peculiarities exert a strong influence on the conduits of ascending mud and gases. If so, this would significantly complicate the task of developing any technological scheme to predict imminent eruptions, because each site would have to be treated to some degree as a separate case.

The Perikushkul mud volcano consists of several about 2-m (6-ft)-high gryphons and gas seeps on the ground on top of a calcareous, vertically tilted formation (Figure 12). The conduit properties are most likely controlled by the rock properties of the underlying formation. In the case of the 226-m (741-ft)-high Dashgil, a significant part of the conduit system rests within the mud volcano edifice, which in itself is the result of hundreds of mudflows. This structural difference offers the most probable explanation for the observed, widely differing gas flux patterns.

CONCLUSIONS

Gas emission data from vents of two mud volcanoes of Azerbaijan, which were collected over extended periods, show strikingly different gas emission characteristics.

On Dashgil, we observe a valve-type gas emission behavior. Typically, gas is emitted for more than several weeks with slowly declining emission rates, to be followed by a substantial drop in emission, which is then followed by renewed emission activity at a high level. An alternative pattern of strong amplitude oscillations was observed in September 2004 and to some degree during the winter of 2005/2006. In strong contrast to the Perikushkul site, no gas burst behavior was observed. The experience from both sites suggests that one time-point measurement results in a very approximate estimate of average gas flux.

The maximum gas output rate from the investigated vent at the Dashgil site is limited to about 3.0 L/min. The monthly volume of gas output has slowly declined at the monitored site between 2004 and 2006.

The gas output at the Perikushkul site appears to be characterized by episodic gas bursts in excess of 5 L/min, which points to a steady buildup of a methane reservoir, which is destroyed by sudden gas discharge to the atmosphere within a few hours.

The characteristics of the radon emission were shown to be entirely independent from the characteristics of the methane flux. The very infrequent but sharp radon

Figure 12. The Perikushkul mud volcano rests on top (slightly to the left of the intersection of the pole with the horizon) of a vertically tilted sedimentary formation.

burst events point to the presence of a radon source, which episodically reaches a critical excess volume with the result of a discharge or spillover of excess radon into the conduit system.

Preliminary evidence tentatively suggests a connection between large-scale Earth movements (Earth's oscillations) and venting activity.

Preliminary evidence suggests that the gas emission pattern at Dashgil is controlled by the vent configuration. In contrast, the properties of the gas reservoir at depth at Perikushkul appear to be the controlling factor for the gas flux.

The gas emission appears to fluctuate between different conduits as time proceeds. This observation requires us to further refine monitoring techniques in the future. Nevertheless, taken together, both investigated vents showed a slow overall decrease in gas output, which points to an increasingly effective blockage of gas flow from depth, with time.

More than 2 yr of experience in monitoring resulted in two detailed data sets on gas emission from two distinct vents of mud volcanoes in Azerbaijan.

CONCLUDING REMARKS

The instrumental setup on the small salse on Dashgil was modernized in June 2007. A second monitoring system was placed on the large salse (Figure 2). The visual impression suggests that the predominant amount of gas on Dashgil emanates from both these systems (with the main fraction from the large salse). Preliminary first results suggest that both instruments record an annual gas output of more than 40,000 m^3 ($\sim$30 t ac^{-1}).

ACKNOWLEDGMENTS

We gratefully acknowledge the Deutsche Luft- und Raumfahrt/Internationales Büro (DLR-IB9 under grant AZE 02/001), the Azerbaijan National Academy of Sciences (ANAS), and Bundesanstalt für Geowissenschaften und Rohstoffe (BGR) for funding for this project. The article has greatly benefited from helpful comments of three reviewers.

REFERENCES CITED

Aliev, Ad. A., I. S. Guliyev, and I. S. Belov, 2002, Catalog of recorded eruptions of mud volcanoes of Azerbaijan: Baku, Azerbaijan, Nafta Press, 88 p.

Etiope, G., and A. V. Milkov, 2004, A new estimate of global methane flux from onshore and shallow submarine mud volcanoes to the atmosphere: Environmental Geology, v. 46, no. 8, p. 997–1002, doi:10.1007/s00254-004-1085-1.

Etiope, G., A. Feyzullayev, C. L. Baciu, and A. V. Milkov, 2004, Methane emission from mud volcanoes in eastern Azerbaijan: Geology, v. 32, p. 465–468, doi:10.1130/G20320.1.

Facchini, U., M. Garavaglia, S. Magnoni, F. Rinaldi, R. Cassinis,

M. Delcourt-Honorze, and B. Ducarme, 1995, Radon levels in a deep geothermal well in the Po plain, *in* C. Dubois, ed., Gas geochemistry: Northwood, UK, Science Reviews, p. 257–279.

Guliev, I., and B. Panahi, 2004, Geodynamics of the deep sedimentary basin of the Caspian Sea region: Paragenetic correlation of seismicity and mud volcanism: Environmental Geology, v. 24, p. 169–176.

Guliev, I. S., 1992, A review of mud volcanism: Report of the Institute of Geology, Azerbaijan National Academy of Sciences, Baku, 65 p.

Hovland, M., A. Hill, and D. Stokes, 1997, Structure and geomorphology of the Dashgil mud volcano, Azerbaijan: Geomorphology, v. 21, p. 1–15, doi:10.1016/S0169-555X(97)00034-2.

Huseynov, D. A., and I. S. Guliyev, 2004, Mud volcanic natural phenomena in the south Caspian Basin: Geology, fluid dynamics and environmental impact: Environmental Geology, v. 46, p. 1012–1023, doi:10.1007/s00254-004-1088-y.

Jakubov, A. A., A. A. Ali-Zade, and M. M. Zeinalov, 1971, Gryazevye vulkany Azerbaidzhanskoi SSR: Atlas (Mud volcanoes of the Azerbaijan SSR: Atlas): Baku, Azerbaijan, Azerbaijan National Academy of Sciences, 258 p.

Kopf, A., 2003, Global methane emission through mud volcanoes and its past and present impact on Earth's climate: International Journal of Earth Sciences, v. 92, p. 806–816, doi:10.1007/s00531-003-0341-z.

Mekhtiev, Sh. F., and E. N. Khalilov, 1988, How the Earth develops: Baku, Azerbaijan, Znanie, 23 p.

Mellors, R., D. Kilb, A. Aliyev, A. Gasamov, and G. Yetirmishli, 2007, Correlations between earthquakes and large mud volcano eruptions: Journal of Geophysical Research, v. 112, B04304, p. 1–11, doi:10.1029/2006JB004489.

Milkov, A. V., R. Sassen, T. V. Apanasovich, and F. G. Dadashev, 2003, Global gas flux from mud volcanoes: A significant source of fossil methane in the atmosphere and in the ocean: Geophysical Research Letters, v. 30, p. 1037, doi:10.1029/2002GL016358.

Planke, S., H. Svensen, M. Hovland, D. A. Banks, and B. Jamtveit, 2003, Mud and fluid migration in active mud volcanoes in Azerbaijan: Geo-Marine Letters, v. 23, p. 258–268, doi:10.1007/s00367-003-0152-z.

Yang, T. F., C. C. Fu, V. Walla, C.-H. Chen, L. L. Chyi, T.-K. Liu, S.-R. Song, M. Lee, C.-W. Lin, and C.-C. Lin, 2006, Seismo-geochemical variation in SW Taiwan: Multi-parameter automatic gas monitoring results: Pure and Applied Geophysics, v. 163, p. 693–709, doi:10.1007/s00024-0β06-0040-3.

13

Battani, A., A. Prinzhofer, E. Deville, and C. J. Ballentine, 2010, Trinidad mud volcanoes: The origin of the gas, *in* L. Wood, ed., Shale tectonics: AAPG Memoir 93, p. 225–238.

Trinidad Mud Volcanoes: The Origin of the Gas

A. Battani, A. Prinzhofer, and E. Deville
Institut Français du Pétrole, Rueil-Malmaison, France
C. J. Ballentine
School of Earth, Atmospheric and Environmental Sciences, Manchester, United Kingdom

ABSTRACT

Bubbling gases from the mud volcanoes of Trinidad and gases associated with oil in deeper reservoirs were sampled and analyzed to understand their possible relationships. Numerous geochemical analyses were performed on the gas samples. The chemical concentrations of organic compounds, CO_2, and noble gases (from He to Xe) were measured, and isotopic signatures ($\delta^{13}C$ from C_1 to C_5, CO_2, and noble gases) were also determined. In the southern part of the island, our data show a typical thermogenic origin for the gases from oil reservoirs. However, the gases from the mud volcanoes, all located in the Southern Range of Trinidad, exhibit intermediate values between thermogenic and bacterial signatures. The hypothesis of simple mixing between these two end members can be discarded because of an incompatible $\delta^{13}C$ for the bacterial end member (ranging from −52 to −33‰). We thus conclude that most hydrocarbon gases found in the mud volcanoes are purely thermogenic gases.

The isotopic $\delta^{13}C$ of CO_2 versus CO_2/C_1 elemental ratios from the oil reservoir gases and the mud volcano gases follow a global trend. We therefore conclude that all gases come from the same source. This is corroborated by the isotopic composition of the hydrocarbons and of CO_2 and by the radiogenic fraction of the associated noble gases.

However, mud volcano gases show an extreme dryness, which cannot be directly related to gas generation but instead to postgenetic processes. A process of migration involving local process of dissolution and diffusion in water and preferential adsorption of the larger organic molecules onto the solid mud particles should be involved. Finally, mud volcano gases have never been trapped but are permanently expelled to the atmosphere and thus have a lower time of residence than oil reservoir gases as indicated by noble gas data.

INTRODUCTION

Gas and oil exploration started by observing the oil and gas seeps above possible deeper reservoirs. A detailed understanding of the subsurface through surface information remains an essential tool in exploration today. Mud volcanism is a phenomenon, which has been described since antiquity, creating surface vents expelling

DOI:10.1306/13231317M933427

mud (water with fine solid mineral particles) and gas flows. Mud volcanoes develop either in offshore regions or in onshore regions (see Higgins and Saunders, 1974; Guliyiev and Feizullayev, 1998; Dimitrov, 2002; Kopf, 2002, for general inventories and references). The areas of mud volcano occurrence are generally characterized by a thick sedimentary pile affected by overpressure conditions. Mud volcanoes occur in a variety of geological settings, mostly in compressional systems (Biju-Duval et al., 1982; Westbrook and Smith, 1983; Brown, 1990; Martinelli, 1999; Kopf and Behrmann, 2000; Kopf et al., 2001), in deltaic settings (Graue, 2000; Sassen et al., 2003; Loncke et al., 2004), and also episodically in hydrothermal areas (Pitt and Hutchinson, 1982; Etiope et al., 2002). From a geochemical point of view, the origin of the expelled fluids is generally poorly known.

Trinidad Island is a well-known target for oil production for more than one century and is a reference area for the study of mud volcanism (Higgins and Saunders, 1974; Dia et al., 1999; Deville et al., 2003a, b). The aim of this study is to define the origin of the gas expelled by the subaerial mud volcanoes of southern Trinidad. Indeed, the main gas escaping from mud volcanoes in Trinidad other than water vapor is methane. The origin (i.e., thermogenic versus bacterial) is still subject to debate even if previous studies on the expelled water and the other fluids showed chemical compositions indicative of a deep origin (Dia et al., 1999; Barboza and Boettcher, 2000; Battani et al., 2001; Deville et al., 2003a). For this purpose, we have analyzed bubbling mud volcano gas, and oil reservoir gas, to identify and understand any possible relationships between these two families.

GEOLOGICAL CONTEXT

Sedimentary volcanism in Trinidad occurs in the plate boundary area between the Caribbean plate and the South American plate at the junction between the Barbados accretionary prism and the transform system of northern Venezuela (Speed, 1985; Stein et al., 1988; Robertson and Burke, 1989; De Mets et al., 1990). Within this compressional and transpressional system, an active belt of mud volcanoes develops from the Barbados tectonic wedge to the fold-thrust belt of northern Venezuela. In this system, the Trinidad mud volcanoes are only the emergent part of a widely developed phenomenon, especially in the offshore area of the Barbados prism (Biju-Duval et al., 1982; Brown and Westbrook, 1987; Deville et al., 2003a, b, 2006).

In Trinidad, the main inland hydrocarbon province is located in the southern part of the island. An isolated dry gas field (Mahaica) is located in the central part of the island within the Caroni basin. All the other hydrocarbon fields are oil fields, containing a small proportion of associated gas. They are accumulated within Tertiary formations and were generated from Cretaceous type II source rocks (Gautier and Naparima Hill formations; Rodriguez, 1988; Talukdar et al., 1989, 1990, 1995; Lawrence et al., 1991; Persad et al., 1993; Requejo et al., 1994; Heppard et al., 1998). In Trinidad, the maturity of the Cretaceous source rocks investigated in outcrops and in the blocks expelled by the mud volcanoes is in the range of the oil window (T_{max} lower than 460°C). The mud volcanoes are also located mainly in the southern part of Trinidad, sometimes exactly at the apex of oil and gas fields, suggesting a possible relationship between them.

The deep structure of the southern part of Trinidad (Central and Southern Ranges) is relatively well known thanks to numerous exploration wells (Kugler, 1953, 1959, 1965, 1996; Ahmad, 1991; Payne, 1991; Tyson and Winston, 1991; Tyson et al., 1991; Higgins, 1996, and many others). A strong subsidence affected the margin during the Tertiary in relation with the flexure of the northern part of the South American plate associated to the eastward relative motion of the Caribbean plate (Pindell et al., 2005). Then, progressively compressional and dextral lateral strike-slip deformations affected southern Trinidad during the Neogene. This deformation is still active today (Saleh et al., 2004). The decollement is inferred to be located within lower Cretaceous strata. The mud volcanoes in outcrop, as well as subsurface mud plugs are located along the Naparima Hill fault system in the Central Range, along the trends of the transpressive thrust anticlines of the Southern Range and along the Los Bajos major transfer fault (Barr, 1953; Birchwood, 1965; Higgins and Saunders, 1974; Yassir, 1989) (Figure 1). The water and mineral compositions of the mud expelled by these mud volcanoes have been extensively studied (Higgins and Saunders, 1974; Dia et al., 1999; Deville et al., 2003a). The vents show a generally persistent modest gas bubbling and some rare catastrophic eruptions with large emissions of mudflows and breccias. The large volume of fossil mud formations indicates that a similar dynamic fluid flow has been occurring for a long time. At the most active visited vents (Lagon Bouffe and Palo Seco), the highest estimated gas fluxes were higher than 10 m^3 (353 ft^3) (CH_4)/min, which is equivalent to 200 tcf per million years. This impressive natural emission of gas must be investigated in terms of source and migration processes. Notice here that during the sampling campaign, we observed many low-pressure oil fields, whereas the mud volcanism within several mud volcanoes is still highly active today. The pressure regime thus seems disconnected between oil fields and mud volcanoes.

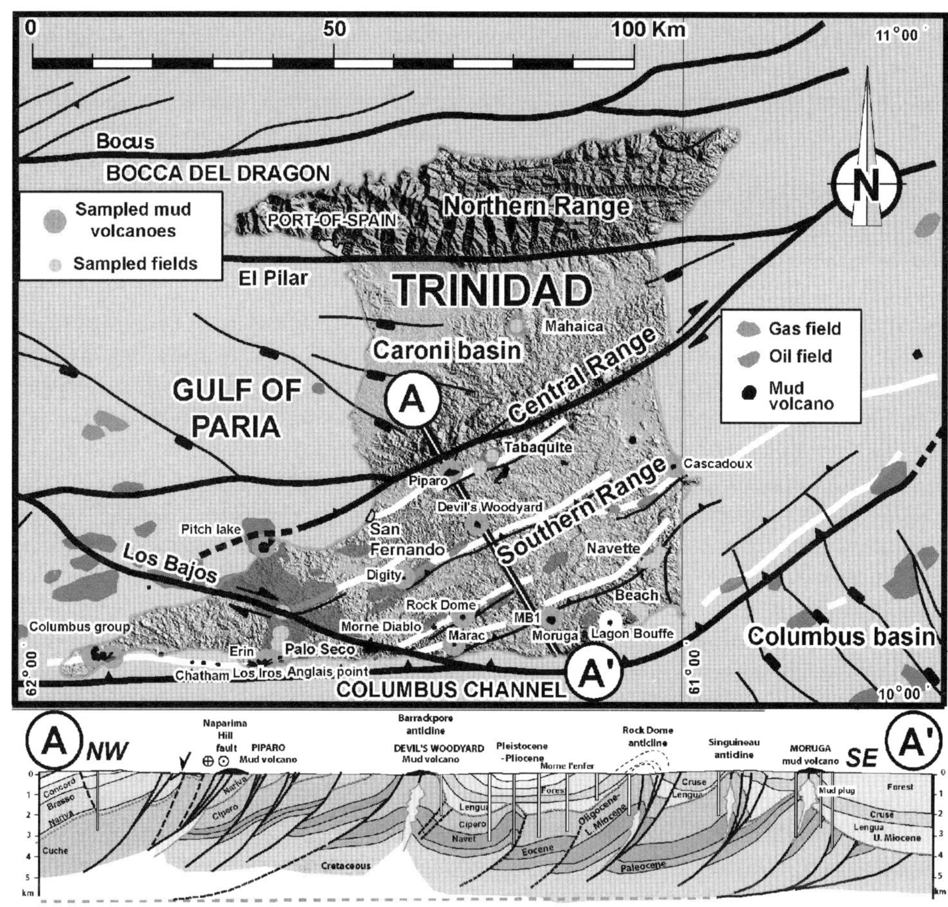

Figure 1. Geological map of Trinidad and sample location (mud volcanoes as well as main hydrocarbons fields). A simplified cross section AA′ from the southern area is also shown. Modified from Deville et al. (2003a). 10 km = 6.2 mi.

SAMPLING AND ANALYTICAL TECHNIQUES

We performed several geochemical analyses on the sampled gas, including major species concentrations, $\delta^{13}C$ of hydrocarbon molecules and CO_2, and elemental and isotopic noble gas determination. All these analyses are used together to determine both the origin of the fluids and the processes that might occur after formation.

GAS SAMPLING

In each sampled locality, a separate gas phase was always present, and we used a funnel to capture the gas emissions. The glass funnel is connected to a stainless steel sampling system. The funnel is positioned in the bubbling gas and submerged into water to avoid any atmospheric contamination during sampling. The gas was flowed through the stainless steel tube for a few minutes to make successive purges before we closed the valves sequentially.

We used two types of container depending on the analysis. For the analysis of the major gases and the carbon isotopic ratios of the different species, the gas was collected in evacuated containers, vacutainers, prepared under secondary vacuum. Noble gases, as they are very volatile, were collected in a stainless steel cylinder fitted with a valve at each end. Sample collection from the gas reservoirs was undertaken directly at the producing well. We therefore used a gas pressure regulator directly connected to the wellhead. Again, we used either vacutainers or stainless steel cylinders according to the type of analysis.

ANALYTICAL TECHNIQUES

Major Elements and Carbon Isotopic Composition

The major compound concentrations (C_1 to C_5, CO_2, N_2, H_2, O_2, He, and H_2S) were determined at the Institut Français du Pétrole (IFP) by chromatography in gas phase using a Varian chromatograph equipped with three columns and with two detectors (flame ionization detector [FID] and thermal conductivity detector [TCD]). Carbon isotopic ratios of C_1, C_2, C_3, iC_4, nC_4, and CO_2 were determined with a Fisons gas-chromatography-combustion-isotope-ratio mass spectrometry (GC-C-IRMS). Every compound was separated by chromatography and then oxidized in H_2O and CO_2 using a copper oxide furnace.

The water was eliminated by the action of a trap cooled with liquid nitrogen, where upon the CO_2 was introduced into the mass spectrometer to determine the $^{13}C/^{12}C$ isotopic ratio.

Abundances and Isotopic Ratios of the Noble Gases

The analyses of the elementary and isotopic compositions of the noble gases were made using two noncommercial mass spectrometers from the Eidgenössische Technische Hochschule Zürich (ETH) noble gas laboratory, equipped with Baur-Signer sources. One mass spectrometer is especially devoted to the measurement of the isotopic ratio of helium and has a resolution of 700. The second mass spectrometer has a separating resolution of about 200, allowing measurements of all other noble gas isotopic ratios as well as concentrations of helium, neon, argon, krypton, and xenon. More details on noble gas analysis can be found in Beyerle et al. (2000). The stainless steel cylinder was connected to an ultrahigh vacuum extraction and purification line before analysis in the mass spectrometers.

Noble gases were separated into two fractions: The noncondensable light helium and neon were expanded and underwent sequential purification on SAES getters and a titanium sponge furnace. The Ar-Kr-Xe fraction was trapped by a cold finger containing zeolite at liquid nitrogen temperature. A split of the He/Ne fraction was taken and expanded into the low-resolution mass spectrometer for the determination of ^{4}He, ^{20}Ne, and ^{22}Ne using an electron multiplier. In the remaining fraction, neon was separated from helium using a cryogenic cold trap adsorbing neon, allowing only the He to be expanded into the high-resolution mass spectrometer for the precise determination of the $^{3}He/^{4}He$ ratio.

The heavy fraction of noble gases is released by heating the zeolite to 180°C and purified using SAES getter pumps. It is then expanded into a calibrated volume, and successive aliquots are taken to allow simultaneous determination of Ar, Kr, and Xe. At the end of the analysis, Xe is separated from Kr and Ar using an activated charcoal trap held at −100°C and reanalyzed to get a precise $^{136}Xe/^{130}Xe$ ratio.

MAJOR ELEMENTS AND CARBON ISOTOPIC RATIOS

The results of analyses of the major elements and the isotopic compositions of carbon are shown in Table 1.

Major Gases: Distinction Between Three Families of Gas

Three groups of samples can be distinguished in Bernard's diagram (Bernard et al., 1976) (Figure 2). One of the groups corresponds to samples from deep reservoirs (Palo Seco, Shuttle, Beach, Marcelle Wali, and Tabaquite fields), with methane contents ranging from 60 to 94% (mean value around 80%) and relatively high contents of C^{2+} (Table 1). These samples have relatively heavy values of methane $\delta^{13}C$ (around −40‰). These characteristics indicate that these samples are thermogenic gases (Figure 2). One of the samples (triangle in Figure 2) shows evidences of biodegradation, outlined in the diagram C_1/C_2 versus iC_4/nC_4 (Figure 3). We distinguish a second group of samples (gas field of Mahaica) that are fairly homogeneous, with the higher proportions of methane (around 99 %) and light values of methane $\delta^{13}C$, which range from −67 to −63‰. These results indicate that these three samples have a bacterial origin.

The samples collected from mud volcanoes are more heterogeneous. They are characterized by a relatively dry gas (low contents in C^{2+}, see Table 1) and a methane isotopic composition ranging from −52 to −33‰. Such values are generally considered as intermediate between a purely bacterial gas and a purely thermogenic gas. However, methane $\delta^{13}C$ may also be affected by transport processes as shown by Prinzhofer and Pernaton (1997), and we will further scrutinize these $\delta^{13}C$ methane following their approach.

The diagram $\delta^{13}C(C_1)$ versus the C_1/C_2 ratio is presented in the Figure 4. Mixing is represented by a straight line. We observe that only three samples have undergone a clear bacterial contamination. It is the case for the Pitch Lake (are asphalt lake) sample for one sample from the Columbus group, and for one sample from Devil's Woodyard. In all other cases, a hypothesis of simple direct mixing between bacterial and thermogenic gas is excluded. As shown in Figure 4, if mixing with bacterial gas occurs, it can only be a late process following the segregation process caused by dissolution-diffusion. However, most of the $\delta^{13}C(C_1)$ isotopic values favor a thermogenic origin of the gas.

Gas of the Mud Volcanoes: Segregation During the Migration

No agreement between the methane $\delta^{13}C$ values and the dryness of the gas is observed if we consider that these signatures are only related to genetic process (too-light gas for this range of dryness, cf. Table 1). This dryness could, however, be explained by a postgenetic process of segregation that could have affected thermogenic hydrocarbons during the migration (Prinzhofer and Pernaton, 1997, Prinzhofer et al., 2000). We present in Figure 5 the chemical proportions (from C_1 to nC_4) of various samples from Trinidad.

Mud volcano gases have systematically lower C^{2+} concentrations than oil reservoir gases. These low concentrations (or depletion) can be related to processes of segregation during migration, such as adsorption on mineral surfaces, and/or interaction between the gas and groundwater. These processes could also explain the isotopic composition of methane slightly enriched in ^{12}C for mud volcano gases (Prinzhofer and Pernaton, 1997; Prinzhofer et al., 2000). The same conclusions regarding the segregation process of alkanes were also reached regarding Italian mud volcanoes by Etiope et al. (2007).

These results suggest a genetic relation between the gases from Trinidad deep reservoirs (except for the Mahaica field, which is bacterial in origin) and the mud volcano gas.

Genetic Relation of Mud Volcanoes and Gas from Oil Reservoirs

The C_1/CO_2 ratio (plotted logarithmically) versus the $\delta^{13}C$ of CO_2 is shown Figure 6a. All the samples from the southern part of the island (including mud volcano gases as well as oil reservoirs) follow the same trend in this diagram. This result gives evidence of a genetic relation between these gases, and both types of gas would come from the same source.

The observed trend cannot be caused by a mixing process (which would give a straight line when plotted using a linear scale; Figure 6b) but corresponds to a phenomenon of distillation. However, we observe that the C_1/CO_2 ratio decreases with increasing $\delta^{13}C(CO_2)$, whereas the opposite behavior is commonly expected during distillation (decreasing C_1/CO_2 ratio correlated with lower values of $\delta^{13}C[CO_2]$). However, we do not know the origin of the very heavy values of $\delta^{13}C(CO_2)$ (reaching almost 30‰). Indeed, none of the potential sources of CO_2 (organic matter, mantle, and carbonates) have such an isotopic signature.

RESULTS OF THE STUDY OF NOBLE GASES

The main advantage of noble gases is that they are chemically inert, and they are thus conservative. Their record in crustal fluids is preserved and provides evidence for the physical processes (partitioning between different fluids, gas phase dissolution into groundwater, diffusion, etc.) occurring in the system (Ballentine et al., 2002). Moreover, noble gases originate from three main sources (atmosphere, crust, and mantle), and the isotopic composition of each source is well characterized. The isotopic composition of helium and to a lesser extent neon provides a strong signal of any mantle-derived contribution to crustal fluids.

Isotopic Ratios of the Noble Gases and Origin of the Fluids

The results (concentrations and isotopic ratios) of the analyses of noble gases are presented in Tables 2 and 3. All data are given to ±1 standard deviation (σ). Isotopic ratios $^3He/^4He$, $^{40}Ar/^{36}Ar$, $^{21}Ne/^{22}Ne$, and $^{21}Ne/^{22}Ne$ allow us to determine the origin of the various fluids (i.e., upper mantle, crust, and atmosphere). The isotopic compositions of helium and neon show the absence of any mantle-derived contribution in this context. This is in agreement with the conclusions obtained during numerous studies of noble gases in other crustal fluids, which show that noble gases from the mantle are present in tectonic contexts of extension, and/or recent magmatism, but rarely in the context of compressive tectonics (Oxburgh et al., 1986; Elliot et al., 1993; Marty et al., 1992; Ballentine and O'Nions, 1994).

In the absence of mantle-derived noble gases, and because they are not produced significantly by nuclear processes, ^{20}Ne, ^{36}Ar, and ^{84}Kr are totally inherited from atmosphere-derived sources (Ballentine et al, 2002). As the thermal peak of maturation of hydrocarbons in Trinidad is recent (probably during the thick Pliocene sedimentation), we also consider that almost all ^{136}Xe is atmosphere derived. The argon isotopic ratio indicates the presence of radiogenic argon-40 (noted $^{40}Ar^*$) when the ratio $^{40}Ar/^{36}Ar$ exceeds the atmospheric ratio of 295.5. As the helium-4 concentration is very low in the atmosphere and the isotopic ratio indicates that no mantle-derived contribution for this gas exists, we conclude that almost all ^4He in the samples is of crustal origin, inherited by radioactive decay of U and Th.

From these results, all our samples can be considered as a mixing between two components, the atmosphere and the crust.

The concentrations in atmosphere-derived noble gases (^{20}Ne, ^{36}Ar, ^{84}Kr, and ^{136}Xe) are higher in the samples of mud volcano gas than in oil reservoir gas samples. This gives evidence of the important interactions between the mud volcano gas and the groundwater, probably during migration. This conclusion is in agreement with the process of gas-phase dissolution into groundwater suggested by the $\delta^{13}C$ of methane measured on the mud volcano samples.

Correlations Between Elementary Ratios of Noble Gases

We present in Figure 7 the $^{84}Kr/^{20}Ne$ ratio versus the $^{136}Xe/^{20}Ne$ ratio. We observe that all the mud volcano gases and the oil reservoir gases align along a straight line. As both axes in this diagram have the same denominator, one explanation would be that the line

Table 1. Gas analyses.*

Well	Hydrocarbon Field	δC_1	δC_2	δC_3	δiC_4	δnC_4	δCO_2	C_1	C_2	C_3	iC_4	nC_4	CO_2	N_2
Reservoirs														
PS1416	Palo Seco	−41.70	−28.88	−26.70	−27.67	−26.09	13.08	81.75	10.24	4.68	0.79	1.23	1.26	0.06
PS1418	Palo Seco	−46.03	−29.88	−28.42	−28.56	−25.88	1.17	81.10	11.87	4.97	0.63	0.98	0.45	0.00
PS1440	Palo Seco	−44.57	−29.17	−27.52	−28.21	−26.22	2.37	78.34	12.88	6.15	0.79	1.21	0.63	0.00
PS11379	Palo Seco	−38.92	−27.89	−25.62	−26.24	−24.94	5.84	89.16	6.21	2.29	0.37	0.48	1.49	0.00
PS1372b	Palo Seco	−40.95	−27.58	−26.57	−26.99	−25.14	9.69	86.86	7.46	3.15	0.40	0.88	1.27	0.00
PS1396	Palo Seco	−39.53	−27.74	−26.70	−27.28	−24.66	−5.72	88.73	7.29	2.81	0.34	0.76	0.08	0.00
PS740	Palo Seco	−47.74	−28.33	−12.71	−27.82	−19.70	11.02	90.98	5.82	0.09	0.43	0.03	2.66	0.00
PS648	Palo Seco	−48.11	−28.68	−26.47	−27.91	−24.75	−13.10	81.63	11.33	4.75	0.87	1.08	0.34	0.00
307/308	Navette	−44.79	−27.35	−25.86	−27.97	−25.44	9.02	76.77	5.87	7.16	1.12	4.08	5.00	0.00
410/556	Navette	−44.30	−28.00	−26.90	−28.60	−26.20	1.40	71.70	12.90	8.50	1.40	2.40	2.30	0.00
410/517	Navette 517	−35.97	−27.18	−26.05	−27.13	−25.95	−25.05	94.09	4.26	1.26	0.14	0.17	0.08	0.00
528/535	Navette	−41.11	−27.36	−26.62	−27.85	−26.36	0.60	73.96	12.20	7.92	1.70	2.77	1.50	0.00
528/525	Navette 525	−39.59	−26.88	−26.14	−25.87	−25.17	−15.13	92.87	4.72	1.61	0.26	0.38	0.17	0.00
528/656	Navette	−41.23	−27.33	−26.24	−28.06	−25.81	−4.98	70.79	13.43	9.01	1.90	3.12	1.73	0.00
85.00	Beach	−38.54	−28.49	−27.38	−27.81	−26.70	13.59	74.96	7.36	5.60	1.16	2.55	7.71	0.66
528/552	Navette	−44.64	−28.11	−26.38	−27.43	−26.56	12.50	60.49	14.45	11.77	2.88	5.06	5.34	0.00
BF629	Marcelle Wali	−39.92	−23.66	−21.15	−25.86	−23.31	22.16	84.77	4.34	1.71	0.68	0.72	7.79	0.21
BF335	Beach	−39.02	−27.07	−26.81	−28.44	−26.88	−4.00	73.95	10.21	8.58	1.99	3.59	1.56	0.12
BF342	Beach	−37.50	−27.50	−27.22	−26.24	−26.07	11.40	81.96	7.68	4.65	0.87	1.64	3.21	0.00
TAB238	Tabaquite Piparo	−43.40	−30.92	−29.45	−30.35	−28.64	21.23	75.80	8.02	6.04	1.52	2.50	6.14	0.00
JR1	Tabaquite Johnston Road	−37.99	−27.60	−27.15	−28.94	−27.50	−13.10	92.89	4.53	1.61	0.32	0.39	0.26	0.00
TA228	Tabaquite Ramdass Trace	−45.03	−29.31	−28.58	−31.53	−27.29	14.73	73.18	5.42	2.31	0.80	0.71	17.58	0.00
TA225	Old Tabaquite	−44.28	−30.31	−28.55	−28.78	−28.62	8.69	80.53	5.59	4.49	1.05	2.13	6.23	0.00
Puits X	Moruga field	−35.95	−27.82	−26.54	−25.83	−25.85	−0.39	91.42	5.06	1.73	0.24	0.38	0.87	0.32
MA1	Mahaica	−64.50	−46.23	−35.28		−30.03	−17.05	99.56	0.13	0.03	0.00	0.04	0.13	0.11
MA2	Mahaica	−63.90	−45.10	−36.90		−30.29	−22.51	98.82	0.13	0.04	0.00	0.04	0.22	0.75
MA3	Mahaica	−67.13	−46.84	−33.81		−30.07	−20.67	98.60	0.10	0.04	0.00	0.04	0.20	0.02
Mud Volcanoes														
	Moruga	−33.96					19.53	87.21	0.06	0.05	0.00	0.00	10.57	2.12
	Palo Seco	−45.70					4.60	99.50	0.00	0.00	0.00	0.00	0.40	0.10
	Palo Seco	−45.10					−11.00	98.37	0.00	0.00	0.00	0.00	0.53	1.11
	Los Iros	−42.27					21.35	93.46	0.06	0.00	0.00	0.00	6.49	0.00
	Lagon Bouffe	−43.47					8.97	89.30	0.07	0.03	0.00	0.00	6.15	4.72
	Pitch Lake	−55.57	−34.65	−32.48	−31.34	−28.64	14.54	71.63	1.49	0.94	0.11	0.07	25.55	0.23
	Pitch 2	−52.58	−32.78	−21.27			14.84	78.07	0.51	0.06	0.00	0.00	20.82	0.54
	Digity	−50.00					11.81	93.19	0.03	0.00	0.00	0.00	6.78	0.00
	Piparo	−47.72	−28.55				10.78	92.44	0.45	0.02	0.03	0.01	6.63	0.42

Devil's Woodyard	−54.98	−27.54	4.03	98.24	0.77	0.00	0.00	0.00	1.00	0.30
MB1	−44.66		19.37	82.64	0.02	0.00	0.00	0.00	16.99	0.36
Rock Dome	−49.26		15.29	89.30	0.00	0.00	0.00	0.00	10.70	0.00
Columbus	−47.94		16.39	94.55	0.03	0.00	0.00	0.00	5.20	0.23
Columbus	−54.65	−32.39	−7.98	96.54	0.13	0.00	0.00	0.00	3.20	0.14
Cascadoux	−44.25		19.42	96.32	0.00	0.00	0.00	0.00	2.57	1.11
Cascadoux	−42.18		18.16	95.47	0.00	0.00	0.00	0.00	2.60	1.92

*The δ values are reported in parts per thousand Peedee Belemnite (‰ PDB) and concentrations are in percent (%).

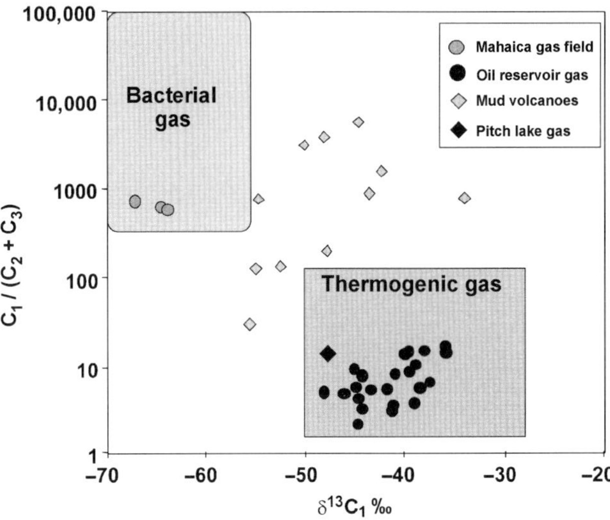

Figure 2. Bernard's diagram (Bernard et al., 1976), $\delta^{13}C_1$ versus $C_1/(C_2 + C_3)$ ratio. The *y* axis is logarithmic. We distinguish three groups of samples. The samples of reservoirs that are of thermogenic origin are in black. Three bacterial gas samples, also from deep reservoirs, taken in the Mahaica field are in gray circles. Almost all mud volcano samples lie in the intermediate range of values in this figure.

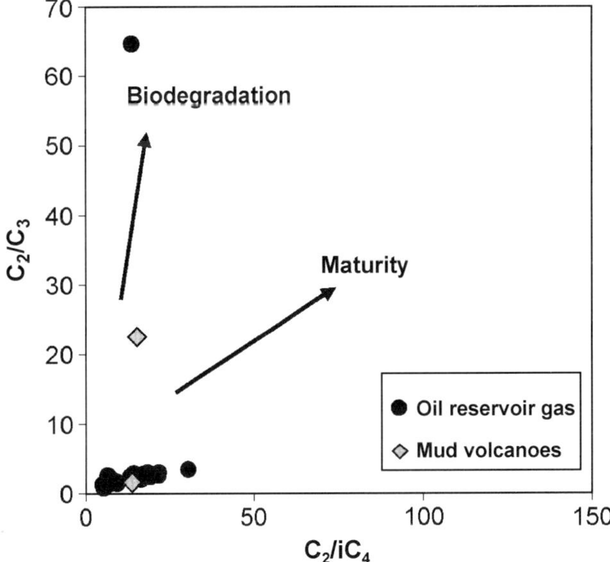

Figure 3. The C_2/iC_4 versus C_2/C_3 diagram. During biodegradation, C_2/C_3 increases much more quickly than C_2/iC_4, the biodegradation preferentially affecting the n-alkanes. One sample from the deep reservoirs shows evidence of biodegradation, whereas the other samples define a trend of maturity.

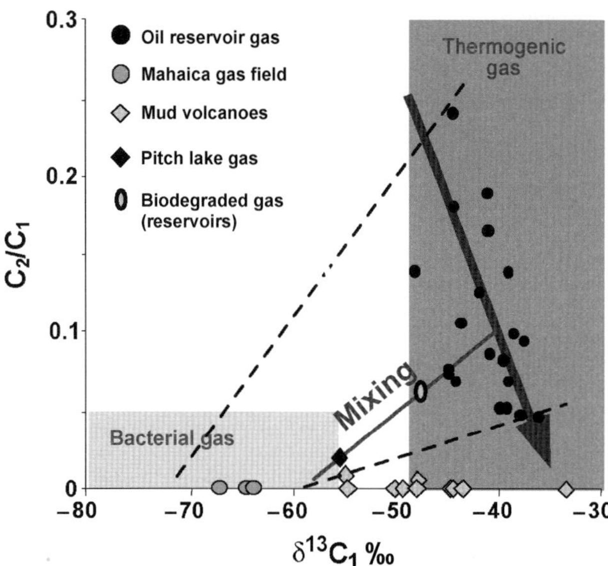

Figure 4. The C_2/C_1 versus $\delta^{13}C_1$diagram. In such a diagram, mixing is represented by a straight line. Only three mud volcano gases can have undergone a simple bacterial contamination, as shown by the mixing lines in this figure. In all other cases, a simple direct process of mixing between bacterial gas and thermogenic gas cannot explain the data because $\delta^{13}C(C_1)$ of the bacterial end member would be too heavy. If a mixing with bacterial gas is involved, it only occurs after the segregation process shown by the arrow.

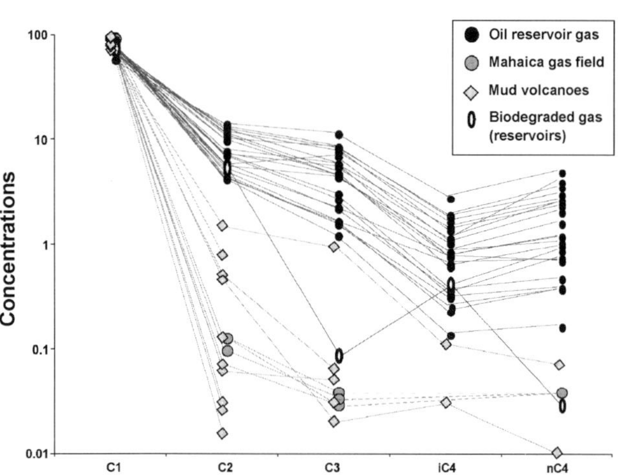

Figure 5. Chemical composition of the various analyzed samples. We distinguish clearly the various families of gas. Gases of mud volcanoes are poor in C^{2+} components, whereas we were able to attribute to them a thermogenic origin. A postgenetic process having affected the gas during the migration is required to explain these results. We also notice the existence of a biodegraded sample, which shows depletion in C_3 and in nC_4, compared to C_2 and iC_4. The last group of samples (Mahaica field) is characterized by a dry gas of bacterial origin (Figure 4).

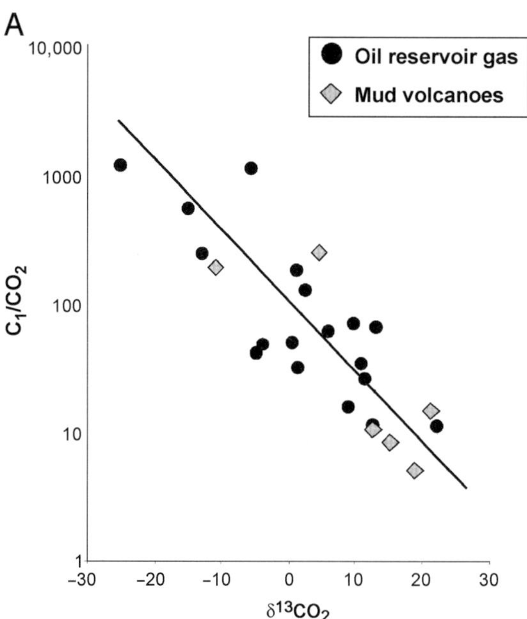

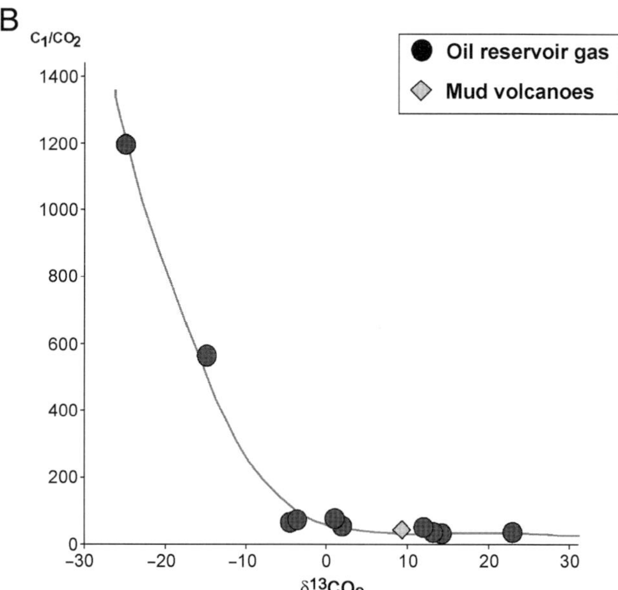

Figure 6. (A) Log (C_1/CO_2) versus $\delta^{13}C(CO_2)$ diagram. There is a correlation between these two geochemical parameters, showing a genetic relationship (the same source) between gas from the reservoirs and gas from the mud volcanoes. The very high values of δ^{13} (CO_2) (approaching 30‰) remain unexplained (see the text). We distinguish two trends; the one corresponding to samples (reservoirs and mud volcanoes) situated in the east of Trinidad and the other one to samples localized in the west of Trinidad. This distinction was also observed on the chemical composition of samples of water (Dia et al., 1999). (B) The C_1/CO_2 ratio versus $\delta^{13}C(CO_2)$. For the legibility of the figure, only the values of samples located in the east were presented in this diagram. We clearly see that, in linear scale, the trend corresponds to a curve. This observation indicates that the correlation cannot correspond to a mixing process.

Table 2. Noble gas isotopic ratios given with associated mean standard deviation 1σ.*

Samples	R/Ra	1 σ	$^{20}Ne/^{22}Ne$	1 σ	$^{21}Ne/^{22}Ne$	1 σ	$^{40}Ar/^{36}Ar$	1 s
MA1	0.0232	0.0008	9.82	0.04	0.0296	0.003	334.45	4.36
MA2	0.0185	0.0033	9.70	0.06	0.0297	0.003	380.65	4.97
MA3	0.0225	0.0012	9.64	0.06	0.0294	0.006	334.27	4.96
JR1	0.0311	0.0013	9.91	0.07	0.0298	0.007	575.74	24.02
Navette 525	0.0282	0.0011	9.84	0.05	0.0302	0.003	623.56	6.15
Navette 525b	0.0280	0.0009	9.89	0.03	0.0305	0.002	611.70	12.64
Navette 517	0.0391	0.0014	9.83	0.04	0.0308	0.003	667.24	11.05
Digity	0.0665	0.0010	9.79	0.03	0.0291	0.002	364.62	4.58
MB1	0.0379	0.0007	9.77	0.03	0.0286	0.001	308.86	1.12
Devil's Woodyard	0.0636	0.0009	9.78	0.03	0.0285	0.001	335.86	2.16
Lagon Bouffe	0.0201	0.0007	9.85	0.03	0.0295	0.001	374.07	1.75
Los Iros	0.0465	0.0009	9.83	0.07	0.0295	0.003	430.07	11.55
Columbus	0.0528	0.0009	9.79	0.03	0.0290	0.001	313.20	1.14
Pitch Lake	0.1668	0.0031	9.80	0.03	0.0290	0.001	294.01	1.35

*The He isotopic ratio is given as R/Ra (Ra is the $^3He/^4He$ atmospheric ratio of 1.4×10^{-6}).

illustrates a two-component mixing between air and atmosphere-equilibrated water. In Figure 8, which shows the $^{84}Kr/^{36}Ar$ ratio versus the $^{136}Xe/^{36}Ar$ ratio, we also observe a linear trend. We notice nevertheless that some samples from the mud volcanoes, as well as all reservoir gases, although on the same trajectory as the mixing line, are beyond the air and atmosphere-equilibrated water end-member values. Figure 9 therefore shows that simple atmosphere-equilibrated water cannot, here, constitute one of the mixing end members.

One explanation might be that the sedimentary source of the hydrocarbons is enriched in sorbed atmosphere-derived Kr and Xe (Torgersen and Kennedy, 1999; Zhou et al., 2005) released into the hydrocarbon phase during primary expulsion and diluted by the normal groundwater-derived signature with increasing water contact (Torgersen and Kennedy, 1999). We propose another explanation involving a two-step process. (1) The water is enriched in atmosphere-derived noble gases because of an excess of dissolved air (Heaton and Vogel, 1981). This phenomenon is commonly, if not systematically, noticed with natural waters. Indeed, in the case of an addition of air, $^{84}Kr/^{20}Ne$ and $^{136}Xe/^{20}Ne$ ratios decrease much more, quantitatively, than $^{136}Xe/^{36}Ar$ and $^{84}Kr/^{36}Ar$ ratios because ^{20}Ne is very sensitive to any addition of air. (2) From this new pole of mixing calculated and represented on Figures 10 and 11, we consider a process of residual distillation inducing a kinetic fractionation, caused by mass difference between the various isotopes. We observe that this model explains the representative samples of reservoir gases, which show an enrichment in heavy isotopes indicating evidence of the residual character of these gases.

The mud volcano gases are enriched in light isotopes consistent with a diffusion process having also affected them during their ascent, the diffused products being enriched in light isotopes, because of their higher mobility. This hypothesis is in good agreement with the isotopic composition of methane, which also suggests the existence of a process of diffusion. We notice, however, in Figures 10 and 11, that mud volcano gases are also interpretable in terms of mixing between gases from the reservoirs and the air, probably because of the existence of an important hydrodynamism within the sedimentary layers. Evidence for the residual character of the gases from the deep reservoirs is shown in Figure 11 by plotting the $^{84}Kr/^{20}Ne$ ratio versus the $^{36}Ar/^{20}Ne$ ratio. We notice in this diagram that all the gases from the hydrocarbon reservoirs exhibit values systematically enriched in heavy isotopes compared to the line of mixing between the air and the atmosphere-derived water where the mud volcano samples plot. In the same figure, we also plotted a line showing the maximum proportion of air that has to be added to atmosphere-equilibrated water to reproduce the composition of the most extreme mud volcano gas (sample Los Iros, composition closest to that of air) from the composition of a reservoir gas (sample JR1, composition farthest from that of air).

Radiogenic Noble Gases and Residence Time of the Gas

The use of radiogenic noble gases provides indication about the respective times of residence of both populations of samples. Figure 11 shows that a 10-ppm addition of air can explain a transition from the composition

Table 3. Noble gas concentrations.*

Samples	^{4}He	1σ	^{20}Ne	1σ	^{40}Ar	1σ	^{84}Kr	1σ	^{136}Xe	1σ
MA1	3.74×10^{-5}	3.34×10^{-7}	2.61×10^{-8}	2.27×10^{-10}	2.88×10^{-5}	5.96×10^{-7}	3.63×10^{-9}	8.57×10^{-11}	1.18×10^{-10}	2.86×10^{-12}
MA2	4.27×10^{-5}	3.81×10^{-7}	2.12×10^{-7}	1.84×10^{-10}	2.51×10^{-5}	5.4×10^{-7}	2.67×10^{-9}	7.14×10^{-11}	1.21×10^{-10}	2.85×10^{-12}
MA3	2.69×10^{-5}	2.40×10^{-7}	9.40×10^{-9}	8.20×10^{-11}	1.41×10^{-5}	3.40×10^{-7}	1.86×10^{-9}	5.89×10^{-11}	4.74×10^{-11}	1.59×10^{-12}
JR1	1.87×10^{-5}	1.67×10^{-7}	6.24×10^{-9}	5.44×10^{-11}	6.41×10^{-5}	6.13×10^{-8}	8.72×10^{-9}	2.96×10^{-11}	2.80×10^{-11}	9.96×10^{-13}
Navette 525	1.22×10^{-5}	1.09×10^{-7}	6.08×10^{-9}	5.29×10^{-11}	1.14×10^{-5}	2.10×10^{-7}	1.30×10^{-9}	3.68×10^{-11}	3.04×10^{-11}	7.96×10^{-13}
Navette 525b	1.21×10^{-5}	1.08×10^{-7}	6.17×10^{-9}	5.36×10^{-11}	1.12×10^{-5}	3.41×10^{-7}	1.18×10^{-9}	5.19×10^{-11}	2.80×10^{-11}	7.96×10^{-13}
Navette 517	1.22×10^{-5}	1.09×10^{-7}	6.49×10^{-9}	5.64×10^{-11}	1.63×10^{-5}	4.30×10^{-7}	1.54×10^{-9}	6.43×10^{-11}	3.83×10^{-11}	1.31×10^{-12}
Digity	1.42×10^{-5}	1.27×10^{-7}	3.95×10^{-8}	3.44×10^{-10}	3.79×10^{-5}	7.84×10^{-7}	4.00×10^{-9}	1.27×10^{-10}	8.97×10^{-11}	5.21×10^{-12}
MB1	8.28×10^{-5}	7.39×10^{-8}	7.85×10^{-8}	6.82×10^{-10}	8.90×10^{-5}	9.62×10^{-7}	1.00×10^{-8}	1.79×10^{-10}	1.92×10^{-10}	4.34×10^{-12}
DW	2.28×10^{-5}	2.04×10^{-7}	7.59×10^{-7}	6.59×10^{-10}	6.93×10^{-5}	9.24×10^{-7}	7.88×10^{-9}	1.76×10^{-10}	1.92×10^{-10}	4.32×10^{-12}
Lagon Trinity	3.93×10^{-5}	3.51×10^{-7}	7.45×10^{-8}	6.47×10^{-10}	4.81×10^{-5}	5.49×10^{-7}	3.74×10^{-9}	1.10×10^{-10}	6.67×10^{-11}	1.71×10^{-12}
Los Iros	6.44×10^{-5}	5.75×10^{-8}	1.35×10^{-8}	1.17×10^{-10}	1.35×10^{-5}	1.26×10^{-7}	1.48×10^{-9}	4.14×10^{-11}	3.05×10^{-11}	1.03×10^{-12}
Columbus	7.06×10^{-5}	6.31×10^{-8}	5.96×10^{-8}	5.17×10^{-10}	6.45×10^{-5}	6.91×10^{-7}	7.89×10^{-9}	1.45×10^{-10}	1.96×10^{-10}	4.32×10^{-12}
Pitch Lake	1.30×10^{-5}	1.16×10^{-8}	1.93×10^{-7}	1.68×10^{-9}	1.85×10^{-4}	2.14×10^{-6}	2.15×10^{-8}	3.80×10^{-10}	4.54×10^{-10}	9.61×10^{-12}

*Noble gas concentrations are in cubic centimeter per cubic centimeter (cm^3/cm^3), with associated mean standard deviation (1σ).

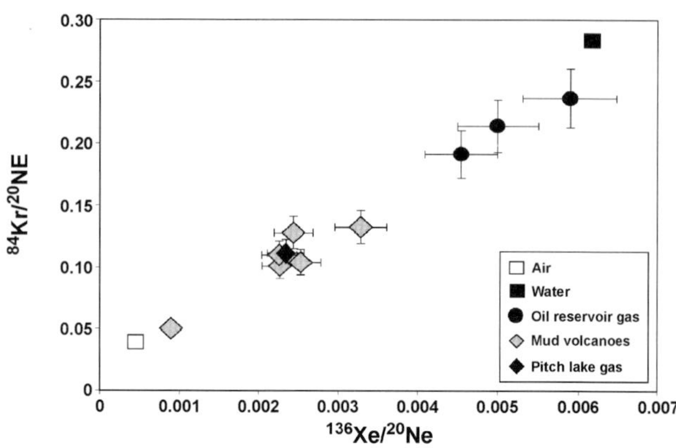

Figure 7. The $^{84}Kr/^{20}Ne$ versus $^{136}Xe/^{20}Ne$ diagram. The origin of these three isotopes results from the contact between the gas and the water that initially contained them because of equilibration with the atmosphere on recharge. In this type of diagram, the denominator is the same on both axes, and mixing is then represented by a straight line. All the experimental points are localized on a line defined between air and water.

of sample JR1 to the composition of the sample of the Los Iros mud volcano. In the same way, Figure 12 shows the analytical points in the diagram $^{40}Ar^{*}/^{20}Ne$ versus $^{4}He/^{20}Ne$, and we observe that a simple air contamination in the proportions calculated previously is not enough to explain the differences of $^{40}Ar^{*}/^{20}Ne$ and $^{4}He/^{20}Ne$ ratios between oil reservoir gas and mud volcano gas. This essential result shows that both populations of gases do not have the same time of residence, and those from mud volcanoes are lower than those from the reservoirs, which are richer in radiogenic isotopes.

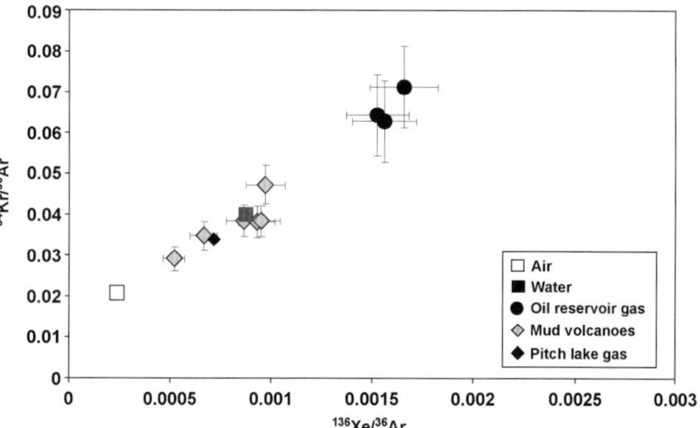

Figure 8. The $^{84}Kr/^{36}Ar$ versus $^{136}Xe/^{36}Ar$ diagram. In this figure, all the analytical points are located along a straight line. The positions of the samples of gas from the deep reservoirs in this diagram, as well as some from the mud volcanoes, clearly show that the water in equilibrium with the atmosphere cannot be considered here as one of the mixing end members.

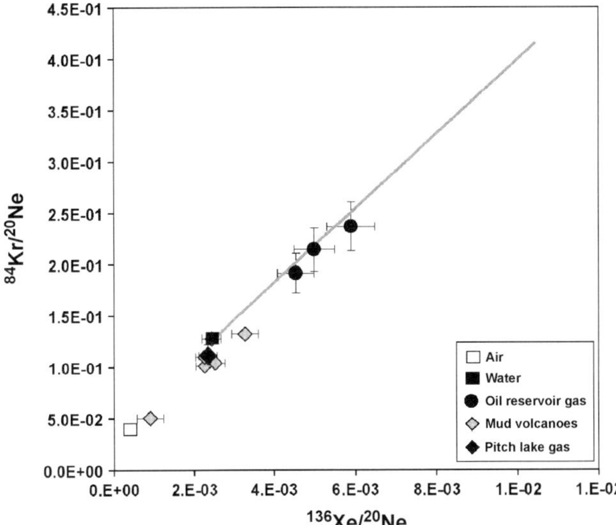

Figure 9. The $^{84}Kr/^{20}Ne$ diagram according to $^{136}Xe/^{20}Ne$, in which we represented a water containing excess air (Heaton and Vogel, 1981) and the model trajectory resulting from residual distillation of the gas. The proposed model explains the compositions of gases from the deep reservoirs, which are systematically enriched in heavy isotopes.

This result shows that although being cogenetic, mud volcano gases cannot result from a simple process of leakage from the reservoirs.

CONCLUSIONS

Most of the gases from mud volcanoes and reservoirs (except the bacterial field of Mahaica) are of thermogenic

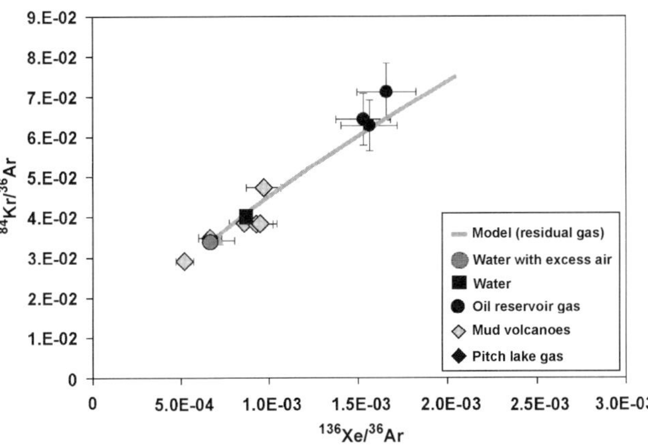

Figure 10. The $^{84}Kr/^{36}Ar$ versus $^{136}Xe/^{36}Ar$ diagram. We observe, as in the previous figure, that all the gases from the deep reservoirs present a residual character with an enrichment in heavy isotopes.

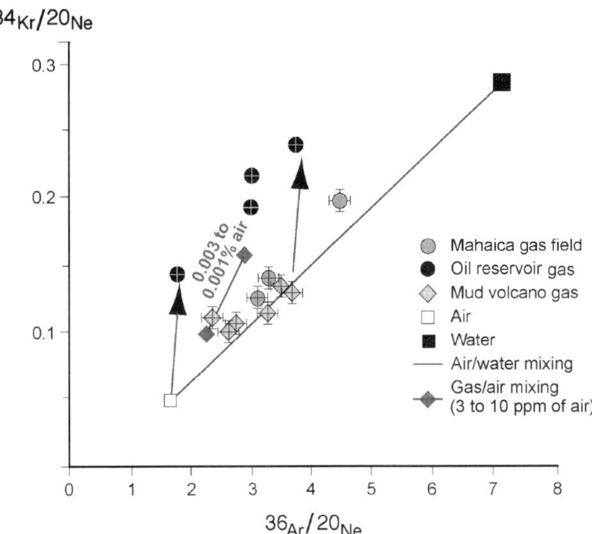

Figure 11. The $^{36}Ar/^{20}Ne$ versus $^{84}Kr/^{20}Ne$ diagram. Arrows represent the evolution of the ratios $^{36}Ar/^{20}Ne$ and $^{84}Kr/^{20}Ne$ during a process of distillation of a gas initially situated on the line of air-water mixing. This diagram provides clear evidence for the residual character of gases of reservoirs, compared to gases of the mud volcanoes. We also calculated, from an equation of mixing, the maximum addition of air from the reservoired gas sample to the composition of mud volcano gas. This contamination consists of 10 ppm of air at maximum.

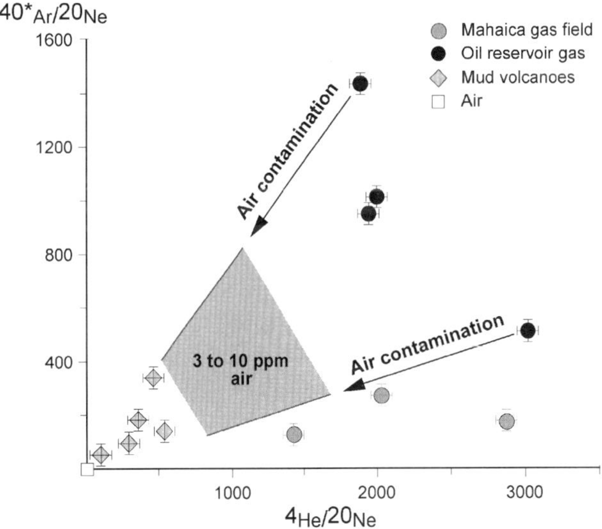

Figure 12. The $^{40}Ar^{*}/^{20}Ne$ versus $^{4}He/^{20}Ne$ diagram. We represented a process of contamination of reservoired gases by air in the same proportions as those calculated and represented in Figure 11 (10 ppm at maximum). We notice that a single phenomenon of contamination of the samples by air cannot explain the differences of these elementary ratios between the gas of reservoirs and the gas of mud volcanoes. This shows that these two populations of samples do not have the same time of residence.

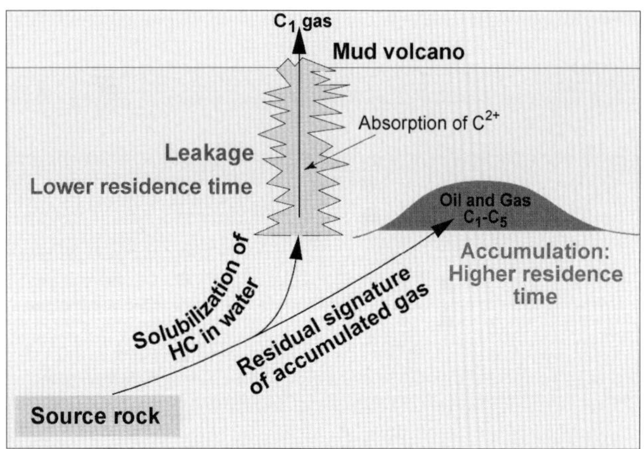

Figure 13. Summary model showing the various stages and the phenomena postgenetic process that affected hydrocarbons (HC). All the HC (except the Mahaica field, see the text) are produced from the same parent rock. A part of the HC, after dissolution and diffusion in water, takes a different pathway of migration toward mud volcanoes, whereas the oil and the residual gas move toward reservoirs for accumulation. Having undergone a process of surface adsorption, the mud volcano gas is almost essentially constituted by methane and expelled regularly in the atmosphere (low time of residence).

origin and are likely produced by the same deep source rock of Cretaceous age. The isotopic and chemical compositions of the major elements, as well as noble gas compositions, reveal, however, the existence of transport processes (adsorption and dissolution or diffusion) that affected locally the mud volcano gas during its ascent. The study of noble gases showed that reservoir gases are enriched in heavy isotopes, whereas mud volcano gases are enriched in light isotopes consistent with diffusive fractionation during migration. The compositions of noble gases indicate that a simple process of leakage of the gas accumulated in reservoirs toward mud volcanoes cannot explain the differences of compositions in radiogenic isotopes. In that case, the residence times of mud volcano gases should be larger than those of reservoir gases, but the contrary is observed (Figure 13). Unlike the oil reservoir gas, the mud volcano gas has never been trapped, but it is expelled in a regular way in the atmosphere and thus has a low time of residence compared to oil reservoir gas. The relatively high flows of gas recorded on the mud volcanoes are issued directly from deep source rocks (the kitchen) and are not issued from the leakage of trapped hydrocarbons. The hydrocarbons fluids expelled from the mud volcanoes thus can be regarded as reflecting the present state of the hydrocarbons produced at depth, but they are not necessary directly representative of the hydrocarbons trapped at depth.

ACKNOWLEDGMENTS

The authors wish to thank Prasanta K. Mukhopadhyay and two anonymous reviewers for critically reading the manuscript and making several useful remarks that improved this article.

We thank J.-L. Faure, X. Guichet, and Ph. Sarda for suggestions and stimulating discussions during the course of this work.

REFERENCES CITED

Ahmad, R., 1991, Structural styles in Trinidad: Field-trip guide, *in* K. A. Gillezeau, ed., Transactions of the 2nd Geological Conference of the Geological Society of Trinidad and Tobago: Geological Society of Trinidad And Tobago, p. 244–265.

Ballentine, C. J., and R. K. O'Nions, 1994, The use of natural He, Ne and Ar isotopes to study hydrocarbon-related fluid provenance, migration and mass balance in sedimentary basins: Geofluids, v. 78, p. 347–361.

Ballentine, C. J., R. Burgess, and B. Marty, 2002, Tracing fluid origin, transport, and interaction in the crust, *in* D. R. Porcelli, C. J. Ballentine, and R. Weiler, eds., Noble gases in geochemistry and cosmochemistry: Reviews in mineralogy and geochemistry: Mineralogical Society of America, v. 47, p. 539–614.

Barboza, S. A., and S. S. Boettcher, 2000, Major and trace element constraints on fluid origin, offshore eastern Trinidad: Geological Society of Trinidad and Tobago and the Society of Petroleum Engineers Special Publication TG03, 11 p.

Barr, K. W., 1953, The mud volcanoes of Trinidad: Caribbean Quarterly, v. 3, no. 2, p. 80–85.

Battani, A., A. Prinzhofer, C. J. Ballentine, and E. Deville, 2001, Identifying the relationship between mud volcanoes, groundwater flow and hydrocarbon reservoirs in Trinidad: A combined noble gas and geochemistry case study: Eos, v. 81, no. 48, p. 442.

Bernard, B. B., J. M. Books, and W. M. Sackett, 1976, Natural gas seepage in the Gulf of Mexico: Earth and Planetary Science Letters, v. 31, p. 48–54, doi:10.1016/0012-821X(76)90095-9.

Beyerle, U., W. Aeschbach-Hertig, D. M. Imboden, H. Baur, T. Graf, and R. Kipfer, 2000, A mass spectrometric system for the analysis of noble gases and tritium from water samples: Environmental Science and Technology, v. 34, no. 10, p. 2042–2050, doi:10.1021/es990840h.

Biju-Duval, B., P. Le Quellec, A. Mascle, V. Renard, and P. Valery, 1982, Multi-beam bathymetric survey and high resolution seismic investigations on the Barbados ridge complex (eastern Caribbean): Tectonophysics, v. 80, p. 275–304, doi:10.1016/0040-1951(82)90070-1.

Birchwood, K. M., 1965, Mud volcanoes in Trinidad: Institute of Petroleum Review, v. 19, p. 164–167.

Brown, K. M., 1990, The nature and hydrologic significance of mud diapirs and diatremes for accretionary systems:

Journal of Geophysical Research, v. 95, no. B6, p. 8969–8982, doi:10.1029/JB095iB06p08969.

Brown, K. M., and G. K. Westbrook, 1987, The tectonic fabric of the Barbados ridge accretionary complex: Marine and Petroleum Geology, v. 4, p. 71–81, doi:10.1016/0264-8172(87)90022-5.

De Mets, C., R. G. Gordo, A. F. Argus, and S. Stein, 1990, Current plate motions: Journal of Geophysical International, v. 101, p. 425–478.

Deville, E., A. Battani, R. Griboulard, S. H. Guerlais, J. P. Herbin, J. P. Houzay, C. Muller, and A. Prinzhofer, 2003a, Mud volcanism origin and processes: New insights from Trinidad and the Barbados prism, *in* P. Van Rensbergen, R. R. Hillis, A. J. Maltman, and C. Morley, eds., Subsurface sediment mobilization: Geological Society (London) Special Publication 216, p. 475–490.

Deville, E., A. Mascle, S.-H. Guerlais, C. Decalf, and B. Colletta, 2003b, Lateral changes of frontal accretion and mud volcanism processes in the Barbados accretionary prism and some implications, *in* C. Bartolini, T. Buffler, and J. F. Blickwede, eds., Mexico and the Caribbean Region: Plate tectonics, basin formation and hydrocarbon habitats: AAPG Memoir 79, chapter 30, p. 1–19.

Deville, E., S. H. Guerlais, Y. Callec, R. Griboulard, P. Huygue, S. Lallemant, A. Mascle, M. Noble, J. Schmitz, and the collaboration of the Caramba working group, 2006, Liquefied vs stratified sediment mobilization processes: Insight from the South of the Barbados accretionary prism: Tectonophysics, v. 428, p. 33–47, doi:10.1016/j.tecto.2006.08.011.

Dia, A. N., M. Castrec-Rouelle, J. Boulegue, and P. Comeau, 1999, Trinidad mud volcanoes: Where do the expelled fluids come from?: Geochimica Cosmochimica Acta, v. 63, p. 1023–1038, doi:10.1016/S0016-7037(98)00309-3.

Dimitrov, L. I., 2002, Mud volcanoes: The most important pathway for degassing deeply buried sediments: Earth Science Reviews, v. 59, no. 1–4, p. 49–76, doi:10.1016/S0012-8252(02)00069-7.

Elliot, T., C. J. Ballentine, R. K. O'Nions, and T. Ricchiuto, 1993, Carbon, helium, neon, and argon isotopes in a Po Basin natural gas field: Chemical Geology, v. 106, p. 429–440, doi:10.1016/0009-2541(93)90042-H.

Etiope, G., A. Caracausi, R. Favara, F. Italiano, and C. Baciu, 2002, Methane emission form the mud volcanoes of Sicily (Italy): Geophysical Research Letters, v. 29, no. 8, p. 2002.

Etiope, G., G. Martinelli, A. Caracuasi, and F. Italiano, 2007, Methane seeps and mud volcanoes in Italy: Gas origin, fractionation and emission to the atmosphere: Geophysical Research Letter, v. 34, no. 14, L14303.1–L14303.5, doi:10.1029/2007GL030341.

Graue, K., 2000, Mud volcanoes in deep water Nigeria: Marine and Petroleum Geology, v. 17, no. 8, p. 959–974, doi:10.1016/S0264-8172(00)00016-7.

Guliyiev, I. S., and A. A. Feizullayev, 1998, All about mud volcanoes: Baku, Azerbaidjan, Nafta Press, Azerbaidjan Publishing House, 52 p.

Heaton, T. H. E., and J. C. Vogel, 1981, Excess air in groundwater: Journal of Hydrology, v. 50, p. 201–208, doi:10.1016/0022-1694(81)90070-6.

Heppard, P. D., H. S. Cander, and E. B. Eggertson, 1998, Abnormal pressure and the occurrence of hydrocarbons in offshore eastern Trinidad, West Indies, *in* B. E. Law, G. F. Ulmishek, and V. I. Slavin, eds., Abnormal pressures in hydrocarbon environments: AAPG Memoir 70, p. 215–246.

Higgins, G. E., 1996, A history of Trinidad oil: Port-of-Spain, Trinidad, Trinidad Express Newspapers Limited, 498 p.

Higgins, G. E., and J. B. Saunders, 1974, Mud volcanoes, their nature and origin: Verhandlungen Naturforschenden Gesselschaft in Basel, v. 84, p. 101–152.

Kopf, A., 2002, Significance of mud volcanism: Reviews of Geophysics, v. 40, no. 2-1, p. 2–52.

Kopf, A., and J. H. Behrmann, 2000, Extrusion dynamics of mud volcanoes on the Mediterranean Ridge accretionary complex, *in* B. Vendeville, Y. Mart, and J.-L. Vigneresse, eds., From the Arctic to the Mediterranean: Salt, shale, and igneous diapirs in and around Europe: Geological Society (London) Special Publication 174, p. 169–204.

Kopf, A., D. Klaeschen, and J. Mascle, 2001, Extreme efficiency of mud volcanism in dewatering accretionary prisms: Earth and Planetary Science Letters, v. 189, no. 3–4, p. 295–313, doi:10.1016/S0012-821X(01)00278-3.

Kugler, H. G., 1953, Jurassic to recent sedimentary environments in Trinidad: Bulletin der Vereinigung Schweizerisches Petroleum-Geologen und-Ingenieur, v. 20, p. 27–60.

Kugler, H. G., 1959, Geological map of Trinidad: Petroleum Association of Trinidad, scale 1:100000, 1 sheet.

Kugler, H. G., 1965, Sedimentary volcanism, *in* J. B. Saunders, ed., Transactions of the 4th Caribbean Geological Conference: Port-of-Spain, Trinidad, Caribbean Printers, p. 11–13.

Kugler, H. G., 1996, The Paleocene to Holocene Formations of Trinidad, *in* H. M. Bolli and M. Knappertsbusch, eds., Treatise on the geology of Trinidad: Basel, Switzerland, National History Museum, 800 p.

Lawrence, S. R., A. Soulsby, and V. H. Childs, 1991, A new framework for hydrocarbons exploration in Trinidad, *in* K. A. Gillezeau, ed., Transactions of the 2nd Geological Conference of the Geological Society of Trinidad and Tobago, p. 170–176.

Loncke, L., J. Mascle, and Fanil Scientific Parties, 2004, Mud volcanoes, gas chimneys, pockmarks and mounds in the Nile deep-sea fan (eastern Mediterranean): Geophysical evidences: Marine and petroleum geology, v. 21, no. 6, p. 669–689, doi:10.1016/j.marpetgeo.2004.02.004.

Martinelli, G., 1999, Mud volcanoes of Italy: a review: Giornale di Geologia, v. 61, no. 3C, p. 107–113.

Marty, B., R. K. O'Nions, E. R. Oxburgh, D. Martel, and S. Lombardi, 1992, Helium isotopes in alpine regions: Tectonophysics, v. 206, p. 71–78.

Oxburgh, E. R., R. K. O'Nions, and R. I. Hill, 1986, Helium isotopes in sedimentary basins: Nature, v. 324, p. 632–635, doi:10.1038/324632a0.

Payne, N., 1991, An evaluation of post-middle Miocene geological sequences, offshore Trinidad, *in* K. A. Gillezeau, ed., Transactions of the 2nd Geological Conference of the Geological Society of Trinidad and Tobago, p. 70–87.

Persad, K. M., S. C. Talukdar, and W. G. Dow, 1993, Tectonic control in source rock maturation and oil migration in Trinidad and implications for petroleum exploration, *in* J. L. Pindell and R. F. Perkins, eds., Mesozoic and early Cenozoic development of the Gulf of Mexico and Caribbean region: A context for hydrocarbon exploration: Proceedings of the 13th Annual Research Conference of the Gulf Coast Section SEPM Foundation, p. 237–249.

Pindell, J., L. Kennan, W. V. Maresh, K.-P. Stanek, G. Draper, and R. Higgs, 2005, Plate-kinematics and crustal dynamics of circum-Caribbean arc-continent interactions: Tectonic controls on basin development in proto-Caribbean margins: Geological Society of America Special Paper 394, p. 7–52, doi:10.1130/2005.2394(01).

Pitt, A. M., and R. A. Hutchinson, 1982, Hydrothermal changes related to earthquake activity at Mud Volcano, Yellowstone National Park, Wyoming: Journal of Geophysical Research, v. 87, p. 2762–2766, doi:10.1029/JB087iB04p02762.

Prinzhofer, A., and E. Pernaton, 1997, Isotopically light methane in natural gas: Bacterial imprint or diffusive fractionation?: Chemical Geology, v. 142, p. 193–200, doi:10.1016/S0009-2541(97)00082-X.

Prinzhofer, A., M. R. Mello, and T. Takaki, 2000, Geochemical characterization of natural gas: A physical multivariable approach and its applications in maturity and migration estimates: AAPG Bulletin, v. 84, p. 1152–1172.

Requejo, A. G., R. Sassen, C. Wielchowsky, and M. J. Klosterman, 1994, Geochemical characterization of lithofacies and organic facies in Cretaceous organic-rich rocks from Trinidad East Venezuela basin: Organic Geochemistry, v. 22, no. 3–5, p. 441–459, doi:10.1016/0146-6380(94)90118-X.

Robertson, R. P., and K. Burke, 1989, Evolution of the southern Caribbean plate boundary vicinity of Trinidad and Tobago: AAPG Bulletin, v. 73, no. 1, p. 490–509.

Rodriguez, K., 1988, Oil source bed recognition and crude oil correlation, Trinidad, West Indies: Organic Geochemistry, v. 13, no. 1–3, p. 365–371, doi:10.1016/0146-6380(88)90057-5.

Saleh, J., K. Edwards, J. Barbate, S. Balkaransingh, D. Grant, J. Weber, and T. Leong, 2004, On some improvements in the geodetic framework of Trinidad and Tobago: Survey Reviews, v. 37, no. 294, p. 604–625.

Sassen, R., A. V. Milkov, E. Ozgul, H. H. Roberts, J. L. Hunt, M. A. Beeunas, J. P. Chanton, D. A. Defreitas, and S. T. Sweet, 2003, Gas venting and subsurface charge in the Green Canyon area, Gulf of Mexico continental slope: Evidence of a deep bacterial methane source?: Organic Geochemistry, v. 34, no. 10, p. 1455–1464, doi:10.1016/S0146-6380(03)00135-9.

Speed, R. C., 1985, Cenozoic collision of the Lesser Antilles arc and continental South America and the origin of the El Pillar fault: Tectonics, v. 4, no. 1, p. 41–69, doi:10.1029/TC004i001p00041.

Stein, S., C. De Mets, R. G. Gordon, J. Brodholt, D. F. Argus, J. Engel, P. Lindgreen, C. Stein, D. Werns, and D. Woods, 1988, A test of alternative Caribbean plate relative motion models: Journal of Geophysical Research, v. 93, p. 3041–3050, doi:10.1029/JB093iB04p03041.

Talukdar, S., W. Gallango, F. Cassani, and C. Vallejos, 1989, Upper Cretaceous source rocks of northern and northwestern South America: AAPG Bulletin, v. 73, no. 3, p. 418.

Talukdar, S., W. Dow, J. Castano, and K. P. Persad, 1990, Geochemistry of oils provides optimism for deeper exploration in Atlantic off Trinidad: Oil & Gas Journal, v. 88 (November), no. 46, p. 118–122.

Talukdar, S., W. Dow, J. Castano, and K. P. Persad, 1995, Origin and evolution of oils and condensates from onshore and offshore Trinidad: Their exploration significance, *in* A. Winston, P. Anthony, and O. Victor Young, eds., Transactions of the 3rd Geological Conference of the Geological Society of Trinidad and Tobago and 14th Caribbean Geological Conference, p. 78.

Torgersen, T., and B. M. Kennedy, 1999, Air-Xe enrichments in Elk Hills oil field gases: Role of water in migration and storage: Earth and Planetary Science Letters, v. 167, p. 239–253, doi:10.1016/S0012-821X(99)00021-7.

Tyson, L., and A. Winston, 1991, Cretaceous to middle Miocene sediments in Trinidad: Field-trip guide, *in* K. A. Gillezeau, ed., Transactions of the 2nd Geological Conference of the Geological Society of Trinidad and Tobago, p. 266–277.

Tyson, L., S. Babb, and B. Dyer, 1991, Middle Miocene tectonics and its effects on late Miocene sedimentation in Trinidad, *in* K. A. Gillezeau, ed., Transactions of the 2nd Geological Conference of the Geological Society of Trinidad and Tobago, p. 26–40.

Westbrook, G. K., and M. J. Smith, 1983, Long decollements and mud volcanoes: Evidence from the Barbados ridge complex for the role of high pore-fluid pressure in the development of an accretionary complex: Geology, v. 11, no. 5, p. 279–283, doi:10.1130/0091-7613(1983)11<279:LDAMVE>2.0.CO;2.

Yassir, N., 1989, Mud volcanoes and the behavior of overpressured clays and silts: Ph.D. dissertation, University of London, London, UK, 249 p.

Zhou, Z., C. J. Ballentine, R. Kipfer, M. Schoell, and S. Thibodeaux, 2005, Noble gas tracing of groundwater/coalbed methane interaction in the San Juan Basin, U.S.A.: Geochimica et Cosmochimica Acta, v. 69, p. 5413–5428, doi:10.1016/j.gca.2005.06.027.